THE
INTERNATIONAL SERIES
OF
MONOGRAPHS ON PHYSICS

GENERAL EDITORS

J. BIRMAN   S. F. EDWARDS   R. FRIEND
C. H. LLEWELLYN SMITH   M. REES
D. SHERRINGTON   G. VENEZIANO

THE INTERNATIONAL SERIES OF MONOGRAPHS ON PHYSICS

98. K. H. Bennemann: *Nonlinear optics in metals*
97. D. Salzmann: *Atomic physics in hot plasmas*
96. M. Brambilla: *Kinetic theory of plasma waves*
95. M. Wakatani: *Stellarator and heliotron devices*
94. S. Chikazumi: *Physics of ferromagnetism*
93. A. Aharoni: *Introduction to the theory of ferromagnetism*
92. J. Zinn-Justin: *Quantum field theory and critical phenomena*
91. R. A. Bertlmann: *Anomalies in quantum field theory*
90. P. K. Gosh: *Ion traps*
89. E. Simánek: *Inhomogeneous superconductors*
88. S. L. Adler: *Quaternionic quantum mechanics and quantum fields*
87. P. S. Joshi: *Global aspects in gravitation and cosmology*
86. E. R. Pike, S. Sarkar: *The quantum theory of radiation*
84. V. Z. Kresin, H. Morawitz, S. A. Wolf: *Mechanisms of conventional and high T_c superconductivity*
83. P. G. de Gennes, J. Prost: *The physics of liquid crystals*
82. B. H. Bransden, M. R. C. McDowell: *Charge exchange and the theory of ion–atom collision*
81. J. Jensen, A. R. Mackintosh: *Rare earth magnetism*
80. R. Gastmans, T. T. Wu: *The ubiquitous photon*
79. P. Luchini, H. Motz: *Undulators and free-electron lasers*
78. P. Weinberger: *Electron scattering theory*
76. H. Aoki, H. Kamimura: *The physics of interacting electrons in disordered systems*
75. J. D. Lawson: *The physics of charged particle beams*
73. M. Doi, S. F. Edwards: *The theory of polymer dynamics*
71. E. L. Wolf: *Principles of electron tunneling spectroscopy*
70. H. K. Henisch: *Semiconductor contacts*
69. S. Chandrasekhar: *The mathematical theory of black holes*
68. G. R. Satchler: *Direct nuclear reactions*
51. C. Møller: *The theory of relativity*
46. H. E. Stanley: *Introduction to phase transitions and critical phenomena*
32. A. Abragam: *Principles of nuclear magnetism*
27. P. A. M. Dirac: *Principles of quantum mechanics*
23. R. R. Peierls: *Quantum theory of solids*
    F. P. Bowden, D. Tabor: *The friction and lubrication of solids*
    J. M. Ziman: *Electrons and phonons*
    M. E. Lines, A. M. Glass: *Principles and applications of ferroelectrics and related materials*

Nonlinear Optics in Metals

Edited by

K. H. BENNEMANN

Institute for Theoretical Physics, Freie Universität Berlin

CLARENDON PRESS • OXFORD
1998

Oxford University Press, Great Clarendon Street, Oxford OX2 6DP

Oxford   New York

Athens   Auckland   Bangkok   Bogota   Bombay   Buenos Aires   Calcutta
Cape Town   Chennai   Dar es Salaam   Delhi   Florence   Hong Kong   Istanbul
Karachi   Kuala Lumpur   Madrid   Melbourne   Mexico City   Mumbai
Nairobi   Paris   São Paolo   Singapore   Taipei   Tokyo   Toronto   Warsaw
and associated companies in
Berlin   Ibadan

Oxford is a trade mark of Oxford University Press

Published in the United States
by Oxford University Press Inc., New York

© Oxford University Press, 1998

All rights reserved. No part of this publication may be
reproduced, stored in a retrieval system, or transmitted, in any
form or by any means, without the prior permission in writing of Oxford
University Press. Within the UK, exceptions are allowed in respect of any
fair dealing for the purpose of research or private study, or criticism or
review, as permitted under the Copyright, Designs and Patents Act, 1988, or
in the case of reprographic reproduction in accordance with the terms of
licences issued by the Copyright Licensing Agency. Enquiries concerning
reproduction outside those terms and in other countries should be sent to
the Rights Department, Oxford University Press, at the address above.

This book is sold subject to the condition that it shall not,
by way of trade or otherwise, be lent, re-sold, hired out, or otherwise
circulated without the publisher's prior consent in any form of binding
or cover other than that in which it is published and without a similar
condition including this condition being imposed
on the subsequent purchaser.

A catalogue record for this book is available from the British Library

Library of Congress Cataloging in Publication Data

Non-linear optics in metals / edited by K. H. Bennemann.
(International series of monographs on physics; 98)
Includes bibliographical references and index.
1. Solids—Surfaces—Optical properties. 2. Metals—Surfaces—Optical properties. 3.
Nonlinear optics. 4. Magnetooptics.
I. Bennemann, K.-H. II. Series: International series of monographs
on physics (Oxford, England); 98.
QC176.8.S8N66   1998   535'.2—dc21   98-8282

ISBN 0 19 851893 5 (Hbk)

Typeset by Technical Typesetting Ireland, Belfast

Printed in Great Britain by Bookcraft (Bath) Ltd., Midsomer Norton, Avon

PREFACE

The interaction of light with matter, in particular metals, has been a classical area of physics. It has advanced our understanding of physics tremendously. Recently, light has been used again rather successfully to investigate the electronic, magnetic and atomic structure of metal surfaces, thin films, multilayers and interfaces. Optical studies represent a nondestructive technique for materials characterization. In particular, the study of magnetism at surfaces, in thin films and of multilayers is of interest, not only for basic research, but also regarding applications like storage of information, magnetic recording, and so on.

For many years the linear Kerr effect, exhibiting typically in metals only Kerr rotations of less than 1°, has been used and developed as a successful tool for solid-state physics research and applications. Only recently has study started on nonlinear optical effects in metals and in particular nonlinear magneto-optical effects. Owing to the high interface sensitivity of nonlinear magneto-optics, in contrast to linear magneto-optics, this is expected to become a successful new tool for studying electronic structure and magnetism at interfaces, in thin films and of multilayer structures. Remarkably, this high sensitivity of nonlinear optics regarding electronic and magnetic structure at metallic surfaces and in thin films has already been observed. In particular, rather large Kerr rotations, much larger than the linear Kerr rotations in metals, for example, have been seen. This is an impressive example demonstrating that Maxwell's equations still offer surprises.

Furthermore, an interesting dependence of the second harmonic light generation (SHG) at interfaces on light polarization has been observed. This is rather different for transition metals than for simple and noble metals. Hence, this too demonstrates clearly that SHG is material-specific and hence a promising new tool for solid-state physics research. Regarding sensitive studies of electronic structure, the observation of quantum-well state effects in SHG of thin films is an illustrative example. Observation of the interdependence of atomic structure and magnetism is another interesting result of SHG. Very recent studies show the importance of using second harmonic light and pump-and-probe spectroscopy for the analysis of the behaviour of systems not at equilibrium.

The first chapter of this book by Qiu and Bader gives a state-of-the-art report of linear Kerr spectroscopy and its application to surface magnetism, which is currently of great interest, and to the analysis of magnetic anisotropy effects and atomic structural changes. This chapter serves to compare linear magneto-optics with nonlinear magneto-optics, probing only interfaces where inversion symmetry is broken as regards the bulk. Comparison of results obtained by linear or nonlinear optics will stimulate both fields and will be particularly useful for future research advances.

The successes of nonlinear magneto-optics so far are described in the following two chapters by Vollmer (of Professor Kirschner's group in Halle) and by Rasing (whose group is in Nijmegen). The results presented for surfaces, thin films and multilayers are partly supplementary and demonstrate that SHG is quite capable of analysing in fine detail electronic and magnetic structure. As an example, the SHG oscillations including their spin dependence due to quantum-well states (QWS) in thin films indicate particularly the potential of nonlinear optics. Furthermore, the various results show that SHG is a very sensitive tool for characterizing material-specific properties including magnetic anisotropy and magnetic structures—see for example the results mentioned by Rasing on antiferromagnetism and on magneto-optical imaging.

The chapter on time-resolved light reflection on metals, in particular of SHG, by Professor Matthias and Hohlfeld, *et al.* (of Professor Matthias's group in Berlin) illustrates the new success of optical studies in metals, perhaps more generally in solid-state physics, and indicates the potential for studying nonequilibrium physics, the dynamics of excitations. This chapter prepares the way for looking at many new problems in the future. Pump-and-probe spectroscopy using nonlinear optics is a very promising new area of physics, in particular for studying the dynamics of magnetism, electronic and atomic structure.

The theory of nonlinear optics and nonlinear magneto-optics (NOLIMOKE, largely developed by the Bennemann group in Berlin), is outlined in the chapter by Hübner. The basic character of SHG and the formulae for nonlinear optics, for the Kerr rotation and for the polarization dependence of SHG at metallic interfaces are described and derived using an electronic theory. This chapter is useful for future theoretical developments and for providing the theoretical basis for analysing the experiments described in the previous chapters.

The last chapter by Luce and Bennemann continues and extends the theory of nonlinear optics. It demonstrates the potential of SHG for analysing fundamental problems regarding the electronic and magnetic structure of surfaces, thin films and multilayer structures. It shows how the interplay of effects due to the electronic energy spectrum and optical transition matrix elements can be used to unfold the complex richness of information contained in SHG. The electronic theory for SHG is also applied to non-equilibrium physics. The time-resolved nonlinear magnetic response of transition metals to light irradiation is analysed. Also, interesting problems to be studied by SHG in the future are discussed.

Future work on electronic and atomic structural phase transitions, on lateral and in-depth resolution of film structure, magnetic contrasts, domain structures, antiferromagnetism and magnetic anisotropy effects, for example, must reveal the full potential of SHG as a new tool for interface and film research. The main purpose of this book is to give a comprehensive introduction to the state of the art in the subject and to lay the ground for further developments in the field of nonlinear optics at metallic interfaces. The optical determination of electronic, atomic and magnetic structures will remain important for basic research as well as for significant technical applications.

Berlin K.H.B.
February 1998

CONTENTS

List of contributors	xiii

1 KERR EFFECT AND SURFACE MAGNETISM   1
Z. Q. Qiu and S. D. Bader

1.1	Introduction		1
	1.1.1	Faraday and the discovery of magneto-optics	1
	1.1.2	Kerr and the role of the magneto-optical Kerr effect in the study of magnetic thin films	2
1.2	Origin of the magneto-optical effect		3
	1.2.1	The search for the origin of the magneto-optical effect	3
	1.2.2	Phenomenological description	5
	1.2.3	Classical model	6
	1.2.4	Quantum description of the magneto-optical effect in a ferromagnet	8
1.3	Macroscopic formalism of magneto-optical Kerr effect		12
	1.3.1	General formulae	12
	1.3.2	Superlattices	17
	1.3.3	Additivity law	18
1.4	Experimental approach		20
	1.4.1	Experimental setup	20
	1.4.2	Verification of the macroscopic formulae	22
1.5	Applications of SMOKE in two-dimensional (2D) magnetic thin films		25
	1.5.1	2D magnetic phase transition	25
	1.5.2	2D spin-reorientation transition	29
	1.5.3	Magnetic anisotropy as a result of symmetry breaking	32
1.6	Summary		38
Acknowledgement			39
References			39

2 MAGNETIZATION-INDUCED SECOND HARMONIC GENERATION FROM SURFACES AND ULTRATHIN FILMS   42
R. Vollmer

2.1	Introduction		42
2.2	Second harmonic generation from metal surfaces		44
	2.2.1	Historical review	44
	2.2.2	Anisotropy of surface SHG	46

	2.2.3	Effects of adsorbates and defects on SHG	49
	2.2.4	SHG spectroscopy	51
2.3	The linear magneto-optical Kerr effect		54
2.4	Instrumentation		59
	2.4.1	The laser system	59
	2.4.2	*In situ* sample preparation	62
	2.4.3	Thermal heating by the laser and damage thresholds	63
2.5	The nonlinear magneto-optical Kerr effect		67
	2.5.1	Phenomenological description	67
	2.5.2	Symmetry analysis of $\chi^{(2,D)}$	68
	2.5.3	Symmetry properties of $\chi^{(2,Q)}$	70
	2.5.4	Nonlinear Kerr angles and asymmetries	71
2.6	MSHG as a surface-sensitive probe of magnetism		75
	2.6.1	MSHG from the Fe(110) surface	77
	2.6.2	Ultrathin Co films on Cu(001)	82
	2.6.3	Fe/Cu(001)	87
	2.6.4	Bulk contributions to SHG	91
	2.6.5	Effects of adsorbates	95
2.7	MSHG studies of thin films		103
	2.7.1	Quantum-well states in thin metallic films	103
	2.7.2	SHG from Cu/Co/Cu(001) and Cu/Fe/Cu(001) sandwich systems	106
	2.7.3	Frequency dependence of the MSHG signal from sandwich systems	112
	2.7.4	MSHG from the Cu/Co/Cu(001) sandwich systems in the longitudinal Kerr geometry	118
2.8	Summary and outlook		120
Acknowledgements			124
References			124

3 NONLINEAR MAGNETO-OPTICAL STUDIES OF ULTRATHIN FILMS AND MULTILAYERS 132
Th. Rasing

3.1	Introduction	132
3.2	Surface second harmonic generation	135
3.3	SHG from multilayers	136
3.4	The magneto-optical Kerr effect	140
	3.4.1 The linear magneto-optical Kerr effect (MOKE)	140
3.5	Magnetization-induced second harmonic generation	143
	3.5.1 The nonlinear magneto-optical tensor components	144
	3.5.2 The nonlinear magneto-optical Kerr effect	148
	3.5.3 Estimates for $\chi^{(2)}(M)$	151
3.6	Instrumentation	152
3.7	Experimental case studies	156

	3.7.1	Demonstration of even/odd tensor components	157
	3.7.2	MSHG as a probe of surface magnetism	160
	3.7.3	MSHG studies of magnetic multilayers	167
	3.7.4	Other MSHG results	171
3.8	Nonlinear Kerr rotations		172
	3.8.1	NOMOKE rotation from Fe	172
	3.8.2	NOMOKE rotation from Ni	177
3.9	MSHG studies of quantum-well states		179
	3.9.1	Quantum-well states in thin metallic films	179
	3.9.2	Cu/Co/Cu(001)	181
	3.9.3	Au/Co/Au(111)	184
	3.9.4	Cu/Co/Au(111)	190
3.10	MSHG studies of antiferromagnetic structures		192
	3.10.1	Cr_2O_3	192
	3.10.2	Other AFM materials	194
3.11	MSGH studies of magnetic garnets		194
	3.11.1	Symmetry properties of thin garnet films	195
	3.11.2	Nonlinear magneto-optical response of thin garnet films	196
	3.11.3	Experimental results and discussion	197
3.12	Nonlinear magneto-optical imaging		201
	3.12.1	Domain and domain-wall contributions in nonlinear magneto-optical microscopy	202
	3.12.2	Nonlinear magneto-optical microscopy of domains in magnetic garnet films	207
	3.12.3	Nonlinear magneto-optical microscopy of antiferromagnetic domains in Cr_2O_3	211
3.13	Summary and future prospects		213
Acknowledgements			214
References			214

4 FEMTOSECOND TIME-RESOLVED LINEAR AND SECOND-ORDER REFLECTIVITY OF METALS 219
J. Hohlfeld, U. Conrad, J. G. Müller, S. S. Wellershoff, and E. Matthias

4.1	Introduction		219
4.2	Transient linear reflectivities		220
	4.2.1	Dynamics of optically excited electrons	220
	4.2.2	Temperature dependence of linear reflectivities	225
	4.2.3	Experimental results	229
	4.2.4	Penetration depth of deposited energy	233
	4.2.5	Saturation of linear reflectivity	238
4.3	Second harmonic generation		239
	4.3.1	Formalism for data analysis	239
	4.3.2	Pump–probe SHG	245
	4.3.3	Polarization-dependent SHG	251

	4.3.4	Electron and magnetization dynamics of Ni	253
4.4	Summary		262

Acknowledgements   264
References   264

5 ELECTRONIC THEORY FOR NONLINEAR MAGNETO-OPTICS   268
W. Hübner

- 5.1 Development of nonlinear magneto-optics   268
- 5.2 Classical theory for nonlinear optics   270
  - 5.2.1 Oscillator model for harmonic generation   274
  - 5.2.2 Nonlinear Mie theory for spherical particles   278
  - 5.2.3 Nonlinear Mie theory for particles of arbitrary shape   293
  - 5.2.4 Magnetic nonlinear Mie theory   297
  - 5.2.5 Magic angle in third harmonic generation   297
- 5.3 Symmetry and nonlinear Kerr rotation   300
  - 5.3.1 Polarization dependence of second harmonic generation   300
  - 5.3.2 Polarization dependence of the nonlinear Kerr effect   311
  - 5.3.3 Enhancement of the nonlinear Kerr rotation   319
  - 5.3.4 Nonlinear magneto-optics in s- and d-wave superconductors   327
  - 5.3.5 Nonlinear magneto-optics in antiferromagnets   335
- 5.4 Microscopic theory for nonlinear magneto-optics   339
  - 5.4.1 Calculation of the nonlinear Kerr susceptibility   340
  - 5.4.2 Second harmonic generation and magnetism   346
  - 5.4.3 Interpretation of the susceptibility tensor   355
  - 5.4.4 Tight-binding theory for nonlinear surface magneto-optics   356
  - 5.4.5 Results: nonlinear Kerr spectra for Ni and Fe surfaces   377
  - 5.4.6 First-principles theory for thin films   398
  - 5.4.7 Results: nonlinear Kerr spectra for thin Fe films   405
- 5.5 Hot-electron dynamics in nonlinear optics   419
  - 5.5.1 Theory   420
  - 5.5.2 Results   422
  - 5.5.3 Conclusions   426
- 5.6 Summary and outlook   427
  - 5.6.1 Summary   427
  - 5.6.2 Outlook   429

Acknowledgements   429
References   430

6 THEORY FOR NONLINEAR OPTICS AT INTERFACES AND IN THIN FILMS OF METALS: SELECTED PROBLEMS   437
T. Luce and K. H. Bennemann

- 6.1 Introduction   437

6.2	General properties of SHG	437
6.3	General theory for SHG oscillations due to quantum-well states in thin films	445
6.4	SHG analysis of magnetic structures	458
6.5	Application of SHG to nonequilibrium physics	467
6.6	Outlook	474
Acknowledgements		474
References		474
Author Index		479
Subject Index		483

CONTRIBUTORS

S. D. Bader
Material Science Division
Argonne National Laboratory
Argonne
Illinois 60439
USA
Fax: + 1 630 252 5219
Tel.: + 1 630 252 5203
e-mail: Bader@anl.gov

K. H. Bennemann
Institute for Theoretical Physics
Freie Universität Berlin
Arnimallee 14
D-14195 Berlin
Germany
Fax: 030 838 6799
e-mail: khb@manuel.physik.fu.berlin.de

U. Conrad
Fachbereich Physik—WE 1
Freie Universität Berlin
Arnimallee 14
D-14195 Berlin
Germany

J. Hohlfeld
Fachbereich Physik—WE 1
Freie Universität Berlin
Arnimallee 14
D-14195 Berlin
Germany
Fax: 030 838 6059
Tel.: 030 838 6234
e-mail: hohlfeld@physik.fu-berlin.de

W. Hübner
Max Planck-Institut für Mikrostrukturphysik
Weinberg 2

D-06120 Halle/Saale
Germany
Fax: +49 345 5582 566
Tel.: +49 345 5582 914
e-mail: huebner@mpi-halle.mpg.de

T. Luce
Institute for Theoretical Physics
Freie Universität Berlin
Arnimallee 14
D-14195 Berlin
Germany
Fax: +49 30 838 3035
Tel.: +49 30 838 6799
e-mail: khb@physik.fu-berlin.de

E. Matthias
Fachbereich Physik—WE 1
Freie Universität Berlin
Arnimallee 14
D-14195 Berlin
Germany
Fax: 030 838 6059
Tel.: 030 838 3340
e-mail: matthias@matth1.physik.fo-berlin.de

J. G. Müller
Fachbereich Physik—WE 1
Freie Universität Berlin
Arnimallee 14
D-14195 Berlin
Germany

Z. Q. Qiu
Materials Science Division
Department of Physics
University of California at Berkeley
Lawrence Berkeley Laboratory
Berkeley
CA 94720
USA
Fax: +1 510 643 8497
Tel.: +1 510 642 2959
e-mail: qiu@physics.berkeley.edu

CONTRIBUTORS

Th. Rasing
Research Institute for Materials
Katholieke Univ. Nijmegen
Postbus 9010
Toernooiveld
NL-6525 ED Nijmegen
The Netherlands
Fax: +31 24 365 2190
Tel.: +31 24 365 3141
e-mail: theoras@sci.kun.nl

R. Vollmer
Max-Planck-Institut für Mikrostrukturphysik
Weinberg 2
D-06120 Halle/Saale
Germany
Fax: +49 345 5511 223
Tel.: +49 345 5582 750
e-mail: Vollme@mpi-halle.mpg.de

S. S. Wellershoff
Fachbereich Physik—WE 1
Freie Universität Berlin
Arnimallee 14
D-14195 Berlin
Germany
Fax: +49 30 838 3340
Tel: +49 30 838 3340
e-mail: matthias@matth1.physik.fu.berlin.de

1
KERR EFFECT AND SURFACE MAGNETISM

Z. Q. Qiu
University of California at Berkeley, Berkeley, USA

and S. D. Bader
Argonne National Laboratory, Argonne, USA

1.1 Introduction

1.1.1 Faraday and the discovery of magneto-optics

The investigation of magneto-optical effects is historically rooted in the early part of the nineteenth century when physicists started to search for relationships between various natural phenomena. Magneto-optical searches date back to 1812 when Domenico Morichini of Rome erroneously claimed that steel needles could be magnetized utilizing violet radiation in the solar spectrum.[1] A null result was reported *circa* 1825, when Sir John Herschel examined the propagation of a beam of polarized light along the axis of a helix carrying an electric current.[1] Then, in 1845, Michael Faraday reported his now famous magneto-optical experiments.[2] He found that the polarization of a beam of light rotates while travelling in a magnetized medium. Let us humble ourselves by looking back at those times of great discovery.

Michael Faraday was born in 1791 the son of a blacksmith. Faraday became an apprentice bookbinder and obtained most of his education through self-learning. From such beginnings he rose to become the director of the British Royal Institute. His keen intuition and rich imagination led him to the discoveries that unambiguously establish him as one of the great scientists of the nineteenth century. Faraday invented the electric motor in 1821, discovered electromagnetic induction in 1831, and discovered both magneto-optics and diamagnetism in 1845. He also laid the foundations of field theory. His last experiment (12 March 1862) was to seek the effect of a magnetic field on the visible spectrum. He obtained a null result, but 35 years later the effect was discovered by Zeeman, and is now known as the Zeeman effect.

Faraday's original search was focused on the relation between light and electricity. On 30 August 1845 he failed to find a change in polarization of light passing through a liquid that was undergoing electrolysis.[3] Then, on 13 September, he substituted magnetic for electric forces, using an electromagnet with an iron core, and discovered the magneto-optical effect. He recorded in his laboratory notebook:[4]

A piece of heavy glass, which was 2 inches by 1.8 inches and 0.5 of an inch thick, being a silicoborate of lead, was experimented with...when contrary magnetic poles were on the same side there *was an effect produced on the polarized ray*, and thus magnetic force and light were proved to have relations to each other. This fact will most likely prove exceedingly fertile, and of great value in the investigation of conditions of nature force.

After acquiring a more powerful electromagnet on 18 September, he continued the experiment with such zeal that he filled 12 pages of his laboratory notebook in one day and concluded with the statement: 'An excellent day's work'. He verified that the effect of the magnet was to rotate the polarization plane of the light by an angle that depended on the strength of the magnet. He subsequently tried experiments on light *reflected* from a metal surface within a magnetic field. While he did obtain an optical rotation from the surface of a polished steel button, his result was inconclusive due to the imperfection of the surface. (Thirty-two years later John Kerr discovered this effect[5], which will become the main focus of this paper: the magneto-optical Kerr effect.) Two months after Faraday's discovery of magneto-optics, he discovered diamagnetism while examining a bar of his 'heavy glass' using an even stronger electromagnet than before. In his description of this discovery, the term 'magnetic field' was used for the very first time.

1.1.2 Kerr and the role of the magneto-optical Kerr effect in the study of magnetic thin films

The magneto-optical Kerr effect (MOKE) was discovered by the Rev. John Kerr in 1877 while he was examining the polarization of light reflected from a polished electromagnet pole. Kerr ultimately received the Royal Medal in 1898 for research that ranked amongst the most important subsequent to Faraday's. Kerr was born in Scotland in 1824. He was educated at the University of Glasgow and received many student prizes, including one that honoured him as the most distinguished student in mathematics and natural philosophy. Though he also was a divinity student, he apparently never assumed official clerical duties, but instead spent his career teaching and conducting research at the Glasgow Free Church Training College for Teachers. He is also best known for his discovery, in 1875, of the first electro-optical effect to be detected: electrical birefringence. Kerr spent many years studying electrical and strain-induced birefringence. In his works he was inspired by the earlier work of Maxwell, and especially of Faraday. When he was presented with the Royal Medal, his presenter said it was a wonder that Kerr learned so much with the 'comparatively simple and ineffectual apparatus at his disposal'. Kerr responded, 'Simple it may be, but not ineffectual; rude, but not crude.' Thus, it might be said about a chapter such as the present one, which introduces linear magneto-optics in a volume devoted to nonlinear effects. Our hope is to remain simple, but send the message that such investigations are not crude, even in comparison to the elegance of nonlinear physics.

Returning to the late nineteenth century for a bit longer, Kerr's initial results suggested that magnetic metals would rotate the polarization plane of transmitted light. In 1884, Kundt succeeded in depositing Fe, Co and Ni films on glass with thicknesses that could still transmit light.[7] The rotation he found was ~30 000 times that produced by the same thickness of glass, with a value of ~200 000°/cm for Fe. Kundt correctly interpreted the rotation was being proportional to the magnetization, rather than to the magnetic field. This work can be viewed as the first investigation of magnetic films using the magneto-optical effect.

Magneto-optical characterizations of surface magnetism began a little over a decade ago. The first surface magneto-optical Kerr effect study, better known by its acronym SMOKE, concerned the magnetic properties of ultrathin Fe films grown epitaxially on Au(100).[8] Hysteresis loops of the Fe film with atomic layer sensitivity were successfully obtained. Since then SMOKE has emerged as a premier surface magnetism technique. SMOKE has been applied to various topics in low-dimensional magnetism ranging from the detection of magnetic order to the characterization of critical behaviour, magnetic surface anisotropies, and the oscillatory antiferromagnetic coupling exhibited by giant-magnetoresistance heterostructures. Additional interest in SMOKE is generated by the recent commercialization of high-density magneto-optical information storage media,[9] and especially by the next-generation candidate material based on Co/Pt superlattices.[10]

This chapter provides background to appreciate some of these developments and highlights examples of contemporary issues that provide a focus for the field. Magneto-optics is presently described in the context of either microscopic quantum theory or macroscopic dielectric theory.[11] Microscopically, the coupling between the electrical field of the propagating light and the electron spin in a magnetic medium occurs through the spin–orbit interaction. Macroscopically, magneto-optical effects arise from the antisymmetric, off-diagonal elements in the dielectric tensor, as discussed in the next section.

1.2 Origin of the magneto-optical effect

1.2.1 The search for the origin of the magneto-optical effect

Early explanations of the Faraday effect were based on an analogy with the kinetics of particle motion. In the *Proceedings of the Royal Society* (June 1856), Sir William Thomson offered a 'microscopic' explanation of the Faraday effect by arguing that the particles in the medium under an external magnetic field follow different circular paths, depending on their direction relative to the magnetic field. From a modern viewpoint, this is the classical explanation of the

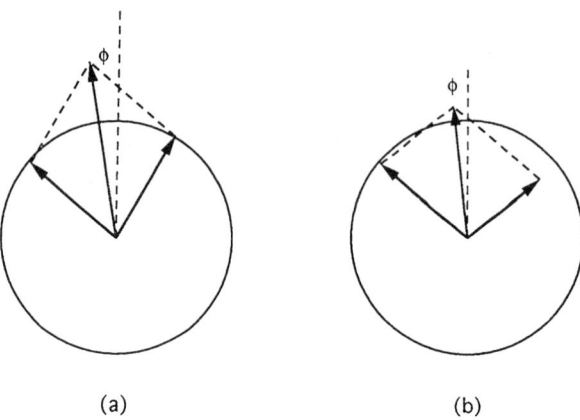

Fig. 1.1. (a) Magneto-optical rotation as a result of different rotation angles of the left-circularly and right-circularly polarized modes. (b) Magneto-optical ellipticity as a result of different absorption rates of the left- and right-circularly polarized modes. Both processes should exist in a real material.

Faraday effect if we regard Thomson's 'particles' as electrons (see sec. 1.2.3). In analogy with mechanical vibrations of a particle, Maxwell expressed linearly polarized light as being a superposition of two circularly polarized components, and realized that the Faraday effect is a consequence of the different propagating velocities of the two circular modes.[12] This explanation remains the phenomenological explanation of the Faraday effect given in introductory physics textbooks. Looked at in greater detail, there are actually two processes taking place for light propagating in a magnetized medium. First, the two circularly polarized modes gain different phase shifts due to their different propagating velocities, resulting in a rotation of the polarization plane. This process is the conventional Faraday rotation and is schematically shown in Fig. 1.1(a). Secondly, the different absorption rates of the medium for the two circularly polarized modes affects the ellipticity, as is shown schematically in Fig. 1.1(b). For transparent materials, the rotation dominates the ellipticity, but for metals, the ellipticity usually dominates the rotation. In general, both effects exist in a magnetized medium.

Macroscopic descriptions of the magneto-optical effect are based on an analysis of the dielectric tensor, which gives different refractive indices for the left- and right-circularly polarized modes. We will give a general analysis of the dielectric tensor in the following section to see which part gives rise to the magneto-optical effect. In sec. 1.2.3, a classical model will be used to describe the origin of the magneto-optical effect. Though the classical model is not adequate to describe real materials, especially ferromagnetic materials, it does provide a simple physical picture of the magneto-optical effect. In sec. 1.2.4, the quantum description of magneto-optical effects in ferromagnetic materials is discussed, from which one can appreciate the role of the spin−orbit interaction in generating the effect.

1.2.2 Phenomenological description

The response of a medium to an external field can be described by the dielectric tensor, ε_{ij}, with $i, j = 1, 2, 3$. This 3×3 tensor reduces to the dielectric constant times the unitary matrix for an isotropic medium. In general, the dielectric tensor can be decomposed into a symmetric part and an antisymmetric part, $\varepsilon_{ij} = (\varepsilon_{ij} + \varepsilon_{ji})/2 + (\varepsilon_{ij} - \varepsilon_{ji})/2$. The symmetric part can be diagonalized by an appropriate rotation of the coordinate system. If the three eigenvalues are the same, the medium is isotropic; otherwise, the medium is anisotropic. Nevertheless, for light propagating along these three principal axes, it is obvious that the polarization plane will remain unchanged, i.e. the symmetric part of the dielectric tensor does not give rise to the Faraday effect. Since the symmetric part of the ε_{ij} is unimportant to the Faraday effect, we will always assume that this part is isotropic with a dielectric constant ε_0. To see the effect of the antisymmetric part of the dielectric tensor, let us consider the special case of a beam of light propagating along the z axis in a medium that has its dielectric tensor of the form

$$\tilde{\varepsilon} = \varepsilon \begin{pmatrix} 1 & iQ & 0 \\ -iQ & 1 & 0 \\ 0 & 0 & 1 \end{pmatrix}, \qquad (1.2.1)$$

where Q is a magneto-optical coupling strength parameter, also known as the Voigt[13] constant.

Then it is easy to show that the two normal modes of the wave are the left-circularly polarized light ($E_y = iE_x$) with $\varepsilon_L = \varepsilon(1 - Q)$, and right-circularly polarized light ($E_y = -iE_x$) with $\varepsilon_R = \varepsilon(1 + Q)$. Thus, the difference in the refractive indices between these two modes, which generates the magneto-optical effect, is proportional to Q (to leading order), i.e. the off-diagonal antisymmetric element Q gives rise to the Faraday rotation. For the general case with the dielectric tensor of the form

$$\tilde{\varepsilon} = \varepsilon \begin{pmatrix} 1 & iQ_z & -iQ_y \\ -iQ_z & 1 & iQ_x \\ iQ_y & -iQ_x & 1 \end{pmatrix}, \qquad (1.2.2)$$

the normal modes are still left- and right-circularly polarized modes with refractive indices $n_L = n(1 - \frac{1}{2}\boldsymbol{Q} \cdot \hat{\boldsymbol{k}})$ and $n_R = n(1 + \frac{1}{2}\boldsymbol{Q} \cdot \hat{\boldsymbol{k}})$, where $n = \sqrt{\varepsilon}$ is the average refractive index, $\boldsymbol{Q} = (Q_x, Q_y, Q_z)$ is called the Voigt[13] vector, and $\hat{\boldsymbol{k}}$ is the unit vector along the light-propagation direction. Thus, the complex Faraday rotation of the polarization plane after travelling a distance L is

$$\theta = \frac{\pi L}{\lambda}(n_L - n_R) = -\frac{\pi L n}{\lambda} \boldsymbol{Q} \cdot \hat{\boldsymbol{k}}. \qquad (1.2.3)$$

The real part of the above formula gives the rotation, and the imaginary part gives the ellipticity. It is interesting to ask why an external magnetic field has a stronger effect on the polarization plane of light than an external electrical field.

Phenomenologically, this can be answered by a simple argument based on time reversal symmetry. Under the time reversal operation, the D and E vectors remain unchanged, but the magnetic field changes sign. Thus, Onsager's relation gives $\varepsilon_{ij}(E, H) = \varepsilon_{ji}(E, -H)$. Here E and H are the external electrical and magnetic fields. By expaning ε_{ij} up to terms linear in the electrical and magnetic fields, it is obvious that the antisymmetric part of ε_{ij} is generated by the magnetic field. The magnetic field is only a special case of time reversal symmetry breaking. In general, any quantity that breaks time reversal symmetry could, in principle, generate antisymmetric elements of the dielectric tensor and, thus, a Faraday rotation. That is why it was magneto-optical-type experiments that were performed[14] on high-temperature superconducting materials to search for the existence of anyons, which are believed to break time reversal symmetry.[15]

1.2.3 Classical model

From the previous section, we saw that the antisymmetric part of the dielectric tensor gives rise to the magneto-optical effect. In this section, we discuss a classical model to see how an external magnetic field generates the antisymmetric part of the dielectric tensor. As an electromagnetic wave is applied to a medium, the electrical field of the wave generates the motion of the electrons in the medium to induce a dipole moment. The coefficient of the electric polarization proportional to the electric field gives the dielectric constant. Thus, the dielectric property of a medium can be obtained by considering the motion of the electrons in the medium. This is the starting point of the microscopic classical model. Consider a medium in the harmonic approximation with natural frequency ω_0. The relaxation time of the electrons is τ. An external magnetic field is applied in the z direction. The equation of motion of an electron in the medium under the influence of an electromagnetic wave is

$$\frac{d^2 r}{dt^2} + m\omega_0^2 r + \frac{m}{\tau}\frac{dr}{dt} = -eE - \frac{e}{c}\frac{dr}{dt} \times B\hat{z}, \qquad (1.2.4)$$

where m and $-e$ are the electron mass and charge, respectively. (The effect of the magnetic field on the wave has been ignored.)

Before doing a quantitative calculation of the dielectric tensor, it is worthwhile to analyse qualitatively the motion of the electrons to provide a physical picture of the magneto-optical effect. Without the external magnetic field, it is obvious that a left-circularly polarized electric field will drive the electron to left circular motion, and a right-circularly polarized electric field will drive the electron to right circular motion. The radii of the electron orbits for left and right circular motion will be the same. Since the electric dipole moment is proportional to the radius of the circular orbit, there will be no difference between the dielectric constants for the left- and right-circularly polarized electromagnetic wave. Thus, there will be no Faraday rotation. After an external magnetic field is applied in the propagation direction of the electromagnetic wave, there will be an additional Lorentz force acting on each electron. This force points towards or away from the circles' centre for left or right circular motion. Thus, the radius for left circular motion will be reduced and the radius for right circular motion will expand. The difference in the radii of the left- and right-circularly polarized modes will give different dielectric constants respectively. Thus, it is the Lorentz force of the external magnetic field that generates the Faraday effect. From the above analysis, it is easy to see that the difference in the dielectric constants for the two circularly polarized modes is proportional to the difference in the radii of the corresponding circular motions. Since this difference is proportional to the Lorentz force, which is proportional to the frequency, then it is obvious from eqn (1.2.3) that the Faraday rotation should be inversely proportional to the square of the wavelength of the light. We will see this also from the following quantitative analysis.

Equation (1.2.4) is linear and can be solved for the electromagnetic wave of frequency ω. The solution is

$$\begin{pmatrix} x \\ y \\ z \end{pmatrix} = -\frac{e}{m} \begin{pmatrix} \dfrac{\gamma}{\omega^2\gamma^2 - \omega_c^2} & \dfrac{i\omega_c/\omega}{\omega^2\gamma^2 - \omega_c^2} & 0 \\ \dfrac{-i\omega_c/\omega}{\omega^2\gamma^2 - \omega_c^2} & \dfrac{\gamma}{\omega^2\gamma^2 - \omega_c^2} & 0 \\ 0 & 0 & \dfrac{1}{\omega^2\gamma} \end{pmatrix} \begin{pmatrix} E_x \\ E_y \\ E_z \end{pmatrix}, \qquad (1.2.5)$$

where $\omega_c = eB/mc$ is the cyclotron frequency, and

$$\gamma = \frac{\omega_0^2}{\omega^2} - 1 - \frac{i}{\omega\tau}$$

is a dimensionless number. Then, from the relation $P_i = -ner_i = (\varepsilon_{ij} - 1)E_j/4\pi$, where n is the electron density, the dielectric tensor can be derived as

$$\tilde{\varepsilon} = 1 + \frac{4\pi n e^2}{m} \begin{pmatrix} \dfrac{\gamma}{\omega^2\gamma^2 - \omega_c^2} & \dfrac{i\omega_c/\omega}{\omega^2\gamma^2 - \omega_c^2} & 0 \\ \dfrac{-i\omega_c/\omega}{\omega^2\gamma^2 - \omega_c^2} & \dfrac{\gamma}{\omega^2\gamma^2 - \omega_c^2} & 0 \\ 0 & 0 & \dfrac{1}{\omega^2\gamma} \end{pmatrix}. \qquad (1.2.6)$$

Thus, for light propagating in the z direction, the Faraday rotation after a distance L is

$$\theta = \frac{L\omega_c}{2n_0 c} \frac{4\pi n e^2}{m(\omega^2\gamma^2 - \omega_c^2)}, \qquad (1.2.7)$$

where

$$n_0 = \sqrt{1 + \frac{4\pi n e^2 \gamma}{m(\omega^2\gamma^2 - \omega_c^2)}}$$

Table 1.1 Experimental values of Faraday rotation

Materials	Temperature (°C)	Rotation (min cm^{-1} Gauss^{-1})
Light flint glass	18	0.0317
Water	20	0.0131
NaCl	16	0.0359
Quartz	20	0.0166

is the average refractive index of the medium. For visible light and a transparent insulator, $\omega_0 \gg \omega \gg \omega_c$ and $\omega\tau \gg 1$, thus $\gamma \approx \omega_0^2/\omega^2$. Therefore, the Faraday rotation is

$$\theta \approx \frac{2\pi n e^3 \omega^2 L B}{m^2 c^2 \omega_0^4 \sqrt{1 + \frac{4\pi n e^2}{m \omega_0^2}}}. \quad (1.2.8)$$

The rotation is proportional to the sample length and the magnetic field, and is inversely proportional to the square of the light wavelength. These results roughly agree with experimental observation. For typical values of $\omega \sim 10^{15}$ s^{-1}, $\omega_0 \sim 10^{16}$ s^{-1}, and $n \sim 10^{23}$ cm^{-3}, eqn (1.2.8) yields a rotation of $\theta \sim 10^{-2}$ min cm^{-1} Gauss^{-1}, which agrees with experimental results (see Table 1.1[16]).

1.2.4 Quantum description of the magneto-optical effect in a ferromagnet

In ferromagnetic materials, the Faraday effect is much stronger than in nonmagnetic materials. Early attempts to explain this assumed that there exists an effective field, rather than the applied field, that determines the Faraday rotation in ferromagnetic materials. In fact, Voigt found that the effective field is of the order of 10^6–10^7 Oe to produce the observed Faraday rotation. This magnitude is of the order of the Weiss field that was postulated to account for the existence of ferromagnetism. The nature of the Weiss field remained unexplained until Heisenberg developed the theory that ascribed the origin of magnetism to the exchange interaction among electrons. Although Heisenberg's exchange interaction correctly reveals the origin of magnetism as an effective magnetic field to align the individual spins, this field alone cannot be used to explain the Faraday effect. This is because it is not coupled to the electron motion, which determines the optical properties of a material. This difficulty was solved in 1932 by Hulme[17] who pointed out that it is the spin–orbit interaction that couples the electron spin to its motion to give rise to the large Faraday rotation in a ferromagnetic. Spin–orbit coupling, $\sim (\nabla V \times \mathbf{p}) \cdot \mathbf{s}$, results from the interaction of the electron spin with the magnetic field the electron 'sees' as it moves through the electric field $-\nabla V$ with momentum $\mathbf{p}$ inside a medium. This interaction couples the magnetic moment of the electron with its motion, and thus connects the magnetic and optical properties of a ferromagnet. Indeed, to a certain extent, the spin–orbit interaction can be thought of as an effective magnetic field vector potential $\sim \mathbf{s} \times \nabla V$ acting on the motion of the electron.

For nonmagnetic materials, this effect is not strong, although the spin–orbit interaction is present, because the equal numbers of spin-up and spin-down electrons cancel the net effect. For ferromagnetic materials, however, the effect manifests itself because of the unbalanced population of electron spins.

Hulme calculated the two refractive indices (right and left polarized) using the Heisenberg model of a ferromagnet, and the Kramers–Heisenberg dispersion formula. This approach represents the refractive index in terms of the eigenenergy and matrix elements of the dipole moment operator with respect to the eigenfunctions of the system. Hulme accounted for the difference of the two refractive indices by the energy splitting due to the spin–orbit interaction. He neglected, however, the change of the wavefunction due to the spin–orbit interaction. This theory is unsatisfying because the quenching of the orbital angular momentum in ferromagnets gives no energy splitting. Kittel[18] showed that it is the change of the wavefunctions due to the spin–orbit interaction that give rise to the correct order of magnitude of the difference of the two refractive indices. Argyres[19] later gave a full derivation of the magneto-optical effect in a ferromagnet using perturbation theory. Subsequent works were performed thereafter to calculate the magneto-optical effect in different regimes.[20–22]

Since the refractive index is related to the complex conductivity by a constant, i.e. $\varepsilon_{ij} = 1 + i4\pi\sigma_{ij}/\omega$, it is only necessary to calculate the conductivity tensor, especially the off-diagonal element, to obtain the magneto-optical effect. There are generally a few assumptions in the microscopic theory. First, the light frequency ω is much greater than the spin–lattice relaxation rate $1/\tau$, i.e. $\omega\tau \gg 1$, so that the effect of spin relaxation is negligible. Secondly, the interaction of the electron with the electrical field of the incident light is much stronger than the interaction of the electron with the magnetic field of the light. This is true in that $v_F/c \ll 1$, where v_F is the Fermi velocity of the electron gas. Thirdly, the wavelength of the incident light is much greater than the atomic spacing, so that one can take the $k \to 0$ long-wavelength limit in the calculation. For short wavelengths, such as for X-rays, where core-level excitations occur, simple models based on atomic wavefunctions and band density of states can be used to give fairly good results.[23] Finally, the interaction of an electron inside a crystal with other electrons and nuclei can be represented by the periodic potential $V(r)$ so that the one-electron approximation can be applied to describe the ground and excited states of the system. With the above approximations, the one-electron Hamiltonian within an electromagnetic field is

$$H = H_0 + H' + H'', \tag{1.2.9}$$

where

$$H_0 = \frac{p^2}{2m} + V(r)$$

$$H' = \frac{\hbar}{4m^2c^2}(\nabla V \times p) \cdot \boldsymbol{\sigma} \tag{1.2.10}$$

$$H'' = \frac{e}{mc} \boldsymbol{A} \cdot \boldsymbol{p}.$$

In this expression, p and $\hbar\boldsymbol{\sigma}/2$ are the momentum and the spin operators of the electron, and A is the vector potential of the electromagnetic field inside the material. Thus, the second term in the Hamiltonian represents the spin–orbit interaction, and the third term describes the interaction of the material with the electromagnetic wave. There are two methods to calculate the conductivity tensor. In the first method, as used by Argyres, one starts with the eigenfunctions of H_0,

$$H_0 \psi_n \sigma(\pm 1) = E_n \psi_n \sigma(\pm 1), \qquad (1.2.11)$$

with

$$\psi_n = \frac{1}{\sqrt{N}} e^{i\mathbf{k}\cdot\mathbf{r}} u_n(r) \qquad (1.2.12)$$

being the Bloch wavefunction of energy E_n and $\sigma = \pm 1$ denotes the quantum state of the electron spin. Then the spin–orbit interaction H' is treated as a perturbation to obtain the wavefunctions of $H_0 + H'$:

$$\phi_{n,\pm 1} = \left(\psi_n \pm \sum_{m \neq n} b_{nm} \psi_m \right) \sigma(\pm 1) \qquad (1.2.13)$$

with

$$b_{nm} = \frac{i\hbar^2/4m^2c^2}{E_m - E_n} \int d\mathbf{r}\, \psi_m^* (\nabla V \times \nabla)_z \psi_n. \qquad (1.2.14)$$

Here z is the quantization axis of the electron spin. Within the electromagnetic field, $\phi_{\alpha,\pm 1}$ will change into $\Psi_{\alpha,\pm 1}$ through the time-dependent perturbation H''. Then, from the current density operator

$$j = \sum_{\alpha,\sigma} \left(\frac{ie\hbar}{2m} (\Psi_{\alpha,\sigma}^* \nabla \Psi_{\alpha,\sigma} - \Psi_{\alpha,\sigma} \nabla \Psi_{\alpha,\sigma}^*) - \frac{e^2}{mc} A \Psi_{\alpha,\sigma}^* \Psi_{\alpha,\sigma} \right), \qquad (1.2.15)$$

the conductivity tensor can be derived. Although the procedure is straightforward, the calculation is tedious. The main conclusion is that the off-diagonal elements are antisymmetric, give rise to the magneto-optical effect and are proportional to the magnetization of the material. Numerical estimates give the order of magnitude of the magneto-optical effect for ferromagnets.

The second approach relates the elements of the conductivity tensor with the quantum transition rates. Macroscopically, the absorption power per unit volume is described by $P = \mathrm{Re}(\sigma_{ij} E_i E_j^*) V/2$. Microscopically, the absorption is due to the transition between different quantum states whose transition rates can be calculated by time-dependent perturbation theory. Then the element σ_{ij} can be deduced. We follow here the method used by Bennett and Stern.[20] Consider the simplest case that a plane wave $\boldsymbol{E}e^{-i\omega t}$ is propagating along the z axis, the absorption rates for linearly polarized ($\boldsymbol{E} = E\boldsymbol{e}_x$), right-circularly [$\boldsymbol{E} =$

$E(\mathbf{e}_x - i\mathbf{e}_y)/\sqrt{2}$], and left-circularly [$\mathbf{E} = E(\mathbf{e}_x + i\mathbf{e}_y)/\sqrt{2}$] polarized light are given by

$$P_x = \frac{VE^2}{2} \operatorname{Re} \sigma_{xx}$$

$$P_R = \frac{VE^2}{2} (\operatorname{Re} \sigma_{xx} + \operatorname{Im} \sigma_{xy}) \quad (1.2.16)$$

$$P_L = \frac{VE^2}{2} (\operatorname{Re} \sigma_{xx} - \operatorname{Im} \sigma_{xy}).$$

The real part of σ_{xx} and the imaginary part of σ_{xy} can then be deduced from P_x, P_R and P_L. The imaginary part of σ_{xx} and the real part of σ_{xy} can be obtained by Kramers–Kronig relations that relate the real and imaginary parts of the refractive index.

Microscopically, the transition rate from state $|\alpha\rangle$ to $|\beta\rangle$ due to the absorption of the electromagnetic wave is

$$W_{\alpha\beta} = \frac{2\pi}{\hbar^2} |H_{\alpha\beta}|^2 \left[\delta(\omega_{\beta\alpha} - \omega) + \delta(\omega_{\beta\alpha} + \omega) \right]. \quad (1.2.17)$$

Here $H_{\alpha\beta}$ are the transition matrix elements of the kinetic momentum

$$\boldsymbol{\pi} = \mathbf{p} + \frac{\hbar}{4mc^2} (\boldsymbol{\sigma} \times \nabla V)$$

for the three polarization modes:

$$H_{\alpha\beta}^x = \frac{E^2 e^2}{4m^2 \omega^2} |\langle \beta | \pi_x | \alpha \rangle|^2$$

$$H_{\alpha\beta}^R = \frac{E^2 e^2}{8m^2 \omega^2} |\langle \beta | \pi^+ | \alpha \rangle|^2, \qquad \pi^+ = \pi_x + i\pi_y \quad (1.2.18)$$

$$H_{\alpha\beta}^L = \frac{E^2 e^2}{8m^2 \omega^2} |\langle \beta | \pi^- | \alpha \rangle|^2, \qquad \pi^- = \pi_x - i\pi_y.$$

Then the conductivity tensor elements can be calculated:

$$\operatorname{Re} \sigma_{xx} = \frac{\pi e^2}{\hbar \omega m^2 V} \sum_\beta{}'' \sum_\alpha{}' |\langle \beta | \pi_x | \alpha \rangle|^2 \left[\delta(\omega_{\beta\alpha} - \omega) + \delta(\omega_{\beta\alpha} + \omega) \right]$$

$$\operatorname{Im} \sigma_{xy} = \frac{\pi e^2}{4\hbar \omega m^2 V} \sum_\beta{}'' \sum_\alpha{}' \left(|\langle \beta | \pi^- | \alpha \rangle|^2 - |\langle \beta | \pi^+ | \alpha \rangle|^2 \right) \quad (1.2.19)$$

$$\times \left[\delta(\omega_{\beta\alpha} - \omega) + \delta(\omega_{\beta\alpha} + \omega) \right],$$

where the α and β summations are over the occupied and unoccupied states, respectively. The imaginary part of σ_{xx} and the real part of σ_{xy} can be obtained from the Kramers–Kronig relations:

$$\text{Im } \sigma_{xx} = \frac{-2\omega e^2}{\hbar m^2 V} \sum_{\beta}{}'' \sum_{\alpha}{}' \frac{|\langle \beta |\pi_x| \alpha \rangle|^2}{|\omega_{\beta\alpha}|(\omega_{\beta\alpha}^2 - \omega^2)}$$

$$\text{Re } \sigma_{xy} = \frac{e^2}{2\hbar m^2 V} \sum_{\beta}{}'' \sum_{\alpha}{}' \left(\frac{|\langle \beta |\pi^-| \alpha \rangle|^2 - |\langle \beta |\pi^+| \alpha \rangle|^2}{(\omega_{\beta\alpha}^2 - \omega^2)} \right).$$

(1.2.20)

Evaluatiuon of these matrix elements requires band-structure calculations. Reference 15 provides some approximations to get a qualitative estimate of the matrix elements.

1.3 Macroscopic formalism of magneto-optical Kerr effect

1.3.1 General formulae

In this section, we discuss a macroscopic formalism for the magneto-optical effect in magnetic multilayers. Since most magnetic materials of interest are metals, which strongly absorb light, it is convenient to measure experimentally the reflected light in order to probe the magneto-optical effect. Therefore, while we will concentrate on the Kerr effect, the formalism can be readily extended to include the Faraday effect. For a given magnetic multilayer the refraction tensor for each layer can be expressed by a 3×3 matrix. The goal is to calculate the final reflectivity along different polarization directions. The general method is to apply Maxwell's equations to the multilayer structure, and to satisfy the boundary conditions at each interface. Zak et al.[24,25] developed a general expression for the Kerr signal based on this method. We will follow that approach.

First, it is important to describe the normal modes of the electromagnetic waves in a magnetic medium. To obtain them, consider a wave $\sim e^{i k \cdot x - i \omega t}$ propagating in a medium whose dielectric tensor is described by eqn (1.2.2). Since the magnetic response of the medium is attributed to the Voigt vector Q in the dielectric tensor, we can assume that the magnetic permeability is 1. Then the relationships between D and E, and B and H, are

$$D = \varepsilon E + i \varepsilon E \times Q \quad \text{and} \quad B = H. \quad (1.3.1)$$

Then Maxwell's equations give

$$k \cdot E + i k \cdot (E \times Q) = 0,$$
$$k \times E = \frac{\omega}{c} H,$$
$$k \cdot H = 0,$$
$$k \times H = -\frac{\omega \varepsilon}{c}(E + i E \times Q).$$

(1.3.2)

MACROSCOPIC FORMALISM OF MOKE

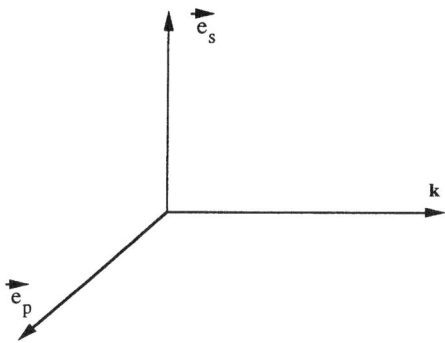

Fig. 1.2. The definition of the s and p directions relative to the wave propagation direction $\mathbf{k}$.

It is easy to see that $\mathbf{D}$, $\mathbf{B}$ (or $\mathbf{H}$) and $\mathbf{k}$ are perpendicular to each other. The $\mathbf{E}$ vector, however, has a component parallel to the wavevector $\mathbf{k}$. Using the familiar s and p polarization modes (see Fig. 1.2), the electric field can be written as

$$\mathbf{E} = E_s \mathbf{e}_s + E_p \mathbf{e}_p + i(-\mathbf{Q} \cdot \mathbf{e}_p E_s + \mathbf{Q} \cdot \mathbf{e}_s E_p) \mathbf{e}_k. \tag{1.3.3}$$

Here $\mathbf{e}_s$, $\mathbf{e}_p$ and $\mathbf{e}_k$ are unit vectors along the s and p and $\mathbf{k}$ directions. E_s and E_p are the s and p components of the electric field, and their equations of motion are

$$\begin{aligned}\left(\frac{\omega^2 \varepsilon}{c^2} - k^2\right) E_s + \frac{i\omega^2 \varepsilon \mathbf{Q} \cdot \mathbf{e}_k}{c^2} E_p &= 0 \\ -\frac{i\omega^2 \varepsilon \mathbf{Q} \cdot \mathbf{e}_k}{c^2} E_s + \left(\frac{\omega^2 \varepsilon}{c^2} - k^2\right) E_p &= 0.\end{aligned} \tag{1.3.4}$$

To first order in Q, it is easy to show that the two normal modes are right (R) and left (L) circularly polarized modes with

$$k_{R,L} = k(1 \pm \tfrac{1}{2}\mathbf{Q} \cdot \mathbf{e}_k) \quad \text{or} \quad n_{R,L} = n(1 \pm \tfrac{1}{2}\mathbf{Q} \cdot \mathbf{e}_k). \tag{1.3.5}$$

Here $k = (\omega/c)\sqrt{\varepsilon}$ and $n = \sqrt{\varepsilon}$ are the wavevector and refractive index, respectively, without the magnetization. After obtaining the two normal modes, any mode of the electromagnetic wave can be viewed as their superposition.

Now, we consider an electromagnetic wave propagating inside a magnetic multilayer structure. At each boundary between two layers, we define the incident and reflected waves as in Fig. 1.3. The boundary conditions involve E_x, E_y, H_x and H_y, but it is more convenient to express these four quantities with the s and p components of the electric field. The x components are easy to write because they are parallel to the s direction:

$$E_x = E_s^i + E_s^r, \tag{1.3.6}$$

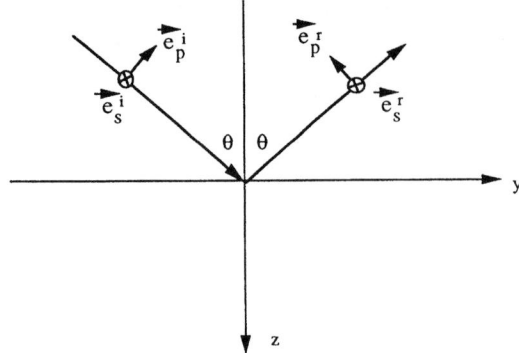

Fig. 1.3. Definitions of the s and p directions for the incident and reflected waves at the boundary between two media.

where the superscripts i and r denote the incident and reflected waves, respectively. For the y components, one has to keep in mind that the electric field has a component $i(-\mathbf{Q} \cdot \mathbf{e}_p E_s + \mathbf{Q} \cdot \mathbf{e}_s E_p)\mathbf{e}_k$ parallel to the $\mathbf{k}$ direction, and that the L and R modes have different refractive indices and incident angles. Then E_y can be expressed as

$$E_y = E_p^{i,L} \cos\theta_L + E_p^{i,R} \cos\theta_R + i\left(-\mathbf{Q} \cdot \mathbf{e}_p^i E_s^i + \mathbf{Q} \cdot \mathbf{e}_s^i E_p^i\right)\sin\theta$$
$$- E_p^{r,L} \cos\theta_L - E_p^{r,R} \cos\theta_R + i(-\mathbf{Q} \cdot \mathbf{e}_p^r E_s^r + \mathbf{Q} \cdot \mathbf{e}_s^r E_p^r)\sin\theta. \quad (1.3.7)$$

Using the relations:

$$E_p^L = +iE_s^L,$$
$$E_p^R = -iE_s^R, \quad (1.3.8)$$
$$n_R \sin\theta_R = n_L \sin\theta_L = n\sin\theta,$$

E_y can be expressed as

$$E_y = \frac{i}{2}\left[-\tan\theta\,(1+\cos^2\theta)Q_y + \sin^2\theta\,Q_z\right]E_s^i + (\cos\theta + i\sin\theta\,Q_x)E_p^i$$
$$+ \frac{i}{2}\left[\tan\theta\,(1+\cos^2\theta)Q_y + \sin^2\theta\,Q_z\right]E_s^r + (-\cos\theta + i\sin\theta\,Q_x)E_p^r. \quad (1.3.9)$$

H_x and H_y can be derived in a similar way from the expression $\mathbf{k} \times \mathbf{E} = (\omega/c)\mathbf{H}$:

$$H_x = \frac{in}{2}(\sin\theta\,Q_y + \cos\theta\,Q_z)E_s^i - nE_p^i$$
$$+ \frac{in}{2}(\sin\theta\,Q_y - \cos\theta\,Q_z)E_s^r - nE_p^r$$
$$H_y = n\cos\theta\,E_s^i + \frac{in}{2}(\tan\theta\,Q_y + Q_z)E_p^i \quad (1.3.10)$$
$$- n\cos\theta\,E_s^r - \frac{in}{2}(\tan\theta\,Q_y - Q_z)E_p^r.$$

Therefore, we obtained the relation between the x and y components of $\mathbf{E}$ and $\mathbf{H}$ with s and p components of the electric field. This relation can be expressed as a matrix product:

$$\begin{pmatrix} E_x \\ E_y \\ H_x \\ H_y \end{pmatrix} = A \begin{pmatrix} E_s^i \\ E_p^i \\ E_s^r \\ E_p^r \end{pmatrix} \qquad (1.3.11)$$

with the 4×4 matrix A known as the *medium boundary matrix*:

$$A =$$

$$\begin{pmatrix} 1 & 0 & 1 & 0 \\ \frac{i}{2}[-\tan\theta(1+\cos^2\theta)Q_y + \sin^2\theta\, Q_z] & \cos\theta + i\sin\theta\, Q_x & \frac{i}{2}[\tan\theta(1+\cos^2\theta)Q_y + \sin^2\theta\, Q_z] & -\cos\theta + i\sin\theta\, Q_x \\ \frac{in}{2}(\sin\theta\, Q_y + \cos\theta\, Q_z) & -n & \frac{in}{2}(\sin\theta\, Q_y - \cos\theta\, Q_z) & -n \\ n\cos\theta & \frac{in}{2}(\tan\theta\, Q_y + Q_z) & -n\cos\theta & -\frac{in}{2}(\tan\theta\, Q_y - Q_z) \end{pmatrix}$$

$$(1.3.12)$$

The inverse matrix of A is

$$A^{-1} =$$

$$\frac{1}{2}\begin{pmatrix} 1 & -\frac{i\sin\theta}{2\cos^3\theta}Q_y & \frac{i}{2n\cos\theta}Q_z & \frac{1}{n\cos\theta} \\ \frac{i}{2}(\sin\theta\, Q_y - \sin\theta\tan\theta\, Q_z) & \frac{1}{\cos\theta} & -\frac{1}{n} + \frac{i\tan\theta}{n}Q_x & \frac{i\sin\theta(1+\cos^2\theta)}{2n\cos^3\theta}Q_y + \frac{i}{2n}Q_x \\ 1 & \frac{i\sin\theta}{2\cos^3\theta}Q_y & -\frac{i}{2n\cos\theta}Q_z & -\frac{1}{n\cos\theta} \\ \frac{i}{2}(\sin\theta\, Q_y + \sin\theta\tan\theta\, Q_z) & -\frac{1}{\cos\theta} & -\frac{1}{n} - \frac{i\tan\theta}{n}Q_x & -\frac{i\sin\theta(1+\cos^2\theta)}{2n\cos^3\theta}Q_y + \frac{i}{2n}Q_x \end{pmatrix}$$

$$(1.3.13)$$

The next step is to derive the propagation matrix that relates $\mathbf{E}$ and $\mathbf{H}$ at the two surfaces of a film of thickness d. Since both the incident and reflected beams are composed of L and R circularly polarized modes, we use 1 and 2 to denote the L and R modes of the incident beam at both surfaces, and 3 and 4 to denote the L and R modes of the reflected beam at both surfaces. Then we have the following relations:

$$E_A^{1,2} = E_B^{1,2} \exp(ik^{1,2}d\cos\theta_{1,2})$$
$$E_A^{3,4} = E_B^{3,4} \exp(-ik^{3,4}d\cos\theta_{3,4}). \qquad (1.3.14)$$

The relation between E_s and E_p at boundaries A and B can then be expressed by a matrix product:

$$\begin{pmatrix} E_s^i \\ E_p^i \\ E_s^r \\ E_p^r \end{pmatrix}_A = D \begin{pmatrix} E_s^i \\ E_p^i \\ E_s^r \\ E_p^r \end{pmatrix}_B, \tag{1.3.15}$$

where D is 4×4 matrix known as the *medium propagation matrix*:

$$D = \begin{pmatrix} U\cos\delta_i & U\sin\delta_i & 0 & 0 \\ -U\sin\delta_i & U\cos\delta_i & 0 & 0 \\ 0 & 0 & U^{-1}\cos\delta_r & -U^{-1}\sin\delta_r \\ 0 & 0 & U^{-1}\sin\delta_r & U^{-1}\cos\delta_r \end{pmatrix} \tag{1.3.16}$$

with

$$U = \exp(-ikd\cos\theta),$$
$$\delta_i = \frac{kd}{2}(Q_y \tan\theta + Q_z), \tag{1.3.17}$$
$$\delta_r = \frac{kd}{2}(Q_y \tan\theta - Q_z).$$

With the A and D matrices, one can calculate the magneto-optical effect under any conditions. Consider a multilayer structure that consists of N individual layers, with a beam of light impinging from initial medium i (Fig. 1.4). After

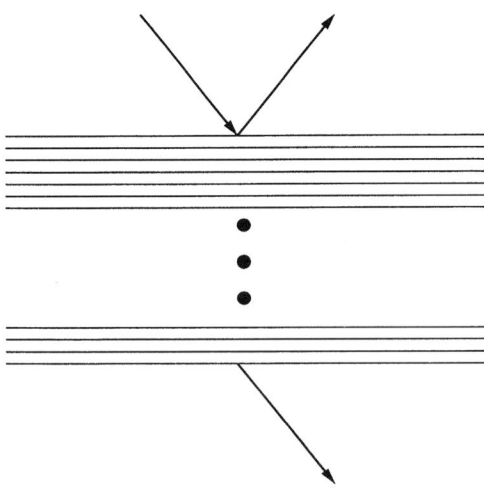

Fig. 1.4. Schematic of a multilayer structure.

MACROSCOPIC FORMALISM OF MOKE

multiple reflections, there will be a reflected beam from the top layer back to medium i and a transmitted beam from the bottom layer into the final medium f. The electric fields in media i and f can be expressed as

$$P_i = \begin{pmatrix} E_s^i \\ E_p^i \\ E_s^r \\ E_p^r \end{pmatrix}_i = \begin{pmatrix} E_s^i \\ E_p^i \\ r_{ss}E_s^i + r_{sp}E_p^i \\ r_{ps}E_s^i + r_{pp}E_p^i \end{pmatrix} \quad \text{and} \quad P_f = \begin{pmatrix} E_s^i \\ E_p^i \\ 0 \\ 0 \end{pmatrix}_f = \begin{pmatrix} t_{ss}E_s^i + t_{sp}E_p^i \\ t_{ps}E_s^i + t_{pp}E_p^i \\ 0 \\ 0 \end{pmatrix}.$$

(1.3.18)

Here r and t are reflection and transmission coefficients of the corresponding components. If P_m is the field component at the bottom surface in the mth layer, the field component at the top surface in the mth layer will be DP_m. Because E_x, E_y, H_x and H_y are related to P by the A matrix, the boundary condition at the interface between the mth and the $(m+1)$th layer is

$$A_m P_m = A_{m+1} D_{m+1} P_{m+1}.$$ (1.3.19)

Since E_x, E_y, H_x and H_y are continuous at each interface, the relation between P_i and P_f can be derived as

$$A_i P_i = A_1 D_1 P_1 = A_1 D_1 A_1^{-1} A_1 P_1 = A_1 D_1 A_1^{-1} A_2 D_2 P_2$$

$$= \cdots = \prod_{m=1}^{N} (A_m D_m A_m^{-1}) A_f P_f.$$ (1.3.20)

If this expression is put in the form $P_i = T P_f$, where

$$T = A_i^{-1} \sum_{m=1}^{N} (A_m D_m A_m^{-1}) A_f \equiv \begin{pmatrix} G & H \\ I & J \end{pmatrix},$$ (1.3.21)

then the 2 × 2 matrices G and I can be used to obtain the Fresnel reflection and transmission coefficients:

$$G^{-1} = \begin{pmatrix} t_{ss} & t_{sp} \\ t_{ps} & t_{pp} \end{pmatrix} \quad \text{and} \quad IG^{-1} = \begin{pmatrix} r_{ss} & r_{sp} \\ r_{ps} & r_{pp} \end{pmatrix}.$$ (1.3.22)

The Kerr rotation ϕ' and ellipticity ϕ'' for s- and p-polarized light are then given by

$$\phi_s = \phi_s' + i\phi_s'' = \frac{r_{ps}}{r_{ss}} \quad \text{and} \quad \phi_p = \phi_p' + i\phi_p'' = \frac{r_{sp}}{r_{pp}}.$$ (1.3.23)

1.3.2 Superlattices

Information about the magneto-optical rotation is contained in the matrix product ADA^{-1}. Using the Pauli matrices $\boldsymbol{\sigma} = (\sigma_x, \sigma_y, \sigma_z)$, it can be shown that $ADA^{-1} =$

$$\begin{pmatrix} \cos\alpha + \dfrac{\sin\alpha \tan\theta}{2} Q_x + \boldsymbol{\rho}\cdot\boldsymbol{\sigma} & \dfrac{[i\alpha\sin\alpha\tan\theta\, Q_y + (\sin\alpha - \alpha\cos\alpha) Q_z]}{2n\cos\theta} + \dfrac{1}{n}\boldsymbol{\eta}\cdot\boldsymbol{\sigma} \\ \dfrac{n[-i\alpha\sin\alpha\tan\theta\, Q_y + (\sin\alpha + \alpha\cos\alpha) Q_z]}{2\cos\theta} - n\boldsymbol{\eta}\cdot\boldsymbol{\sigma} & \cos\alpha - \dfrac{\sin\alpha\tan\theta}{2} Q_x + \boldsymbol{\rho}'\cdot\boldsymbol{\sigma} \end{pmatrix},$$
(1.3.24)

where

$$\alpha = \frac{2\pi nd\cos\theta}{\lambda},$$

$$\rho_x = \rho'_x = \frac{[\alpha\cos\alpha\sin^2\theta - (\cos^2\theta + 1)\sin\alpha]\tan\theta\, Q_y}{4\cos^2\theta}$$
$$- \frac{i\alpha\sin\alpha\tan^2\theta\, Q_z}{4},$$

$$\rho_y = \rho'_y = \frac{i[-\sin\alpha\sin^2\theta + (\cos^2\theta + 1)\alpha\cos\alpha]\tan\theta\, Q_y}{4\cos^2\theta}$$
$$+ \frac{\alpha\sin\alpha(1 + \cos^2\theta) Q_z}{4\cos^2\theta},$$
(1.3.25)

$$\rho_z = -\rho'_z = -\frac{\sin\alpha\tan\theta\, Q_x}{2},$$

$$\eta_x = -\frac{i\sin\alpha}{2}\left(\frac{1}{\cos\theta} - \cos\theta\right),$$

$$\eta_y = \frac{\sin\alpha}{2}\left(\frac{1}{\cos\theta} + \cos\theta\right), \qquad \eta_z = 0.$$

To obtain quantitative results, one must calculate the matrix product numerically.

1.3.3 Additivity law

For ultrathin films where the $\sum_i n_i d_i \ll \lambda$, i.e. the total optical thickness of the film is much less than the wavelength of the light, the ADA^{-1} matrix can be simplified to

$$ADA^{-1} = \begin{pmatrix} 1 & 0 & 0 & -\dfrac{i2\pi d}{\lambda} \\ -\dfrac{2\pi d}{\lambda} n\sin\theta\, Q_y & 1 + \dfrac{2\pi d}{\lambda} n\sin\theta\, Q_x & \dfrac{i2\pi d}{\lambda}\cos^2\theta & 0 \\ \dfrac{2\pi d}{\lambda} n^2 Q_z & \dfrac{i2\pi d}{\lambda} n^2 & 1 - \dfrac{2\pi d}{\lambda} n\sin\theta\, Q_x & 0 \\ -\dfrac{i2\pi d}{\lambda} n^2\cos^2\theta & \dfrac{2\pi d}{\lambda} n^2 Q_z & -\dfrac{2\pi d}{\lambda} n\sin\theta\, Q_y & 1 \end{pmatrix}.$$
(1.3.26)

Then the product $\prod_{m=1} A_m D_m A_m^{-1}$ can be evaluated:

$$\prod_{m=1} A_m D_m A_m^{-1} =$$

$$\begin{pmatrix} 1 & 0 & 0 & -\frac{i2\pi}{\lambda}\sum_m d_m \\ -\frac{2\pi}{\lambda} n \sin\theta \sum_m d_m Q_y^{(m)} & 1 + \frac{2\pi}{\lambda} n \sin\theta \sum_m d_m Q_x^{(m)} & \frac{i2\pi}{\lambda}\left(\sum_m d_m - n^2 \sin^2\theta \sum_m d_m/n_m^2\right) & 0 \\ \frac{2\pi}{\lambda}\sum_m d_m n_m^2 Q_z^{(m)} & \frac{i2\pi}{\lambda}\sum_m d_m n_m^2 & 1 - \frac{2\pi}{\lambda} n \sin\theta \sum_m d_m Q_x^{(m)} & 0 \\ -\frac{i2\pi}{\lambda}\left(\sum_m d_m n_m^2 - n^2 \sin^2\theta \sum_m d_m\right) & \frac{2\pi}{\lambda}\sum_m d_m n_m^2 Q_z^{(m)} & -\frac{2\pi}{\lambda} n \sin\theta \sum_m d_m Q_y^{(m)} & 1 \end{pmatrix}$$

(1.3.27)

If the initial and final media are nonmagnetic, then the 2×2 matrices G and I in eqn (1.42) can be calculated to give the following reflection coefficients:

$$r_{ss} = \frac{n_i \cos\theta_i - n_f \cos\theta_f}{n_i \cos\theta_i + n_f \cos\theta_f},$$

$$r_{pp} = \frac{n_f \cos\theta_i - n_i \cos\theta_f}{n_f \cos\theta_i + n_i \cos\theta_f},$$

$$r_{ps} = -\frac{4\pi}{\lambda} \frac{n_i \cos\theta_i}{(n_i \cos\theta_i + n_f \cos\theta_f)(n_f \cos\theta_i + n_i \cos\theta_f)}$$

$$\times \left(\cos\theta_f \sum_m d_m n_m^2 Q_z^{(m)} - n_f n_i \sin\theta_i \sum_m d_m Q_y^{(m)}\right),$$

$$r_{sp} = -\frac{4\pi}{\lambda} \frac{n_i \cos\theta_i}{(n_i \cos\theta_i + n_f \cos\theta_f)(n_f \cos\theta_i + n_i \cos\theta_f)}$$

$$\times \left(\cos\theta_f \sum_m d_m n_m^2 Q_z^{(m)} + n_f n_i \sin\theta_i \sum_m d_m Q_y^{(m)}\right),$$

(1.3.28)

where n_i, θ_i and n_f, θ_f are the refractive indices and incident angles of the initial and final media, respectively. Then an additivity law is obtained which states that the total Kerr signal is a simple summation of the Kerr signals from each magnetic layer, and is independent of the nonmagnetic spacer layers. This additivity law is true only in the ultrathin limit where the total thickness of the layered structure is much less than the wavelength of the incident light. For thicker films, interference effects will cause deviations from the additivity law. Ultimately the Kerr signal saturates at its traditional 'bulk' value that is independent of thickness.

There are three important Kerr configurations: polar, longitudinal and transverse. In the polar Kerr effect H is oriented normal to the film plane (z in Fig. 1.5(a)); thus, it is sensitive to the perpendicular component of the magnetization. The longitudinal case has H applied in the film plane and in the plane of the incident light (y axis in Fig. 1.5(b)); thus, it is sensitive to the in-plane

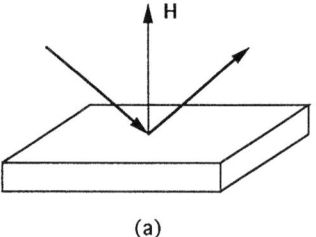

(a)

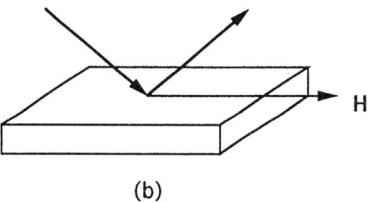

(b)

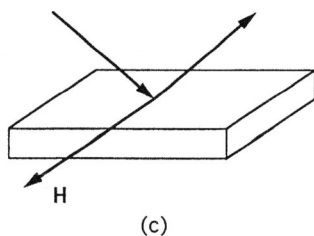

(c)

Fig. 1.5. Three common configurations for the SMOKE measurements: (a) polar, (b) longitudinal, and (c) transverse Kerr effect.

component of the magnetization. The transverse Kerr effect also has H applied in the film plane, but perpendicular to the incident plane of the light (x axis in Fig. 1.5(c)). The transverse Kerr effect depends on terms that are second order in Q, and manifests itself by a change of reflectivity. The polar signal is usually an order of magnitude greater than the longitudinal signal due to an n^2 enhancement factor in its numerator.

1.4 Experimental approach

1.4.1 Experimental setup

An experimental SMOKE setup has the advantage of simplicity. A linear

EXPERIMENTAL APPROACH

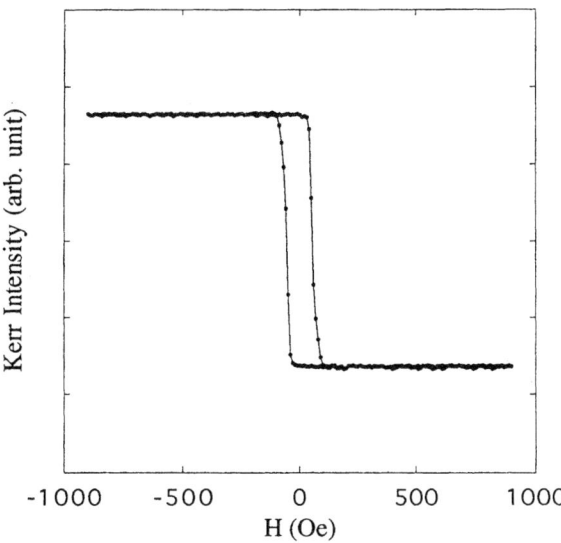

Fig. 1.6. A SMOKE loop taken from a six-monolayer Fe film grown on an Ag(100) substrate.

p-polarized (or s-polarized) laser serves as the light source. After reflection from the sample surface, the light intensity is detected by a photodiode with a linear (analysing) polarizer in front of it. The analysing polarizer is set at a small angle δ ($\sim 1-2°$) from the extinction condition to provide a d.c. bias. Then the reflected intensity as a function of the external magnetic field can be used to generate a magnetic hysteresis loop. Figure 1.6 shows an example of a hysteresis loop measured by SMOKE for a 6 ML Fe/Ag(100) film, where ML denotes monolayer. It is quite apparent that SMOKE can readily achieve monolayer sensitivity.

For a quantitative analysis, the influence of the optical conditions must be considered. The Kerr intensity measured by the photodiode after the analysing polarizer is

$$I = |E_p \sin \delta + E_s \cos \delta|^2 \approx |E_p \delta + E_s|^2. \qquad (1.4.1)$$

Recall that the expression $E_s/E_p = \phi' + i\phi''$ gives the Kerr rotation ϕ' and ellipticity ϕ''. Then eqn (1.49) becomes

$$I = |E_p|^2 |\delta + \phi' + i\phi''|^2 \approx |E_p|^2 (\delta^2 + 2\delta\phi') = I_0\left(1 + \frac{2\phi'}{\delta}\right) \qquad (1.4.2)$$

with

$$I_0 = |E_p|^2 \delta^2 \qquad (1.4.3)$$

representing the intensity at zero Kerr rotation. Thus, the measured intensity as a function of H yields the magnetic hysteresis loop. The saturation Kerr rotation ϕ'_m can be determined by the relative change of the Kerr intensity ΔI obtained upon reversing a field value that is equal to or greater than its saturation value:

$$\phi'_m = \frac{\delta}{4} \frac{\Delta I}{I_0}. \tag{1.4.4}$$

For *in situ* measurements, the ultrahigh-vacuum (UHV) windows (w) used as viewports usually produce a birefringence, $\phi'_w + i\phi''_w$, that prevents the realization of the optical extinction condition. In this situation, a quarter-wave plate is usually placed before the analysing polarizer to cancel the window birefringence. It can be shown that the extinction condition can be realized if the principal axis of the quarter-wave plate makes an angle ϕ'_w to the p axis, and the polarization axis of the analysing polarizer makes an angle of $\pi/2 + \phi'_w - \phi''_w$ to the p axis. Then the measured Kerr intensity becomes

$$I = |E_p|^2 (\delta^2 + 2\delta\phi'') = I_o\left(1 + \frac{2\phi''}{\delta}\right), \tag{1.4.5}$$

i.e. the relative Kerr intensity determines the Kerr ellipticity rather than the rotation in this case. The effect of the quarter-wave plate is to produce a $\pi/2$ phase difference between the s and p components so that the analysing polarizer will 'see' $i(\phi' + i\phi'') = -\phi'' + i\phi'$ instead of $\phi' + i\phi''$, i.e. the rotation and ellipticity are interchanged. Then to measure the rotation, a half-wave plate could be used to replace the quater-wave plate.

1.4.2 Verification of the macroscopic formulae

Experimental verification of the macroscopic formulae was sought via the investigation of various series of films of Co layers and Co/Cu superlattices. The films were grown and measured *in situ* via SMOKE utilizing a He−Ne laser. The single crystals of the Co overlayers and Co/Cu superlattices were grown at room temperature by the UHV evaporation process known as molecular-beam epitaxy (MBE) onto single-crystalline substrates of Cu(100) and Cu(111). The film growth took place in a UHV chamber equipped with high-energy electron diffraction (RHEED), low-energy electron diffraction (LEED) and Auger electron spectroscopy. The Cu substrate single-crystal disc was ∼10 mm in diameter. Its surface was first mechanically polished down to a ∼0.25 μm paste finish. Then the Cu was ultrasonically cleaned in methanol before it was put into the UHV chamber. Cycles of 3 keV Ar$^+$ sputtering and annealing at 650°C were used to clean the Cu substrate surface *in situ*. After this treatment, a well ordered Cu surface was formed as identified by RHEED and LEED. The RHEED intensity also was monitored during the growth of the film on the Cu(100) substrate in order to follow the process and to calibrate the thickness monitor. Over 200 RHEED oscillations were observed during the growth of

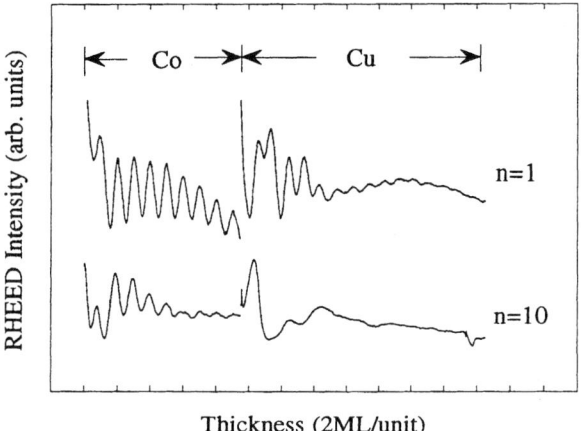

Fig. 1.7. RHEED oscillations taken during the growth of a [Co(9.5 ML)/Cu(16 ML)]$_n$ superlattice on a Cu(100) substrate.

Co/Cu superlattice (Fig. 1.7). Each oscillation represents the growth of an atomic layer. The persistence of the oscillations indicates a stable, well defined growth mode.

The Kerr ellipticities of the films were measured *in situ* as described in the last section. The results are plotted in Fig. 1.8. The ellipticity for Co on polycrystalline Cu is also plotted for comparison. We first concentrate on the magneto-optical behaviour of a single Co layer on the Cu substrate. The ellipticity data for the overlayers increase linearly in the ultrathin regime, reach a maximum at ~120 Å of Co, and approach a constant value for >400 Å of Co. The initial rise is expected since the Kerr effect is sensitive to the increasing amount of Co. In the thick regime, >400 Å of Co, the signal becomes constant since the absorption of light limits the depth sensitivity. In the intermediate regime the maximum in the ellipticity at ~120 Å of Co is attributed to an optical effect: the reflectivity changes from being dominated by Cu to Co. Since Cu has a higher reflectivity than Co, it acts as a mirror to enhance the signal. Similar behaviour is also observed in the Fe/Au system.[26] It is also interesting to note that the ellipticity is independent of crystal orientation in the thickness range studied.

To analyse the data quantitatively, we applied the formalism described in the last section to simulate the results. The refractive indices used were obtained from tabulations in the literature[27]: $n_{Cu} = 0.249 + 3.41i$ and $n_{Co} = 2.25 + 4.07i$. The values of Q_1 and Q_2, where $Q = Q_1 + iQ_2$, for Co were left as free parameters to best fit the experimental curves; the values $Q_1 = 0.043$ and $Q_2 = 0.007$ were obtained. The calculated curves, depicted as the solid lines in Fig. 1.8, are in good overall agreement with the experimental data. In particular, the peaked behaviour of the overlayers is faithfully reproduced. The ellipticities of three epitaxial Co/Cu superlattices were also measured *in situ* after each

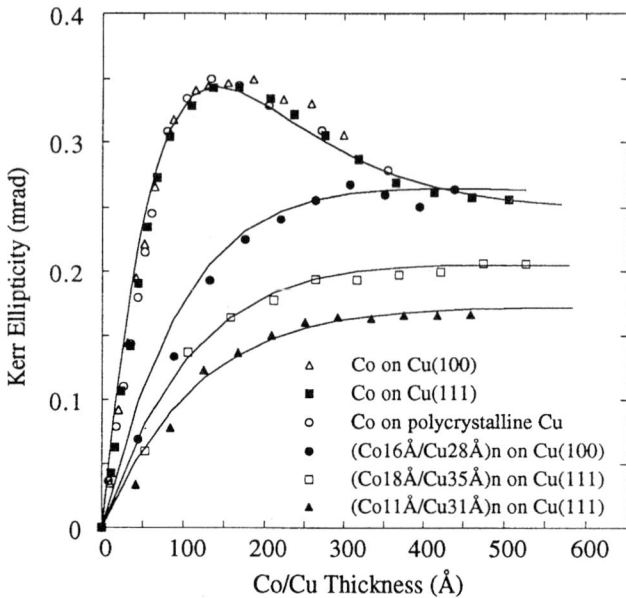

Fig. 1.8. The Kerr ellipticity measured for different samples. The solid lines are theoretical calculations.

Co/Cu bilayer was grown. The superlattices used were $[\text{Co}(16\,\text{Å})/\text{Cu}(28\,\text{Å})]_n$ grown on Cu(100), and $[\text{Co}(11\,\text{Å})/\text{Cu}(31\,\text{Å})]_n$ and $[\text{Co}(18\,\text{Å})/\text{Cu}(35\,\text{Å})]_n$ both grown on Cu(111). The ellipticity results for the superlattices also appear in Fig. 1.8 as a function of the total superlattice thickness. Again the ellipticities initially increase linearly in the ultrathin region, and then saturate in the thick regime, although there is no maximum at intermediate thickness as for the overlayer cases. The lack of a maximum in the intermediate thickness regime is because the reflectivity is not evolving from that of Cu to that of Co, as in the overlayer cases above. Instead, the reflectivity maintains itself at an average value between the two limits, since both Co and Cu remain within the penetration depth of the light, no matter how thick the superlattice becomes. Using the Q value obtained from the Co overlayers, the Kerr ellipticities for the superlattices were calculated and plotted in Fig. 1.8. The agreement with the experimental data is obvious.

To test the additivity law, the experimental data in Fig. 1.8 were replotted in Fig. 1.9 as a function of the thickness of only the magnetic Co layers, as opposed to the total superlattice thickness. All the data in the ultrathin regime then fall onto a single straight line. This result confirms the additivity law that the total Kerr signal in the ultrathin regime is a summation of the Kerr signal from each individual magnetic layer and is independent of the thickness of the nonmagnetic spacer layers.

Despite the good overall agreement, the calculated ellipticity can be seen to

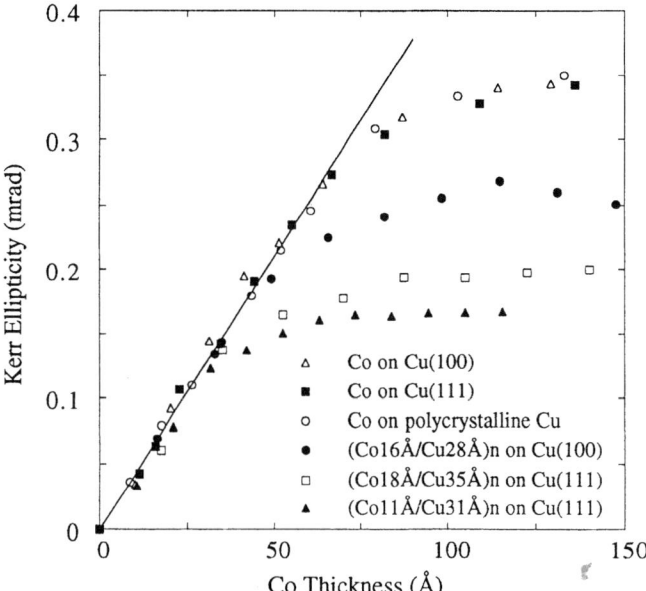

Fig. 1.9. The additivity law shows that the Kerr signal in the ultrathin regime depends only on the thickness of the magnetic layers.

exceed the experimental values in the ultrathin regime. For example, the calculated linear slope is 6.6 μrad Å^{-1}, while the experimental result yields only 4.3 μrad Å^{-1}. This systematic deviation can be due, for instance, either to the breakdown of the macroscopic description in the ultrathin region, or to optical parameters (n and Q) that deviate from their bulk values.

1.5 Applications of SMOKE in two-dimensional (2D) magnetic thin films

The high sensitivity and the local nature of the SMOKE technique have made it one of the more important *in situ* magnetic measurement techniques in the study of magnetic thin films. Since its debut in 1985 SMOKE has been applied to a variety of topics in thin film magnetism. Rather than summarize all important results achieved by SMOKE, we focus on three topics in the Fe/Ag system.

1.5.1 *2D magnetic phase transition*

As is well known, the magnetic exchange interaction in transition metals can be well described by a Heisenberg model. Mermin and Wagner[28] proved in 1966 that quantum fluctuations in a 2D Heisenberg lattice destroys long-range magnetic order at finite temperature. Experimentally, however, the Curie temperature (T_C) in most 2D magnetic films is finite. This seeming contradiction indicates that, in addition to the Heisenberg exchange interaction, there must exist another energy term in magnetic thin films. The Heisenberg Hamiltonian is invariant under rotation. In a real lattice, however, the electrons that contribute to the magnetization (3d electrons, for example) usually obey the lattice symmetry in their wavefunctions due to the crystal field. Thus, the spin–orbit interaction can transfer the lattice symmetry from the electron radial wavefunctions to the electron spins, breaking the spin isotropy. In other words, in addition to the isotropic Heisenberg exchange interaction, there also exist other energy terms that favour special directions for the electron spins. Such energy terms are referred to as the magnetic anisotropies. Generally, the magnetic anisotropy energy can be expanded in even-order terms of the magnetization:

$$E_{\text{ani}} = \sum_{\alpha,\beta} A_{\alpha\beta} u_\alpha u_\beta + \sum_{\alpha,\beta,\mu,\nu} B_{\alpha\beta\mu\nu} u_\alpha u_\beta u_\mu u_\nu + \cdots, \qquad (1.5.1)$$

where u_α is the αth component of the unit vector of the magnetization, and the $A_{\alpha\beta}$ and $B_{\alpha\beta\mu\nu}$ are the anisotropy coefficients. In a lattice with cubic symmetry (fcc and bcc, for example), it can be shown that the lowest-order term in the anisotropy energy is $\sim u^4$. However, when translational symmetry along one direction is broken, the square term in the anisotropy is no longer zero. Since this type of broken translational symmetry occurs at the surface of a crystal, the square term in the anisotropy is often called the *surface anisotropy* while the u^4 term is referred to as the *volume* or *bulk anisotropy*. Remember there are only two independent components in the three components of a unit vector, so the surface anisotropy can be written in the form

$$E_{\text{surf}} = K_1 u_z^2 + K_2 u_x^2, \qquad (1.5.2)$$

where the z axis is along the surface normal, and the x axis is along one of the two principal axes within the surface plane. Based on symmetry, K_2 vanishes if the surface normal is more than a 2-fold rotation axis. The sign of K_1 (K_2) will tell if the z (x) axis is the easy axis of the magnetization. Lattice distortions in thin films also can break the cubic symmetry to give rise to quadratic anisotropy. For some systems, this strain-induced anisotropy even dominates the total magnetic anisotropy.[29] The appropriate model includes the Heisenberg exchange interaction plus the magnetic anisotropy. The interesting question is: What is the effect of the magnetic anisotropy on the 2D long-range magnetic order? There are several approaches to this topic as discussed in the following.

Subsequent to the pioneering work of Polyakov,[30] who applied renormalization group theory to a 2D Heisenberg lattice, Bander and Mills examined the effect of the magnetic surface anisotropy on a 2D Heisenberg system.[31] Renormalization group theory cannot answer the question of whether or not long-range order exists. What it can answer is how the magnetization will behave if such a phase transition occurs. The results of Bander and Mills are twofold:

1. If the uniaxial magnetic anisotropy triggers a 2D long-range magnetic order,

then the Curie temperature should be $T_C \sim T_3/\ln(4\pi T_3/3K)$, where T_3 is the Curie temperature of the corresponding 3D lattice, and K is the effective anisotropy, which includes both the surface and the shape anisotropy terms.

2. The transition at T_C should belong to the 2D Ising universality class. Using experimental values of the surface anisotropy, the estimated magnitude of T_C from this theory is about 10^2 K, which is of the order of experimentally observed values. This suggests that the long-range magnetic order of thin films could originate from the magnetic surface anisotropy.

Experimental investigations of this subject usually involve temperature-dependent studies of the magnetization to determine its critical exponent β according to the power law $M(T) \sim (1 - T/T_C)^\beta$ for $T < T_C$ as T approaches T_C. Examples of systems studied include Fe/Au(100),[32] Fe/Pd(100),[33] Fe/Ag(111),[34] Co/Cu(111),[35] Ni/Cu(100),[36] Ni/Cu(111),[37] Ni/W(110),[38] Tb/W(110)[39] and Fe/W(110).[40] The β values determined for the various systems usually fall into two groupings: one has β close to the Ising value of 1/8, and the other has β close to 1/4. Films with $\beta \approx 1/8$ tend to have either a perpendicular easy axis, or an in-plane easy axis with an in-plane uniaxial anisotropy. Therefore, their phase transitions belong to the 2D Ising class. However, there are a few exceptions, for example, Ni(111)/W(110). Ni(111)/W(110) has an in-plane easy axis, but the Ni(111) surface has 6-fold rotational symmetry, so the in-plane surface anisotropy should have vanished. However, the W(110) substrate has only 2-fold rotational symmetry. Thus, we anticipate that the Ni(111) overlayer is strained by its substrate in one in-plane direction, which destroys its 6-fold symmetry and results in the in-plane uniaxial anisotropy.

On the other hand, films in the $\beta \approx 1/4$ grouping,[41] without exception, possess in-plane easy axes and zero in-plane magnetic surface anisotropy. Although this grouping exhibits universal behaviour, the underlying mechanism of the phase transition is still not well understood. These films should be described by an XY model plus a volume anisotropy. It is well known that the 2D XY model does not possess a finite T_C, but exhibits instead a Kosterlitz–Thouless transition for which the magnetic susceptibility diverges from a finite value.[42] After adding a volume anisotropy to a 2D XY system, theoretical studies show that a finite T_C could exist, but the transition at T_C exhibits non-universal behaviour.[43] Experimental investigations, however, always indicate a universal value of β, independent of film thickness in the 2D regime. An improvement to the theoretical calculation would be to consider a finite-size effect, since a real film of monolayer thickness usually breaks into finite-size domains or islands. Recent Monte Carlo simulations of the 2D XY model[44] show that magnetic fluctuations in a finite-size system yield a finite T_C with the critical exponent $\beta = 3\pi^2/128 \sim 0.231$, which is similar to the experimental value. But in this work the volume anisotropy is neglected. Therefore, it is still not clear what the role of the volume anisotropy is in a 2D magnetic thin film. More investigations are warranted.

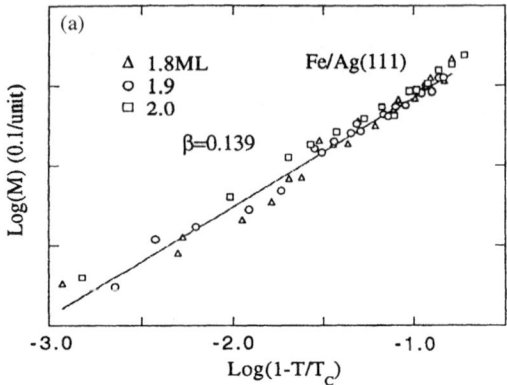

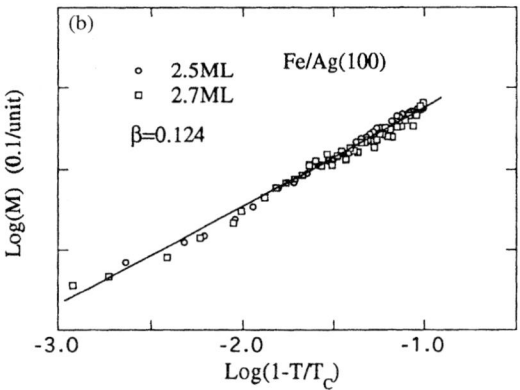

Fig. 1.10. Log–log plots of the magnetic remanence vs reduced temperature for (a) the Fe/Ag(111) and (b) the Fe/Ag(100) systems. The straight lines indicate a universal behaviour, $M \sim (1 - T/T_C)^\beta$, with exponent β close to the 2D Ising theoretical value of 1/8.

The SMOKE technique has been playing an important role in the study of 2D magnetic phase transitions. In particular, SMOKE has been used to determine experimental β values. This is achieved by measuring magnetic remanences at different temperatures and fitting of the data to the expression: $M \propto (1 - T/T_C)^\beta$. As an example, we show results obtained for two systems: (i) Fe/Ag(111), which has an in-plane magnetization and a weak in-plane surface anisotropy, and (ii) Fe/Ag(100), which has a strong perpendicular surface anisotropy. Log–log plots of the remanent values of M vs $(1 - T/T_C)$ are shown in Fig. 1.10 for the two systems. The solid lines represent the scaling law, and the slopes give experimental approximations to the critical exponent β. The β values obtained are quite close to the 2D Ising value of 1/8. This result conforms with theoretical expectation that the magnetic surface anisotropy, even if vanishingly small, makes the 2D phase transition Ising-like.

It is worthwhile to discuss the methodology used in the determination of β. There are two difficulties in the power-law fitting procedure. First, T_C has to be taken as a fitting parameter since it cannot be determined independently due to the magnetization tail that extends above T_C. A small change in the T_C value chosen in the analysis will, in general, affect the value derived for β. The tail above T_C is caused either by the remanent magnetic field (the Earth's field, for example) or by the finite-size effect, which prohibits the divergence of the correlation length at the transition. Recently, Farle et al. approached the problem differently by determining T_C independently by monitoring the divergence of the magnetic susceptibility in *thick* Gd films.[45] In this way, they determined the T_C value within an accuracy of ± 0.1 K, and then they analysed their magnetization data with β as the only fitting parameter. Their result agrees with the 3D Heisenberg value. Unfortunately, the method does not work well for ultrathin films because of background problems from the substrate due to its diamagnetic response and alignment problems that introduce a tilt in the hysteresis loop, which limits the accuracy of the T_C determination.

A second difficulty is that the power law should be valid only within a certain range ΔT within T_C. While the exponent β is a universal parameter, ΔT is not. In general, ΔT is larger for a system with a shorter correlation length. But it is not clear what the ΔT should be for a 2D magnetic thin film. Instead of fitting the data to the power-law expression, Kohlhepp et al. utilized the exact solution of the 2D Ising model over a wide range of temperature.[35] But the rigorous physical meaning of this method is not clear since the systems studied are Heisenberg systems whose anisotropy makes their critical behaviour Ising-like. Hence, it is not necessary for such systems to behave in an Ising-like fashion for temperatures outside the critical region. For example, for $T \ll T_C$, while the 2D Ising solution gives an exponential temperature dependence of the magnetization, spin-wave excitations in 2D usually introduce a quasi-linear temperature dependence to the magnetization. The subject of the transition width ΔT would benefit from further investigations.

1.5.2 2D spin-reorientation transition

The investigation of the 2D spin-reorientation transition (SRT) was originally motivated as a test of the Mermin–Wagner theorem (see sec. 1.5.1). A real magnetic thin film can usually be described by an isotropic Heisenberg exchange, a magnetic surface anisotropy (K_S) and a shape anisotropy ($-2\pi M^2$) that originates from the short-range part of the dipole–dipole interaction. The direction of the easy axis of magnetization is determined by the sign of the effective surface anisotropy $K_{\text{eff}} = K_S/d - 2\pi M^2$, where d is the film thickness. For systems with $K_S > 0$, a magnetization perpendicular to the film plane can be stabilized at low temperature and below certain thicknesses. Changing temperature or thickness can cause K_{eff} to vanish at some point below T_C, and for the film to approach an isotropic 2D Heisenberg system. At $K_{\text{eff}} = 0$ the spin-reorientation transition should occur wherein M changes its direction from perpendicular to in-plane. The question of interest regards the presence or absence of long-range magnetic order at the transition. The possibility of an SRT below the Curie temperature was analysed theoretically using renormalization group theory by Pescia and Pokrovsky.[46] Since quantum fluctuations are strong in the vicinity of the SRT, one has to revert to the original dipole interaction to replace the shape anisotropy term $2\pi M^2$, which provides an adequate description only if all the moments are aligned in the same direction. This is because, as will be seen later, $2\pi M^2$ only represents the short-range part of the dipole interaction. The long-range part becomes crucial to understand the formation of stripe domains. The basic result of the renormalization group theory is that the uniaxial anisotropy and the strength of the dipole interaction are renormalized differently. Under certain conditions, these two terms could cross over at a temperature below T_C to result in the SRT.

Several groups have carried out experiments on this subject. The first were reported by Pappas et al.[47] using spin-polarized electron spectroscopy to characterize the systems Fe/Cu(100) and Fe/Ag(100). They found that at low temperature the easy axis was normal to the surface plane, at high temperature it was in-plane, and in the SRT region there was a temperature gap ~20 K wide within which the magnetic remanence vanished. With more precise measurements made possible via SMOKE, Qiu et al.[48] studied the Fe/Ag(100) system as a function of both temperature and film thickness. They found that the magnetization is not identically zero in the transition region, but is markedly reduced and exhibits structure in a 'pseudo-gap' that resembles an asymmetric ramping towards zero with increasing temperature or film thickness. Thus, the gap, if it exists, must be at least an order of magnitude smaller than the ~20 K reported by Pappas et al. Also, it should be mentioned that the SRT as a function of film thickness occurs within about one atomic layer thickness.

It is almost impossible to explore the detailed features of the SRT by growing many samples with different film thicknesses. This is because of sample-to-sample variations that cannot be adequately controlled. This difficulty was overcome by the advent of wedge-shaped samples. Such magnetic wedges were used earlier in the study of oscillatory magnetic coupling of giant-magnetoresistance systems. Wedged samples provide an essentially continuous change of film thickness so that, as a local probe scans across the sample, the magnetization can be measured for different thicknesses in a systematic manner. SMOKE is a local-probe technique that measures the magnetization within the confines of the laser spot. When applied to the SRT, its thickness resolution can achieve the 0.04 ML level for a typical wedge with 2 ML/mm slope. Figure 1.11 shows the SMOKE results for the SRT of the Fe/Ag(100) system. It is obvious that the magnetization in the gap region is non-vanishing. The interesting question to ask concerns the description of the 'gap' region.

Yafet and Gyorgy were first to recognize that a stripe domain structure has a lower energy than a single domain structure in a 2D system with perpendicular uniaxial anisotropy.[49] The dipole interaction can be separated into the short- and long-range parts. The short-range part behaves like the shape anisotropy $2\pi M^2$ and can be combined with the perpendicular surface anisotropy to give

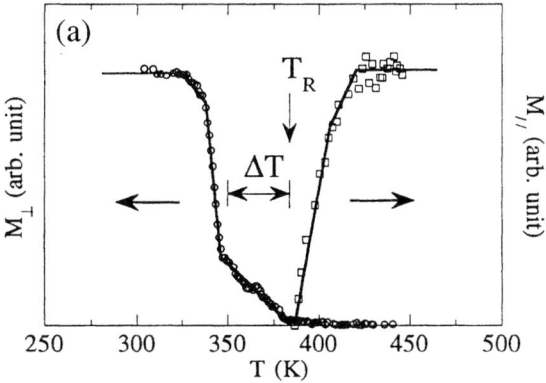

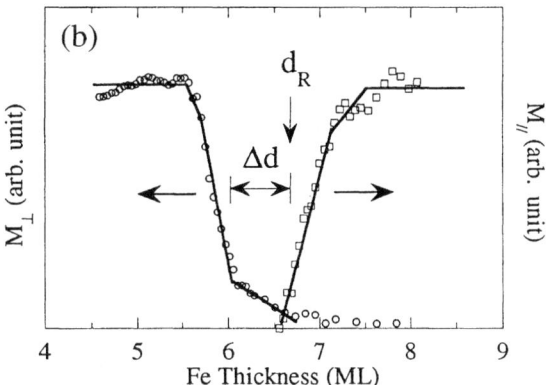

Fig. 1.11. Spin-reorientation transition as a function of (a) temperature and (b) film thickness.

the effective surface anisotropy. The long-range part of the dipole interaction tends to align neighbouring magnetic moments antiparallel, thus favouring the formation of magnetic domains. For a stripe domain structure of wavelength k ($k = 2\pi/L$, where L is the domain size), Yafet and Gyorgy showed that the energy is lowered, compared with a single domain structure, by an amount proportional to the wavevector k. The formation of stripe domains can be likened to the creation of a spin wave of wavevector k. As is well known, the spin wave will increase the total energy by Dk^2 due to the magnetic exchange interaction, where D is the spin-wave stiffness. Thus, the size of the stripe domains can be obtained by minimizing the total energy, which consists of (i) a positive k^2 term from the exchange interaction, (ii) a negative term linear in k from the long-range part of the dipole interaction, and (iii) the effective surface anisotropy term. It was found that the domain size increases almost exponentially as the effective surface anisotropy departs from zero. Therefore, the stripe domains are observable only in the vicinity of the SRT where the effective surface anisotropy is nearly zero and the domain size is less than the sample size. This explains why there is a gapped region in the SRT within which the magnetic remanence is greatly suppressed. It should be mentioned here that an equivalent picture of stripe domains was also proposed by Erickson and Mills[50] when they examined the effect of the dipole interaction on the spin-wave excitations in a 2D magnetic system. They found that the spin-wave energy becomes imaginary within the SRT region. The imaginary energy implies the 'death' of the dynamic spin wave, which converts into the static stripe domain structure. Stripe domains form a 1D ordered state, which itself is unstable against thermal fluctuations. Indeed, Kashuba and Pokrovsky[51] found that the stripe domain structure is equivalent to a 2D liquid-crystal system in that it possesses orientational order but no spatial order.

Experimental observation of the stripe domains in the SRT region was accomplished by Allenspach and Bischof who applied the SEMPA technique (scanning electron microscopy with polarization analysis) to study the SRT in the Fe/Cu(100) system.[52] They observed that the single domain structure of the film breaks into stripe domains ($\sim 1\,\mu$m size) in the gap region. Results of dynamic properties of the SRT are also consistent with a stripe domain structure.[53] The effect of higher-order magnetic anisotropies on the SRT is also a topic of interest. Under certain conditions, the higher-order anisotropy can alter the nature of the SRT to result in a continuous rotation of the magnetization.[54]

1.5.3 Magnetic anisotropy as a result of symmetry breaking

The magnetocrystalline anisotropy is one of the most important properties of 2D magnets. The magnetic anisotropy energy originates from the spin−orbit interaction,[55] and thus must obey the symmetry of the lattice. In an effort to understand how symmetry breaking induces a uniaxial magnetic anisotropy, a few groups have performed experiments on magnetic thin films grown on stepped (100) substrates. Atomic steps on a (100) surface break the 4-fold rotational symmetry of the film surface, and therfore should induce a uniaxial anisotropy within the film plane. This type of step-induced uniaxial anisotropy has been observed in (i) Ni/Fe and Fe films on Ag(3 1 100),[56] (ii) Co films on stepped Cu(100),[57,58] and (iii) Fe films on stepped W(100).[59] The unusual (3 1 100) indexing for the Ag substrate specifies the step direction [311] and average terrace width (100 lattice spacings). Experimentally, stepped surfaces consist of low-Miller-index terraces uniformly separated by atomic steps, and are created by polishing a sample surface that is misaligned by a few degrees from the terrace normal direction. Such surfaces are also referred to as *vicinal* surfaces, because crystallographically they are oriented in the *vicinity* of fundamental, low-Miller-index faces.

In the above listed cases the step density is fixed. To explore experimentally the relationship between induced anisotropy and step density, many substrates with different vicinal angles would be needed. In practice, it is difficult to

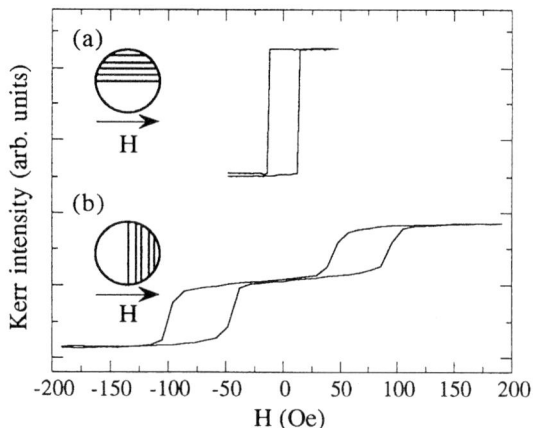

Fig. 1.12. Longitudinal hysteresis loops taken for a 25 ML Fe film grown on a stepped Ag(001) substrate with 6° vicinal angle. The external magnetic field is applied (a) parallel and (b) perpendicular to the step edges. The different loops indicate a uniaxial magnetic anisotropy within the film plane.

prepare multiple surfaces under identical conditions. To overcome this difficulty a recent study was performed for which a single 'curved' substrate was prepared with a graded step density. A 1 cm diameter Ag(001) substrate was used. Half of the surface was polished to retain its [001] orientation, and served as the reference, while the other half was polished with a curvature such that the vicinal angle varied continuously from 0° to 10°. Another sample was prepared for comparison purposes with a fixed angle of ~6°. In both cases, the atomic steps on the fcc Ag surface were parallel to the [110] direction, so that the steps on the bcc Fe(100) overlayer are parallel to its [100] direction.

To show the step-induced, in-plane magnetic anisotropy, an Fe film of ~25 ML thickness was grown onto the fixed-angle substrate. At this thickness, the magnetization is fully in-plane as indicated by a zero polar SMOKE signal. Longitudinal hysteresis loops were taken by applying the magnetic field parallel and perpendicular to the step edges. Figure 1.12 shows the results. For a field parallel to the step edges, the hysteresis loops has a square shape with full remanence. For a field perpendicular to the step edges, the loop shears into two separate ones and exhibits zero remanence. This behaviour shows that the atomic steps induce an in-plane, uniaxial magnetic anisotropy with the easy axis parallel to the step edges. The split loops are characterized by a shift field H_s, which is the field required to switch the magnetization from the easy to the hard axis, and thus is proportional to the step-induced anisotropy.

To explore the relation between this induced anisotropy and the step density, a 25 ML Fe film was grown onto the curved substrate. The SMOKE laser beam was then used to serve both as the magnetic probe and also as the reflection beam to determine the local vicinal angle α of the substrate. Therefore, as the laser beam scanned across the curved substrate, the hysteresis loop and the

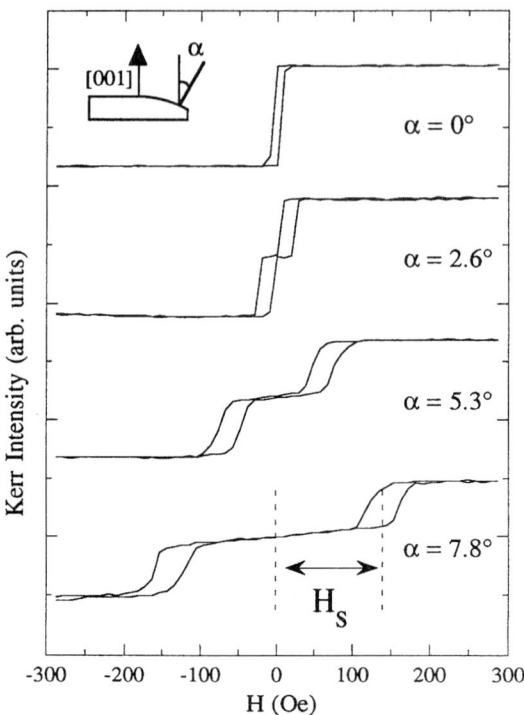

Fig. 1.13. Hysteresis loops taken at different positions along a 25 ML Fe film grown on a curved Ag(001) substrate. The entire relationship between the step-induced anisotropy and the step density is obtained on a single sample.

vicinal angle were measured simultaneously. Figure 1.13 shows representative longitudinal SMOKE loops at four vicinal angles for H along the hard axis. It is obvious that the induced anisotropy increases as the step density increases. Figure 1.14 shows the relation between H_s and α. The error bars in Fig. 1.14 account for the fact that the laser spot covers a finite range of step densities. The linear dependence of H_s on α in the log–log plot (Fig. 1.14(b)) indicates a power-law relation. Fitting $H_s \sim \alpha^n$ (the solid lines in Fig. 1.14) yields an exponent $n = 1.97 \pm 0.07$ (the slope in the log–log plot). Thus, the step-induced uniaxial anisotropy depends quadratically on step density. It should be mentioned that, as the laser scans across the sample, the magnetic field H also makes an angle α to the sample surface so that the in-plane magnetic field component should be $H \cos \alpha$ instead of H. However, the difference is negligible for the small values of α used ($< 1.2\%$ for $\alpha < 9°$).

To understand step-induced anisotropy better, and especially its quadratic relationship to the step density, we consider a phenomenological model based on Néel's pair-bonding mechanism. In Néel's model, the magnetic anisotropy is determined by the spin–orbit interaction due to nearest-neighbour electronic hybridization; therefore, it should be accurate for localized electrons. In transition-metal magnets, the magnetic 3d electrons are not fully localized, so the

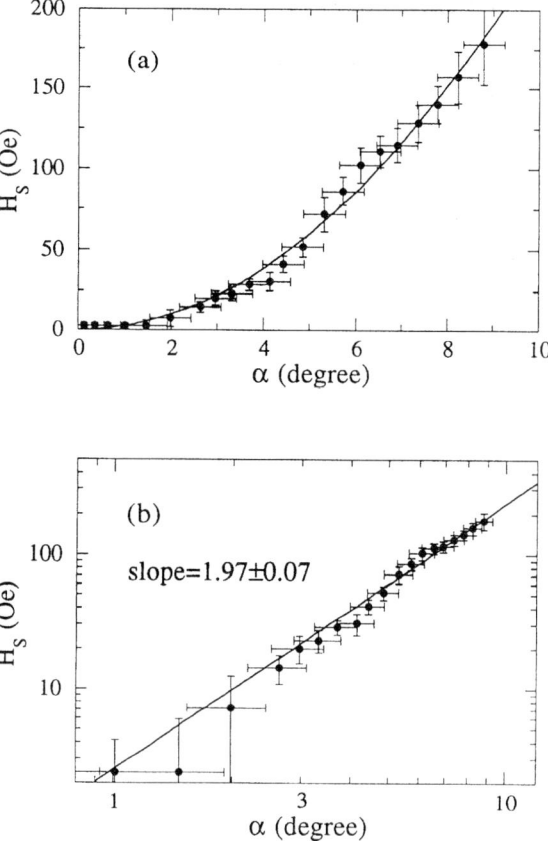

Fig. 1.14. The relation between H_s, which is proportional to the step-induced anisotropy, and the vicinal angle α, which is proportional to the step density. The solid lines represent a quadratic relation $H_s \sim \alpha^2$.

magnetic anisotropy should be derived with realistic wavefunctions from an energy-band calculation. If we disregard the numerical values of the anisotropy constants from the Néel model, but only keep its functional form, which reflects the symmetry of the lattice, the energy density of a stepped [001] bcc film of thickness d is

$$E = -\frac{K_S}{d} u_\zeta^2 + 2\pi M^2 u_z^2 + K_V\left(u_\xi^2 u_\eta^2 + u_\eta^2 u_\zeta^2 + u_\zeta^2 u_\xi^2\right) - \frac{K_{sp}}{dL} u_\eta u_\zeta, \quad (1.5.3)$$

where $\boldsymbol{u}$ is the unit vector of the magnetization $\boldsymbol{M}$, and ξ, η and ζ are the [100], [010] and [001] axes, respectively. The steps are on the (001) surface with edges parallel to the [100]. The average terrace length between steps is L. The normal direction (z axis) of the stepped surface makes an angle α to the [001] axis, so that $\alpha \approx a/L$ is proportional to the step density (a is the layer spacing in the [001]). The first three terms in eqn (1.5.3) represent the surface, shape and volume anisotropy, respectively. (The demagnetization factor is taken to be 1.) The last term is the anisotropy generated by the atomic steps on a [001] bcc surface.[60] The effect of strain is ignored in this model. It should be kept in mind, however, that the lattice distortion (even a small amount) could sometimes generate an appreciable amount of volume-type uniaxial anisotropy, and also could make K_V significantly different from its bulk value.[24] Therefore, K_S and K_V in eqn (1.5.3) should be viewed as operationally defined surface and volume anisotropies, and should in general be thickness-dependent. After a coordinate transformation from the crystal ξ,η,ζ frame to the film x,y,z frame with x and y axes in the plane of the film and parallel and perpendicular to the step edges, respectively, the energy density of eqn (1.56) transforms (to order α^2) into

$$E = \left[-\frac{K_S + (K_{sp}/a - K_S)\alpha^2}{d} + 2\pi M^2\right]u_z^2$$
$$+ \frac{(K_{sp}/a - K_S)\alpha^2}{d}u_y^2 - \frac{(K_{sp}/a - 2K_S)\alpha}{d}u_y u_z$$
$$+ K_V\left[u_x^2 u_y^2 + u_y^2 u_z^2 + u_z^2 u_x^2 - 2\alpha u_y u_z(u_y^2 - u_z^2) + \alpha^2(u_y^4 + u_z^4 - 6u_y^2 u_z^2)\right]. \tag{1.5.4}$$

We first consider the case of thick films where strong shape anisotropy forces the magnetization into the film plane ($u_z = 0$). The energy density is then

$$E = \frac{(K_{sp}/a - K_S)\alpha^2}{d}u_y^2 + K_V(u_x^2 u_y^2 + \alpha^2 u_y^4), \tag{1.5.5}$$

which has in-plane, uniaxial anisotropy $(K_{sp}/a - K_S)\alpha^2/d$. The quadratic dependence of this anisotropy on the step density is consistent with our experimental observations. For $K_{sp}/a > K_S$, the hard axis will be along the y axis, i.e. perpendicular to the step edges. For H applied along the hard axis, it can be shown that the hysteresis loop will split into two loops with an offset field

$$H_S = \frac{2(K_{sp}/a - K_S)\alpha^2}{Md}.$$

A numerical estimate of K_{sp} can be obtained by fitting the data in Fig. 1.14(a) with this formula. Taking the values[61] $d = 25$ ML, $a = 1.435$ Å, $M = 1.71 \times 10^3$ G and $K_S = 1.6$ erg cm^{-2}, the fit yields $K_{sp} \approx 5.73 \times 10^{-8}$ erg cm^{-1}.

We now turn our attention to the thin-film regime. For a flat surface ($\alpha = 0$), the quadratic energy terms in eqn (1.5.4) reduce to the familiar form $(-K_S/d + 2\pi M^2)u_z^2$, which gives rise to a sharp spin-reorientation transition between $u_z = 1$ and $u_z = 0$ at $d_R = K_S/(2\pi M^2) \approx 6$ ML. The formation of stripe domains and the effect of the volume anisotropy usually make the transition less abrupt.[62,63] Under certain circumstances, the volume anisotropy can even alter the nature of the SRT from a first-order switching to a continuous rotation. For a stepped surface, the competition among the three quadratic terms in the

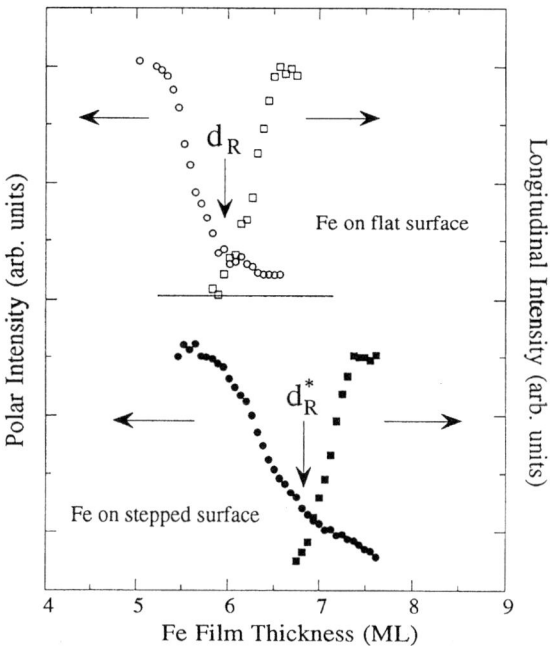

Fig. 1.15. Spin reorientation for Fe films on (a) flat and (b) stepped Ag(001) surfaces. A higher critical thickness for the transition is obtained on the stepped surface.

anisotropy energy will result in a continuous rotation of the magnetization towards the film plane in the yz plane upon increasing film thickness. At a critical thickness d_R^*, the magnetization will switch via a first-order transition. The z components of the magnetization and d_R^* are

$$u_z^2 = \frac{1}{2} - \frac{2\pi M^2 d - K_S - 2(K_{sp}/a - K_S)\alpha^2}{2\sqrt{(K_{sp}/a - 2K_S)^2 \alpha^2 + [2\pi M^2 d - K_S - 2(K_{sp}/a - K_S)\alpha^2]^2}},$$

$$d_R^* = \frac{K_{sp}^2/a^2}{8\pi M^2(K_{sp}/a - K_S)} + \frac{(K_{sp}/a - K_S)\alpha^2}{2\pi M^2}. \qquad (1.5.6)$$

For $\alpha = 6°$ (corresponding to our fixed-angle substrate), eqn (1.5.6) gives $d_R^* \approx$ 6.4 ML and $u_z \approx 0.5$ (at $d = d_R^*$). Thus, a larger SRT thickness is expected for Fe films grown on the fixed-angle vs the flat substrate.

The SRT as a function of film thickness was investigated by growing wedged Fe films (slope ~ 1 ML/mm) on the fixed-angle substrate. The polar and longitudinal remanences on the flat and stepped surfaces were measured at 150 K and are plotted in Fig. 1.15. For the stepped surface, the longitudinal SMOKE loop was taken along the in-plane easy axis. As expected, Fig. 1.15 shows that the SRT thickness has shifted from ~6 ML on the flat surface to ~6.8 ML on the stepped surface. The discrepancy between the calculated 6.4 and the observed 6.8 ML values could be attributed to several causes. Experimentally, the error in the zero-thickness registration point in the wedged samples can cause an apparent shift in the film thickness (estimated to be within ~0.2 ML). Theoretically, d_R^* was calculated using the value of $K_{sp} = 5.73 \times 10^{-8}$ erg cm^{-1}, which was derived from the 25 ML Fe film. If we allow K_{sp} to vary as a function of film thickness,[64] a value of $K_{sp} = 6.7 \times 10^{-8}$ erg cm^{-1} would give the observed 6.8 ML SRT thickness. It is also interesting to note that the polar signal in the SRT region on the stepped surface decreases with increasing film thickness more slowly than on the flat surface. This may be indicative of the continuous rotation of the magnetization in the yz plane on the stepped surface. (A significant y component in the longitudinal signal when applying H along the y (hard) axis is revealed in a detailed investigation on the shape of the SRT curve to be published elsewhere.)

The quadratic relation between step-induced anisotropy and step density originates from the $\sim u_\eta u_\zeta$ term in eqn (1.5.3). In general, the steps on a (001) surface of a cubic-symmetry lattice should induce both $\sim u_\eta u_\zeta$ and $\sim u_\xi^2$ anisotropies. The $\sim u_\xi^2$ term will result in a linear relation between the induced anisotropy and the step density. For a bcc lattice, the u_ξ^2 term should be absent because the missing atoms at the step contribute only to the $\sim u_\eta u_\zeta$ term. For an fcc lattice, the steps could induce both $\sim u_\eta u_\zeta$ and $\sim u_\xi^2$ anisotropies, so a superposition of linear and quadratic dependences of the induced anisotropy on the step density is expected. In a real experimental system, strain (there is a large lattice mismatch in the normal direction of the Fe/Ag(001) system) and defects at the steps could also give some residual u_ξ^2 anisotropy. By fitting the data with the inclusion of both a linear and a quadratic term, we find that the linear term comprises only 2.5, 0.5 and 0.3% of the overall anisotropy at $\alpha = 1°$, 5° and 10°, respectively. Hence, the dependence of the step-induced anisotropy on step density in the present case is nearly purely quadratic.

1.6 Summary

We showed select applications of the SMOKE technique in magnetic thin films. Although SMOKE is a powerful technique, it has certain weaknesses. For example, it cannot distinguish surface or interface magnetism from that arising from the interior layers. This is an area where nonlinear MOKE has major advantages, as will be described in other chapters of this book. SMOKE also cannot, in general, distinguish an antiferromagnetic phase from a nonmagnetic phase. These drawbacks leave many experimental challenges for the future. Also, concerning theoretical challenges, a microscopic understanding of magneto-optics in the monolayer regime is needed since macroscopic continuum theory must ultimately break down. Experimentally, it is also important to enhance both spatial and time resolution so that small-scale processes, such as domain wall dynamics, can be investigated. Possible ways to realize this goal involve combining SMOKE with other techniques, such as near-field optical spectroscopy, scanning tunnelling microscopy, and/or pump-and-probe methods. Another direction is in the investigation of magneto-optical effects with synchrotron radiation, such as in magnetic circular dichroism and related studies. The rapid development to understand and exploit the nonlinear magneto-optical effect is discussed throughout this book. The purpose of the present chapter was to stimulate interest in the impact of SMOKE on contemporary thin-film and surface magnetism. It is hoped that the reader is left with some sense of the history of the field and of the broad opportunities that remain to be explored.

Acknowledgement

This work was supported by the US Department of Energy, Basic Energy Sciences—Materials Sciences under contract DE-AC03-76SF00098 (at Berkeley) and W-31-109-ENG-38 (at Argonne).

References

1. See the preface of the book, *The Effects of a Magnetic Field on Radiation—Memoirs by Faraday, Kerr and Zeeman*, edited by E. P. Lewis (American Book Company, New York, 1900) p. v.
2. M. Faraday, *Trans. Roy. Soc.* (*London*) **5**, 592 (1846).
3. Faraday, *Diary*, 30 August 1845, **4**, 7434, edited by Thomas Martin (London, 1932–1936); *ibid*, 7437–7444.
4. *Ibid*, 13 September 1845, **4**, 7504.
5. J. Kerr, *Phil. Mag.* **3**, 339 (1877); **5**, 161 (1878).
6. See *Molecular Electro-Optics*, Part I, *Theory and Methods*, edited by C. T. O'Konski (Marcel Dekker, New York, 1976) p. 517.
7. Kundt, *Phil. Mag.* (5), **18**, 308 (1884).
8. E. R. Moog and S. D. Bader, *Superlattices and Microstructures* **1**, 543 (1985); S. D. Bader, E. R. Moog, and P. Grünberg, *J. Magn. Magn. Mater.* **53**, L295 (1986).
9. S. Klahn, P. Hansen, and F. J. A. M. Greidanus, *Vacuum*, **41**, 1160 (1990).
10. K. Nakamura, S. Tsunashima, S. Iwata, and S. Uchiyama, *IEEE Trans. Magn.* **25**, 3758 (1989); S. Hashimoto, H. Matouda, and Y. Ochiai, *Appl. Phys. Lett.* **56**, 1069 (1990); S. Hashimoto, Y. Ochiai, and K. Aso, *J. Appl. Phys.* **67**, 2136 (1990).
11. L. D. Landau and E. M. Lifshitz, *Electrodynamics of Continuous Media* (Pergamon, London, 1960).
12. James Clerk Maxwell, *Electricity and Magnetism*, Vol. ii, Chap. xxi.
13. W. Voigt, *Magneto- und Elektro-optic* (B. G. Teubner, Leipzig, 1908); and *Handbuch der Elektrizität und des Magnetismus*, Vol. IV:2, p. 393 (J. A. Barth, Leipzig, 1915).
14. R. F. Kiefl, *et al.*, *Phys. Rev. Lett.* **64**, 2082 (1990); K. B. Lyons, *et al.*, *ibid*, **64**, 2949 (1990); S. Spielman, *et al.*, *ibid*, **65**, 123 (1990).
15. F. Wilczek, *Scientific American*, May 1991, p. 58.

16. Table 8.2 of *Optics* by E. Hecht and A. Zajac (Addison-Wesley, Reading, Mass., 1974).
17. H. R. Hulme, *Proc. Roy. Soc.* **A135**, 237 (1932).
18. C. Kittel, *Phys. Rev.* **83**, 208 (A) (1951).
19. P. N. Argyres, *Phys. Rev.* **97**, 334 (1955).
20. H. S. Bennett and E. A. Stern, *Phys. Rev.* **137**, A448 (1965).
21. Y. R. Shen, *Phys. Rev.* **133**, A51 (1964).
22. J. E. Erskine and E. A. Stern, *Phys. Rev.* B **8**, 1239 (1973).
23. J. E. Erskine and E. A. Stern, *Phys. Rev.* B **12**, 5016 (1975).
24. J. Zak, E. R. Moog, C. Liu, and S. D. Bader, *J. Magn. Magn. Mater.* **89**, 107 (1990).
25. J. Zak, E. R. Moog, C. Liu, and S. D. Bader, *Phys. Rev.* B **43**, 6423 (1991).
26. E. R. Moog, S. D. Bader, and J. Zak, *Appl. Phys. Lett.* **56**, 2687 (1990).
27. J. H. Weaver, in *CRC Handbook of Chemistry and Physics*, 69th edition, edited by R. C. Weast, M. J. Astle, and W. H. Beyer (CRC Press, Boca Raton, 1988) p. E-387ff.
28. M. D. Mermin and H. Wagner, *Phys. Rev. Lett.* **17**, 1133 (1966).
29. K. Baberschke, *Appl. Phys.* A **62**, 417 (1996).
30. A. M. Polyakov, *Phys. Lett.* **59**B, 79 (1975).
31. M. Bander and D. L. Mills, *Phys. Rev.* B **38**, 12015 (1988).
32. W. Dürr et al., *Phys. Rev. Lett.* **62**, 206 (1989).
33. C. Liu and S. D. Bader, *J. Appl. Phys.* **67**, 5758 (1990).
34. Z. Q. Qiu, J. Pearson, and S. D. Bader, *Phys. Rev. Lett.* **67**, 1646 (1991).
35. J. Kohlhepp, H. J. Elmers, S. Cordes, and U. Gradmann, *Phys. Rev.* B **45**, 12287 (1992).
36. F. Huang, G. J. Mankey, M. T. Kief, and R. F. Willis, *J. Appl. Phys.* **73**, 6760 (1993).
37. C. A. Ballentine et al., *Phys. Rev.* B **41**, 10175 (1990).
38. Y. Li and K. Baberschke, *Phys. Rev. Lett.* **68**, 1208 (1992).
39. C. Rau, *Appl. Phys.* A **49**, 579 (1989).
40. C. H. Back, C. Würsch, A. Vaterlaus, U. Ramsperger, U. Maier, and D. Pescia, *Nature*, 378, **597** (1995).
41. W. Dürr, M. Taborelli, O. Paul, R. Germar, W. Gudat, D. Pescia, and M. Landolt, *Phys. Rev. Lett.* **62**, 206 (1989); F. Huang, G. J. Mankey, M. T. Kief, and R. F. Willis, *J. Appl. Phys.* **73**, 6760 (1993); R. L. Fink, C. A. Ballentine, J. L. Erskine, and J. A. Araya-Pochet, *Phys. Rev.* B **41**, 10175 (1990).
42. J. M. Kosterlitz and D. J. Thouless, *J. Phys.* C **6**, 1181 (1973).
43. J. V. José, L. P. Kadanoff, S. Kirkpatrick, and D. R. Nelson, *Phys. Rev.* B **16**, 1217 (1977).
44. S. T. Bramwell and P. C. W. Holdsworth, *J. Phys. Condens. Matter.* **5**, L53 (1993).
45. M. Farle, W. A. Lewis, and K. Baberschke, 'A detailed analysis of the magneto-optic Kerr signal in UHV of gadolinium thin films near the Curie temperature' (preprint).
46. D. Pescia and V. L. Pokrovsky, *Phys. Rev. Lett.* **65**, 3179 (1990).
47. D. P. Pappas, K.-P. Kämper, and H. Hopster, *Phys. Rev. Lett.* **64**, 3179 (1990).
48. Z. Q. Qiu, J. Pearson, and S. D. Bader, *Phys. Rev. Lett.* **70**, 1006 (1993).
49. Y. Yafet and E. M. Gyorgy, *Phys. Rev.* B **38**, 9145 (1988).
50. R. P. Erickson and D. L. Mills, *Phys. Rev.* B **46**, 861 (1992).
51. A. Kashuba and V. L. Pokrovsky, *Phys. Rev. Lett.* **70**, 3155 (1993); *Phys. Rev.* B **48**, 10335 (1993).
52. R. Allenspach and A. Bischof, *Phys. Rev. Lett.* **69**, 3385 (1992).
53. A. Berger and H. Hopster, *Phys. Rev. Lett.* **76**, 519 (1996).

REFERENCES

54. A. Berghaus, M. Farle, Yi Li, and K. Baberschke, in *Magnetic Properties of Low-Dimensional Systems II*, ed. by L. M. Falicov, F. Mejia-Lira, and J.-L. Moran-Lopez, *Springer Proc. Phys.*, Vol. 50 (Springer, Berlin, Heidelberg, 1990) p. 61.
55. D. Wang, R. Wu, and A. J. Freeman, *Phys. Rev. Lett.* **70,** 869 (1993) and references therein.
56. B. Heinrich, S. T. Purcell, J. R. Dutcher, K. B. Urquart, J. F. Cochran, and A. S. Arrott, *Phys. Rev.* B **38,** 12879 (1988).
57. A. Berger, U. Linke, and H. P. Oepen, *Phys. Rev. Lett.* **68,** 839 (1992).
58. W. Weber, C. H. Back, A. Bischof, C. Würsch, and R. Allenspach, *Phys. Rev. Lett.* **76,** 1940 (1996).
59. J. Chen and J. Erskine, *Phys. Rev. Lett.* **68,** 1212 (1992).
60. D. S. Chuang, C. A. Ballentine, and R. C. O'Handley, *Phys. Rev.* B **49,** 15084 (1994).
61. K. B. Urquhart, B. Heinrich, J. F. Cochran, A. S. Arrot, and K. Myrtle, *J. Appl. Phys.* **64,** 5334 (1988).
62. A. Kasuba and V. L. Pokrovsky, *Phys. Rev. Lett.* **70,** 3155 (1993).
63. B. Shulz and K. Baberschke, *Phys. Rev.* B **50,** 13467 (1994).
64. W. Wulfhekel, S. Knappmann, and H. P. Oepen, *J. Appl. Phys.* **79,** 988 (1996).

2
MAGNETIZATION-INDUCED SECOND HARMONIC GENERATION FROM SURFACES AND ULTRATHIN FILMS

R. Vollmer

Max-Planck-Institut für Mikrostrukturphysik, Halle, Germany

2.1 Introduction

Optical second harmonic generation (SHG) and nonlinear optics in general became a rapidly growing field of research soon after the invention of the laser by Maiman in 1960. In the following years it was concentrated on nonlinear light generation in bulk materials. Starting in the 1980s the potential of second harmonic and sum frequency generation was explored in the investigation of surfaces and adsorbed molecules. However, it was not before 1989 when Pan *et al*.[1] and Hübner and Bennemann[2] proposed to use SHG as a means to probe surface magnetism. From the qualitative estimates of Pan *et al*. and the calculations of Hübner and Bennemann it seemed feasible to observe the magneto-optical response from a ferromagnetic surface.

Some years earlier the linear magneto-optical Kerr effect (MOKE) was applied successfully in the study of surface magnetism of ultrathin films.[3,4] Since then MOKE has become a premier surface magnetism technique because of its simplicity and relatively high sensitivity. Submonolayer ferromagnetic films could be detected with this method in favourable circumstances.[5] It was shown that the individual magneto-optical contributions from a multilayer structure add linearly.[6] By measuring the MOKE signal at different wavelengths a separation of the Kerr signal from the individual layers is possible.[7]

However, although the application of MOKE to ultrathin films sometimes was called surface magneto-optical Kerr effect (SMOKE) it lacks an intrinsic surface sensitivity and only information integrated over a (single) ferromagnetic layer can be obtained. In view of problems like exchange coupled ferromagnetic layers, where the coupling strength depends critically on the magnetic structure at the interfaces of the layers or surface and interface anisotropies of ultrathin films, an easy-to-use method having the capabilities of MOKE but intrinsic surface and interface sensitivity was highly desirable. Already in 1990 it was written in a review paper by Falicov *et al*...[8]

The Kerr effect is not an inherently surface sensitive probe. ... It is of interest to use complementary techniques with different probing depth to understand coupled magnetic layers, for instance. It should be possible to develop the Kerr effect into such a probe by using nonlinear optical processes ...

The first experimental evidence for the Kerr effect in the SH light generated from a surface was given by Reif et al. in 1991.[9] Magnetization-induced second harmonic generation (MSHG) at buried interfaces was demonstrated by Spierings et al. in 1993 on Au/Co multilayers. In general much larger relative Kerr effects in the SHG were found compared to the corresponding linear case. Reif et al. found Kerr angles as large as 76° in PtMnSn(111).[10] However, because of the lack of inversion symmetry the SHG was not restricted to the surface in this case.

A big step forwards was made after ultrashort-pulse solid-state laser systems like the self-mode-locked Ti:sapphire laser became commercially available. Despite the rather large relative Kerr signals the absolute SHG from surfaces is quite low. The SHG yields in the above-mentioned experiments were of the order of 10^{-15} SH photons per incident photon, making data acquisition times very long (of the order of an hour). This has been greatly improved with the new generation of lasers. Full hysteresis loops can be taken now within a few seconds, making MSHG a fast analytic tool for surface and interface magnetism.

Magneto-optical effects in bulk SHG from rare-earth garnet films were already observed by Aktsipetrov et al. in 1990[11] and theoretically described by Borisov and Lybchansky.[12] Recently, also this magneto-optical bulk SHG has received much interest because of the possibility of domain imaging of antiferro-magnetic materials like Cr_2O_3.[13-16] The observability of the domains results from an interference of magnetic dipole transitions and weakly allowed electric dipole transitions, the latter one being forbidden by symmetry above the Néel temperature. The observation of SHG in garnet films results from breaking the inversion symmetry of the whole film. Therefore electric dipole SHG is allowed and the magnetic contrast results from the interference of SHG due to the breaking of the inversion symmetry by the lattice distortion and SH light due to the further reduction of the symmetry of the magnetic field.[17,18] Nonlinear magneto-optical effects in these ferromagnetic garnet films are essentially the same as those discussed above resulting only from the symmetry breaking at surfaces and interfaces. They can yield additional information about the (magnetic) structure of the film. In this chapter, however, the discussion is restricted to materials with inversion symmetry and SHG at the *surface* and *interfaces*. Bulk SHG may occur as well in these cases but only due to higher-order terms.

One word on nomenclature: Besides MSHG different acronyms have been introduced in the literature, NLMOKE, NMOKE, NOLIMOKE or NOMOKE (nonlinear magneto-optical Kerr effect)[19-22] or SH-MOKE (second harmonic magneto-optical Kerr effect).[23] They all mean the same, namely the magneto-optical Kerr effect in the second harmonic light generated at the surface or the interfaces of the sample. The term 'nonlinear' in NLMOKE, etc., addresses the nonlinear electric polarization response (generating light at the second harmonic frequency of the incident light field) of the sample. The magneto-optical Kerr effect in the SH light is linear in the magnetization to first order.[24]

In this chapter SHG and MSHG are discussed with emphasis on the influence of magnetization to surface and interface SHG from well characterized surfaces and ultrathin films. Section 2.2 gives a review of ordinary SHG on metal surfaces with a discussion of azimuthal anisotropy of SHG, the effects of adsorbates and wavelength dependence. Section 2.3 describes briefly the magneto-optical Kerr effect (MOKE) in the linear reflected light. Polar, longitudinal and transverse Kerr geometry are introduced. Because the efficiency of surface SHG is very low, pulsed lasers generating high power densities on the sample have to be used. After a short description of the experimental equipment used for the experiments described in later sections, estimates of temperature rise of the sample caused by the laser pulse are given in sec. 2.4. A phenomenological description of MSHG is given in sec. 2.5. Symmetry properties of the nonlinear susceptibility are described and formulae for nonlinear Kerr angles and Kerr asymmetries are given for high-symmetry geometries that correspond to the longitudinal, transverse and polar Kerr geometries in the linear case. In sec. 2.6 experimental results of MSHG from surfaces and ultrathin films of ferromagnetic materials are presented. The influences of adsorbates on MSHG and the contribution of higher-order bulk SHG are explored. Quantum size effects in the SHG from multilayer structures like Cu/Co/Cu(001) are described in sec. 2.7. The chapter is closed with a summary and an outlook to future application of MSHG in the investigation of magnetic thin films.

2.2 Second harmonic generation from metal surfaces

2.2.1 Historical review

Soon after the invention of the laser by Maiman and the demonstration of optical second harmonic generation (SHG) from a quartz crystal in 1960 by Franken et al.,[25] optical SHG from metal surfaces were first observed in 1965 by Brown et al.[26] The historical development of experiment and theoretical interpretation of the SHG until 1992 has been described in the excellent review article of Janz and van Driel.[27] Therefore, we will give here only a very brief summary and focus on the latest developments. Some earlier reviews have addressed the investigation of surface polaritons,[28] the application of second harmonic generation as a tool for surface science,[29] and on SHG from the metal/liquid interface.[30]

The first experiments were analysed by Jha[31] in terms of magnetic dipole and electric quadrupole contributions

$$P(2\omega) = \alpha \left(E \times \frac{\partial H}{\partial t} \right) + \beta E(\nabla \cdot E), \quad (2.2.1)$$

with E and H the electric and magnetic field components of the incident light and α and β parameters depending on the electronic structure of the metal. Electric dipole contributions from the surface were neglected in this approach. A surface contribution arises only due to the second term in eqn (2.2.1) because of the rapid change of the normal component of the electric field vector at the surface or interface.

The first work that specifically addressed the SH response at an interface was published by Rudnick and Stern.[32] The SH radiation is produced by three different nonlinear current sources, the bulk current and two surface currents, parallel and perpendicular to the surface. The last one had been neglected in the earlier work. To estimate the amplitude of the SH light generated from the perpendicular and parallel surface currents, the authors introduced the coefficients a and b:

$$P_z^s(2\omega) \propto a(\omega)E_z(\omega)E_z(\omega) \qquad (2.2.2)$$

$$P_x^s(2\omega) \propto b(\omega)E_x(\omega)E_z(\omega) \qquad (2.2.3)$$

with $a \approx 1$ and $|b| \approx 1$. This parametrization has been used in subsequent publications on the subject.[33-37] It turned out that the original estimates by Rudnick and Stern of $a \approx 1$ strongly underestimated the surface contribution to the SHG. Using a quantum-mechanical theory, Weber and Liebsch[34] showed that the sharp cutoff of the charge density at the surface introduces serious errors in the calculation of the perpendicular nonlinear surface currents. Their calculation predicted a 1-2 orders of magnitude larger surface SH polarization. The early experiments by Quail and Simon[38] on the Al/glass and Ag/glass interfaces gave quite low values for a of the order of 1. However, this has to be attributed to the reduced SHG at the metal/glass interface compared to the metal/vacuum interface. On the other hand, by quantitative measurements of the SH light from an Al surface in ultrahigh vacuum (UHV) at a laser wavelength of $\lambda = 1.17$ eV, Murphy et al.[39] found a strong enhancement of the SHG at large angles of incidence. Using the theoretical results of ref. 37 they derived $a = -36 - 9i$, in very good agreement with the theoretical prediction. Furthermore, these experiments proved that for larger incident angles the SHG is determined mainly by the perpendicular nonlinear surface current for the vacuum/metal interface.

Despite the good agreement of theory and experiment on the second harmonic response normal to the surface, significant discrepancies became evident at smaller incident angles where signs of an anisotropic SH response were detected. Shortly after the observation of strong azimuthal anisotropy in centrosymmetric semiconductors[40-42] it was also discovered for metal surfaces.[43] Its was concluded that this anisotropy can only arise from interband transitions, i.e. from the periodic lattice potential, which had not been included in the theory. However, the observed azimuthal anisotropy for even the simple metal aluminium was surprising.[39,44] Janz et al.[45] suggested that the observed anisotropy is not an intrinsic property of an atomically flat surface but caused to a large extent by surface steps present at all surfaces. The authors derived this from the behaviour of the SH yield on O_2 exposure and temperature dependence of the SH. In both cases, with increasing O_2 coverage and with increasing temperature the anisotropic contribution decreased and finally fell below the detection limit of the experiment. It was shown on a vicinal Al(001) surface that the presence of steps significantly enhanced the SHG and introduced a strong anisotropy in the SH yield.[46] A theoretical study by Ishida and Liebsch[47] using a stepped jellium surface model did not find a strong step-induced anisotropy of the SH response. Ying et al.[48] did not find a strong temperature dependence of the azimuthal anisotropy as found previously.[49] Therefore, the microscopic origin of the observed anisotropy is open again. The isotropic part of the SHG from clean metal/vacuum interfaces can be described by the theoretical model of Liebsch and Schaich[37] quite well for not too high frequencies.

Still a number of questions remain. Quantitative theoretical description of SHG from metal surfaces is only possible in the simplest case. On the other hand, only a small fraction of the experiments were performed in a clean well defined environment. Questions regarding the influence of adsorbates or the influence of the surface or interface structure are not solved or only on a phenomenological level. Some of the more recent developments addressing these questions will be discussed in more detail below.

2.2.2 *Anisotropy of surface SHG*

Strong azimuthal anisotropy has now been observed on a number of metal surfaces: from the (111) surfaces of aluminium,[39,45,48] copper,[43,50] nickel,[51] silver[52-56] and gold,[54,57-59] from the (110) surfaces of copper,[60,61] silver[58,62-64] and gold,[65] and even from the (001) surfaces of aluminium,[44,62] copper,[62,66] silver[62] and gold.[62] Special interest should be paid to the Al(111) surface. Because aluminium is a nearly-free-electron metal and the (111) surface is extraordinarily smooth, the strong observed SH anisotropy is surprising at first view. This shows that even in that case SHG senses the lattice potential. In some of the earlier work surface resonances were invoked for the strong observed SH anisotropy from (111) surfaces.[39,43] However, even in the absence of a surface resonance strong anisotropy was observed.[55] Recently, Petukhov and Liebsch[67-69] developed a microscopic model which includes the Al crystal potential. Amplitude and frequency variation of the anisotropic surface SH component described by the χ_{xxx} tensor element was found to be in good agreement with the experimental results of refs 39 and 45. Moreover, they found an extraordinarily large penetration depth of the nonlinear tangential surface current $j_{2x}(z, \omega)$ of several tens of ångströms contrary to the normal component which is restricted to the surface within a few ångströms. According to the authors the physical origin of the large penetration depth of the tangential current is the delocalized nature of the Bloch states in aluminium. This is illustrated in Fig. 2.1. The introduction of a surface couples bulk states with evanescent surface states in the bandgap. New eigenstates can be constructed from a linear combination of

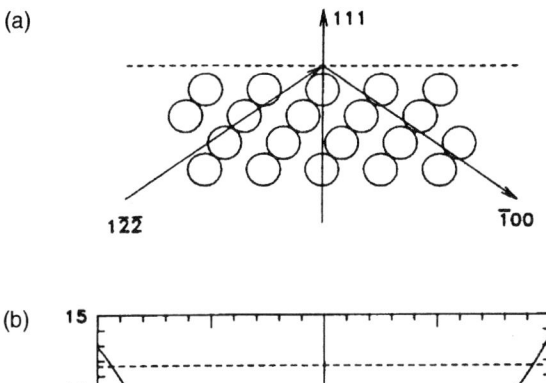

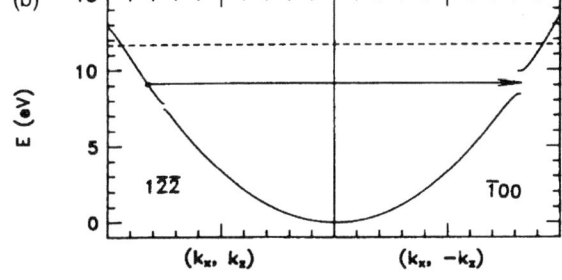

Fig. 2.1. Illustration of the Bragg reflections in a semi-infinite Al(111) crystal. In (a) the cut through the surface region in the $(1\bar{1}0)$ plane is shown. In (b) the bulk band structure in the $[\bar{1}00]$ (right) and $[1\bar{2}\bar{2}]$ (left) directions is plotted. The arrows indicate an electron coming from the $[1\bar{2}\bar{2}]$ direction which, after reflection from the surface, appears within the energy gap created by the V_{100} Fourier component of the lattice potential. After ref. 69.

bulk states with wavevector $k_1 = (K, k_z)$ and $k_2 = (K, -k_z)$, i.e. the specularly reflected wave. For energies in the range of the energy gap created in the $[\bar{1}00]$ direction by the V_{100} Fourier component, eigenstates along $[1\bar{2}\bar{2}]$ have no corresponding propagating states in the $[\bar{1}00]$ direction but only evanescent states in the gap. In this way noncentrosymmetric electron states in the vicinity of the surface are created that may contribute to the anisotropic SH response. The length to which the evanescent waves penetrate into the crystal can be calculated from the imaginary part of their wavevector and amounts to 16 Å in the middle of the gap and diverges towards the band edges.

Despite the large penetration depth of the parallel surface currents the anisotropic SH response can be quite sensitive to modifications of the surface as it is evident from the experimental results of ref. 45. In particular, it was found that oxygen adsorption strongly decreases the SH anisotropy. In the theoretical model of Petukhov and Liebsch the adsorbates act as scattering centres. Therefore, the electrons are scattered more diffusely in all directions, reducing the relative weight of the evanescent surface states to the SH anisotropy.

A similar theory can be applied for stepped (100) surfaces.[68] As for the (111) surfaces, not the nonsymmetric outer profile of the electric density but the nonsymmetric stacking of the atomic planes parallel to the macroscopic surface

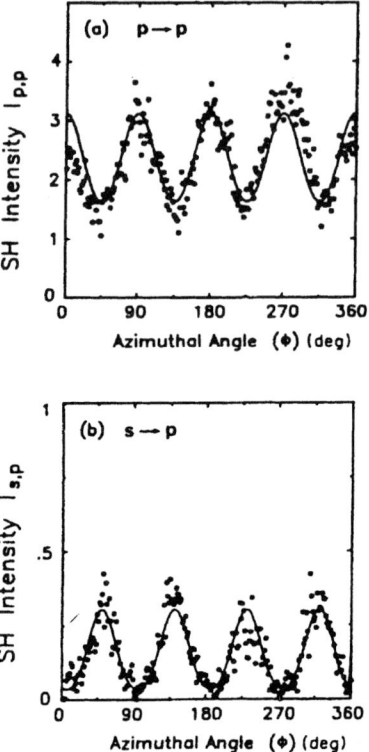

Fig. 2.2. The p-polarized second harmonic intensity from the Al(100)/air interface as a function of the azimuth angle ϕ: (a) for p-polarized incident fundamental light, (b) for s-polarized incident light. $\lambda = 1.06$ μm, $\theta_i = 32°$. After ref. 62.

plane is responsible for the observed anisotropy in ref. 46. More recently Jacobsen et al.[70] found a strong azimuthal dependence of the SH signal even on the oxide-covered stepped Al(001) surface.

The above model cannot account for the observed anisotropy on the (100) surfaces. The C_{4v} symmetry of the surface allows only an isotropic SH response. Nevertheless a strong anisotropy was observed.[44,62] Figure 2.2 shows the p-polarized SH response from an Al(100)/air interface. The surface was electrochemically polished, removing the plastically deformed layer from the mechanical polishing process. However, ambient oxygen produces an amorphous Al_2O_3 surface layer, which should, however, only modify the isotropic part of the SHG. As shown by Petrukhov and Liebsch[71] the bulk contribution to the anisotropy for the (100) surfaces is small compared to the surface response mainly determined by $\chi^{(2)}_{zzz}$. However, for the oxidized Al surfaces $\chi^{(2)}_{zzz}$ is greatly reduced. In addition the experiments were carried out at an incident angle of $\theta = 32°$, further suppressing the surface contribution. Unfortunately, up to now there exist no experimental data about the anisotropic SH response from clean Al(100) surfaces.

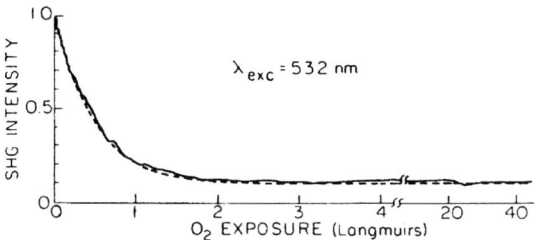

Fig. 2.3. Second harmonic signal from the Rh(111) surface as a function of the O_2 exposure. Solid curve, experimental results; dashed curve, theoretical fit using eqn (2.2.4). After ref. 72.

2.2.3 Effects of adsorbates and defects on SHG

While for clean metal surfaces a microscopic theory is developed, the understanding of how adsorbates affect SHG is still only on a phenomenological level. The first experiments on the subject were driven by interest in catalytical reactions on surfaces. Generally it was found that in most cases the SH yield dropped with increasing adsorbate coverage for adsorbates like O_2 and CO. It was shown in a number of publications that SHG can be used in this way to measure the adsorbate coverage on metal surfaces with submonolayer sensitivity in a very fast and clean way.[72-78] Qualitatively, the strong reduction of the surface SHG by strongly bound molecules like oxygen is explained by the reduced polarizability of the conduction electrons of the metal substrate. Figure 2.3 shows the p-polarized SH intensity of p-polarized incident light at $\lambda = 532$ nm from a Rh(111) surface as a function of oxygen exposure.[72] An exponential drop of the SH intensity is observed. In the case of noninteracting adsorption sites the nonlinear susceptibility can be written as:

$$\chi^{(2)} = A + B \frac{\theta}{\theta_S} \qquad (2.2.4)$$

with A and B the contribution from the bare metal surface and the net contribution from the oxygen-saturated surface having coverage θ_S. Assuming a simple Langmuir adsorption kinetics,[79]

$$\frac{d\theta}{dt} = Kp\left(1 - \frac{\theta}{\theta_S}\right), \qquad (2.2.5)$$

the constant K, which is proportional to the sticking coefficient and the ratio B/A, can be determined by using eqns (2.2.4) and (2.2.5) and fitting the resulting SH intensity to the experimental data in Fig. 2.3.

The behaviour of the SH intensity as function of the adsorbate coverage can be much more complex than in the example discussed above. As an example in Fig. 2.4 the SH intensities $\lambda = 532$ nm and $\lambda = 1.06\ \mu$m of the incident light from a Pt(111) surface are compared as a function of CO exposure.[74] While at the longer wavelength ($\lambda = 1.06\ \mu$m) the SH intensity drops by about one order

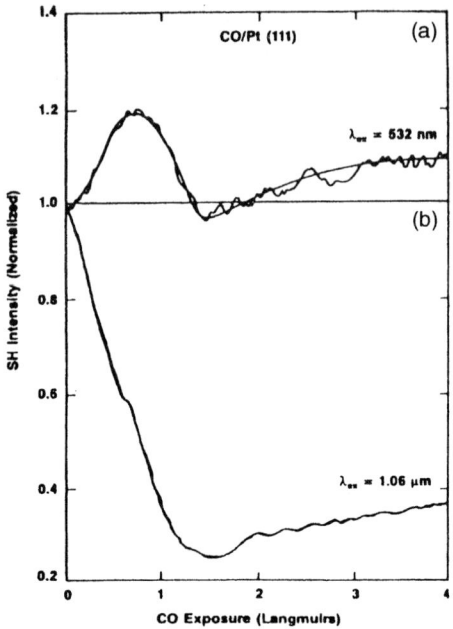

Fig. 2.4. Total second harmonic signal from a Pt(111) surface as a function of the Co exposure for an incident wavelength of $\lambda = 532$ nm (a) and $\lambda = 1.06$ μm (b). The incident angle $\theta = 67°$ from the surface normal. The azimuthal angle is 30° away from the projection of the [211] direction onto the surface. The curves are normalized to the SH intensity from the clean Pt surface for each wavelength. After ref. 74.

of magnitude it increases at $\lambda = 532$ nm. The authors propose that resonance enhancement via interband transitions or surface states in the vicinity of the 1.17 eV (1.06 μm) excitation energy may be responsible for the observed strong frequency dependence. A similar effect is observed on the Ag(110) surface. While Heskett et al.[75] found an exponential decrease of the SH intensity with oxygen coverage up to 0.5 monolayers for an incident wavelength $\lambda = 532$ nm and a plane of incidence along the [001] direction, Reiff and Block[64] found after a minor initial drop an increase of the SH intensity by a factor 9.5 for the 0.5 monolayers coverage compared to the clean surface for $\lambda = 1.06$ μm and the optical plane along the [1̄10] direction. This dramatic dependence of the SH upon oxygen exposure of the azimuth is easily understood if one considers the fact that the Ag(110) surface reconstructs upon oxygen exposure to an added row-type structure. This strongly modifies the silver states besides the appearance of oxygen-derived states. Quoting surface band-structure calculations, the authors of ref. 64 attribute this behaviour to silver surface states, which are shifted in energy upon oxygen adsorption. Therefore, with the oxygen adsorbed the silver surface states come closer to resonance with the fundamental incident light while for the clean silver surface resonances are unimportant. This interpretation is also supported by spectroscopic measurements that will be discussed in sec. 2.2.4. However, an increase of the SH yield with oxygen exposure has been observed also for other surfaces such as Cu(100)[66] and Fc(110),[80] where not such large surface reconstructions are known to be induced by oxygen exposure. Therefore, the system under investigation has to be well investigated by other methods before SHG can be used as an adsorbate coverage monitor. Nevertheless, the advantages of SHG, like the speed of measurement, and the nondestructiveness and cleanness, and the very high sensitivity, outweigh the possible difficulties in the interpretation of the data. To monitor the surface coverage SHG has been applied up to now successfully in thermal desorption spectroscopy experiments[73,76] and surface diffusion.[81] Since the SH yield depends not only on the amount of adsorbate but also on the adsorption site, it is possible to monitor adsorbate phase transition.[75,77,82]

For the adsorption of alkali atoms on semiconductors[83] and metal surfaces[72,84,85] generally a large increase of the SHG was found. For $\lambda = 1.06\ \mu$m laser excitation the SH intensity increases by more than two orders of magnitude within an alkali coverage of $1/3$.[84] The authors of ref. 84 deduce from a simple model that this strong SH enhancement is caused by a resonance of the SH light with the effective plasma frequency of the interfacial layer. Using the time-dependent density-functional approach Liebsch[36] showed that the nonlinear dipole moment shows a pronounced resonance-like structure near $2\omega \approx 0.8\omega_p(\text{ads.})$, with $\omega_p(\text{ads.})$ the bulk plasma frequency of the adsorbed alkali metal. The calculation showed that this surface local-field effect can enhance the SH intensity by several orders of magnitude.

2.2.4 SHG spectroscopy

Until recently SHG studies were restricted to one or a few fixed wavelengths by the availability of high-energy short-pulse lasers. In most investigations a Nd:YAG laser was used either at $\lambda = 1.06\ \mu$m or at the doubled frequency, $\lambda = 532$ nm. The first spectroscopic SHG measurements on metal surfaces were performed by Janz et al. on low-indexed aluminium surfaces[45] using a synchronously pumped dye laser emitting a train of 3 ps pulses with a repetition rate of 76 MHz. The wavelength of the laser could be tuned in the range between 565 and 870 nm. In Fig. 2.5 the wavelength dependence of the isotropic part of SH intensity for three incident angles $\theta = 67.5°$, $45°$ and $22.5°$ is compared to the theoretical results of Liebsch and Schaich.[37] Since the theory uses a jellium model for the aluminium surface, lattice effects like the anisotropic SH response and differences in the SH response from different surfaces of the same material cannot be expected to be described well. However, especially for the (111) surface the theory describes the isotropic SH response quite accurately except for a region around 800 nm (1.5 eV), which is attributed to an interband transition between parallel bands. Larger differences were observed for the (100) surface. An overall three times smaller SH intensity compared to the (111) surface is attributed to the larger effect of the lattice potential for the (100) surface. This is consistent with the larger workfunction for this surface, $\Phi = 4.41$ eV compared to $\Phi = 4.24$ eV for the (111) surface. The strong rise of the

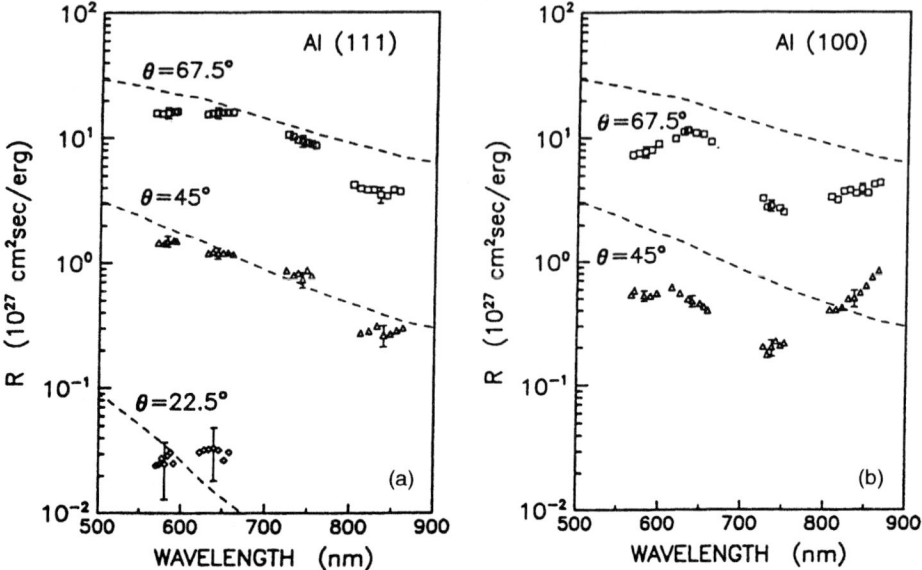

Fig. 2.5. The variation of the isotropic component of the p-polarized SH reflectivity $R = I_{2\omega}/I_\omega^2$ with the incident wavelength for (a) Al(111) and (b) Al(100) surfaces at incident angles of 67.5° (squares), 45° (triangles) and 22.5° (diamonds). The dashed lines represent the theoretical SH reflectivity obtained from the LDA-RPA calculation of ref. 37. After ref. 45.

SH signal at wavelengths above 800 nm (at $2\omega < 3$ eV) may be attributed to the surface state $\bar{\Gamma}$ on the Al(100) surface, which is located 2.75 eV below the Fermi energy.

Already in 1988 Hicks et al.[86] observed on the Ag(110) surface a strong resonance enhancement of SHG when 2ω is near the interband absorption edge at ≈ 3.9 eV (see Fig. 2.6). The main purpose of that paper was to demonstrate the use of the SH response as a local ultra-fast thermometer. The surface was heated by a 8 ns Nd:YAG laser at $\lambda = 1.06$ μm and the surface was probed by a picosecond dye laser. The SH response at the peak wavelength $\lambda_{SH} = 314$ nm was found to be proportional to the temperature at the surface. The temporal development of the 8 ns pump pulse can be described entirely by a thermal diffusion model. It was pointed out first by Liebsch and Schaich[37] that the observed strong enhancement at $\lambda_{SH} = 314$ nm is mainly due to a minimum in the linear dielectric function $\varepsilon(2\omega)$ rather than due to a change in the nonlinear susceptibility $\chi^{(2)}$ and therefore determined mainly by the bulk properties of Ag(110). Consequently, this resonance appears at the (111) surface as well.[87] Even on the Ag(111) surface in aqueous solution this enhancement was observed.[52] In a detailed study Li et al.[53] showed that both the frequency dependence of the azimuthal SH anisotropy and the polarization dependence can be fully described by the frequency dependence of the linear optical constants.

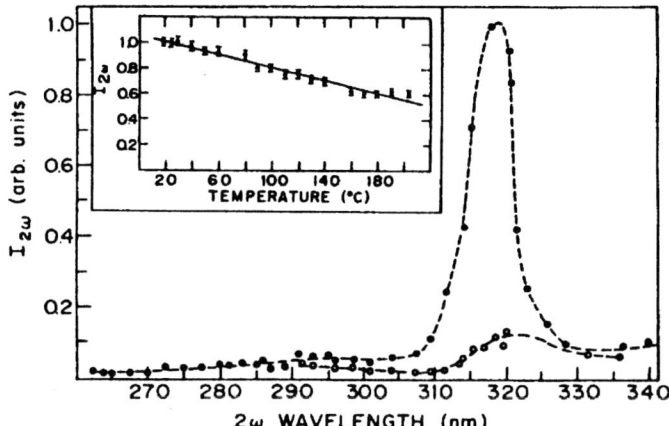

Fig. 2.6. Frequency dependence of p-polarized SHG from the Ag (110) surface for p-polarized fundamental light at an incident angle of $\theta_i = 60°$. Filled circles for 94 K and open circles for 573 K surface temperature. The inset shows the temperature dependence at the peak at 628 nm fundamental light with $\theta_i = 55°$. The straight line is a linear fit of the experimental points, $I_{2\omega} + 1.059 - (2.494 \times 10^{-3})T$. After ref. 86.

The first clear evidence of surface state resonance in $\chi^{(2)}$ was given by Urbach et al.[63] The SHG spectra obtained for two different planes of incidence are shown in Fig. 2.7. The peak denoted 'A' is due to the silver interband transition already discussed above. At a somewhat lower energy a second peak denoted as 'B' can be assigned to a transition from an occupied surface state to an unoccupied surface state at $\overline{Y}$. The symmetry of the initial and final surface state require the polarization of the incident light to have a component parallel to the $\overline{\Gamma Y}$ ([001]) direction. Therefore, the resonance enhancement is seen for the incident plane along the [1$\overline{1}$0] direction only for s-polarized incident light and for the incident plane along [001] only for p-polarized incident light. The relative intensities for the two azimuthal directions are determined mainly by the linear optical constants.

A similar strong enhancement as for silver has been observed for polycrystalline copper at the onset of interband transitions at 2.1 eV ($\omega = 620$ nm).[88] The authors pointed out that in this case the enhancement of SHG could not be described by the frequency dependence of the linear optical constants, which alone lead to a reduction rather than to an enhancement of SHG. From the polarization dependence of the SH signal the authors deduce that the enhancement is caused by a resonance in the bulk and/or surface nonlinear susceptibility at the onset of single-photon interband transitions. Because this study was performed on a polycrystalline copper sample in ambient air, the effects of the oxide layer and other adsorbates could not be separated.

The Cu(110) surface and the effect of oxygen on the SHG were investigated under more well defined conditions by Woll et al.[61] As shown in Fig. 2.8 they found a maximum in the SH intensity of a fundamental wavelength at about

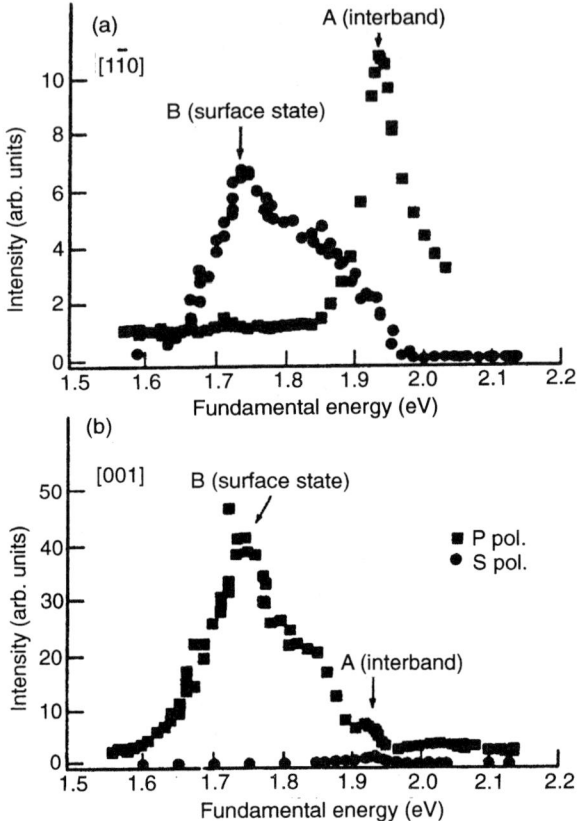

Fig. 2.7. Photon energy dependence of the SHG from an Ag(110) surface. The squares represent p-polarized incident light, the circles s-polarized incident light. (a) The plane of incidence is parallel to [1$\bar{1}$0]. (b) The plane of incidence is parallel to [001]. Note the intensity units are arbitrary but the same for (a) and (b). After ref. 63.

600 nm for s-polarized incident light and the incident plane along the $\overline{\Gamma Y}$ direction as found in ref. 88 for p-polarized incident light from a polycrystalline surface. However, Woll et al. attribute the observed maximum at $\lambda = 600$ nm to a resonant transition at ω between two surface states along the $\overline{\Gamma Y}$ direction. This interpretation was corroborated by the observed dependence of the SH intensity on the polarization of the incident light and on the azimuthal direction of the plane of incidence. For the $\overline{\Gamma X}$ azimuth the resonance enhancement is observed for p-polarized incident light. On oxygen exposure the SH intensity at the resonance is decreased because of an energy shift in the surface states.

2.3 The linear magneto-optical Kerr effect

Magneto-optical effects of ferromagnetic materials are produced by the combined effect of spin−orbit coupling and exchange interaction. As shown by

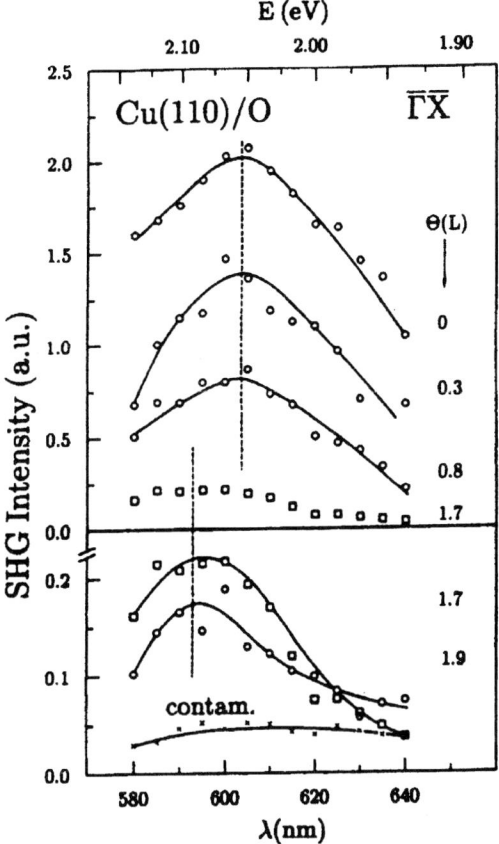

Fig. 2.8. Wavelength dependence of s-polarized SHG from the Cu(110) surface for s-polarized fundamental light at an incident angle of 45° for different oxygen exposures $\theta(L)$. The energy scale refers to the fundamental frequency. After ref. 61.

Argyres[89] the resulting action on the polarization of the electrons by an electric field can be described by an effective magnetic field. This acts as a 'Lorentz force' on the electron currents induced by the incident electromagnetic wave. A rotation of the induced electron current results, which in turn leads to a rotation of the electric field vector of the reflected wave depending linearly on the magnetization ***M***. This leads to the Faraday effect for the transmitted light and to the magneto-optical Kerr effect for the reflected light. An overview of experimental data collected on the Kerr effect can be found in recent review articles of Buschow[90] and Schoenes[91] for bulk materials and in the article of Bader[92] for ultrathin films. Recently, the calculation of the magneto-optical Kerr effect from first principles became possible.[93-96]

For the Faraday effect (also called circular magnetic birefringence) (Fig. 2.9(a)) only the component of the magnetization parallel to the propagation

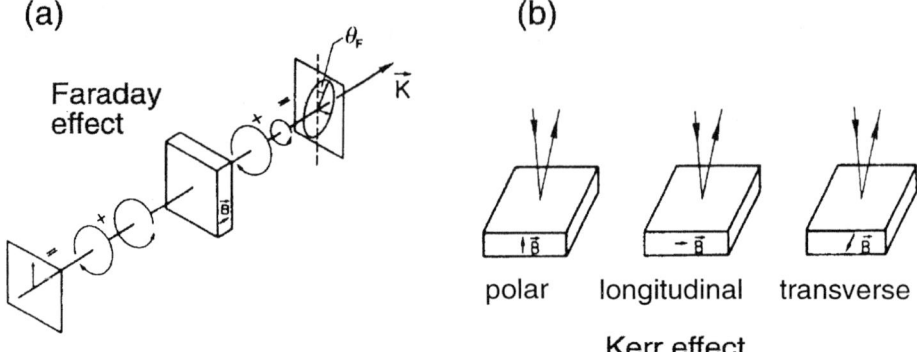

Fig. 2.9. Geometrical setup (a) for the Faraday effect and (b) for the magneto-optical Kerr effect for the three different configurations which differ in the relative orientation of surface normal and direction of the magnetization relative to the optical plane.

direction of the light leads to an effect linear in M. The magnetic field introduces a difference in the index of refraction for left and right polarized light. The polarization direction of a linearly polarized light beam is rotated by an angle Φ_F proportional to the length of the medium. Besides the circular birefringence a circular dichroism can be observed if the absorption of right and left circularly polarized light differs. The component of M perpendicular to the propagation direction of light gives rise to linear magnetic birefringence which is to lowest order quadratic in M (Voigt effect or Cotton–Mouton effect). Therefore these effects do not break the time inversion symmetry and do not change sign upon reversal of the magnetization direction.

For the magneto-optical Kerr effect (Fig. 2.9(b)) three configurations can be distinguished. In the polar configuration, the magnetization is parallel to the surface normal. This configuration is very close to the Faraday effect in transmission. Similar holds for the longitudinal geometry, where the magnetization is parallel to the surface but still in the optical plane. In these two geometries the magnetization has a component parallel to the propagation direction and therefore the above picture of the magnetization-induced circular birefringence can be applied for the reflected light as well. In the transverse geometry the magnetization is perpendicular to the optical plane. Still a magneto-optical effect linear in M results, as will be shown below.

A phenomenological description of the magneto-optical Kerr effect has been given by a number of authors for the problem of reflection from a magnetic surface[97-104] and layered structures.[6, 105-107] The most general formulae of the problem including crystalline birefringence and dichroism have been derived by Yeh[108] and Mansuripur.[106] Here only the very basic problem of reflection from a ferromagnetic surface of a cubic crystal is considered. In the formulae all contributions of quadratic and higher order in M are ignored. It is the purpose of this section to give a clear introduction to the Kerr effect to serve as a basis for understanding the magneto-optical effects in the frequency-doubled light discussed in subsequent sections.

As derived in ref. 101 the solution of Maxwell's equations inside a cubic crystal in the presence of a magnetization is linear combinations of the right (+) and left (−) circularly polarized plane waves

$$D^{\pm}(r,t) = D^{\pm} e^{i\omega(t - n^{\pm} \cdot r/c)}. \quad (2.3.1)$$

$n^{\pm}$ points in the propagation direction and $|n^{\pm}|$ is the index of refraction, which is slightly different for the two waves. The D field vector is related to the electric field vector E by

$$E = \varepsilon^{-1} D \approx \frac{1}{\varepsilon_0} D - \frac{1}{\varepsilon_0^2} g \times D \quad (2.3.2)$$

with the dielectric tensor

$$\varepsilon = \begin{pmatrix} \varepsilon_0 & ig_z & -ig_y \\ -ig_z & \varepsilon_0 & ig_x \\ ig_y & -ig_x & \varepsilon_0 \end{pmatrix} \quad (2.3.3)$$

Here ε_0 is the scalar dielectric constant of the medium in the absence of a magnetization, g_x, g_y and g_z are the components of the gyrotopic vector $g = \varepsilon_0 Q \hat{g}$, which is parallel to the magnetization direction. (Q is the Voigt constant, $\hat{g}$ the unit vector in the direction of g.) In eqns (2.3.2) and (2.3.3) terms higher than linear in g have been neglected. Up to linear order in g the refractive index obeys the following equation:

$$n^{\pm} \approx n \pm \tfrac{1}{2} \hat{n} \cdot g \quad (2.3.4)$$

with $n = \sqrt{\varepsilon_0}$ and the vector $n = n\hat{n}$ points in the direction the wave would propagate for $g = 0$. The $D^{\pm}$ waves are transverse waves with

$$D_p^{\pm} = \pm i D_s^{\pm} \quad (2.3.5)$$

From eqns (2.3.2), (2.3.4) and (2.3.5) together with Maxwell's equation $H = n \times E$ the x, y components of E and H, which are continuous across the surface boundary, can be determined and thus the components of the reflected light $E_s^{(r)}$ and $E_p^{(r)}$ can be expressed by $E_s^{(i)}$ and $E_p^{(i)}$ via the reflectivity tensor:

$$\begin{pmatrix} E_s^{(r)} \\ E_p^{(r)} \end{pmatrix} = \begin{pmatrix} r_{ss} & r_{ps} \\ r_{sp} & r_{pp} \end{pmatrix} \begin{pmatrix} E_s^{(i)} \\ E_p^{(i)} \end{pmatrix}. \quad (2.3.6)$$

The complex Kerr angle for s- and p-polarized light is defined as

$$\Phi_{Ks} = \Phi'_{Ks} + i\Phi''_{Ks} = \frac{E_p^{(r)}}{E_s^{(r)}},$$

$$\Phi_{Kp} = \Phi'_{Kp} + i\Phi''_{Kp} = \frac{E_s^{(r)}}{E_p^{(r)}}. \quad (2.3.7)$$

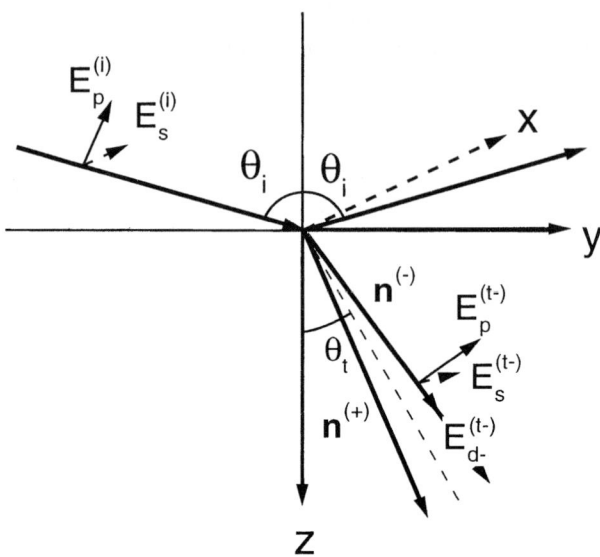

Fig. 2.10. Components of the electromagnetic field of the incident light wave $E_p^{(i)}$ and $E_s^{(i)}$, the reflected light wave $E_p^{(r)}$ and $E_s^{(r)}$, and the two transmitted waves $E_p^{(t\pm)}$ and $E_s^{(t\pm)}$. The propagation direction of the two transmitted waves is in the yz plane at an angle of $\theta_{t\pm}$ from the z direction.

With the definition of the coordinate system as shown in Fig. 2.10 for the polar Kerr configuration M is parallel to the z axis. The Kerr angles are given by[102]

$$\Phi_{Ks}^{(P)} = -\frac{inQ}{4\cos\theta_i}\frac{t_{ss}t_{pp}}{r_{ss}}$$

$$= -inQ\frac{\cos\theta_i}{(n\cos\theta_i + \cos\theta_t)(\cos\theta_i + n\cos\theta_t)},$$

$$\Phi_{Kp}^{(P)} = -\frac{inQ}{4\cos\theta_i}\frac{t_{ss}t_{pp}}{r_{pp}}$$

$$= -inQ\frac{\cos\theta_i}{(n\cos\theta_i + \cos\theta_t)(\cos\theta_i + n\cos\theta_t)}.$$

(2.3.8)

Here t_{ss} and t_{pp} are the usual transmission coefficients

$$t_{ss} = \frac{2\cos\theta_t}{\cos\theta_i + n\cos\theta_t} \qquad t_{pp} = \frac{2\cos\theta_t}{n\cos\theta_i + \cos\theta_t} \qquad (2.3.9)$$

and r_{ss} and r_{pp} are the reflection coefficients

$$r_{ss} = \frac{\cos\theta_i - n\cos\theta_t}{\cos\theta_i + n\cos\theta_t} \qquad r_{pp} = \frac{n\cos\theta_i - \cos\theta_t}{n\cos\theta_i + \cos\theta_t}. \qquad (2.3.10)$$

The angle θ_t is determined by Snell's law $n\sin\theta_t = \sin\theta_i$.

In the longitudinal Kerr configuration the magnetization direction is parallel to the y axis and the following formulae result:

$$\Phi_{Ks}^{(L)} = \frac{inQ}{4\cos\theta_i} \tan\theta_t \frac{t_{ss} t_{pp}}{r_{ss}}$$

$$= inQ \tan\theta_t \frac{\cos\theta_i}{(n\cos\theta_i + \cos\theta_t)(\cos\theta_i - n\cos\theta_t)},$$

$$\Phi_{Kp}^{(L)} = -\frac{inQ}{4\cos\theta_i} \tan\theta_t \frac{t_{ss} t_{pp}}{r_{pp}}$$

$$= -inQ \tan\theta_t \frac{\cos\theta_i}{(n\cos\theta_i - \cos\theta_t)(\cos\theta_i + n\cos\theta_t)}.$$

(2.3.11)

The formulae for the Kerr angles in the longitudinal configuration differ only by an additional factor $\tan\theta_t$. Since for metals in the frequency range of visible light the magnitude of the refractive indices n is usually much larger than $\sqrt{2}$, this leads to a reduction of the magnitude of the complex Kerr angle in the longitudinal geometry with respect to the polar geometry. For iron at 800 nm the refractive index is $n = 3.02 \pm 3.67i$[109] and therefore for $\theta_i = 45°$ the factor $|\tan\theta_t| = 0.15$.

For the transverse Kerr geometry no Kerr rotation occurs because the magnetization direction is perpendicular to the optical plane. In the simple picture mentioned above the Lorentz force causes a rotation of the D field in the optical plane. For p-polarized incident light a change in the reflectivity results (magnetic dichroism). This does not occur for s-polarized incident light because M and E are parallel. The asymmetry

$$A_K^{(T)} = \left| \frac{E_p^{(r)}(M) - E_p^{(r)}(-M)}{E_p^{(r)}(M) + E_p^{(r)}(-M)} \right|$$

(2.3.12)

can be expressed as

$$A_K^{(T)} = \left| Q \frac{\sin 2\theta_i}{\cos^2\theta_t - n^2\cos\theta_i} \right|$$

(2.3.13)

Figure 2.11 illustrates the dependence of the magneto-optical effect for the polar, longitudinal and transverse Kerr effect on the angle of incidence calculated for iron at 800 nm ($n = 3.02 + 3.67i$[109] and $Q = 0.0032 + 0.043i$[110]).

2.4 Instrumentation

2.4.1 The laser system

In special cases the output of a 20 mW CW laser diode is sufficient for the

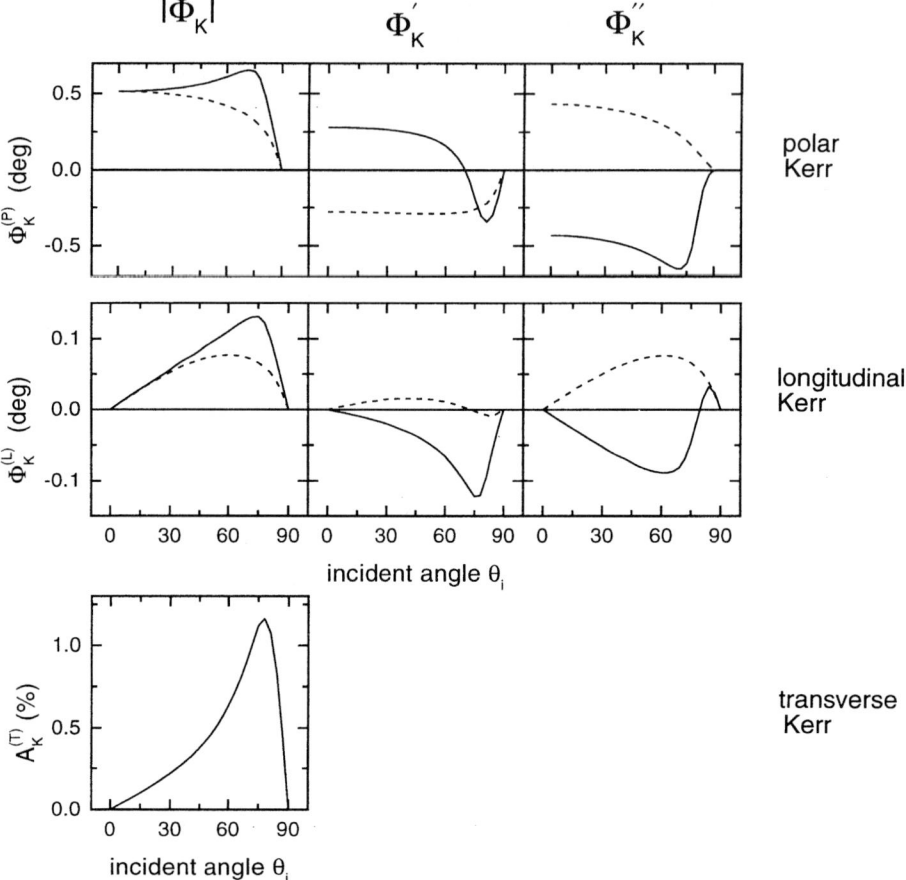

Fig. 2.11. Calculated Kerr angles for iron at 800 nm for the polar, longitudinal and transverse geometry. For the polar and longitudinal geometry the magnitude of the complex Kerr $|\Phi_K|$ (left panels), the Kerr rotation Φ'_K (middle) and the Kerr ellipticity Φ''_K is plotted. Curves corresponding to p-polarized incident light are drawn as solid lines, those for s-polarized light with dashed lines. For the transverse Kerr only the asymmetry $A_K^{(T)}$ is shown.

detection of surface SHG as demonstrated by Shen.[111] In general, however, without the local field enhancement on rough surfaces, pulsed lasers are necessary to create any detectable SHG from clean metal surfaces. In the past Nd:YAG lasers (pulse duration about 10 ns)[9,51,70,82,112] and synchronous pumped dye lasers (several picoseconds)[45] have been used. More recently mode-locked Ti:sapphire lasers (pulse duration about 100 fs) have become commercially available, and these have been used for the experiments described in the subsequent sections. This type of laser is tunable over a wavelength range from 720 nm down to about 1 μm with a maximum pulse energy of 10 nJ at about 790 nm. The repetition rate is 80 MHz.

Absolute intensity measurements of the surface SHG from the nearly-free-electron metal aluminium have been performed by Janz et al.[45] In the wavelength range of 500 to 850 nm ratios $R = I_{2\omega}/I_{\omega}^2$ of the order of 10^{-20} to 10^{-19} cm² s J⁻¹ have been measured. For the experiments the Gaussian shaped laser beam is focused onto the sample (in a UHV chamber) by a telescope down to a diameter ranging between 25 and 100 μm. Taking the above-mentioned ratio $I_{2\omega}/I_{\omega}^2 = 10^{-19}$ cm² s J⁻¹, and a pulse energy of 5 nJ (400 mW average power), this would result in an output of SH light of about 5×10^{-14} to 3×10^{-15} W or 2×10^5 to 10^4 SH photons per second. This is the value estimated from the nonlinear susceptibility of aluminium.

For Cu below the d-band excitation threshold similar ratios can be expected. Because of the high density of states of the d electrons in transition metals like Fe, Co and Ni, a much larger SHG yield is expected. Therefore, instead of photon counting we used a lock-in technique by chopping the incident beam in our experiments.

Figure 2.12 shows schematically the optical setup. The polarization of the incident beam can be rotated by a Babinet–Soleil compensator. In front of the telescope a Schott OG570 red transmitting edge filter is placed to filter out any SH light possibly generated in the rearward light pass. The SH light generated at the sample in the UHV chamber is detected by a photomultipler. A dichroic filter/mirror separate the SH light from the fundamental light, which is also monitored during the experiments. After the dichroic filter a calcite polarizer

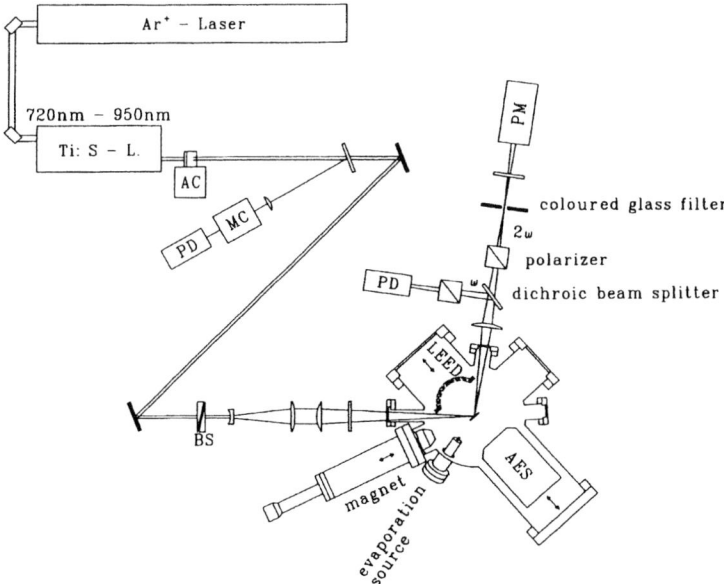

Fig. 2.12. Schematic drawing of the optical setup used for the *in situ* measurements: AC, autocorrelator; MC, monochromator; PD, photodiode; BS, Babinet–Soleil compensator; PM, photomultiplier.

can be placed to analyse the polarization of the outgoing SH light. A collimator is placed after the analyser to minimize any stray light from the optical components. To remove the residual fundamental light a Schott OG39 blue transmitting edge filter (3–5 mm thick) is placed in front of the multiplier. This additionally reduces the fundamental light by at least 15 orders of magnitude. Figure 2.12 shows schematically the experimental setup for the longitudinal Kerr geometry. Measurements in transverse or polar Kerr configuration can be performed by using additional viewports of the UHV chamber for incident and outgoing light beams not shown in the figure.

In comparison to the first experiment of Reif *et al.*[9] the SH signal for the present experimental setup is increased by 2–3 orders of magnitude.

2.4.2 *In situ sample preparation*

The sample temperature can be varied between 120 K and ≈ 1000 K. A magnetic field is supplied by a dipole magnet. The maximum field at the position of the sample is about 400 Oe. For ferromagnetic bulk samples an iron yoke on the manipulator is used on which the sample crystal is mounted to reach the necessary fields for magnetic saturation of the sample. The UHV chamber is equipped with standard surface science instruments like a rear-view low-energy electron diffraction (LEED) system, cylindrical mirror analyser for Auger electron spectroscopy (AES), quadrupole mass spectrometer, ion sputter gun, etc. Ultrathin metal films are grown *in situ* by electron evaporation sources. The growth can be monitored by using the electron gum of the AES system and the fluorescent screen of the LEED system as a medium-energy electron diffraction (MEED) system. Analogous to the more generally used RHEED (reflection high-energy electron diffraction) the reflected intensity of a 3 keV electron

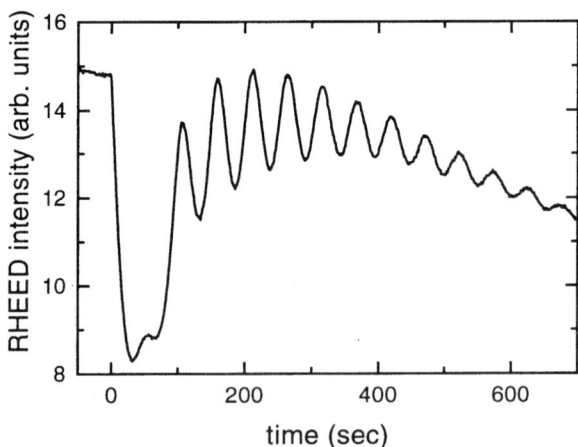

Fig. 2.13. MEED oscillations during the growth of Co on Cu(100) grown at room temperature.

beam incident nearly parallel to the surface is observed during the growth. A detailed discussion of RHEED can be found for example in ref. 113. As an example the MEED oscillations during the growth of Co on Cu(100) are shown in Fig. 2.13. The incident angle of the electron beam is chosen close to the 'in-phase' condition in which the electron waves scattered from adjacent (100) lattice planes interfere constructively. The initial reduction of the intensity and the following oscillatory behaviour of the reflected intensity are mainly caused by the periodic change of the surface roughness. At low coverages the island borders scatter more electrons in nonspecular directions. In the case of layer-by-layer growth the layers are nearly completely filled before nucleation in the next layer occurs. Therefore, the surface becomes smoother again when the islands start to coalesce and the MEED intensity increases again. This is repeated each time a monolayer is completed, resulting in the oscillatory intensity seen in Fig. 2.13. The first maximum corresponding to the completion of the monolayer is greatly suppressed because double layer growth occurs for the first two layers.[114]

2.4.3 Thermal heating by the laser and damage thresholds

The observation of nonlinear effects on surfaces makes it necessary to use rather high-power lasers. Therefore, one has to consider unwanted side-effects beside the generation of frequency-doubled light. The most important of them are unwanted heating and possible damage of the sample by the incident laser light. Damage created by a (single ultrashort) laser pulse may be not detectable unless time-resolved measurements are performed. Damage thresholds given in $J\,cm^{-2}$ depend strongly on the material, the wavelength and the pulse duration of the incident laser light. In the following we discuss the effect of thermal heating of a copper sample heated by a 6 ns Nd:YAG laser at $\lambda = 1064$ nm and by a 80 fs Ti:sapphire laser.

Thermal heating of a bulk copper sample by a Nd:YAG laser

The laser light incident on the sample surface is partially reflected, partially transmitted and partially absorbed in the solid. The energy is absorbed by excitation of electrons and is finally transferred via inelastic scattering of these excited electrons to the lattice. In metals for not too low temperatures the mean free time between collisions of electrons is of the order of 100 fs and therefore fast compared to the duration of the laser pulse which is a few nanoseconds. Therefore, it is justified to assume thermal equilibrium of electrons and lattice.

The temporal and spatial temperature profile in the solid caused by the laser pulse is influenced by a number of factors such as heat losses by radiation, convection and heat diffusion. In addition the thermal conductivity and the heat capacity depend on the temperature. For a relatively small temperature increase we can restrict ourselves to heat diffusion with temperature-independent heat conductivity and heat capacity. Usually the absorption in metals is large, i.e. the penetration depth of the light is of the order of a few hundred ångströms, which is much smaller than the lateral extensions of the laser beam, usually of the order of 1 mm. Therefore, for not too long times lateral heat diffusion can be neglected and the heat flow can be described by a one-dimensional heat diffusion equation

$$-K\frac{\partial^2 T}{\partial z^2} + \rho c \frac{\partial T}{\partial t} = A(z,t) \qquad (2.4.1)$$

with K the thermal heat conductivity, ρ the density and c the specific heat. $A(z,t)$ is the energy source term given by

$$A(z,t) = I_0(1-R)\alpha \exp(-\alpha z)q(t) \qquad (2.4.2)$$

with I_0 the maximum laser intensity, R the reflectivity and α the absorption coefficient. For the temporal shape of the laser pulse $q(t)$ we assume a Gaussian shape

$$q(t) = \frac{1}{\sqrt{\pi}\sigma} \exp\left(-\left(\frac{t}{\sigma}\right)^2\right). \qquad (2.4.3)$$

The resulting temperature profile for the effect of a Nd:YAG laser pulse ($\sigma = 6$ ns, $\lambda = 1064$ nm) on a Cu surface is shown in Fig. 2.14. For the calculation the following parameters were used: initial surface temperature $T_0 = 300$ K, heat conductivity $K = 3.7$ W cm^{-1} K^{-1}, heat capacity $\rho c = 3.5$ J cm^{-3} K^{-1}, reflectivity $R = 0.97$, absorption coefficient $\alpha = 0.0087$ Å^{-1}. The incident fluence

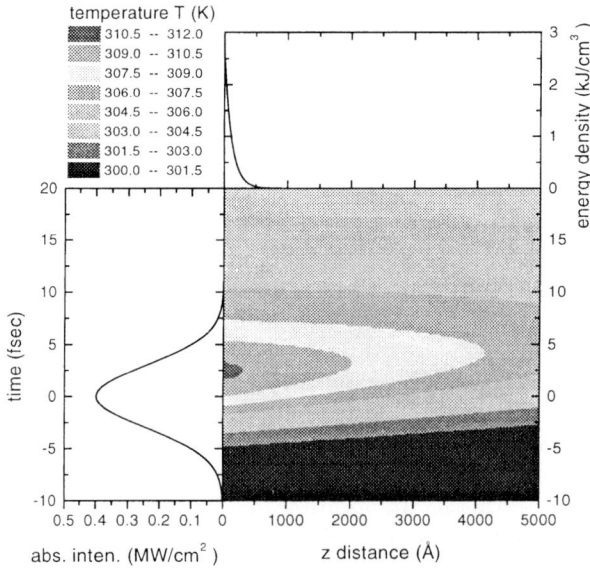

Fig. 2.14. Temperature profile generated by a 6 ns laser pulse on a Cu surface. For parameters, see text.

was chosen to be $F = 100\,\text{mJ cm}^{-2}$. The maximum heating even for this quite high fluence is only about 10 K. First, this is because only 3% of the incident energy is absorbed in the solid, and secondly during the pulse the energy is distributed by thermal diffusion over a distance of about 1 μm, much larger than the penetration depth of the light. Such high laser fluence can be used on Cu samples and on ultrathin transition-metal films on Cu as long as the film thickness is much smaller than the penetration depth of the light. Using a Co sample instead of Cu would cause a maximum temperature increase of about 200 K.

Thermal heating of a thin film sample by a Nd:YAG laser

In contrast to bulk samples with high heat conductivity, using a thin Cu or Co film on a substrate with low heat conductivity like glass causes a very large temperature increase. For example in a 3000 Å thick Cu film there is no large temperature gradient through the film. In the one-dimensional model, discussed here, the temperature increase is simply given by the heat capacity and the absorbed energy. For $R = 0.7$, $F = 100\,\text{mJ cm}^{-2}$, one gets a temperature increase of about 270 K. Therefore, one has to use in this or similar cases incident fluences as low as $7\,\text{mJ cm}^{-2}$.[115] Usually the laser spot area on the sample is of the order of 1 mm². Therefore, lateral diffusion even in the case of highly conductive materials is not important for the temperature increase generated by a single laser pulse.

Thermal heating of a Cu sample by a Ti:sapphire laser

The pulse duration of the Ti:sapphire laser of about 100 fs is much smaller than the time needed to transfer the energy stored in the electronic system by the laser pulse to the lattice. For this reason thermal diffusion as discussed above is unimportant no matter whether the sample is a thin film or a bulk sample. The pulse energies in a single laser pulse are of the order of a few nanojoules. However, to reach a significant SH signal the beam is usually focused much stronger than the Nd:YAG laser beam. Depending on the focus size, ranging from 100 to 25 μm, an incident peak fluence between $F \approx 50\,\mu\text{J cm}^{-2}$ and $F \approx 2\,\text{mJ cm}^{-2}$ is obtained. At $\lambda = 800$ nm the reflectivity of Cu is about $R = 0.97$ and the absorption coefficient $\alpha = 0.0084\,\text{Å}^{-1}$. The assumption of instantaneous heat deposition into the lattice would result in a temperature increase of

$$\Delta T(z) = \frac{(1-R)F}{\rho c}\, \alpha \exp(-\alpha z). \qquad (2.4.4)$$

Even for the extreme case of a peak fluence of $2\,\text{mJ cm}^{-2}$ this would result only in a maximum temperature increase at $z = 0$ of $\Delta T = 14.4$ K. Therefore, single-shot damage due to lattice heating can be ruled out. Lattice heating due to multishot heating can be treated as if a CW laser is used.

However, for these short times the electronic system and the lattice are not in thermodynamic equilibrium as shown by Eesley.[116,117] While for nanosecond laser pulses a one-temperature model is still adequate even for large pulse energies, as has been shown by Hicks et al.,[86] this model is not sufficient for the description of laser excitation with femtosecond laser pulses. Corkum et al.[118] described the heat transport inside the metal for such ultrashort pulses by a two-temperature model:

$$C_e \frac{\partial}{\partial t} T_e = \frac{\partial}{\partial z} K_e \frac{\partial}{\partial z} T_e - g(T_e - T_i) + A(z,t),$$

$$C_i \frac{\partial}{\partial t} T_i = g(T_e - T_i),$$

(2.4.5)

where the electronic heat capacity C_e is given by $C_e = C'_e T_e$, $C'_e = 96.6\,\text{J}\,\text{m}^3$ for copper. K_e is the electronic heat conductivity and its temperature dependence can be described approximately by $K_e = K_{e0} T_e / T_i$, $K_{e0} = 400\,\text{W}\,\text{m}^{-1}\,\text{K}^{-1}$. $C_i = 3.5 \times 10^6\,\text{J}\,\text{m}^{-3}\,\text{K}^{-1}$ is the lattice heat capacity and $g \approx 2 \times 10^{17}\,\text{W}\,\text{m}^{-3}\,\text{K}^{-1}$ the electron–lattice coupling constant. $A(z,t)$ is the energy source term as defined in eqn (2.4.1). The resulting electron temperature profile shown in Fig. 2.15 is calculated for an incident fluence $F = 2\,\text{mJ}\,\text{cm}^{-2}$, reflectivity $R = 0.97$ and absorption coefficient $\alpha = 0.0084\,\text{Å}^{-1}$ corresponding to a laser pulse with wavelength $\lambda = 800\,\text{nm}$ on a Cu surface. Despite the fact that the pulse energy does not heat the lattice significantly, the electron system is heated by more than 200 K. Although this temperature is not sufficient to cause any damage to the crystal, it may change the optical properties, i.e. the refractive indices and $\chi^{(2)}$.

Although lattice heating by a single laser shot can be neglected, the high repetition rate of about 80 MHz of the Ti:sapphire laser may still limit the

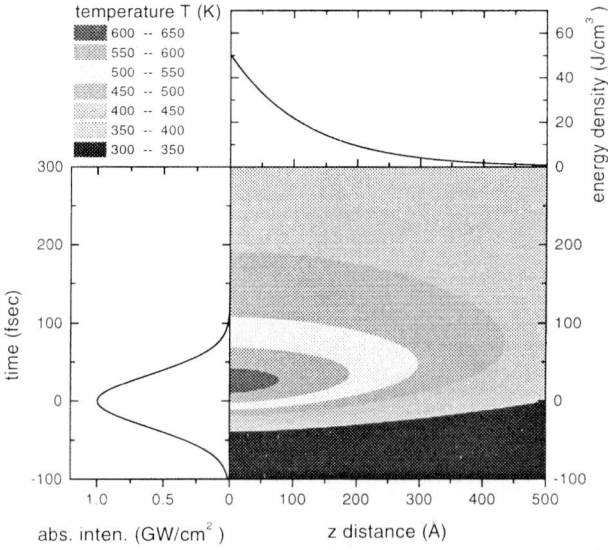

Fig. 2.15. Electron temperature profile generated by a 80 fs (FWHM) laser pulse on a Cu surface. For parameters, see text.

average incident laser power. This is a minor concern for highly conductive single-crystal substrates, but for thin films on insulating or semiconducting materials this may easily lead to damage. The temperature in the film is largely determined by the heat transition resistance at the film boundaries and depends strongly on the specific kind of sample. Details of model calculations can be found, for example, in the book of Ready.[119]

2.5 The nonlinear magneto-optical Kerr effect

2.5.1 Phenomenological description

The presence of a magnetization changes the polarization state and/or the intensity of the reflected light as described in sec. 2.3. Similar can be observed in the frequency-doubled light. This effect is named *nonlinear* Kerr effect in the literature.[2,120,121] However, to first approximation the effect is *linear* in the magnetization.[24] It turns out that essentially the same qualitatively different configurations as for the linear magneto-optical Kerr effect can be distinguished in highly symmetric configurations. Then, in the polar and longitudinal configurations the polarization of the SH light rotates in opposite direction on reversal of the magnetization. However, (M)SHG is described by a third-rank tensor instead of a second-rank tensor as for the linear reflected light. Instead of a single magneto-optical constant Q up to five independent magnetic tensor elements contribute. In general the situation is as complicated as in the case of the linear Kerr effect in optically active media. This more complicated behaviour, on the other hand, can be used in principle for the determination of the magnetization direction without the necessity to rotate the external magnetic field. However, in the case of a highly symmetric configuration—the magnetization along a crystallographic mirror plane and the optical plane parallel to a mirror plane as well—a quite simple behaviour is found for entirely s- or p-polarized incident fundamental light. To give some more insight into the symmetry properties of the nonlinear Kerr effect, a symmetry analysis of the nonlinear susceptibility tensor is given below.

The polarization with frequency 2ω generated in the medium by the incident light wave

$$E(\omega) = E^{(\omega)} e^{i k^{(\omega)} \cdot r - i \omega t} \tag{2.5.1}$$

at frequency ω can be expanded in a series

$$P_i^{(2\omega)} = \chi_{ijk}^{(2,D)} E_j^{(\omega)} E_k^{(\omega)} + i \chi_{ijkl}^{(2,Q)} E_j^{(\omega)} k_k^{(\omega)} E_l^{(\omega)} + \cdots . \tag{2.5.2}$$

For systems with inversion symmetry like cubic crystals in the first term (and all higher terms with an even power of $E^{(\omega)}$) in eqn (2.5.2) is nonzero only at the surface or at buried interfaces where the inversion symmetry is broken. This is the term of main interest and it is in many cases the dominant one, as will be

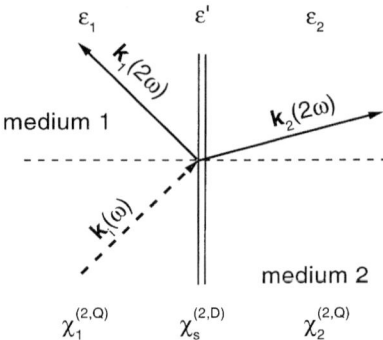

Fig. 2.16. Schematic representation of the phenomenological second harmonic effects on the interface between two centrosymmetric media. The pump wave at frequency ω and the two SH waves generated by the pump beam at the interface by the nonlinear susceptibility $\chi_s^{(2,D)}$ are shown. (The SH beams generated by the $\chi_{1,2}^{(2,Q)}$ bulk susceptibilities are not shown.)

shown in the subsequent sections. The second term arises from the nonlocal response of the medium to the incident light wave and contains contributions of magnetic dipole and electric quadrupole character. These contributions are described by a fourth rank tensor $\chi^{(2,Q)}$ and are therefore not forbidden in the bulk but usually quite small. The total contribution of this term, however, can become comparable to the surface dipolar term.

Most phenomenological descriptions[122-125] use the model illustrated in Fig. 2.16. The effect of the surface nonlinearity is modelled by an infinitesimal thin sheet characterized by the nonlinear susceptibility $\chi_s^{(2,D)}$ and a dielectric constant ε'. The bulk linear dielectric response is assumed to be isotropic and is determined by $\varepsilon_1(\omega)$, $\varepsilon_1(2\omega)$ and $\varepsilon_2(2\omega)$, $\varepsilon_2(2\omega)$ respectively. There is an ambiguity in the choice of ε_s. One can choose for ε_1 the value for medium 1, ε_1, or that for medium 2, ε_2, or even that for the vacuum. This, however, only results in a redefinition of $\chi_s^{(2,D)}$:

$$\chi_{s1}^{(2)} = \frac{\varepsilon_2(\omega)^2 \varepsilon_1(2\omega)}{\varepsilon_1(\omega)^2 \varepsilon_2(2\omega)} \chi_{s2}^{(2)}. \qquad (2.5.3)$$

In the following subsections we give a symmetry analysis of the tensor $\chi^{(2,D)}$ and some general remarks about the $\chi^{(2,Q)}$ tensor. For the high-symmetry configurations explicit formulae for the nonlinear Kerr rotation and asymmetry will be given as well.

2.5.2 Symmetry analysis of $\chi^{(2,D)}$

The superscript D in $\chi^{(2,D)}$ is suppressed for simplicity in this subsection. In the presence of a magnetization the symmetry of the crystal is reduced. Compared

Table 2.1 Nonvanishing elements of $\chi^{(2)}(M)$ for the (001) surface of a cubic crystal. $M \parallel [100]$ and $[001]$ (x is along the [100] direction, z is along the surface normal)

Direction of M	Nonvanishing tensor elements of $\chi^{(2)}(M)$	
	Even elements	Odd elements
[100]	xzx = xxz, yzy = yyz zxx, zyy, zzz	xyx = xxy, yxx, yyy yzz, zyz = zzy
[001]	xzx = xxz = yzy = yyz zxx = zyy, zzz	xyz = xzy = −yzx = −yxz zxy = zyx

to the nonmagnetic case more tensor elements of $\chi^{(2)}$ become nonzero. Generally the tensor elements of $\chi^{(2)}$ can be split into two contributions:

$$\chi^{(2)}_{ijk} = \chi^{(2,\text{nm})}_{ijk}(M) + \chi^{(2,\text{m})}_{ijk}(M) \qquad (2.5.4)$$

with $\chi^{(2,\text{nm})}_{ijk}(M)$ even and $\chi^{(2,\text{m})}_{ijk}(M)$ odd under reversal of the magnetization:

$$\chi^{(2,\text{nm})}_{ijk}(-M) = \chi^{(2,\text{nm})}_{ijk}(M), \qquad (2.5.5)$$

$$\chi^{(2,\text{m})}_{ijk}(-M) = -\chi^{(2,\text{m})}_{ijk}(-M). \qquad (2.5.6)$$

For crystals having high symmetry it can be shown that the elements $\chi^{(2,\text{nm})}_{ijk}(M)$ and $\chi^{(2,\text{m})}_{ijk}(M)$ are mutually exclusively nonzero. For example, if the magnetic point group of the crystal contains the combined symmetry element $\underline{m} = R \circ m$ of spin inversion R and mirror plane m, then for the $\chi^{(2,\text{nm})}$ part of the tensor m alone is a symmetry operation and $\underline{m}$ transforms $\chi^{(2,\text{m})}$ into its negative:

$$\begin{aligned} m \circ \chi^{(2,\text{nm})}_{ijk}(M) &= \chi^{(2,\text{nm})}_{ijk}(M), \\ \underline{m} \circ \chi^{(2,\text{m})}_{ijk}(M) &= -\chi^{(2,\text{m})}_{ijk}(-M). \end{aligned} \qquad (2.5.7)$$

This is the case for all low-index surfaces of cubic crystals. Tables of the nonzero tensor elements have been derived by Pan et al.[1] and they are reproduced here

Table 2.2 Nonvanishing elements of $\chi^{(2)}(M)$ for the (110) surface of a cubic crystal. $M \parallel [001]$, $[\bar{1}10]$ and $[110]$ (x is along the [001] direction, z is along the surface normal)

Direction of M	Nonvanishing tensor elements of $\chi^{(2)}(M)$	
	Even elements	Odd elements
[001]	xzx = xxz, yzy = yyz zxx, zyy, zzz	xxy = xyx, yxx, yyy yzz, yzz = zzy
[$\bar{1}$10]	xzx = xxz, yzy = yyz zxx, zyy, zzz	xxx, xyy, xzz yxy = yyx, zxz = zzx
[110]	xzx = xxz, yzy = yyz zxx, zyy, zzz	xyz = xzy, yzx = yxz zxy = zyx

as Tables 2.1 and 2.2 for the (001) and (110) surfaces and the relevant direction of magnetization.

2.5.3 Symmetry properties of $\chi^{(2,Q)}$

For the nonmagnetic case the properties of $\chi^{(2,Q)}$ have been discussed by a number of authors for either the isotropic case[123,124,126,127] or the case of cubic symmetry.[125] A symmetry analysis of $\chi^{(2,Q)}$ including magnetism has been given by Koopmans et al.[128] for the isotropic case. A full treatment of $\chi^{(2,Q)}$ for cubic media has not been worked out yet for the magnetic case.

In the case of a nonmagnetic cubic crystal the elements of $\chi^{(2,Q)}(M)$ can be written in the following form:[125]

$$\chi^{(2,Q)}_{ijkl} = \chi^Q_1 \delta_{ijkl} + \chi^Q_2 \delta_{ij}\delta_{kl} + \chi^Q_3 \delta_{ik}\delta_{jl} + \chi^Q_4 \delta_{il}\delta_{jk} \qquad (2.5.8)$$

with χ^Q_1, χ^Q_2, χ^Q_3, χ^Q_4 constants. For the case of a plane wave incident on a single interface the second and fourth terms are zero due to $\nabla E = 0$ and $n \times E = 0$ in isotropic or cubic media. The SH field generated from the first term χ^Q_1 generates the anisotropic bulk response for cubic materials. For isotropic materials it vanishes as well. The nonlinear bulk polarization, therefore, can be written for the case of an isotropic medium as

$$P^{(2,Q)} = \tfrac{1}{2}\chi^Q_3 \nabla(E \cdot E). \qquad (2.5.9)$$

From this term only p-polarized SHG results. (The anisotropic contribution $\chi^Q_1 \delta_{ijkl}$ would also generate s-polarized SH light.) Including the effect of magnetization eqn (2.5.9) can be generalized to[128]

$$P^{(2,Q)} = \tfrac{1}{2}\chi^{(Q,nm)}_3 \nabla(E \cdot E) + \tfrac{1}{2}\chi^{(Q,m)}_3 \hat{M} \times \nabla(E \cdot E) \qquad (2.5.10)$$

with $\hat{M}$ the unit vector in the direction of the magnetization. The part even in the magnetization $\chi^{(Q,nm)}_3$ has exactly the same dependence on the fundamental light as in the nonmagnetic case. The magnetization-induced part $\chi^{(Q,m)}_3$ is perpendicular to the even part. Table 2.3 lists the nonvanishing elements of $\chi^{(2,Q)}(M)$ which belong to the χ^Q_1 and χ^Q_3 subexpressions in eqn (2.5.8). A full list of all elements of $\chi^{(2,Q)}(M)$ can be found in ref. 128 for the isotropic case.

Table 2.3 Nonvanishing elements of $\chi^{(2,Q)}(M)$. Only the contributions from the χ^Q_1 and the χ^Q_3 terms are listed (x is along the [100] direction)

Direction of M		Nonvanishing tensor elements of $\chi^{(2,Q)}(M)$	
		Even elements	Odd elements
[010]	χ^Q_1	$xxxx = zzzz$	$zxxx = -xzzz$
		$yyyy$	
	χ^Q_3	$zxzx = xzxz$	$xxzx = -zzxz$
		$xyxy = zyzy$	$xyzy = -zyxy$
		$yxyx = yzyz$	$yxyz = -yzyx$

2.5.4 Nonlinear Kerr angles and asymmetries

In this and the following two subsections explicit formulae of the nonlinear Kerr asymmetry or the nonlinear Kerr angle are given for certain highly symmetric configurations. Contributions from the bulk due to $\chi^{(2,Q)}(M)$ are neglected. Furthermore, the linear magneto-optical effects on the incident fundamental as well as on the outgoing SH light are ignored as well. The nonlinear magneto-optical effects (Kerr rotation, Kerr asymmetry) are usually orders of magnitude larger than the linear effects. Calculations by Pustogowa et al.[20,129] explicitly including these effects revealed that they do not lead to significantly different results. Then the field $E(2\omega)$ of the outgoing SH light can be expressed as:[21,125,130,131]

$$E^{(2\omega)}(\Phi,\phi) = \frac{4\pi i}{\cos\theta_i}\frac{\omega}{c}|E^{(\omega)}|^2 \begin{pmatrix} T_p F_c \cos\Phi \\ T_s \sin\Phi \\ T_p N^2 F_s \cos\Phi \end{pmatrix}^T \chi^{(2)} \begin{pmatrix} f_c^2 t_p^2 \cos^2\phi \\ t_s^2 \sin^2\phi \\ f_s^2 t_p^2 \cos^2\phi \\ 2 f_s t_p t_s \cos\phi \sin\phi \\ 2 f_c f_s t_p^2 \cos^2\phi \\ 2 f_c t_p t_s \cos\phi \sin\phi \end{pmatrix}$$

(2.5.11)

with Φ and ϕ the angles of the polarization direction of $E^{(\omega)}$ and $E^{(2\omega)}$ with respect to the optical plane. $\chi^{(2)}$ is written in the form

$$\chi^{(2)} = \begin{pmatrix} \cdot & \cdot & \cdot & \cdot & \cdot & \cdot \\ \cdot & \chi^{(2)}_{ijj} & \cdot & \cdot & \chi^{(2)}_{ijk} & \cdot \\ \cdot & \cdot & \cdot & \cdot & j \neq k & \cdot \end{pmatrix}.$$

In this expression it was assumed that the nonlinear sheet is just below the surface with the dielectric constant ε' that of the crystal. The f_s, f_c are defined as

$$f_s = \frac{\sin\theta_i}{n} \qquad f_c = \sqrt{1-f_s^2} \qquad (2.5.12)$$

with n the refractive index of the medium at frequency ω, and t_s and t_p are the linear transmission coefficients t_{ss} and t_{pp} of eqn (2.3.9). The definitions of F_s, F_c, T_s, T_p and N are similar with N the refractive index at 2ω, $N = N(2\omega)$.

Transverse Kerr geometry

In the transverse Kerr configuration, the magnetization is perpendicular to the optical plane. With the definition of the coordinate system indicated in Fig. 2.17 the optical plane lies in the yz plane and the magnetization is parallel to the x-axis. To be specific the surface is chosen to be the [001] surface, i.e. the surface

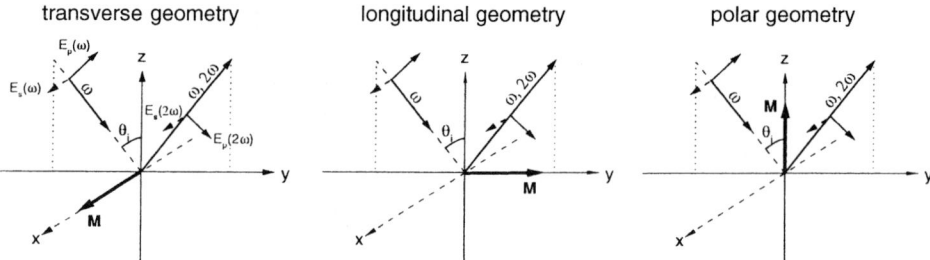

Fig. 2.17. Definition of the coordinate system for the transverse, longitudinal and polar Kerr geometries.

has C_{4v} symmetry. However, essentially the same is true for the [110] surface. With the magnetization parallel to the x axis $\chi^{(2)}$ has the following form:

$$\chi^{(2)} = \begin{pmatrix} 0 & 0 & 0 & 0 & \chi^{(2,\text{nm})}_{xzx} & \chi^{(2,\text{m})}_{xxy} \\ \chi^{(2,\text{m})}_{yxx} & \chi^{(2,\text{m})}_{yyy} & \chi^{(2,\text{m})}_{yzz} & \chi^{(2,\text{nm})}_{yzy} & 0 & 0 \\ \chi^{(2,\text{nm})}_{zxx} & \chi^{(2,\text{nm})}_{zyy} & \chi^{(2,\text{nm})}_{zzz} & \chi^{(2,\text{nm})}_{zyz} & 0 & 0 \end{pmatrix}. \quad (2.5.13)$$

Generally, there is no s-polarized SH light in this configuration for the incident light, for either entirely p- or s-polarized incident fundamental light, as can be verified with the aid of Table 2.2. For p-polarized incident light there are three even ($yzy = yyz$, zyy and zzz) and three odd (yyy, yzz and $zyz = zzy$) contributing tensor elements. For s-polarized incident light there is only one even (zxx) and one odd (yxx) tensor element. The field amplitude of the outgoing SH light can be written as

$$E^{(2\omega)}(s_{\text{in}}, p_{\text{in}})(\pm M) = 4\pi i \frac{\omega}{c} \frac{T_{s,p}}{\cos\theta_i} \left[\chi^{(2,\text{nm})}_{s,p}(\pm M) \pm \chi^{(2,\text{m})}_{s,p}(\pm M) \right] \quad (2.5.14)$$

with the effective susceptibilities

$$\chi^{(2,\text{nm})}_s = t_s^2 F_s N^2 \chi^{(2)}_{zxx} \quad (2.5.15)$$

$$\chi^{(2,\text{m})}_s = t_s^2 F_c \chi^{(2)}_{yxx}, \quad (2.5.16)$$

$$\chi^{(2,\text{nm})}_p = t_p^2 \left(f_c^2 N^2 F_s \chi^{(2)}_{zyy} + f_s^2 N^2 F_s \chi^{(2)}_{zzz} + 2 f_c f_s F_c \chi^{(2)}_{yyz} \right) \quad (2.5.17)$$

$$\chi^{(2,\text{m})}_p = t_p^2 \left(f_c^2 F_c \chi^{(2)}_{yyy} + f_s^2 F_c \chi^{(2)}_{yzz} + 2 f_c f_s N^2 F_s \chi^{(2)}_{zyz} \right), \quad (2.5.18)$$

which now depend on the incident angle θ_i through the f_s, F_s, etc.

For comparison with experimental data it is useful to define an asymmetry for p- and s-polarized incident light

$$A_p = \frac{I_{pp}(+M) - I_{pp}(-M)}{I_{pp}(+M) + I_{pp}(-M)} \qquad A_s = \frac{I_{sp}(+M) - I_{sp}(-M)}{I_{sp}(+M) + I_{sp}(-M)} \quad (2.5.19)$$

with $I_{pp}(M) = |E_p^{(2\omega)}(M)|^2$ for p-polarized incident light and $I_{sp}(M) = |E_p^{(2\omega)}(M)|^2$ for s-polarized incident light. Using eqn (5.14) $A_{s,p}$ can be expressed as

$$A_{s,p} = \frac{2R_{s,p}}{1+R_{s,p}^2} \cos \varphi_{s,p} \quad \text{with} \quad R_{s,p} = \left| \frac{\chi_{s,p}^{(2,m)}}{\chi_{s,p}^{(2,nm)}} \right| \quad (2.5.20)$$

and $\varphi_{s,p}$ the phase angle between $\chi_{s,p}^{(2,nm)}$ and $\chi_{s,p}^{(2,m)}$. Note that for s-polarized incident light only one odd and only one even tensor element contributes.

It is interesting to remember that there is *no* transverse linear magneto-optical Kerr effect for s-polarized incident light. In the nonlinear case this can happen because of the presence of the 'cross'-tensor elements like $\chi_{zxx}^{(2)}$ which generate p-polarized SH light for s-polarized incident fundamental light.

Longitudinal Kerr geometry

In the longitudinal geometry the magnetization M is parallel to the y axis. An inspection of Table 2.2 reveals that in this geometry for p- as well as for s-polarized incident light the electric field of the SH light has p components generated entirely by even tensor elements and s components generated entirely by odd tensor elements. $\chi^{(2)}$ has the following form:

$$\chi^{(2)} = \begin{pmatrix} \chi_{xxx}^{(2,m)} & \chi_{xyy}^{(2,m)} & \chi_{xzz}^{(2,m)} & 0 & \chi_{xzx}^{(2,nm)} & 0 \\ 0 & 0 & 0 & \chi_{yzy}^{(2,nm)} & 0 & \chi_{xxy}^{(2,m)} \\ \chi_{zxx}^{(2,nm)} & \chi_{zyy}^{(2,nm)} & \chi_{zzz}^{(2,nm)} & 0 & \chi_{zxx}^{(2,m)} & 0 \end{pmatrix}. \quad (2.5.21)$$

With eqn (2.5.11) the s- and p-polarized SH light $E_{s,p}^{(2\omega)}$ is found to be

$$E_p^{(2\omega)}(p_{in}) = 4\pi i \frac{\omega}{c} |E^{(\omega)}|^2 \frac{T_p}{\cos \theta_i} t_p^2 \left[2f_c f_s F_c \chi_{yyz}^{(2)} + N^2 F_s \left(f_c^2 \chi_{zyy}^{(2)} + f_s^2 \chi_{zzz}^{(2)} \right) \right] \quad (2.5.22)$$

$$E_s^{(2\omega)}(p_{in}) = 4\pi i \frac{\omega}{c} |E^{(\omega)}|^2 \frac{T_s}{\cos \theta_i} t_p^2 \left(f_c^2 \chi_{xyy}^{(2)} + f_s^2 \chi_{xzz}^{(2)} \right) \quad (2.5.23)$$

for p-polarized incident light, and

$$E_p^{(2\omega)}(s_{in}) = 4\pi i \frac{\omega}{c} |E^{(\omega)}|^2 \frac{T_p}{\cos \theta_i} t_s^2 N^2 F_s \chi_{zxx}^{(2)} \quad (2.5.24)$$

$$E_s^{(2\omega)}(s_{in}) = 4\pi i \frac{\omega}{c} |E^{(\omega)}|^2 \frac{T_s}{\cos \theta_i} t_s^2 \chi_{xxx}^{(2)} \quad (2.5.25)$$

for s-polarized incident light.

The nonlinear Kerr angle for p-polarized incident light can be defined now as

$$\Phi_{Kp}^{(2,L)} = \Phi_{Kp}^{\prime(2,L)} + i\Phi_{Kp}^{\prime\prime(2,L)} = \arctan\left(\frac{E_s^{(2\omega)}(p_{in})}{E_p^{(2\omega)}(p_{in})}\right). \quad (2.5.26)$$

For s-polarized incident light the nonlinear Kerr angle is defined by the *same* ratio $E_s^{(2\omega)}/E_p^{(2\omega)}$:

$$\Phi_{Ks}^{(2,L)} = \Phi_{Ks}^{\prime(2,L)} + i\Phi_{Ks}^{\prime\prime(2,L)} = \arctan\left(\frac{E_s^{(2\omega)}(s_{in})}{E_p^{(2\omega)}(s_{in})}\right) \quad (2.5.27)$$

with the only difference that now the fields $E_s^{(2\omega)}$ and $E_p^{(2\omega)}$ are generated by the surface with s-polarized incident light. The nonmagnetic contribution of the outgoing SH light is for both p- and s-polarized incident light entirely p-polarized, and the magnetization-induced part is always s-polarized in this longitudinal geometry, which is different from the linear case. There the reflected light is polarized in the same direction as the incident light in the absence of magnetization. Since nonlinear Kerr angles can be very large, the small-angle approximation used in eqn (2.3.7) for the definition of the linear Kerr angles cannot be applied in general. Using eqns (2.5.22) to (2.5.25) it follows that

$$\tan \Phi_{Ks}^{(2,L)} = \frac{T_s}{T_p} \frac{\chi_{xxx}^{(2)}}{N^2 F_s \chi_{zxx}^{(2)}}, \quad (2.5.28)$$

$$\tan \Phi_{Kp}^{(2,L)} = \frac{T_s}{T_p} \frac{f_c^2 \chi_{xyy}^{(2)} + f_s^2 \chi_{xzz}^{(2)}}{2 f_c f_s F_c \chi_{yyz}^{(2)} + N^2 F_s \left(f_c^2 \chi_{zyy}^{(2)} + f_s^2 \chi_{zzz}^{(2)}\right)}. \quad (2.5.29)$$

For the comparison with experiments it is useful to give an expression for the asymmetry measured at a certain polarization direction Φ:

$$A_{s,p}(\Phi) = \frac{I(+M,\Phi) - I(-M,\Phi)}{I(+M,\Phi) + I(-M,\Phi)} = \frac{2 R_{s,p} \tan \Phi}{1 + R_{s,p}^2 \tan^2 \Phi} \cos \varphi_{s,p} \quad (2.5.30)$$

with

$$R_{s,p} = \left|\frac{E_s^{(2\omega)}(s_{in}, p_{in})}{E_p^{(2\omega)}(s_{in}, p_{in})}\right| \quad (2.5.31)$$

and $\varphi_{s,p}$ the complex phase angle between $E_s^{(2\omega)}(s_{in}, p_{in})$ and $E_p^{(2\omega)}(s_{in}, p_{in})$. Φ measures the angle of the $E_p^{(2\omega)}$ vector to the optical plane, i.e. $\Phi = 0$ for $E_p^{(2\omega)}$ entirely p-polarized. In the small-angle approximation $\Phi_{Ks,p}^{(2,L)} \approx \tan \Phi_{Ks,p}^{(2,L)}$, the curves $A_{s,p}$ assume their minimum and maximum values $A_{s,p}(\Phi) = \pm \cos \varphi_{s,p}$ at angles $(\pi/2 \pm |\Phi_{Ks,p}^{(2,L)}|)$. Then the nonlinear Kerr rotation $\Phi_{Ks,p}^{\prime(2,L)}$ and the nonlinear Kerr ellipticity $\Phi_{Ks,p}^{\prime\prime(2,L)}$ are given by

$$\Phi_{Ks,p}^{\prime(2,L)} = |\Phi_{Ks,p}^{(2,L)}| \cos \varphi_{s,p}, \quad (2.5.32)$$

$$\Phi_{Ks,p}^{\prime\prime(2,L)} = |\Phi_{Ks,p}^{(2,L)}| \sin \varphi_{s,p}. \quad (2.5.33)$$

Polar Kerr geometry

In the polar geometry the magnetization M is parallel to the surface normal, the z axis. $\chi^{(2)}$ has the following form:

$$\chi^{(2)} = \begin{pmatrix} 0 & 0 & 0 & \chi_{xzy}^{(2,m)} & \chi_{xxz}^{(2,nm)} & 0 \\ 0 & 0 & 0 & \chi_{yzy}^{(2,nm)} & -\chi_{xzy}^{(2,m)} & 0 \\ \chi_{zxx}^{(2,nm)} & \chi_{zyy}^{(2,nm)} & \chi_{zzz}^{(2,nm)} & 0 & 0 & 0 \end{pmatrix}. \quad (2.5.34)$$

The p-polarized components of the SH light $E_p^{(2\omega)}(s_{in})$ and $E_p^{(2\omega)}(p_{in})$ are the same as in the longitudinal geometry and given by eqns (2.5.22) and (2.5.24). For p-polarized incident light the s-polarized component $E_s^{(2\omega)}(p_{in})$ is

$$E_s^{(2\omega)}(p_{in}) = 4\pi i \frac{\omega}{c} |E^{(\omega)}|^2 \frac{T_s}{\cos\theta_i} t_p^2 f_c f_s \chi_{xzy}^{(2)}. \quad (2.5.35)$$

There is no s-polarized SH light for s-polarized incident light.

The nonlinear Kerr angle for the polar geometry can now be defined in the same way as in eqn (2.5.26) for the longitudinal geometry:

$$\tan\Phi_{Kp}^{(2,P)} = \frac{T_s}{T_p} \frac{2f_c f_s \chi_{xzy}^{(2)}}{2f_c f_s F_c \chi_{yyz}^{(2)} + N^2 F_s \left(f_c^2 \chi_{zyy}^{(2)} + f_s^2 \chi_{zzz}^{(2)}\right)}. \quad (2.5.36)$$

Because of $E_s^{(2\omega)}(s_{in}) = 0$ there is *no* Kerr rotation for s-polarized incident light in the polar Kerr geometry, which is different from the linear case.

It should be noticed that the above expressions for the nonlinear Kerr angles are derived under the assumption of a C_{4v} (or C_∞) symmetry of the surface with the optical plane containing a mirror plane of the crystal. They are still valid for a C_{2v} symmetry for the indicated high-symmetry configurations of either entirely p- or s-polarized incident light. For other surface symmetries not containing two orthogonal mirror planes, the situation in general is more complicated. For example, in the case of a C_{3v} symmetry the *even* tensor element $\chi_{xxx}^{(2)}$ will lead to 'nonmagnetic' s-polarized outgoing light in addition to the 'magnetic' light generated by the odd tensor element $\chi_{yxx}^{(2)}$. Therefore the nonlinear Kerr angle cannot be defined as in eqn (2.5.27) because the s-polarized component does not simply change sign on reversal of magnetization.

2.6 MSHG as a surface-sensitive probe of magnetism

The first experimental proof of a nonlinear magneto-optical Kerr effect from surfaces of media having inversion symmetry and its surface sensitivity was given Reif *et al.* by SHG measurements from the Fe(110) surface.[9] For the experiments a frequency-doubled Nd:YAG laser working at $\lambda = 532$ nm with a repetition rate of 20 Hz was used providing pulses of 6 ns duration with a peak power density of about 5×10^6 W cm^{-2} at the sample. Measurements were performed

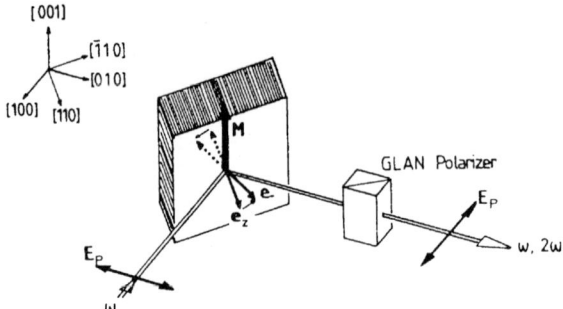

Fig. 2.18. Schematic representation of the experimental setup for the MSHG measurements of Reif *et al.*[9]

in the transverse geometry with the [$\bar{1}$10] direction in the optical plane using p-polarized incident light (see Fig. 2.18). An external magnetic field was applied to switch the magnetization of the sample parallel or antiparallel to the [001] direction. The angle of incidence was 45°. A typical result is shown in Fig. 2.19. The SH intensity was measured for 250 laser shots with the magnetization parallel to the [100] direction ('up'). Then the laser was blocked by a UV filter for 100 laser shots to determine the background light level. After reversal of the magnetization of the SH signal was measured for another 250 laser shots ('down'). This cycle was repeated several hundred times to improve the statistics. Despite the quite low signal level a clear intensity difference was observed (Fig. 2.19(a)). The background base pressure in the vacuum chamber was about

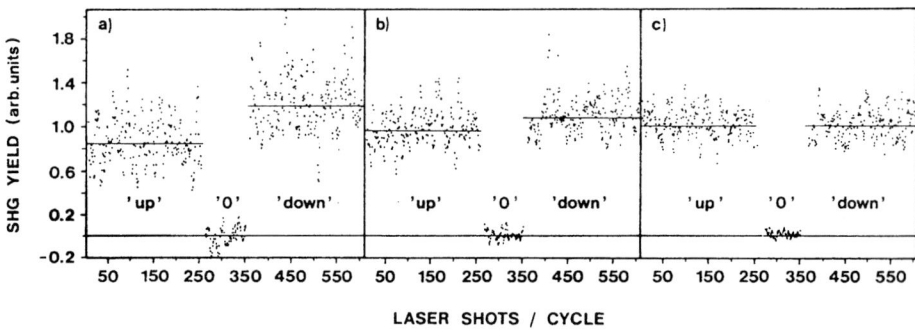

Fig. 2.19. SHG signal in the transverse Kerr geometry from an Fe(110) surface generated by a frequency-doubled Nd:YAG laser operating at $\lambda = 532$ nm. The SH signal is shown for three different times elapsed since sample preparation ((a) ≈ 45 min, (b) ≈ 60 min, (c) ≤ 180 min). The data represent an averaged superposition of cycles consisting of 250 laser pulses with the magnetization direction in the $-M$ direction ('up'), 100 pulses with blocked beam (background, '0'), and 250 pulses in the M direction ('down'). The signals are normalized to the average value of the signal for the magnetization in the two opposite directions. The solid lines represents the average over the respective regions of interest.

10^{-9} mbar. Therefore, the Fe surface was covered with adsorbates, mainly CO, and the ratio $R_p = |\chi_p^{(2,m)}/\chi_p^{(2,nm)}| \approx \frac{1}{2}A_p$ defined in eqn (2.5.20) dropped with a time constant of about half an hour ($\cos\varphi = 1$ is assumed). In this paper the assumption was made that only the magnetization-induced part of the nonlinear susceptibility $\chi_p^{(2,m)}$ is cancelled by the adsorbates. This assumption is opposite to what one would expect under the assumption that the adsorbate simply reduces the polarizability of the surface and therefore the total SHG. However for the transition metals very often a different behaviour is found. For example Hamilton et al.[132] found a strong reduction of SHG for the Ni(111) surface but only a very weak one for the Ni(110) surface at a fundamental wavelength of 532 nm. The above assumption is further corroborated by the later investigations on the Fe(110) surface, ultrathin Fe films on Cu(001), and Co films on Cu(001) measured with the Ti:sapphire laser at about $\lambda \approx 800$ nm (see sect. 2.6.5). A ratio of

$$\left|\chi_p^{(2,m)}/\chi_p^{(2,nm)}\right| = 0.25$$

was derived. Since the nonlinear magneto-optical effect should roughly scale with the magnetic moments,[19] a simple scaling of the value for Ni calculated by Hübner[133] results in

$$\left|\chi_p^{(2,m)}/\chi_p^{(2,nm)}\right|_{\text{theory}} = 0.27,$$

in excellent agreement with the experiment. Comparing this value with that of the linear magneto-optical constant $|Q|$, which is of the order of $|Q| \approx 0.03$[110] in the optical wavelength range, the 'nonlinear magneto-optical Kerr constant' $|\chi_p^{(2,m)}/\chi_p^{(2,nm)}|$ would be enhanced by about a factor of 10. It should be noted, however, that neither $\chi_p^{(2,m)}$ nor $\chi_p^{(2,nm)}$ are constants but they are a linear superposition of three odd and three even tensor element, $\chi_{ijk}^{(2,m)}$ and $\chi_{ijk}^{(2,nm)}$, respectively, with coefficients depending on the incident angle θ_i. As discussed in sec. 2.5.4, on going to grazing incidence the ratio $|\chi_p^{(2,m)}/\chi_p^{(2,nm)}|$ would approach infinity. Nevertheless, generally large magneto-optical Kerr effects are observed and it was found that the magnetization-induced tensor elements are of the same order of magnitude as the nonmagnetic tensor elements. This will be illustrated for some examples below.

2.6.1 MSHG from the Fe(110) surface

With the new Ti:sapphire laser setup the experiments on Fe(110) discussed above could be continued with much better signal statistics as described in sec. 2.4 but at a different wavelength of about 800 nm. Besides the laser system the experimental setup (Fig. 2.20) was similar to that used in ref. 9. An elliptically shaped Fe(110) sample of 2 mm thickness was clamped onto a soft iron yoke in such a way that the magnetic field is applied along the [001] direction of the iron crystal, the easy axis of bcc iron. A current of a few hundred milliamps running through the coil wound around the iron yoke was sufficient for magnetization

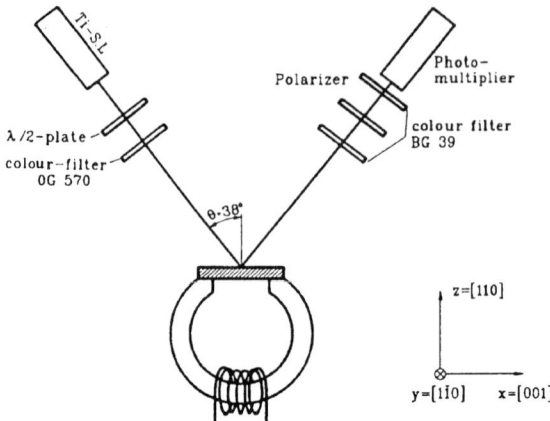

Fig. 2.20. Schematic drawing of the experimental setup used for the experiment on Fe(110). The longitudinal geometry is shown. For the transverse Kerr geometry the iron yoke with the sample is rotated by 90° about the axis normal. The angle of incidence was $\theta_i = 38°$ in all experiments.

saturation. The standard cleaning procedure was applied as described in ref. 134. All contaminants were below 1 at % as checked by Auger electron spectroscopy.

Transverse geometry

Figure 2.21 shows two hysteresis loops taken simultaneously in the transverse geometry with p-polarized incident light. In (a) the SH intensity and in (b) the reflectivity $R_{pp} = |r_{pp}|^2$ of the fundamental light is shown as a function of the applied magnetic field. While the reflectivity changes only by about 0.25% (in good agreement with the calculated value in Fig. 2.11) there is a change of the SH light intensity by a factor of 5.3 corresponding to an asymmetry $A_p = 0.68$ or a ratio $R_p = 0.4$ as defined in eqns (2.5.19) and (2.5.20) (assuming $R_p < 1$). This is a surprisingly large value if one compares this value with the value of Reif *et al.* of only $R = 0.25$ for the shorter wavelength of 523 nm. The theory of Pustogowa *et al.*[19] predicts a lower value at longer wavelength for the quantity Im($\chi^{(2)}_{yzz}(\omega)$). However, for p-polarized incident light the 'effective susceptibilities' $\chi^{(2,\text{nm})}$ and $\chi^{(2,\text{m})}$ are a linear combination of three tensor elements each with coefficients depending on the linear optical constants and geometry. The ratio of individual magnetic tensor elements of $\chi^{(2)}$ cannot be derived in this case from an experiment at a single incident angle because several even and several odd components contribute simultaneously. From eqns (2.5.17) and (2.5.18) it can be deduced that the contributions from the $\chi^{(2)}_{zzz}$ and the $\chi^{(2)}_{yzz}$ tensor elements, which should be the largest elements, are strongly suppressed by the Fresnel coefficients for the actual experimental situation. For the next largest pair of tensor elements $\chi^{(2)}_{yyz}$ and $\chi^{(2)}_{zyz}$ the coefficients deviate by a factor

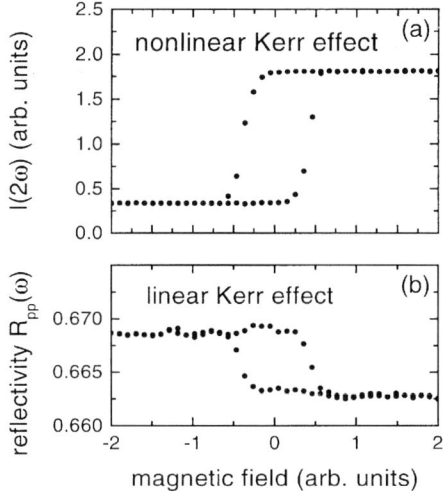

Fig. 2.21. Hysteresis loops of the SH light and the reflected light from the Fe(110) crystal taken simultaneously in the transverse Kerr geometry. The optical plane was along the [$\bar{1}$10] azimuth.

$N^2 F_s/F_c$, which leads to an enhancement of the magnetic component by a factor of about 2.3 for 800 nm and only 1.7 for 532 nm. Therefore the ratio R_p is enhanced for the longer wavelength due to the larger refractive index N at 2ω for 800 nm compared to 532 nm incident light.

For s-polarized light the situation is much simpler. Only one even and one odd tensor elements contribute. The measured asymmetry is lower compared to the case of p-polarized incident light and depends critically on the cleanliness of the Fe surface. An average value of 0.15 is obtained for the asymmetry A_s corresponding to an $R_s = 0.07$. From eqns (2.5.15) and (2.5.15) it follows that R_s equals the ratio $\chi^{(2)}_{yxx}/\chi^{(2)}_{zxx}$ *divided* now by the above-mentioned factor $N^2 F_s/F_c$. The nonlinear magneto-optical effect is therefore *reduced* by linear optical effects. Now, after all linear optics has been properly taken into account the ratio

$$\left| \frac{\chi^{(2)}_{yxx}}{\chi^{(2)}_{zxx}} \right|_{Fe} = 0.15$$

is in remarkably good agreement with the theoretical estimate given by Pustogowa *et al.*[19] who estimated 0.175 for the ratio. However, this agreement might be accidental since Pustogowa *et al*, did not take into account the tensor character of $\chi^{(2)}$ and this ratio may be rather different for different corresponding pairs of magnetic and nonmagnetic elements. Furthermore, in the above estimates phase factors have been ignored by setting $\varphi_{s,p} = 0$. However, they can dramatically change the SH light output for a given geometry as pointed out by Hübner *et al.*[131]

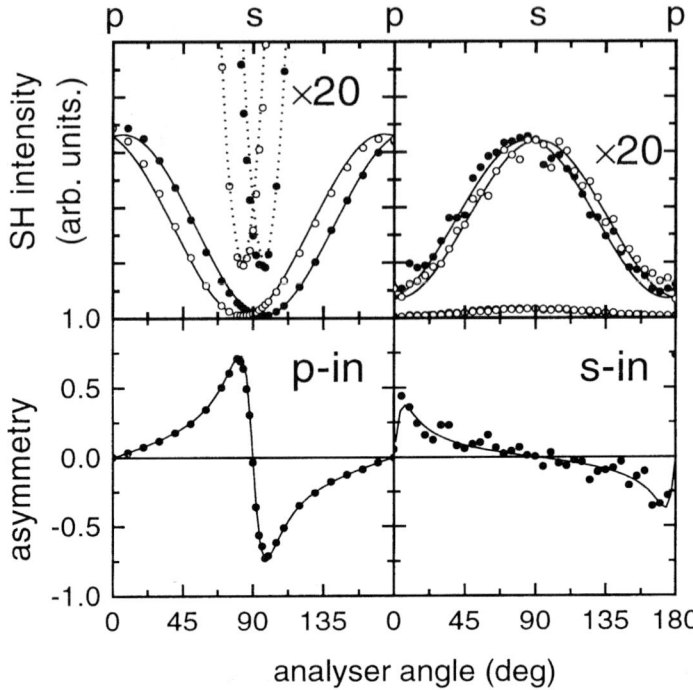

Fig. 2.22. SH intensity from the Fe(110) surface in the longitudinal geometry as a function of the analyser angle Φ. The optical plane is along the [001] direction. Left panels are for p-polarized incident light and right panels are for s-polarized incident light. The bottom two panels show the asymmetry calculated from the intensities in the top panels. The solid lines are curves fitted to the data by eqn (2.5.30).

Longitudinal geometry

In the longitudinal geometry the phase factor between even and odd tensor components of $\chi^{(2)}$ can be determined because they lead to orthogonal polarizations of the outgoing SH light. Measurements were performed with the magnetization still in the [001] direction, but the optical plane now parallel to the [$\bar{1}$10] azimuth direction. Figure 2.22 shows the SH light as a function of the analyser angle Φ placed in the outgoing beam path for p-polarized incident light (left panels) and for s-polarized incident light (right panels). In the bottom two panels the asymmetry $A_{s,p}(\Phi)$ as defined in eqn (2.5.30) is calculated. The solid lines are curves fitted to the data using the expression in (2.5.30) with $R_{s,p}$ and $\varphi_{s,p}$ as fitting parameters. For p-polarized incident light $R_p = 0.17 \pm 0.01$ and $\varphi_p = 43° \pm 3°$ are obtained, corresponding to $|\Phi_{Kp}^{(2,L)}| = 10°$, $\Phi_{Kp}^{\prime(2,L)} = 7°$ and $\Phi_{Kp}^{\prime\prime(2,L)} = 7°$. For s-polarization of the incident light there is barely any asymmetry visible in the measured intensity curve. However, a clear SH intensity is observed for s-polarized outgoing SH light, which according to the symmetry analysis in sec. 2.5 is entirely caused by the magnetic contribution. On the other hand the p-polarized 'nonmagnetic' SH intensity almost vanishes. This situation corresponds to a Kerr angle $|\Phi_{Kp}^{(2,L)}|$ close to 90°. By gas adsorption, which increases the p-polarized component of the SH light, it was confirmed that the s-polarized SH intensity is indeed of magnetic origin. The increasing p component reduced the Kerr angle from close to 90° down to lower values and then a strong asymmetry was observed. This is shown later in Fig. 2.36 where the subject of gas adsorption on MSHG is further discussed. The fit of the asymmetry function $A_{s,p}(\Phi)$ to the measured data gives $R_s \approx 6$ and $\varphi_s \approx 67°$. ($|\Phi_{Ks}^{(2,L)}| = 87°$, $\Phi_{Ks}^{\prime(2,L)} = 86°$ and $\Phi_{Ks}^{\prime\prime(2,L)} = 9°$).

For p-polarized incident light three nonmagnetic, $\chi_{zxx}^{(2)}$, $\chi_{xxz}^{(2)}$ and $\chi_{zzz}^{(2)}$, and two magnetic tensor elements, $\chi_{yxx}^{(2)}$ and $\chi_{yzz}^{(2)}$, contribute and a clear separation of linear optical and nonlinear effects is difficult. (Here the same coordinate system is used as for the transverse geometry. Therefore the indices x and y are exchanged with respect to eqns (2.5.24) and (2.5.25). The incident plane is now in the xz plane.) For s-polarized incident light there is, as in the transverse geometry, only one even tensor element, $\chi_{zyy}^{(2)}$, and one odd element, $\chi_{yyy}^{(2)}$. From eqns (2.5.24) and (2.5.25) a ratio

$$\left|\frac{\chi_{yyy}^{(2)}}{\chi_{zyy}^{(2)}}\right| \approx 14$$

can be deduced. Because of the smallness of $\chi_{zyy}^{(2)}$, however, the absolute size or the relative size compared to tensor elements expected to be large such as $\chi_{zzz}^{(2)}$ could not be determined in these experiments.

It is interesting to compare the results of the clean Fe(110) surface with MSHG results of the Nijmegen group[22,135] on the Fe/Cr interface and the surface of an Fe(100) whisker either capped with Cr or Au or oxidized.[136] The Fe/Cr sample consisted of a 2 nm Fe film covered with a 2 nm Cr film on a Si(100) surface covered with an amorphous SiO_2 layer of 525 nm. For s-polarized incident light on the Fe/Cr interface a relatively large nonlinear Kerr angle was found. For an angle of incidence of 45° the magnitude of the complex Kerr angle was measured to be 17°. The authors pointed out that for this multilayer system the local field at the Fe/Cr interface is strongly influenced by the thickness of the SiO_2 layer, which in turn influences the nonlinear Kerr angle. A simulation of the Cr/Fe interface with a semi-infinite Fe bulk substrate gave a Kerr angle of about 10° for $\theta_i = 38°$. For the Fe(100) whisker for s-polarized incident light a Kerr angle between 10° and 20° was found. However, this strong disagreement between their and our results is not a surprise because the large Kerr angle of close to 90° in Fe(110) is caused by an almost vanishing nonmagnetic p-polarized SH component. A cover layer like Cr or Au very likely will change the total nonmagnetic contribution from the sample, leading to a decrease of the Kerr angle. For the oxidized Fe surface this has been shown[137] and will be discussed further in sec. 2.6.5. From this discussion it can be concluded that the Kerr angle alone is not a good measure of the strength of the magnetization-induced part of the SH light in general because the nonmagnetic SH yield also varies strongly with surface treatment.

For p-polarized incident light the nonmagnetic p-polarized SH component is quite large for the Fe(110) sample. Therefore, small additional nonmagnetic contributions from the cover layer should not lead to such drastic effects as for the s-polarized incident light. Moreover it was proven experimentally that a Cr cover layer on Fe(110) does not change the magnetic s-polarized component of the outgoing SH light (although changes of the 'nonmagnetic' p-polarized component are observed).[138] Still a large discrepancy of the Kerr angles for Fe(110) ($\Phi_{Kp}^{(2,L)} = 10°$) and Cr/Fe(100) ($\Phi_{Kp}^{(2,L)} = 1.2°$) is found. A possible reason could be the azimuthal anisotropy of the nonlinear Kerr angle. While there should be no azimuthal dependence of the Kerr angle on Fe(100) in the dipole approximation, there is clearly an anisotropy for the Fe(110) surface as can be seen most clearly from the fact that the nonmagnetic contribution for s-polarized incident light in the longitudinal geometry is close to zero while there is definitely a considerable nonmagnetic contribution in the transverse geometry for the same polarization of the incident light—about 1/10 of the SH intensity for p-polarized incident light. Therefore, $\chi_{zxx}^{(2)} \neq \chi_{zyy}^{(2)}$.

Another point was neglected in the above discussion. Only contributions from the surface term $\chi^{(2,D)}$ to the SH signal were considered. It was suggested that bulk contributions by the higher-order term $\chi^{(2,Q)}$ might be important.[80] For the above discussion this would mean that the nonlinear Kerr signal is not entirely generated at the surface but to a certain part from a depth region corresponding to the penetration depth of the incident light of about 200 Å. The role of bulk contributions will be further discussed in secs 2.6.4 and 2.6.5.

2.6.2 Ultrathin Co films on Cu(001)

While the MSHG experiments on the Fe(110) surface and the sensitivity of the magnetization-induced signal on the adsorbate coverage gave a first experimental indication of the surface sensitivity of MSHG, a clear proof was given by the MSHG experiments on ultrathin Co films on Cu(001).[120] This work was a close collaboration of the groups of Rasing in Nijmegen and Kirschner in Halle. Because the experimental results are also reviewed by Th. Rasing in this book these results are discussed only briefly and we focus on the new results obtained in Halle.

To investigate the dependence of MSHG on the thickness of an ultrathin ferromagnetic film the system Co on Cu(001) has been chosen for several reasons. The growth of Co on Cu(001) is close to the ideal layer-by-layer Frank–van der Merve growth.[114,139–142] It grows pseudomorphically onto the Cu(001) substrate in a face-centred cubic (fcc) structure in contrast to the natural room-temperature structure of Co which is hexagonal close-packed (hcp). Therefore complications in the interpretation of the data arising from the change of roughness of the film during growth are largely avoided. Secondly, the magnetic properties are well investigated.[139,143–148] For a thickness up to at least 20 monolayers (ML) the easy axis of magnetization is parallel to the surface plane along the ⟨110⟩ azimuthal direction. The Curie temperature for films thicker than 2 ML is well above room temperature. A 3 ML film already has a

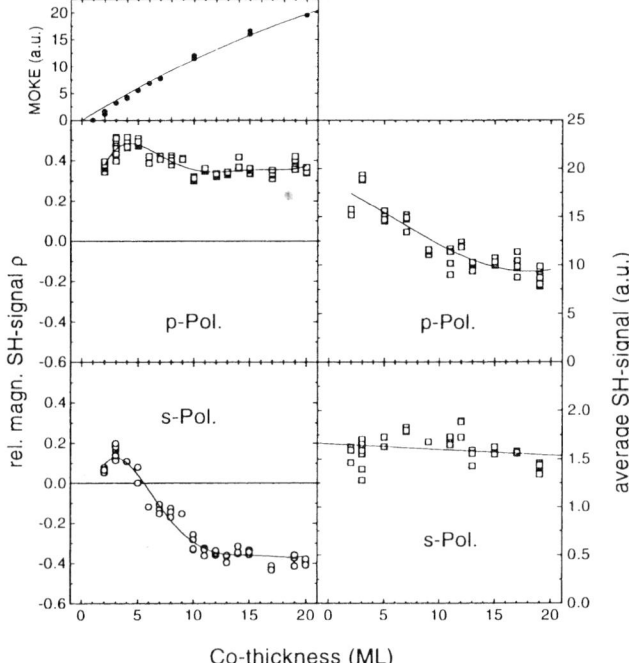

Fig. 2.23. Dependence of the linear MOKE (left topmost panel) and the MSHG (bottom four panels) from Co films on Cu(001) as a function of the layer thickness. The linear measurements were taken in the longitudinal geometry, while for the MSHG measurements the transverse geometry was chosen. After ref. 80.

Curie temperature of about 550 K.[114] Fcc Co(001) has a 4-fold in-plane anisotropy. However, small miscuts introduce a small uniaxial anisotropy in the surface plane.[147,148] Along the easy axis, however, square-shaped hysteresis loops are observed in all cases for thickness larger than 2 ML.

Preparation and characterization of the Co films as well as the experimental setup is described in sec. 2.4. Figure 2.23 shows the dependence of the SH signal from a Co film on a Cu(001) single crystal as a function of the Co film thickness measured in the transverse Kerr geometry. The incident angle was about 35° and the wavelength of the incident laser light for the MSHG measurements was 800 nm. On the left-hand side the asymmetry A_p and A_s as defined in eqn (2.5.19) is plotted for p- and s-polarized incident light (denoted as ρ in the figure). On the right-hand side correspondingly the arithmetic average of $I_{pp}(M)$ and $I_{pp}(-M)$ and $I_{ss}(M)$ and $I_{ss}(-M)$, respectively, is shown. For comparison the linear Kerr signal measured with a HeNe laser at 633 nm in the longitudinal geometry is depicted in the right top panel of the figure. While the linear Kerr signal increases nearly linearly with the film thickness, the asymmetry for p-polarized incident light, A_p, has already at 2 ML Co film thickness the same value of about 0.4 as for 20 ML, clearly proving the surface/interface sensitivity of MSHG with an effective depth of less than 2 ML. Also the average SH intensity is essentially constant. Actually the SH intensity at 2 ML is even slightly larger than for a thicker sample. This and the bump in the asymmetry in the thickness range from 2 to 6 ML is attributed to changes of the electronic structure of the films in this thickness range, as will be discussed below. The SH signal from the Cu substrate (for this [110] azimuth) is almost two orders of magnitude lower. Therefore the overwhelming part of the signal comes from the Co layer. It should be mentioned that the measured asymmetries A_p in refs 120 and 149 are slightly larger than that shown in Fig. 2.23. This is attributed to a smaller contamination of the surface with adsorbates in the present case. Very small amounts of adsorbates like CO present in the residual gas of the UHV chamber influence the SHG from the surface, leading to an *enhanced* asymmetry. This will be discussed further in sec. 2.6.5. The extreme sensitivity of the MSHG signal on adsorbates gives an additional indication of its large surface sensitivity.

The asymmetry for s-polarized incident light shows a somewhat more complicated behaviour. While for 2 to 4 ML it has a small but constant value, it changes sign at 5 ML until it reaches a constant negative value at around 10 ML. The average SH intensity is constant in the full investigated range from 2 to 20 ML. From theoretical arguments it is expected that for s-polarized incident light the effective thickness in which the SH light is generated should be larger because strong screening effects are present mainly for the perpendicular components of the polarization and not for the tangential part if mainly delocalized electronic states are involved.[69] However, the average SH signal is constant from 2 ML up to 20 ML. Therefore s-polarized incident light generates SH light from a strongly localized region at the interfaces, too. We believe that this behaviour is caused by changes of the electronic structure at the surface and/or the buried Co/Cu interfaces. Generalizing the model of a infinitely thin nonlinear sheet embedded in the crystal just below the surface described in sec. 2.5 to the case of a thin film as illustrated in Fig. 2.24, there are four contributions to the nonlinear susceptibility: the nonmagnetic contribution from the surface $\chi_s^{(2,nm)}$ and from the interface $\chi_i^{(2,nm)}$ as well as the corresponding magnetic ones $\chi_s^{(2,m)}$ and $\chi_i^{(2,m)}$. A free-standing film of a cubic material possesses inversion symmetry. Therefore, the total SH yield should be zero if the

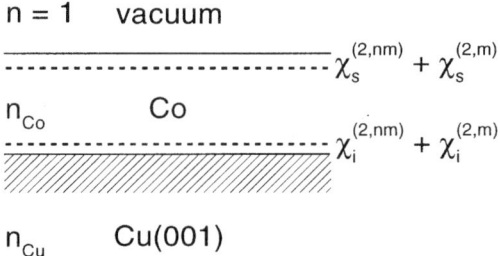

Fig. 2.24. Contributions to MSHG from a film on a substrate.

thickness of the film is much smaller than the penetration depth of the light so that the fields at the two interfaces are the same. From this one can conclude that $\chi_s^{(2,\text{nm})} = -\chi_i^{(2,\text{nm})}$ and $\chi_s^{(2,\text{m})} = -\chi_i^{(2,\text{m})}$. In reality all these four elements are complex quantities and depend on the individual electronic structure at the surface and at the interface and therefore the cancellation is not perfect. For the nonmagnetic part this cancellation is obviously far from perfect, because comparable SH intensities were measured for the Co films as for the Fe(110) surface. This does not need to be the case for the magnetic components, which involve different tensor elements. Small electronic changes in the film with thickness should cause only small variations in the elements of the nonlinear susceptibility tensors. Because of the relatively large net magnetic contribution for p-polarized incident light, the effective susceptibilities $\chi_s^{(2,\text{m})}$ and $\chi_i^{(2,\text{m})}$ are far from cancellation, while for s-polarized incident light the asymmetry at low thickness is small, indicating a nearly equal magnitude of $\chi_s^{(2,\text{m})}$ and $\chi_i^{(2,\text{m})}$ for this polarization. Minor changes in the electronic structure now can lead to relatively large changes in the asymmetry. This is further enhanced by the about one order of magnitude smaller amplitude of the average SH intensity compared to the SH intensity for p-polarized incident light.

The electronic nature of the bump in A_p between 2 and 6 ML and the sign change in A_s rather than a change in the magnetic moment is further supported by the strong dependence of the asymmetry on the frequency of the incident light as depicted in Fig. 2.25. While the above data were taken at individual layers for each thickness the measurements in Fig. 2.25 are measured on a wedge-like Co film on Cu(001). The incident angle $\theta_i = 38°$ was slightly larger than that for the measurements in Fig. 2.23. The SH signal for the magnetization in opposite directions is shown for two wavelengths of the incident light, 790 nm and 740 nm, in the top part of the figure for p-polarized incident light and for s-polarized incident light in the bottom part of the figure. For p-polarized incident light the bump in the asymmetry curve is significantly reduced compared to the 790 nm curve. For s-polarized incident light the asymmetry is close to zero at low thickness but decreases faster to negative values at shorter wavelength.

In ref. 149 the relative strength of surface to interface contributions for p-polarized incident light have been calculated under the *ad hoc* assumption that the adsorbate-covered surface does not contribute to the SH signal. It results that the magnitude of the surface contribution exceeds the interface contribution. The assumption that mainly the magnetic components are affected and the nonmagnetic components remain nearly constant during gas exposure leads to the opposite result that $\chi_s^{(2,\text{m})} < \chi_i^{(2,\text{m})}$. Under the latter assumption the observed frequency dependence indicates a reduced surface contribution of the magnetic tensor elements at shorter wavelength in the lower thickness range. In any case, the contributions from surface and buried interface are found to be of similar size. Which contribution dominates depends strongly on the specific system. For, example, in the case of fcc Fe/Cu (001) the magnetic surface contribution will dominate, as will be discussed in secs 2.6.3 and 2.6.5.

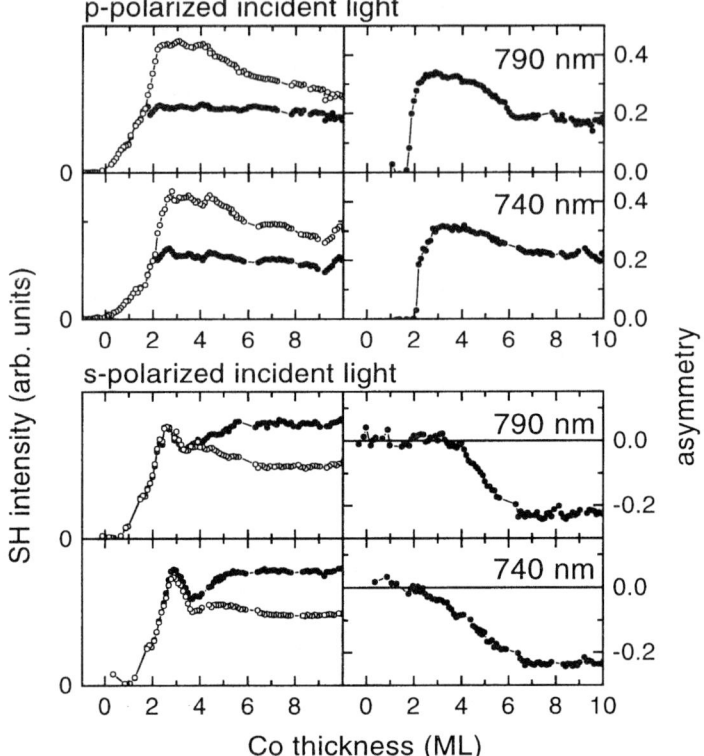

Fig. 2.25. Frequency dependence of the MSHG signal from a Co wedge on Cu(001). Top four panels, p-polarized incident light; bottom four panels, s-polarized incident light. In each of these two groups the SH intensity for the magnetization in the $+M$ direction (open symbols) and in the $-M$ direction (solid symbols) versus the Co film thickness is plotted on the left side. The calculated asymmetries for 790 nm wavelength of the incident light (top) and 740 nm (bottom) are plotted on the right side. $\theta_i = 38°$.

Similar measurements have been taken in the longitudinal geometry. Figure 2.26 shows the SH light from a 9 ML thick Co film on Cu(001) as a function of the analyser angle Φ for p-polarized and s-polarized incident light together with the calculated asymmetry curves $A_p(\Phi)$ and $A_s(\Phi)$. The fit gives

$$|\Phi_{Kp}^{(2,L)}| = 2.5° \qquad \Phi_{Kp}'^{(2,L)} = 2° \qquad \text{and} \qquad \Phi_{Kp}''^{(2,L)} = -1.5°$$

$$|\Phi_{Kp}^{(2,L)}| = 12.5° \qquad \Phi_{Kp}'^{(2,L)} = -10° \qquad \text{and} \qquad \Phi_{Kp}''^{(2,L)} = -7.4°$$

for p-polarization and for s-polarization, respectively. Compared to the result from the Fe(110) surface the Kerr angles are found to be smaller in this case. This is very likely an effect of the partial cancellation of surface and interface contributions as discussed above. For fcc iron films on Cu(001) a Kerr angle of about 3° to 4° was measured (see sec. 2.6.3) for p-polarized incident light in the polar geometry, quite similar to the result found here for the Co films.

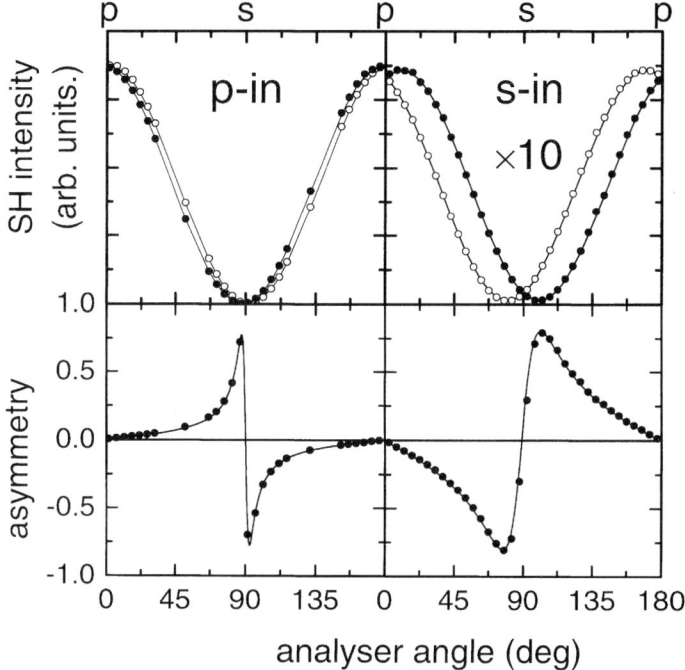

Fig. 2.26. SH intensity from a 9 ML Co film on Cu(001) in the longitudinal geometry as a function of the analyser angle Φ. Left panels are for p-polarized incident light and right panels are for s-polarized incident light. The bottom two panels show the asymmetry calculated from the intensities in the top panels. The solid lines are curves fitted to the data by eqn (2.5.30). $\theta_i = 39°$.

2.6.3 Fe / Cu(001)

Ultrathin fcc Fe films grown on Cu(100) have received considerable interest continuously during the past 15 years.[121,150–161] For coverages up to 11 ML Fe grows at room temperature in the fcc-like structure while for thicker films it transforms into the bcc structure of bulk iron.[155] Below a coverage of 4 ML the Fe grows epitaxially on the Cu(001) substrate with a tetragonal distorted fcc structural phase (phase I). In the thickness range from 5 to 11 ML the interior of the film exists as an undistorted fcc structure and only the interlayer distance between the surface layer and the subsequent layer is expanded[159] (phase II). The magnetic behaviour of the Fe films reflects the structural changes as the magnetic structure of fcc iron films depends on the unit-cell volume.[153] This dependence of the magnetic structure on the lattice constant has been demonstrated experimentally in comparison with different Cu/Au alloys[160,162] and Fe/Cu$_3$Au(100)[163,164] or carbon incorporation into the Fe film.[165] Theoretically now not less than six coexisting phases are predicted.[166]

Experimentally a high-moment ferromagnetic phase was found in the thickness region of phase I and a paramagnetic or an antiferromagnetic low-moment phase was found in the thickness range of phase II.[160,167,168] Antiferromagnetic ordering in the phase II of Fe/Cu(001) at a Néel temperature T_N of about 200 K was reported by was Li *et al.*[157] An antiferromagnetic low-moment phase has also been observed in Fe precipitates in Cu matrix, but with a Néel temperature of only $T_N \approx 65$ K.[169] For fcc films on Cu(001) in phase I generally the high-moment phase was found while in phase II linear MOKE measurements showed a reduced Kerr signal compared to the MOKE signal at 4 ML.[154] This reduced Kerr signal was found to be independent of the film thickness in phase II. The absolute value corresponds to a magnetically 'live layer' of 2 ML. It was proposed that only the two layers closest to the surface, which have an enlarged interlayer spacing, are in the high-moment phase while the rest of the film, which has the undistorted fcc lattice structure, is in a paramagnetic or low-moment antiferromagnetic phase. This view, that there is a 2 ML 'live layer' in the magnetic high-moment phase on top of the residual film, which is in a paramagnetic phase at room temperature, is supported now by site-selective conversion electron Mössbauer spectroscopy (CEMS)[170] and by spin-polarized appearance potential spectroscopy experiments.[171] The fcc Fe films grown at room temperature on Cu(001) show perpendicular magnetic anisotropy, in both phase I and phase II, i.e. the easy axis of magnetization is perpendicular to the film surface. At the fcc to bcc transition at about 11 ML the easy axis flips into the surface plane.

In this section we want to show how similar information about the location of the magnetic 'live layers' can be obtained from the measurements of the nonlinear Kerr effect without the drawbacks of CEMS having very long measuring times. Compared to the spin-polarized appearance potential spectroscopy experiments in MSHG the signal sources are more strongly localized at the interfaces as shown in sec. 2.6.2.

The Fe films were grown in a UHV system with a base pressure of 4×10^{-11} mbar. Growth temperature was 300 K, growth rate 1 ML min^{-1}. The Ti:sapphire laser system described in sec. 2.4 was used. Because the Fe films are perpendicularly magnetized in the thickness range up to 11 ML, the polar Kerr geometry was chosen for the measurements. A schematic view of the experimental setup is given in Fig. 2.27. The angle of incidence was about 40° and the [110] azimuthal direction was in the optical plane. Figure 2.27(b) shows the asymmetry A as a function of the analyser angle denoted as α here from a 7 ML thick Fe film on Cu(001). A nonlinear Kerr rotation of $\Phi'^{(2,P)}_{Kp} = 2.5°$ is obtained for p-polarized incident light. ($|\Phi^{(2,P)}_{Kp}| \approx 3°$ to 4°.) Because of the perpendicular direction of the magnetization, no Kerr rotation is observed for s-polarized incident light.

To investigate the thickness dependence of the MSHG signal a wedge-like Fe film was grown onto the Cu(100) substrate ranging from 0 to 12 ML. At a fixed analyser angle of $\alpha = 105°$ the SH signal was recorded for $+M$ and $-M$ magnetization directions while scanning with the laser beam over the sample. The result is shown in Fig. 2.28. Measurements were taken at 220 K in remanence, i.e. with no external magnetic field. The upper part shows the SH

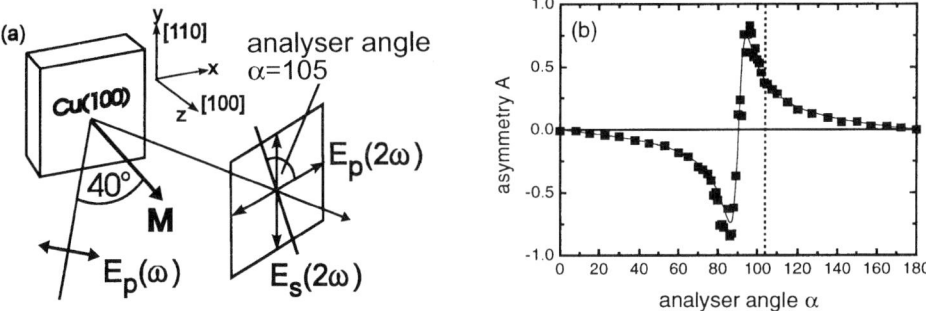

Fig. 2.27. (a) Experimental setup. (b) Asymmetry A as a function of the analyser angle α for a 7 ML thick Fe film on the Cu(100) for p-polarized incident light at $\lambda = 769$ nm. $\alpha = 0$ corresponds to p-polarized outgoing SH light. The vertical dashed line indicates the analyser angle at which the Fe thickness dependence of the MSHG signal on the iron wedge sample was measured. After ref. 121.

intensities $I(+M)$ and $I(-M)$ and the asymmetry calculated after eqn (2.5.19). For comparison in (c) the liner Kerr ellipticity $\Phi_K''^{(1,P)}$ is plotted measured with a HeNe laser at 633 nm.

Four different thickness ranges can be distinguished. Below ≈ 1.3 ML the Fe film is not ferromagnetically ordered and therefore neither a linear Kerr signal nor an SH asymmetry is observed. The SH signal increases monotonical from the very low value of the Cu(100) substrate. (As mentioned in sec. 2.6.2 the SH single from the Co films was about two orders of magnitude larger than that of the Cu substrate. The signal levels from the Fe film and the Co film are of comparable magnitude.) After the onset of ferromagnetic order in the second thickness range named 'phase I' in Fig. 2.28 the linear Kerr signal increases nearly linearly with the film thickness while the SH asymmetry is nearly constant. There is a minor increase between 1.3 and about 2.5 ML. A similar increase has been found by Detzel et al.[171] from spin-polarized appearance potential spectroscopy measurements in the same thickness range and the authors attributed this increase to an increasing magnetic moment of the Fe layer. The constant asymmetry reflects the surface/interface sensitivity of MSHG as discussed in sec. 2.6.2.

While the linear Kerr signal drops rapidly at about 4 ML at the transition from phase I to phase II there is little change in the SH asymmetry and this constant asymmetry gives a direct indication that the magnetization in phase II is indeed localized at the surface and/or the Fe/Cu interface. There is a strong drop in the SH signal itself at this thickness. This can be attributed to different surface reconstructions. LEED investigations showed a 1×4 superstructure at 2 ML and 1×5 superstructure at 3 to 4 ML surface reconstruction of the film below 5 ML.[141, 172–174] A LEED structure analysis by Müller et al. of the 1×5 superstructure revealed a complex reconstruction pattern with sinusoidal in-plane shifts of the atoms and vertical buckling of the layers with an amplitude as

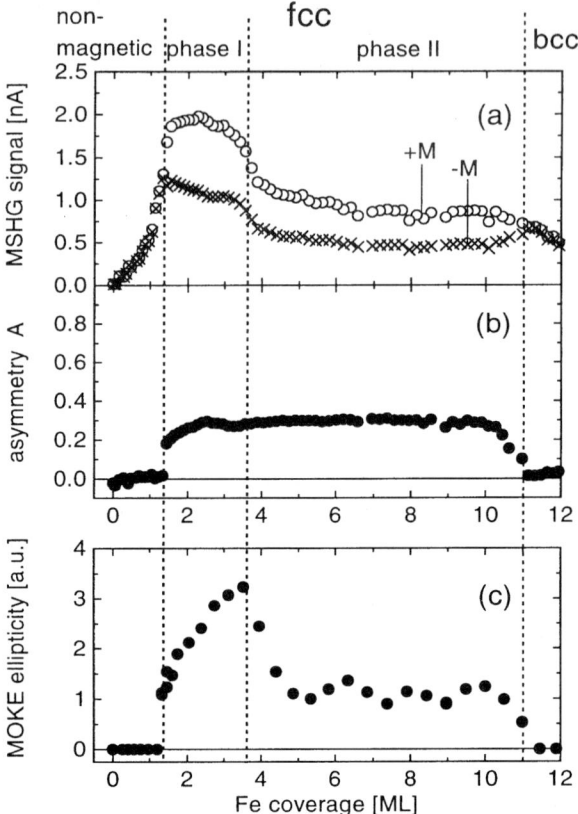

Fig. 2.28. Thickness dependence of (a) of the SH light for magnetization in opposite directions and (b) the resulting asymmetry calculated from the SH intensities in (a). For comparison the linear MOKE signal (ellipticity) is shown in (c). Both the linear and the nonlinear Kerr measurements were measured in remanence at $T = 220$ K.

large as 0.28 Å at the surface.[158,159] The Fe film in thickness range II shows only a 2×1 reconstruction with smaller in-plane shifts.[175–177] However there was no need to assume a vertical buckling of the layers to fit the experimental data. SHG is very sensitive to surface reconstruction.[57,65,178] Because of the screening effects present for vertical displacements SHG should be more sensitive to vertical displacements and the larger SH intensity in phase I may be explained by the large buckling present in this phase but not in phase II.

The fact that the asymmetry in both phases is nearly identical might be accidental. In fact, taking the different Curie temperatures for phase I and phase II into account, a 10% higher asymmetry results in phase II with respect to phase I from the extrapolation of the measured data to 0 K. Since MSHG does not measure magnetic moments directly and the SH intensity as well as the MSHG asymmetry are strongly influenced by any change in the electronic structure, it cannot determine quantitatively the magnitude of magnetic moments. However, qualitatively the nearly constant asymmetry measured in phase I and phase II can be understood if a nearly equal surface magnetization in both phases is assumed. As mentioned in sec. 2.6.1 the 'nonmagnetic' tensor element that contributes most strongly to the outgoing SH light for not too large incident angles is the $\chi^{(2)}_{yyz}$ element; the only 'magnetic' tensor element is the $\chi^{(2)}_{xyz}$ element. Therefore in both cases the SH polarization is generated by the 'mixed' components of the incident light and it is parallel to the surface. Therefore one can expect that these two tensor elements are affected by the change in surface reconstruction in a similar way and it can be expected that the Kerr angle (or the asymmetry at a fixed analyser angle), which is proportional to the ratio of $\chi^{(2)}_{xyz}$ and $\chi^{(2)}_{yyz}$, is a good measure of the surface magnetization. A nearly constant surface magnetization in the two phases is also consistent with a nearly constant enlarged interlayer spacing of the two topmost layers of about 1.86 Å.[159]

In the fourth thickness region in Fig. 2.28 the film transforms fully into the bcc phase via a martensitic phase transition.[155] In the polar Kerr geometry used in the experiments no Kerr signal is detected because the magnetization flips into the film plane. Very small amounts of bcc precipitates already present at a thickness as low as 4.6 ML obviously do not affect the SHG signal.

Strictly speaking, the experimental result only proves that the magnetically 'live' layers are located at the surface *and/or* the buried Fe/Cu interface. In secs 2.6.2 and 2.6.5 it is shown that presumably in the case of Co films on Cu(100) the buried interface has a larger 'magnetic' contribution than the surface at a wavelength of about 800 nm of the fundamental light. This was concluded from the behaviour of the SH signal on CO or O_2 absorption on the Co film. The asymmetry increased upon gas adsorption. For Fe/Cu(001) just the opposite was observed, a decrease of the magnetic component with increasing O_2 coverage. Under the same assumptions as made for the Co/Cu(001) case it follows that in this case the surface contribution dominates. A comparison of the influence of O_2 adsorption at $T_s = 170$ K on the linear Kerr signal and the nonlinear Kerr signal shows a nearly 40% decrease of the nonlinear Kerr signal at an exposure of 0.5 L (Langmuirs) while only a drop of 7% was observed in the linear Kerr signal, further corroborating the dominant surface contribution to the 'magnetic' component of the SH light. In phase II less than 0.2 L suppress both the linear and the nonlinear Kerr signal.

2.6.4 Bulk contributions to SHG

While in many cases the dipolar surface contribution $\chi^{(2,D)}$ dominates the SHG, in the case of materials with inversion symmetry,[123] this is not guaranteed in general and depends also on the incident wavelength and angle of incidence, θ_i, used in the experiments. While for larger angle of incidence the surface response tends to dominate due to the large $\chi^{(2,D)}_{zzz}$ contribution to the SHG, for smaller angle of incidence bulk contributions may become more important.[39] A clear experimental separation of bulk and surface contributions of SHG from a surface is difficult and in general not possible, as discussed by a number of authors[126-128,179,180]. For example, while for the isotropic case the s-polarized outgoing SH light should be generated entirely due to the dipolar surface susceptibility[122], this is already not true for the less symmetric cases of a (001) surface of a cubic crystal.

Although adsorption experiments may give some indication about the presence of bulk contributions as suggested, for example, in the case of the Fe(110) surface and discussed in secs 2.6.1 and 2.6.5, this kind of analysis is not free from assumptions about the effect of the adsorbates on the surface SHG. For the (001) surface of a cubic crystal the magnitude of the anisotropic part of the bulk-like contributions, $\chi^{(2,Q)}$, can be obtained easily from the azimuthal dependence of the SHG. Because of the 4-fold axis the dipolar surface part of the nonlinear polarization has to be isotropic with respect to an azimuthal rotation as can be verified with the aid of Table 2.2. According to Sipe et al.[125] the anisotropic part of the bulk contribution can be written as

$$E_p^{(2\omega,Q)}(p_{in}) = C \frac{n\zeta}{8(nf_c + NF_c)} (a_p + c_p \cos 4\psi) \qquad (2.6.1)$$

with the same prefactor

$$C = 4\pi i \frac{\omega}{c} |E^{(\omega)}|^2 \frac{T_{s,p}}{\cos \theta_i} \qquad (2.6.2)$$

as for the dipolar part and

$$a_p = t_p^2 (3F_c f_s f_c^2 + 4F_s f_s^2 f_c) \quad \text{and} \quad c_p = t_p^2 F_c f_s f_c^2. \qquad (2.6.3)$$

The f_s, f_c, etc. are the same as defined in sec. 2.5; ψ is the azimuthal angle measured from a $\langle 100 \rangle$ direction. The anisotropy parameter ζ is defined through the four independent tensor elements of $\chi^{(2,Q)}$ defined in eqn (2.5.8) as

$$\zeta = \chi_1^Q - (\chi_2^Q + \chi_3^Q + \chi_4^Q). \qquad (2.6.4)$$

Therefore the total SH intensity from the surface can be expressed as

$$I_p^{(2\omega)}(p_{in}) = |A + B \cos 4\psi|^2. \qquad (2.6.5)$$

The factors A and B can be found by comparison of eqn (2.6.5) with eqns (2.6.1) and (2.5.14). Therefore, in the presence of significant (anisotropic) bulk contributions a 4-fold azimuthal anisotropy is expected. In Fig. 2.29 an azimuthal scan of the SH intensity from a 10 ML thick Co film on Cu(001) is shown. The transverse Kerr geometry was used, i.e. an external field of about 50 Oe was applied always perpendicular to the optical plane. The SH intensities $I(+M)$ (solid circles) and $I(-M)$ (open circles) together with the arithmetic average $I_{av} = [I(+M) + I(-M)]/2$ (squares) are plotted. A magnetization reduces the symmetry of the (001) surface to 2mm. Now even the surface SH response exhibits an azimuthal anisotropy. However, if the magnetization-induced part of

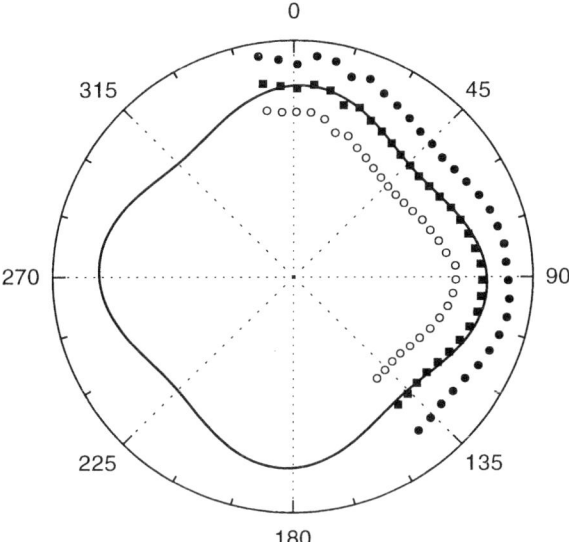

Fig. 2.29. Azimuthal dependence of the SH signal from a 10 ML Co film on Cu(001) for p-polarized incident light in the transverse Kerr geometry at $\lambda = 790$ nm incident light. $\theta_i = 38°$. The solid and open circles denote the SH intensities $I(+M)$ and $I(-M)$. The squares are the arithmetic average $I_{av} = [I(+M) + I(-M)]/2$ and the solid line is a fit to I_{av} according to eqn (2.6.5).

the SHG is comparably small—in this case $A_p = 0.2$ and therefore $\chi^{(2,D,m)}/\chi^{(2,D,nm)} \approx 0.1$—the magnetic contribution to I_{av} should be of the order of 1%. I_{av} shows a small azimuthal dependence in Fig. 2.29 of the order of 10% indicating an anisotropy of the nonlinear susceptibility of more than 5%. This is more than one would expect from the above rough estimate due to the symmetry reduction by the magnetization. (The 4-fold symmetry, however, of the SH intensity would be expected, because of the 4-fold magnetic anisotropy of the (001) Co film.) The main reason for the observed azimuthal anisotropy in the SHG, however, is probably not the film itself but the SHG from the substrate, as will be discussed below.

While for the Co film on Cu(001) the azimuthal anisotropy is small, this is not the case for the bare substrate as can be seen in Fig. 2.30. Here the SH intensity as a function of the azimuthal angle is shown for the Cu(001) substrate with otherwise identical experimental conditions to the results shown in Fig. 2.29 for the Co film. An intensity variation of more than a factor of 5 is observed indicating that the bulk contributions are at least of the same size as the surface contributions. A quantitative analysis of the s-polarized SH light revealed for $\lambda = 790$ nm a ratio between $\zeta' = n\zeta/[8(nf_c + NF_c)]$ and $\chi_{yxy}^{(2,D)}$ [66] of

$$\frac{\zeta'}{\chi_{yzy}^{(2,D)}} = 0.3. \qquad (2.6.6)$$

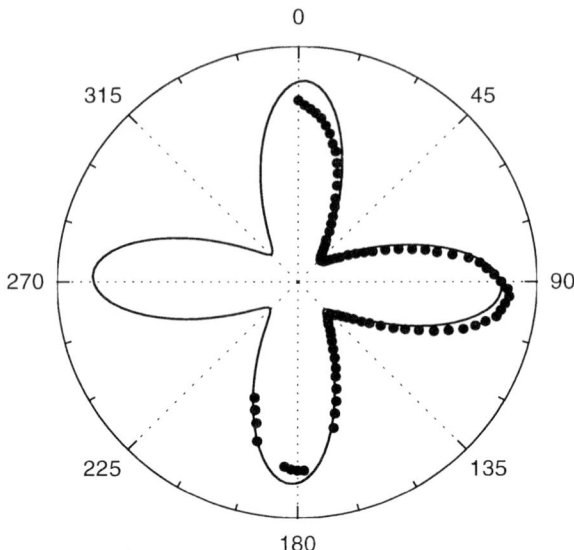

Fig. 2.30. Azimuthal dependence of the SH signal from the Cu(001) surface for p-polarized incident light and p-polarized outgoing SH light ($I_{pp}^{(2\omega)}$, circles). $\lambda = 790$ nm. The solid line is a fit to $I_{pp}^{(2\omega)}$ according to eqn (2.6.5).

Because the SH intensity from the Co film is larger by a factor of 20 to 100 for p-polarized incident light, depending on the azimuth direction, this large anisotropy of the substrate is strongly suppressed but may still account completely for the residual anisotropy observed in Fig. 2.29 since substrate and film contributions of the SH light amplitude add *coherently*. The SH intensity for s-polarized incident light is lower by a factor of about 7. Consequently the observed azimuthal anisotropy is larger compared to the case of p-polarized incident light as can be seen in Fig. 2.31. $I_{av}(\psi)$ varies by about 50% which almost scales with the ratio of the SH intensity of p- and s-polarized incident light corroborating that the azimuthal anisotropy observed for the Co films is indeed mainly caused by the bulk contributions of the Cu substrate.

The bulk contributions from the Co film itself are probably less important in the case of ultrathin films simply because the thickness of the film is much smaller than the penetration depth of the light, which is of the order of 200 Å in the cases discussed here. An (anisotropic) bulk contribution from the film itself should produce an SH signal increasing proportionally with the film thickness, which is not observed in Figs 2.23 and 2.25. The nonferromagnetic substrate contributes only to the nonmagnetic part of SHG and therefore the magnetization-induced part comes entirely from the ferromagnetic film. For the Fe(110) surface discussed in sec. 2.6.1 the bulk may also contribute to the magnetization-induced SH signal. However, because of the lower symmetry of the (110) surface, bulk contributions cannot be derived directly from azimuthal scans in this case because $\chi^{(2,D)}$ itself is anisotropic. The effect of adsorbates discussed in sec. 2.6.5 gives some indication that the bulk contributions to the

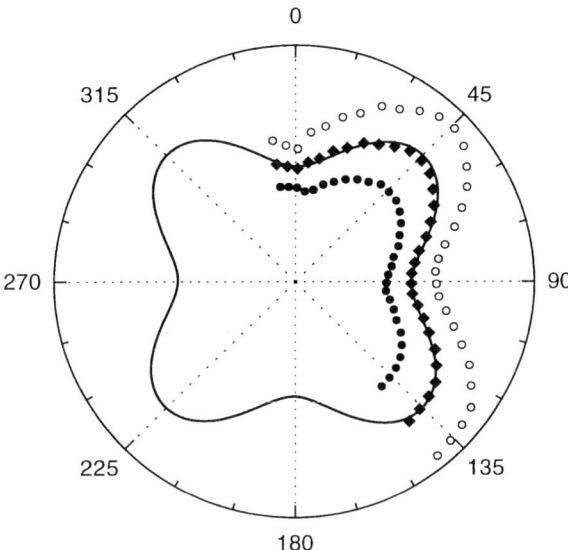

Fig. 2.31. Same as Fig. 2.29 for s-polarized incident light.

magnetic SHG might be significant.

2.6.5 Effects of adsorbates

As already mentioned in the previous sections SHG and magnetization-induced SHG as well are very surface-sensitive. SHG studies under ambient air always face the problem that one interface, the surface, is not well characterized, because the adsorbate film present at the surface may alter the SH signal considerably. It is the purpose of this section to demonstrate the quite dramatic changes of the SHG from surfaces with adsorbate coverage. The influence of adsorbates was already discussed briefly in sec. 2.2.3. Here it is focused on the influence of the adsorbates on the magnetization-induced SHG.

In case of the linear Kerr effect the reflectivity is only very weakly altered by adsorbed molecules. Changes in the Kerr angle directly reflect changes in the magneto-optical Voigt constant Q, i.e. in the off-diagonal elements of the linear susceptibility tensor $\chi^{(1)}$. This is not true for the nonlinear case. Because of its surface sensitivity not only the 'magnetic' part but also the 'nonmagnetic' part of SHG is strongly affected by adsorbates. In certain cases the SH yield can change by orders of magnitude with a monolayer coverage of adsorbates.[46,61,84,132] Another complication arises from the fact that the magnetization-induced tensor elements of $\chi^{(2)}$ are of similar size compared to the 'nonmagnetic' elements of $\chi^{(2)}$. Therefore, the Kerr angle is no longer simply proportional to a ratio $\chi_{\rm m}^{(2)}/\chi_{\rm nm}^{(2)}$ but a nonmonotonic functional dependence on $\chi_{\rm m}^{(2)}/\chi_{\rm nm}^{(2)}$ results.

In the following we give some recent experimental results that demonstrate the above-mentioned influences of adsorbates on the MSHG signal.

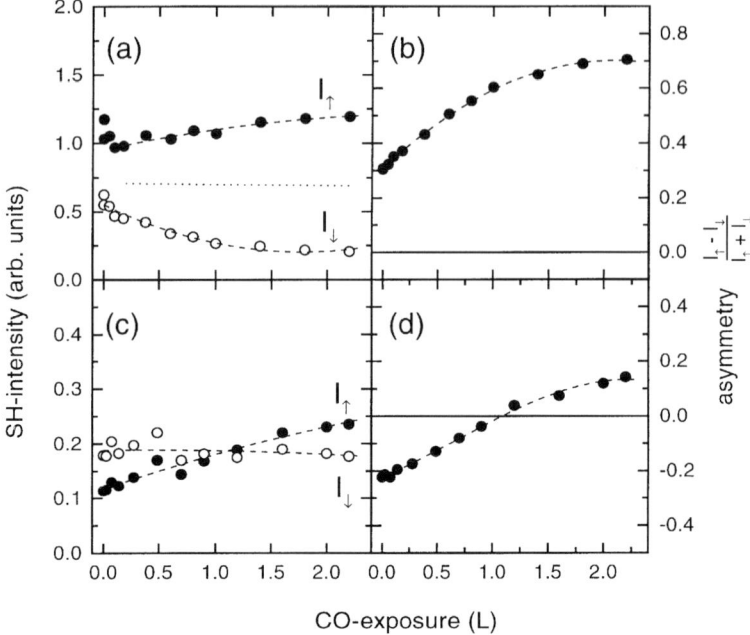

Fig. 2.32. The SH intensity of a 10 ML Co film as a function of the CO exposure for p-polarized incident light (top panels) and s-polarized incident light (bottom panels). The measurements were done in the transverse Kerr geometry. In (a) and (c) open and solid symbols represent the measured SH intensities for the magnetization M in opposite directions. In (b) and (d) the asymmetry A_p and A_s are shown, respectively, calculated from the measured SH intensities in (a) and (c). Dashed lines are guidelines to the eye while the dotted line in (a) represents the arithmetic average of $I_\uparrow$ and $I_\downarrow$. The wavelength of the incident light was $\lambda = 790$ nm.

CO and O_2 adsorption on Co and Fe films on Cu(001)

In Fig. 2.32 the p-polarized SH intensity from a 10 ML thick Co film on Cu(001) is measured as a function of the CO adsorption for p- as well as for s-polarized incident light in the transverse Kerr geometry ($M \parallel y$). While in both cases the arithmetic average of the SH light for the magnetization in opposite directions, $I_{av} = (I_\uparrow + I_\downarrow)/2$, does not change much with the CO coverage, a strong change is observed in the magnetization-dependent signal. For p-polarized incident light a monotonic *increase* of the asymmetry is observed, while for s-polarized incident light the absolute value of the asymmetry decreases, changes sign around 1 L Co exposure and increases again. An obvious explanation for this behaviour can be given by assuming that mainly the magnetization-induced (odd) tensor components of the nonlinear susceptibility are affected by the Co adsorption, while the even tensor components remain nearly unchanged as assumed in the paper by Reif *et al.*[9] for the Fe(110) surface. Opposite to the case in ref. 9 where only one surface contributes, in the case of a film always two interfaces, the surface and the film/substrate interface, contribute to the SH intensity. These two contributions tend to compensate each other although in the present case the two interfaces are not equal and only a partial compensation is expected. Differences in the local fields are unimportant because of the very low thickness of the films. Furthermore only the surface is modified in a way that almost does not affect the linear fields. Therefore, changes in the SH intensity directly reflect changes in the nonlinear susceptibility. Assuming a reduction of the magnetization-induced tensor elements at the surface with CO adsorption, it follows that for p-polarized incident light the magnetization-induced SH from the buried Co/Cu interface always exceeds that of the vacuum/Co interface. The opposite is true for s-polarized incident light. In that case the surface contribution initially exceeds that from the buried interface. Similar results were also obtained from experiments with O_2 exposure.

It is interesting to compare the above finding for an fcc Co film on Cu(001) with that of an fcc Fe film on Cu(001). Because of its complicated magnetic behaviour it received large interest as discussed in sec. 2.6.3. Here we want to restrict the discussion to the homogeneous magnetized phase I at Fe film thickness below 4 ML and the behaviour of the SHG on O_2 coverage. Figure 2.33 shows the SH intensity from a 3 ML thick fcc Fe film on Cu(001) as a function of the O_2 exposure. At 3 ML thickness the easy axis of magnetization is perpendicular to the surface. Therefore the polar Kerr geometry was chosen for the measurements. The experimental setup was similar to that shown in Fig. 2.14 except that the magnetic field was applied (nearly) perpendicular to the crystal surface. For the SH light the analyser was set to an angle of 15° from s polarization. For the linear reflected light simultaneously measured with the same Ti:sapphire laser the analyser was set at an angle of 2° from the minimum of transmission. In contrast to the case of the Co films, O_2 exposure of fcc Fe films leads to a reduction of the asymmetry measured at a fixed analyser angle. Assuming the same action of O_2 on the SHG from the Fe film as on the Co film, it can be concluded that the surface outweighs the buried interface in the magnetization-induced part of the SHG. The magnitude of the (complex) Kerr angle Φ_K of the clean Fe film was determined to approximately 3° to 4°. At an analyser angle of 15° from the s polarization, therefore, the measured SH intensity is dominated by the nonmagnetic components and the arithmetic average $(I_\uparrow + I_\downarrow)/2$ can be taken as a measure for the nonmagnetic component alone. Again this average remains essentially constant as in the case of CO adsorption on the Co film. A small maximum around 2–3 L O_2 exposure can be attributed to an ordered $c(2 \times 2)$ oxygen superstructure seen in LEED at that coverage.

For completeness we note that around 10 L O_2 exposure the linear Kerr signal becomes zero within experimental uncertainty while the asymmetry for the frequency-doubled light changes sign and stays constant afterwards. At this oxygen coverage the easy axis of magnetization changes from perpendicular to in-plane. For that direction of the magnetization no linear Kerr rotation can be measured in this geometry and a linear transverse Kerr signal is too small to be detected. The asymmetry measured in the nonlinear Kerr signal above 10 L O_2 exposure is caused by the nonlinear transverse Kerr effect.

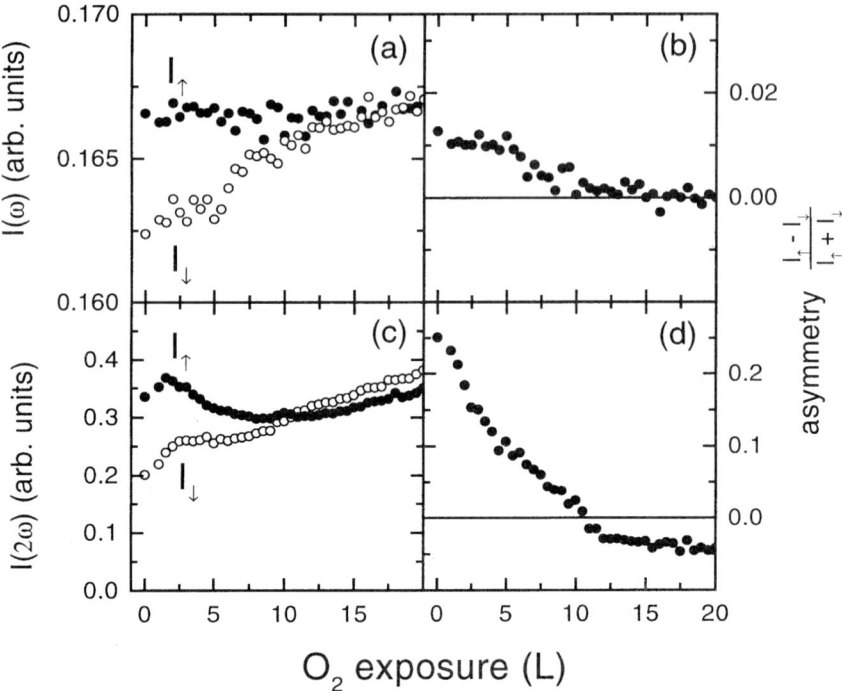

Fig. 2.33. The SH intensity of a 3 ML Fe film as a function of the O_2 exposure measured in the polar Kerr geometry. (c) The SH intensity for the magnetization M in opposite directions. (d) The asymmetry calculated from the intensities $I_\uparrow$ and $I_\downarrow$. The analyser was set to an angle of 15° from s polarization. For comparison the linear reflected light simultaneously measured with the same Ti:sapphire laser is shown in the top two panels. The analyser for the linear light was set to an angle of 2° from the minimum of transmission. The incident light was p-polarized. $T_{\text{sample}} \approx 280$ K. Wavelength of the incident light $\lambda = 790$ nm.

O_2 adsorption on Fe(110)

Using a ferromagnetic single crystal the problem of having more than one contributing interface can be avoided and one may hope to study the influence of adsorbates without a hypothesis about the action of the adsorbates on the (magnetic) SHG. However, it will be shown below that the effect of O_2 depends strongly on the geometry and light polarization used in the particular experiments. Very different behaviour for the individual cases is found, indicating that single tensor elements of $\chi^{(2,D)}$ are affected quite differently by the oxygen adsorption.

In Fig. 2.34(b) the total SH intensity as a function of the O_2 exposure is shown for p-polarized incident light in the transverse Kerr geometry. As for the Co films on Cu(001) the arithmetic average of $I_\uparrow$ and $I_\downarrow$ does not decrease with oxygen exposure. Instead a pronounced maximum in the average SH signal

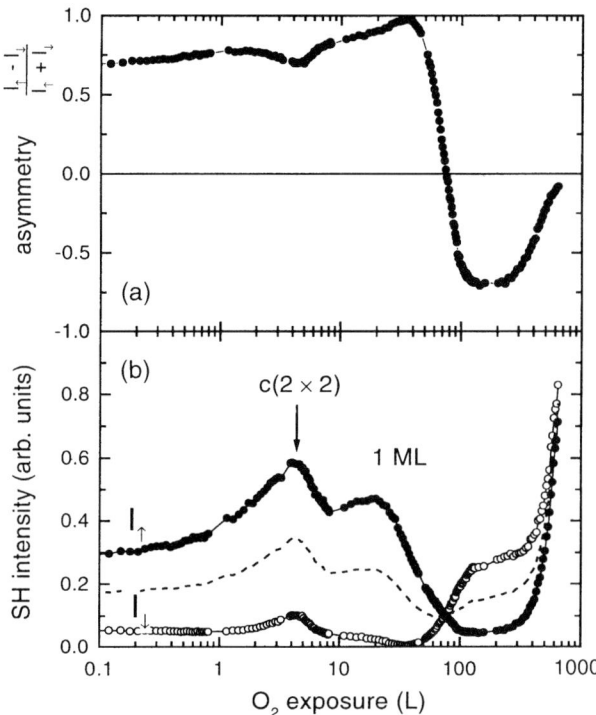

Fig. 2.34. Total SH intensity from a Fe(110) surface for the transverse Kerr geometry as a function of oxygen exposure for p-polarized incident light. In (b) $I_\uparrow$ (filled circles) and $I_\downarrow$ (open circles) denote the SH intensity for the magnetization in opposite directions. The dashed line represents the arithmetic average of $I_\uparrow$ and $I_\downarrow$. In (a) the asymmetry calculated from $I_\uparrow$ and $I_\downarrow$ is shown. $T_{\text{sample}} \approx 330$ K.

occurs for an O_2 exposure of 3–4 L. There are two known superstructures commonly denoted as c(2 × 2) and c(3 × 1) at coverages of 0.25 and 0.33 ML.[181] These superstructures are maximally developed at 3 L and ≈5 L and are not resolved as individual peaks in Fig. 2.34. A second maximum at around 25 L nearly coincides with the completion of a monolayer coverage. On further O_2 exposure the chemical structure of the surface becomes quite complicated because of oxide formation.[182,183] Interestingly for the asymmetry the same behaviour is observed as for the CO adsorption on Co films on Cu(001), namely an increase of the asymmetry with adsorbate coverage. But now only one interface, the surface, contributes to the SHG and the increase cannot be explained by destructive interference between contributions from two interfaces. While it cannot be excluded directly that in this particular geometry O_2 actually enhances the magnetization-induced SHG the possibility of bulk contributions is discussed in ref. 80. With the assumption that O_2 reduces the magnetization-induced part of surface SHG, qualitatively the same behaviour as for CO adsorption on Co films can be postulated. But now instead of contributions from

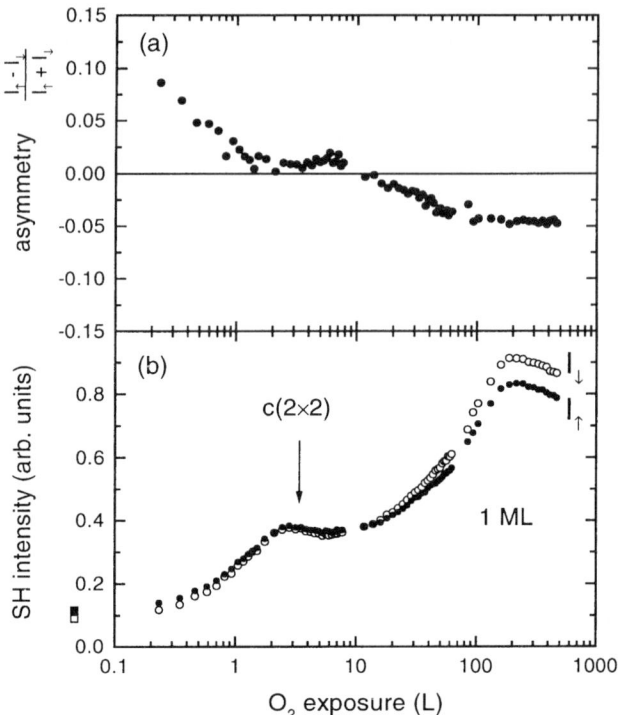

Fig. 2.35. Same as in Fig. 2.34 for s-polarized incident light. $T_{\text{sample}} \approx 470\,\text{K}$.

a buried interface the bulk contributions interfere with the surface contribution.

For s-polarized incident light (Fig. 2.35) the asymmetry decreases with increasing O_2 exposure and almost vanishes at around 0.25 ML coverage while the SH intensity itself increases strongly with O_2 coverage. Therefore part of the decrease in the asymmetry is caused by an increase of the 'nonmagnetic' contribution but still a net reduction of the magnetic component results assuming a constant phase between 'magnetic' and 'nonmagnetic' parts. As for the p-polarized incident light the c(2 × 2) oxygen superstructure creates a maximum in the (average) SH intensity. (Monolayer coverage happens at larger O_2 exposure compared to Fig. 2.34 because of the larger surface temperature of about 470 K in this experiment.)

Summarizing the results from the oxygen coverage dependence of SHG in transverse geometry it can be concluded that oxygen strongly affects the 'nonmagnetic' part of SHG. The change of the magnetic component alone cannot be determined directly in this geometry, because both 'magnetic' and 'nonmagnetic' components generate p-polarized SH light and the phase factor φ is not known. In the longitudinal geometry this is easily achieved because of the orthogonal polarization of the two components. Figure 2.36 shows a series of plots of the SH intensity versus polarizer angle Φ in the longitudinal geometry for p-polarized

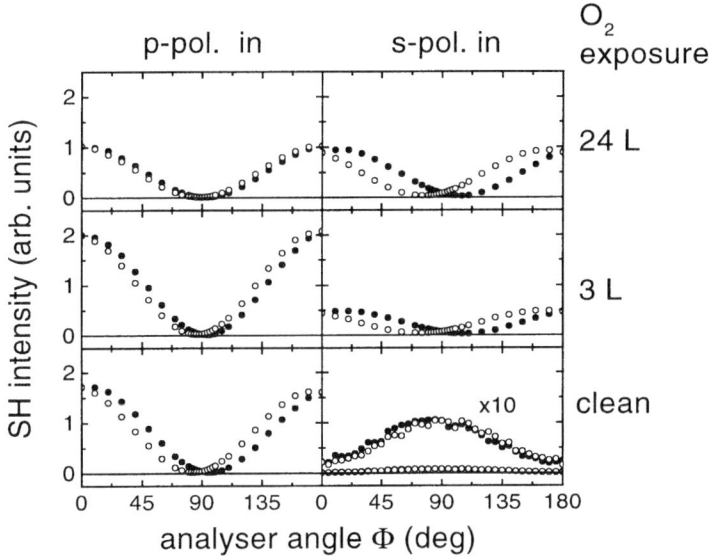

Fig. 2.36. SH intensities $I(+M)$ and $I(-M)$ from the clean and oxygen-covered Fe(110) surface in the longitudinal geometry as a function of the analyser angle Φ, on the left for p-polarized and on the right for s-polarized incident light. 0° corresponds to p-polarized outgoing SH light. $T_{\text{sample}} \approx 340\,\text{K}$.

incident light (left side) and s-polarized incident light (right side) for different amounts of O_2 exposure. Now the optical plane is parallel to a different azimuthal direction, the [100] direction, which is also the direction of the magnetization. Because of the only 2-fold symmetry of the (110) surface also for the 'nonmagnetic' part of SHG different tensor elements contribute compared to the transverse geometry. While for p-polarized incident light the curves for $I_\uparrow$ and $I_\downarrow$ approach each other for increasing O_2 exposure, for s-polarized incident light the curves $I_\uparrow$ and $I_\downarrow$ are already nearly identical for the clean Fe(110) surface. As discussed in sec. 2.6.1 this is caused by the almost vanishing *even* p-polarized component of the SH light. All light is of 'magnetic', odd origin and therefore no asymmetry can occur. Oxygen exposure increases the p-polarized part and consequently an asymmetry is observed for the oxygen-covered surface. The clean surface corresponds to a Kerr angle $|\Phi^{(2,L)}_{\text{Ks}}| = 87°$ (see sec. 2.6.1). This angle drops down to $|\Phi^{(2,L)}_{\text{Ks}}| = 22°$ at 3 L O_2 ($R_s = 0.39$, $\varphi_s \approx 45°$). After the initial fast drop the Kerr angle decreases only very slowly. At an exposure of 24 ML the Kerr angle is $|\Phi^{(2,L)}_{\text{Ks}}| = 17°$ ($R_s = 0.31$, $\varphi_s \approx 38°$). As noted by Pan *et al.*[1] the phase angle between even and odd tensor elements of $\chi^{(2,D)}$ should be 90° in the absence of dissipation. Since the indices of refraction $n(\omega)$ and $N(2\omega)$ are complex, dissipation in the linear E fields is large. However, at large O_2 exposures the deviation of the phase φ from 90° is caused mainly by the linear optical constants and not by the nonlinear tensor elements themselves. Assuming only dipolar contributions the phase between $\chi^{(2,D)}_{xxx}$ and $\chi^{(2,D)}_{zxx}$

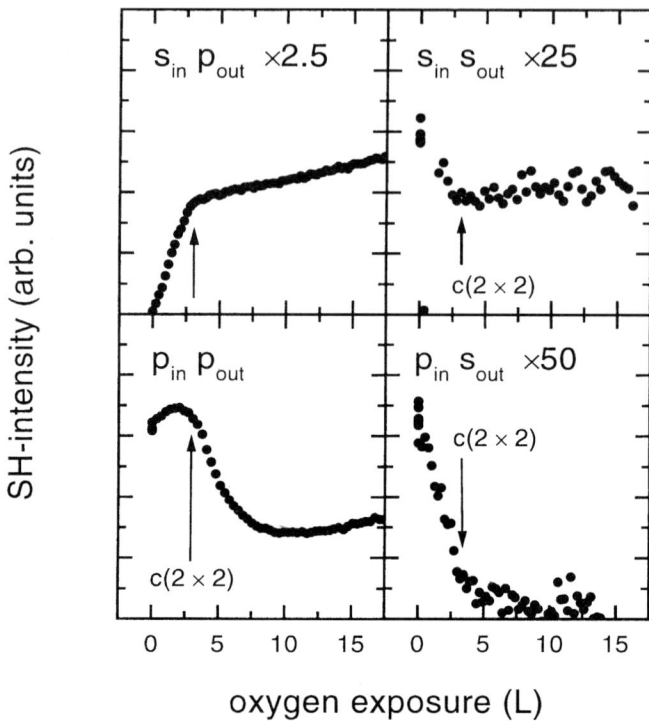

Fig. 2.37. SH intensity as a function of O_2 exposure in the longitudinal Kerr geometry for all four combinations of the polarization of the incident fundamental light and the outgoing SH light. The SH intensity in the left panels is generated entirely by the nonmagnetic tensor elements $\chi_{nm}^{(2)}$, while the SH intensity in the right panels is generated entirely by magnetic tensor elements $\chi_m^{(2)}$. $T_{sample} \approx 470\,\mathrm{K}$. The intensities in the four panels are drawn approximately to scale.

deviates only by 5° from 90° if the factor $N^2 F_s/F_c$ is taken into account as described in sec. 2.6.1. For the clean surface there are larger deviations of the phase from 90°.

While for s-polarized incident light O_2 causes much larger changes in the Kerr angle, stronger changes in the magnetization-induced signal occur for p-polarized incident light. This is illustrated in Fig. 2.37. The SH intensities for all four polarization combinations of incident fundamental and outgoing SH light are plotted as a function of the O_2 exposure. p-polarized SH light is generated from the 'nonmagnetic' tensor elements and s-polarized SH light is generated entirely from the 'magnetic' tensor elements. For s-polarized incident light the magnetic part of the SH light, $I_s(s_{in})$, is little influenced by oxygen adsorption. There is only a small drop of $I_s(s_{in})$ at low O_2 exposure. The 'nonmagnetic' part, $I_p(s_{in})$, on the other hand is strongly *enhanced* by O_2. This behaviour is responsible for the observed strong dependence of the nonlinear Kerr angle $|\Phi_{Ks}^{(2,L)}|$ and therefore its strong surface sensitivity noted in ref. 80. However, this strong change is not due to a change in the magnetic properties of the surface but to electronic changes. $I_p(s_{in})$ increases from a value close to the background level almost two orders of magnitude with increasing oxygen coverage up to 1/4, the coverage of the c(2 × 2) superstructure (3 L exposure), and increases further but slower on further increase of the oxygen coverage while the change in $I_s(s_{in})$ is much less than a factor 2. The surprisingly fast rise of $I_p(s_{in})$ may indicate quite larger and nearly equal SH light contributions from $\chi^{(2,D)}$ and $\chi^{(2,Q)}$ which compensate each other at the clean surface. Small amounts of O_2 reduce the dipolar part also by a small amount, which however because of the initially very low signal causes a strong *relative* increase of the SH light intensity. The opposite is true for p-polarized incident light. $I_s(p_{in})$ is decreased by about one order of magnitude after 3 L of O_2 while the 'nonmagnetic' part $I_p(p_{in})$ is nearly unchanged. At further O_2 exposure $I_p(p_{in})$ starts to drop. Therefore, a strong surface sensitivity of the *magnetization*-induced SHG is found for p-polarized incident light despite the fact that the Kerr angle, which is proportional to $\sqrt{[I_s(p_{in})/I_p(p_{in})]}$, does not change much with oxygen adsorption.[80] After 3 L O_2 exposure $I_s(p_{in})$ definitely is not zero as can be seen from Fig. 2.37. Even at the highest oxygen coverage shown in the figure a clear symmetry is still visible. While the asymmetry $A(\Phi)$ depends linearly on $\chi_m^{(2)}$, the intensity $I_s(s_{in})$ depends *quadratically* on $\chi_m^{(2)}$.

In summary it can be concluded that gas adsorption experiments can give qualitative information about the relative strength of 'magnetic' susceptibilities $\chi_s^{(2,m)}$ and $\chi_i^{(2,m)}$. However, also the influence of adsorbates on the 'nonmagnetic' part of the susceptibilities $\chi_s^{(2,nm)}$ and $\chi_i^{(2,m)}$ have to be taken into account. Even in the apparently simpler case of a single ferromagnetic crystal surface the effect of the adsorbates depends strongly on the (Kerr) geometry and polarization of the incident light. Some experimental results like the increase of the asymmetry with O_2 adsorption and the strong increase of $I_p(s_{in})$ might indicate significant contributions of bulk SHG.

2.7 MSHG studies of thin films

2.7.1 Quantum-well states in thin metallic films

Periodic quantum size effects in ultrathin metallic layers have been observed in a number of different physical properties, in the periodic variation of the workfunction, the surface energy,[184,185] transport phenomena like tunnelling currents[186,187] or in the growth properties of thin films.[188] More recently quantum size effects in magnetic films were investigated extensively. Antiferromagnetic exchange coupling of two magnetic films through a paramagnetic layer has been observed by Grünberg.[189] This antiferromagnetic coupling gives rise to phenomena like giant magnetoresistance (GMR)[190,191] and spin valve effects. Subsequently oscillations in the interlayer exchange coupling were observed as a function of the interlayer thickness.[192,193] Theoretically this coupling across nonmagnetic spacer layers has been described by concepts similar to the

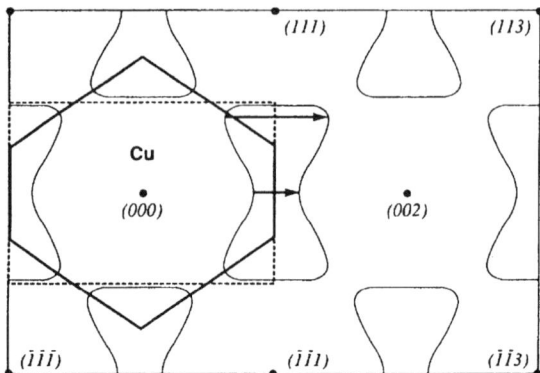

Fig. 2.38. Cross-section of the Fermi surface for a Cu(001) spacer. The first bulk Brillouin zone and film adapted Brillouin zone are represented as bold solid and dashed contours. The extremal wavevectors giving rise to the oscillatory interlayer coupling are indicated by two horizontal arrows.

Ruderman–Kittel–Kasuya–Yoshida (RKKY) theory for localized magnetic moments in a nonmagnetic matrix.[194] In these models a ferromagnetic layer interacts with the conduction electrons of the spacer layer and induces a spin polarization in that layer. The spin polarization extends throughout the spacer layer and interacts with the second ferromagnetic layer mediating an indirect exchange coupling between the two ferromagnetic layers.[195–197] Alternatively, the exchange coupling is described by a quantum-well model.[198–202] In both cases, however, the oscillation period is determined by the 'nesting' vectors at extremal points of the Fermi surface (so-called callipers) as shown in Fig. 2.38 for the case of a Cu(001) spacer. For this case of Cu(001) two oscillation periods with $\Lambda_1 = 5.9$ ML and $\Lambda_2 = 2.6$ ML were predicted by the theory[197] and confirmed by experiments on Co/Cu/Co/Cu(001) multilayer structures.[203–206] The existence of confined quantum-well states and their spin polarization in the Cu spacer layer have been proven by photoemission and inverse photoemission[199,207,208] and spin-polarized photoemission.[209–212] Meanwhile also the phase and amplitude of the oscillations in the interlayer coupling are well understood by means of first-principles calculations.[213–215]

Quantum size effects were found in the magnitude of the Kerr angle measured on ultrathin ferromagnetic layers.[216,217] An oscillatory-like behaviour was observed either as a function of the thickness of the ferromagnetic layer in Au/Fe/Au(001)[218] or as a function of the paramagnetic cover layer in Au/Co/Au(111).[219] In the latter case a period of about 7.7 ML was observed largely independent of the wavelength of the incident light. (The corresponding period for interlayer coupling is 6.4 ML.[220] The oscillatory dependence of the Kerr effect on the thickness of the cover layer was explained by Suzuki and Bruno[221,222] as an induced Kerr effect in the paramagnetic layer due to the existence of spin-polarized quantum-well states in the cover layer. The mechanism leading to a Kerr effect is illustrated in Fig. 2.39 for a bulk ferromagnet

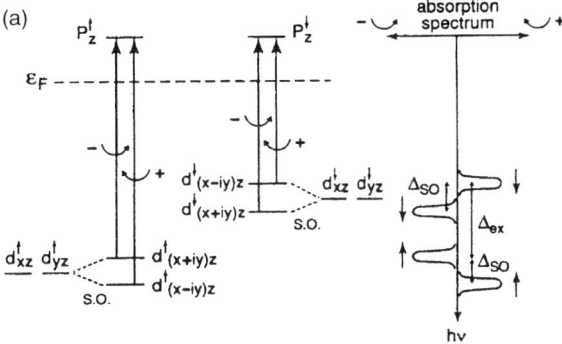

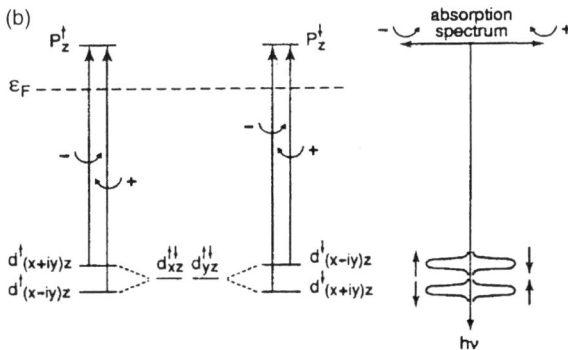

Fig. 2.39. Sketch of the energy levels in the bulk ferromagnet (a) and a bulk paramagnet (b) showing the electrical dipolar optical transitions for the left- and right-circularly polarized light. After ref. 222.

and a bulk paramagnet. Optical transitions allowed according to the dipolar selection rules ($\Delta l = \pm 1$, $\Delta m_l = 1$ for left-circularly polarized light and $\Delta m_l = -1$ for right-circularly polarized light) are indicated by vertical lines. Because of the simultaneous presence of spin–orbit splitting Δ_{SO} and exchange splitting Δ_{ex}, an inequality in the absorption for right and left polarized light is obtained and therefore a Kerr rotation results in an appropriate geometry. In a paramagnet, because of the missing exchange splitting, no asymmetry in the absorption results. In thin films, however, spin-dependent confinement can lead to different densities of states for the final state, for example ($p_z^\uparrow$ and $p_z^\downarrow$ in Fig. 2.39). The amplitude of this induced Kerr effect varies with the difference in the density of states for spin-up (majority) and spin-down (minority) electrons in the final states of the optical transitions. Now a similar oscillatory dependence can be expected as a function of the thickness of the paramagnetic layer as for the case of interlayer coupling though a paramagnetic layer. The difference is that now not the density of states near E_F but the energy of the final state (which is assumed to be unpolarized) and the photon energy together determine the oscillation period. For the system Au/Fe(001) an oscillation period strongly varying with the photon energy is obtained ranging from 8 ML at $\hbar\omega \approx 1.7$ eV to infinity at $\hbar\omega = 2.63$ eV. This finding does not agree with the nearly constant period found in Au(111) films on hcp Co(0001) mentioned above. However, the electronic structure, especially the extremal wavevectors, are completely different for the (111) orientation of the film compared to the (001) surface that this discrepancy is not unexpected. A theoretical treatment of the Au(111) orientation has not been published up to now. For bcc Fe/Au/Fe(100) sandwich systems Katayama et al.[223] found an oscillatory behaviour in the saturation Kerr rotation as a function of the Au spacer layer thickness of about 5 ML in a wide photon energy range between 2.5 to 3.8 eV. Currently the oscillations in the Kerr angle are not understood as well as the oscillations in the interlayer coupling.

In the case of linear reflected light the influence of the quantum-well states is only observable in the (small) variations of the Kerr angle. On the other hand, the sources of SHG are strongly localized at the surface and the interfaces. Therefore, small changes in the electronic structure of the film significantly modify the SH intensity, as already mentioned in the preceding sections. The quantum-well states affect not only the magnetization-induced part of SHG in analogy to the Kerr angle in the linear case but also the average or magnetization-independent part of SHG. This will be discussed below on the example of Cu/Fe/Cu(001) and Cu/Co/Cu(001) sandwich systems.

2.7.2 SHG from Cu/Co/Cu(001) and Cu/Fe/Cu(001) sandwich systems

Figure 2.40 shows the average SH intensity and the calculated asymmetry for Cu/10 ML Co/Cu(100) and Cu/3 ML Fe/Cu(001) as a function of the Cu cover layer. In secs 2.6.2 and 2.6.3 it was mentioned that the easy axis of magnetization for ultrathin fcc Co films is in the surface plane along the $\langle 110 \rangle$ azimuth and for ultrathin Fe layers it is perpendicular to the surface. In the case of Co films this does not change when a Cu cover layer is grown on top. For the Fe films the easy axis flips in-plane at a certain Cu thickness. This thickness depends strongly on minute differences in step density, growth temperature, etc., which are difficult to control. For the measurements shown in Fig. 2.40 the Fe film was magnetized perpendicular to the film plane in the whole Cu thickness range shown. The measurements for the Cu/Co/Cu(001) sandwiches were performed in the transverse Kerr geometry at $\lambda = 790$ nm. All other experimental conditions were identical to those of the measurements shown earlier in Fig. 2.23. In the case of the Fe sandwich the polar geometry was chosen as in Fig. 2.28 ($\lambda = 770$ nm). In contrast to the measurements on the Co sandwiches, which were done on individual films, a Cu wedge was grown on top of the 3 ML Fe film, so that all data shown result from a single sample. The average SH signal can be taken to a good approximation as being proportional to $|\chi^{(2)}_{nm}|^2$. Assuming $\varphi = 0°$ for the complex phase angle between $\chi^{(2)}_{nm}$ and, $\chi^{(2)}_{m}$, the 'magnetic' contribution to the average SH intensity, $|\chi^{(2)}_{m}|^2$, is below 10% of

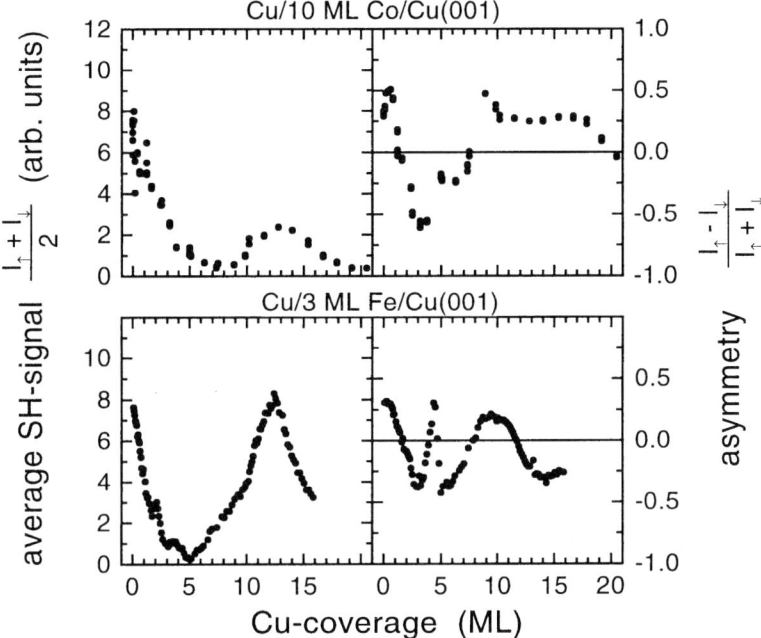

Fig. 2.40. Average SH intensity (left side) and asymmetry (right side) as a function of the Cu cover layer thickness for Cu/10 ML Co/Cu(001) (top) and Cu/3 ML Fe/Cu(001) (bottom) sandwiches. The data on the Co sandwich were measured in the transverse geometry at $\lambda = 790$ nm and with p-polarized incident light; the data on the Fe sandwich were measured in the polar geometry at $\lambda = 770$ nm with the analyser set at an angle $\Phi = 105°$.

$|\chi^{(2)}_{nm}|^2$. For the polar Kerr measurements the analyser was set to an angle of 15° from the s polarization, so that the average SH intensity $(I_\uparrow + I_\downarrow)/2$ is largely determined by the 'nonmagnetic' contribution as well.

In both cases the average SH intensity drops with Cu coverage from the first few monolayers. This behaviour can be easily understood with the aid of a symmetry argument. By Cu adsorption on the Fe or Co surface the two interfaces, the buried Cu/transition-metal interface and the transition-metal surface, which changes now to a transition-metal/Cu interface, become more similar. Finally, for very thick Cu layers the inversion symmetry of the whole system Cu/transition-metal/Cu is restored and no net SH radiation is allowed in the dipole approximation. As long as the transition-metal films are much smaller than the optical wavelength, optical phase differences between the two equivalent interfaces can be ignored and the SH light wave generated at the two individual interfaces interfere destructively. A new interface, the Cu surface, is created by the Cu film growth which, however, produces much less SH light than the two other interfaces, as can be seen in Figs 2.23 and 2.28 respectively.

While the drop of the average SH intensity up to a coverage of about 4 ML can be understood as an optical destructive interference effect, the following increase and oscillatory behaviour is unexpected if the Cu overlayer is considered as a homogeneous film with bulk electronic structure. The maximum at about 12 to 13 ML in Fig. 2.40 is definitely not caused by optical interference in the multilayer structure as is the case for the systems studied in refs 224 and 225, for example, on much thicker multilayer structures. (A calculation of the SH intensity using bulk optical constant but including all multiple reflections from the individual boundaries yields an essentially constant SH signal.[120]) Therefore, these fast variations of the SH intensity with the Cu cover layer thickness must be of electronic origin.

In ref. 120 it was already speculated that the variations in the average SH intensity and in the asymmetry are caused by the presence of confined electronic states in the Cu layer. However, it was not clear what is the origin of the observed oscillations. There is no direct relation to the oscillation periods observed for interlayer coupling because not only electronic states close to the Fermi energy contribute but more or less all states with $\pm 2\hbar\omega$ around the Fermi energy. The situation could be described similarly to the case of the linear Kerr effect discussed above. Optical two-photon transitions from the Cu 3d band into the quantized Cu 4s band are resonantly enhanced if the energy difference between a (discrete) d-band state and an s-band state equals the energy of the SH photon. This is illustrated in Fig. 2.41. For films of finite thickness k_z is no longer a 'good quantum number' and the continuous bulk bands along the k_z direction split up into a set of discrete levels (quantum-well states, QWS) as indicated in Fig. 2.41. These discrete states derived from the Cu 4s band are indicated as black dots in Fig. 2.41 for a 12 ML thick Cu film. The vertical arrow (a) describes the above-mentioned resonant transition for a

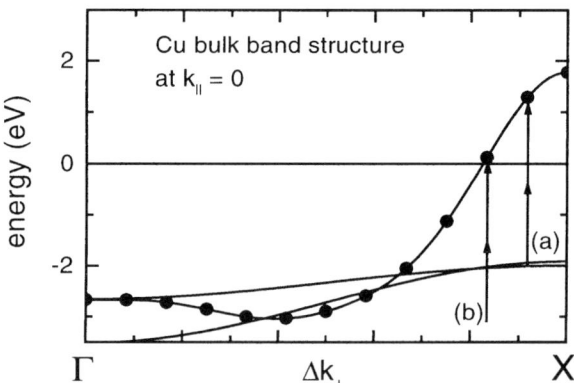

Fig. 2.41. Band structure of bulk Cu(001) in the [001] direction and the 'quantum-well states' derived from the bulk band structure. The quantized states derived from the 4s branch are indicated for a 12 ML thickness of the Cu film. Arrow (a) indicates the transition between a Cu 3d band and the Cu 4s band, which becomes resonant at this thickness. Arrow (b) indicates a transition into a Cu 4s state close to the Fermi energy.

photon energy of 1.6 eV ($2\hbar\omega = 3.2$ eV). If the thickness of the film is increased the QWS is moved more towards the X point and the optical transition becomes less resonant. Therefore a maximum at 12 ML could be expected from this picture. At 24 ML, 36 ML, etc., there is again a resonant optical transition into a QWS and a periodic intensity variation of the SH intensity results with a period of 12 ML. A second possible mechanism is indicated by the transition (b). On increasing film thickness each time a QWS runs through the Fermi energy, it becomes a new possible final state for an optical transition. This transition is nonresonant along the k_z direction but may become resonant at a nonzero $k_\parallel$. Because this kind of optical transition is related to the Fermi energy, a periodic SH intensity oscillation with the same period observed for the interlayer coupling is expected, about 5.5 ML.

The periodicity with layer thickness in transitions of kind (b) in Fig. 2.41 is independent of the photon energy while the period of transitions (a) is strongly dependent on it. For higher photon energy the resonance shifts to larger $\Delta k_\perp$ values and therefore it occurs at a larger film thickness.

These two mechanisms of SH enhancement by transition into QWS are also possible for linear reflected light at 2ω. However, the weight these kinds of transitions contribute to the total SH light (or the reflected light) depends on the density of states (DOS) in the initial and final states involved in the transition. The DOS of the quantum-well states, i.e. the sharpness of the state, is defined by the degree of confinement in the Cu layer. At the vacuum side all electron waves are reflected but at the transition-metal side this is only true for some states, namely Cu states at energies at which no corresponding (same symmetry, same spin direction) states in the transition metal are present. In Fig. 2.42 the band structure along [001] for Cu, minority- and majority-spin states of Co are shown. Only the Cu states at 2.2 eV below the Fermi energy for

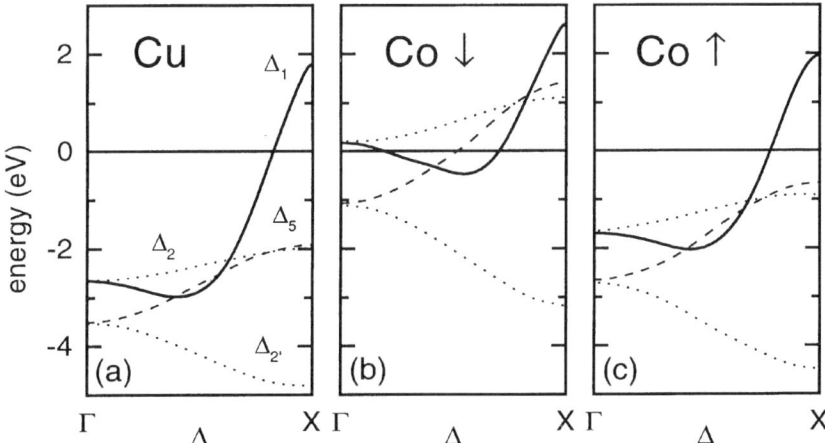

Fig. 2.42. Band structure of (a) Cu, (b) minority- and (c) majority-spin fcc Co along the δ direction ([001]). Bold solid lines correspond to Δ_1 symmetry. After [220].

majority-spin states and below 0.55 eV for minority-spin states are fully confined in the Cu layer. However, because of the difference in density of states at the other energies, still a nonzero reflection coefficient results.[220] The relatively weak confinement of the QWS, above the Fermi energy, is also found in a first-principles calculation by van Geldern et al.[215] While below E_F δ-like spikes in the DOS clearly indicate the confined QWS, above E_F the DOS shows only broad maxima with a width of the order of several hundred meV. It is therefore rather unlikely that the variations of such broad structures in the DOS with the Cu layer lead to the observed strong intensity variations in the SH light without any further enhancement mechanism. Such a mechanism was proposed by Luce et al.[226] and a model calculation was performed for the Cu/Fe/Cu(001) sandwich system. Their model was based on two considerations: (i) SH is generated only in a very narrow range around the interfaces. (ii) Because of (i) it is quite sensitive to hybridization of Fe states and Cu states. Therefore, SHG can be enhanced by resonances of the intermediate state with Fe-derived states. Using the bulk Cu band structure and an *ab initio* band structure for an Fe monolayer, they calculated the nonlinear susceptibility tensor $\chi^{(2)}$ and the resulting SH intensity. Only the SHG from the Cu film/Fe interface is considered. Because of the contribution of the Fe d states the SH amplitude from this interface overwhelms by far (factor 50!) the contribution from the Cu surface. The Fe/Cu substrate interface would give only a constant contribution to the total SH intensity. (For details see Chapter 5 by Hübner in this book.) For $\hbar\omega = 1.61$ eV the SH intensity shows strong maxima at ≈ 11 ML and ≈ 22 ML. These maxima are caused mainly by transitions of kind (a) in Fig. 2.41 with the intermediate state (nearly) in resonance with a 3d Fe-derived state.

I want to emphasize that hybridization of Cu states with Fe states is an essential prerequisite. Without this hybridization the Cu layer would be symmetric with respect to inversion and no SH light is generated in the dipole approximation. At least one of the three involved states, the final state, the intermediate state or the initial state, must be hybridized with substrate states. The k_z conservation in optical transitions involving QWS is relaxed due to the confinement of the state in the Cu layer. Therefore optical transitions between confined layer states corresponding to different k_z values of the bulk band structure are possible. While for QWS well above E_F this may be restricted to neighbouring k_z values because of the relatively weak confinement, (virtual) optical transitions between QWS below E_F with larger differences in k_z values still have a significant transition probability because of the nearly complete confinement. This is also true for the Cu 3d-derived QWS, but because of the small dispersion of these bands compared to the Cu 4s band it is less important.

Besides the two kinds of transitions drawn in Fig. 2.41 a large number of additional transitions involving Fe-derived states and nonvertical transitions are possible at the Cu film/Fe interface. This is illustrated in Fig. 2.43. All QWS derived from the Cu bulk bands are plotted in an energy versus Cu layer thickness diagram as bars. The dense accumulation of QWS below -2 eV results from the Cu 3d bands while the QWS above -2 eV are derived from the sp-like

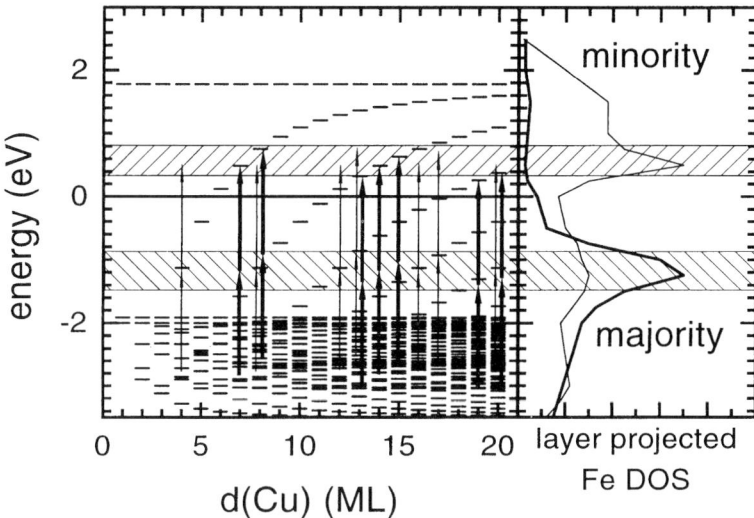

Fig. 2.43. Illustration of the model describing the observed oscillations in the (M)SHG from the Cu/Fe/Cu(001) sandwich. For explanation, see text.

Cu 4s band. On the right-hand side the layer projected Fe DOS at the Cu/Fe interface is drawn schematically after a calculation of ref. 227. The important point is that there are two pronounced maxima in the Fe DOS at the interface, one for the minority-spin electrons at about 0.5 eV above E_F and one for the majority-spin electrons at about 1.1 eV below E_F. While in the bulk band structure of Fig. 2.41 the energy of the initial state for transitions of kind (a) is constrained to a very narrow interval because of the k_z conservation and the very small dispersion of the Cu 3d bands in the relevant part of the Brillouin zone, the width is considerably larger if one considers the relaxed k_z conservation in thin films and the hybridization of Cu and Fe states at the interface. For example, taking half of the width of the Δ_5 band as the effective width of the initial state, all oscillations would be smeared out. Therefore, only sharper features in the spectral density of the initial, final, or intermediate states produce intensity oscillations in the SHG.

The maxima in the minority and majority DOS of Fe are sharp features. Two families of two-photon transitions are drawn in Fig. 2.43. The bold arrows indicate transitions from Cu or Fe 3d states into a Cu 4s QWS above E_F. These transitions are enhanced if the intermediate state has an energy in the range of the maximum of the Fe majority DOS. This is only the case at certain Cu layer thicknesses. A period Λ_1 of about 6 to 7 ML results. The period is somewhat larger compared to the period observed in interlayer coupling because the QWS are probed at an energy ≈ 0.5 eV above E_F at $\hbar\omega = 1.6$ eV. Because of the dispersion of the Cu 4s band the periodicity at which a QWS crosses a certain energy level increases with that energy.

In the second family of two-photon transitions, indicated as thin arrows in Fig.

2.43, the initial state is a Cu or Fe-derived 3d state as for the first family but now the intermediate state is a Cu 4s QWS. The transition is enhanced if the final-state energy is in the range of the maximum in the minority Fe DOS at the interface. Now occupied QWS about 1.1 eV below E_F are probed and therefore a shorter period Λ_2 of about 3 to 4 ML results. The total SH light amplitude is generated by both families of transitions. Therefore the SH intensity oscillates with the sum frequency $1/\Lambda_1 + 1/\Lambda_2$ and the oscillation amplitude is modulated with the beating frequency $1/\Lambda = 1/\Lambda_2 - 1/\Lambda_1$. Because of the finite energy width of the peaks in the Fe DOS and the roughness of the Cu film, the fast oscillations may be strongly suppressed, so that only the long-period oscillation Λ is observed. Taking $\Lambda_1 = 6.5$ and $\Lambda_2 = 4.2$ a period of $\Lambda \approx 12$ ML results in agreement with the experimental result shown in Fig. 2.40.

This model also accounts for the observed strong variations in the asymmetry in Fig. 2.40. The transitions involving empty QWS are enhanced strongly only for the majority-spin states because the spin direction is conserved in the optical transitions. Only the majority Fe DOS shows the strong maximum at -1.1 eV while the minority Fe DOS is almost flat in this energy region and therefore the transition probability is much less enhanced by the intermediate state for transitions between minority-spin states. For the transitions involving occupied QWS the situation is reversed. Now transitions between minority-spin states are much stronger than those between majority-spin states because of the larger Fe DOS at 0.5 eV. The interpretation of the asymmetry curves, however, is complicated by the fact that the asymmetry is not a unique measure of the magnetization-induced part of SHG. The asymmetry is proportional to the ratio $\chi_m^{(2)}/\chi_{nm}^{(2)}$ for not too large $\chi_m^{(2)}$. When the nonmagnetic part of the SH light amplitude changes sign at a certain Cu layer thickness, the magnetization-induced part does not necessarily do so, too. Therefore, in the thickness range close to the sign change of $\chi_{nm}^{(2)}$ the magnetization-induced part $\chi_m^{(2)}$ becomes becomes larger compared to $\chi_{nm}^{(2)}$ and for the asymmetry the full expressions (2.5.20) and (2.5.30), respectively, have to be used. Therefore, some of the sign changes may be just caused by sign changes of the nonmagnetic SH light amplitude and an apparently faster oscillation in the asymmetry is observed. This will be discussed further in section 2.7.4.

2.7.3 Frequency dependence of the MSHG signal from sandwich systems

In the simple model, discussed in section 2.7.2, it is difficult to quantify the relative contributions of different kinds of transitions to the SHG without a detailed calculation including a realistic band structure of the sandwich structure. Besides the indicated transitions in Fig. 2.43 several other transitions are possible as well including those of types (a) and (b) indicated in Fig. 2.41. However, some additional information about the relative contributions of all these transitions can be obtained from measurements of the SHG at different incident wavelengths. All three models involving transitions of types (a) and (b) or the transitions drawn in Fig. 2.43 predict a different dependence on the

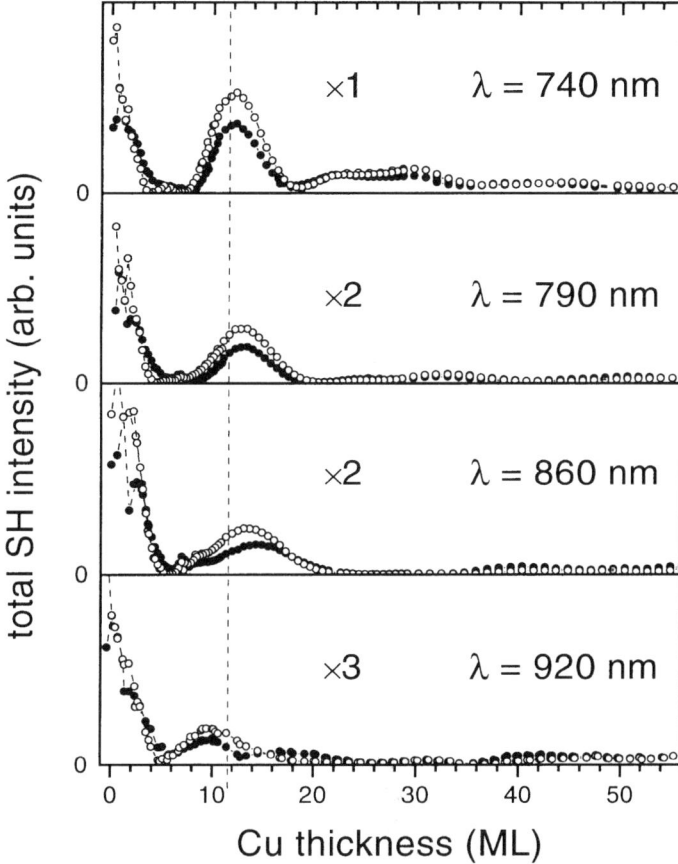

Fig. 2.44. SH intensity from a Cu/10 ML Co/Cu(001) sandwich as a function of the Cu cover layer thickness for various wavelengths of the incident light. The measurements were performed in the transverse Kerr geometry with p-polarized incident light. The intensities are normalized against a quartz reference and scaled by the indicated factors for the different incident wavelengths.

wavelength of the incident light. The frequency dependence of the SHG from the x Cu/10 ML Co/Cu(001) sandwich systems is discussed in the following. Similar results were obtained from x Cu/Fe/Cu(001) sandwiches.

Figure 2.44 shows the SH intensity from a Cu/10 ML Co/Cu(001) wedge as a function of the Cu cover layer in the transverse geometry with p-polarized incident light. Open and solid circles represent the total SH intensity for the magnetization in opposite directions. The measurements were performed in remanence. Because of a small miscut of the Cu sample the uncovered Co films show a small uniaxial magnetic anisotropy as is known from Co films on stepped Cu(001) surfaces.[228,229] On Cu adsorption the magnetic anisotropy changes, causing a flip of the easy axis at very low Cu coverage, and the easy axis flips back again below 1 ML.[230] In this range no intensity difference in the total SH is observed because the magnetization component perpendicular to the optical plane becomes zero. (The zero asymmetry is more clearly seen in Fig. 2.46.) With the exception of the range below 1 ML no change in the easy axis is observed in the full Co thickness range from 1 ML to more than 50 ML. The Co film and the Cu wedge were grown at 300 K with a rate of about 1 ML/min. The angle of incidence was about 38°.

Experimental conditions for the curve at $\lambda = 790$ nm were nearly identical to those for the measurements at individual layers shown in Fig. 2.40 and a very similar result is obtained, a strong maximum of the SH intensity at about 13 ML. At larger Cu thickness the SH intensity drops. Nevertheless two more maxima at about 24 ML and a slightly larger one at 33 ML can be observed. From that an average oscillation period of 11 ML results. For the shorter wavelength $\lambda = 740$ nm a shorter period is found. (Maxima at about 12 ML, 22 ML and 30 ML, average period 9 ML.) This behaviour is opposite to that expected for the transitions of type (a) in Fig. 2.41 but is consistent with the beating frequency model illustrated in Fig. 2.43. In that model the fixed energy points of the transitions are the energies at which the maxima in the minority- and majority-spin DOS occur. For the transitions involving majority-spin states (bold arrows) the QWS are probed now at higher energies because of the higher photon energy. The period Λ_1 of the occurrence of a QWS at this higher energy is longer compared to the lower energy case. For the other transitions involving minority-spin states the QWS are probed now at lower energies resulting in a shorter period Λ_2. Because the shorter period Λ_2 becomes shorter and the longer period Λ_2 becomes longer, the period of the beating frequency Λ becomes longer.

The increase of the period is observed in the experimental curves in Fig. 2.45 from $\Lambda = 740$ nm ($\lambda \approx 9$ ML) at $\lambda = 860$ nm ($\lambda \approx 13$ ML, the second maximum is missing). It should be noted, however, that the assignment of oscillation periods from the experimental data is not unique because of complications discussed below. A larger period of about 12 ML for $\lambda = 740$ nm can also be derived from the data if sign changes of the SH field amplitude are considered. Nevertheless an increasing period with increasing wavelength results.

At even longer wavelength $\lambda = 920$ nm, however, a shorter period is measured again. The reason for that behaviour is not clear. It could be just because the $E(2\omega)$ field amplitude generated by the non-oscillatory background from all three contributing interfaces has an opposite phase to the field amplitude of the uncovered Co film. Because the intensities and not the $E(2\omega)$ field amplitude are measured, the first observed maximum in the intensity actually corresponds to a minimum in the oscillatory part of the $E(2\omega)$ amplitude. Therefore, maxima and minima are interchanged. Assuming this interpretation then, besides the minimum at 5 ML where the $E(2\omega)$ amplitude changes its sign, the first minimum occurs at about 15 ML. Also the apparently shorter period could be explained in this way.

For s-polarized incident light shown in Fig. 2.45 a similar result as for the

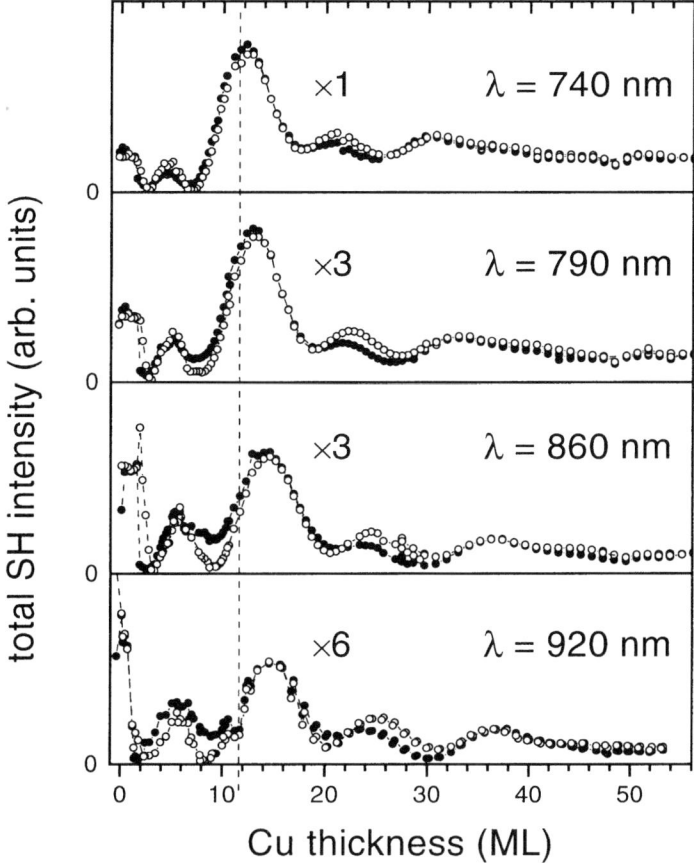

Fig. 2.45. Same as Fig. 2.44 for s-polarized incident light.

p-polarized incident light is obtained. The strongest maximum occurs at nearly the same positions as for the case of p-polarized incident light. Note that now this peak shifts monotonically with increasing wavelength to larger thicknesses even for $\lambda = 920$ nm. The average oscillation period increases from about 9 ML at $\lambda = 740$ nm to about 11 ML at $\lambda = 920$ nm. The periods are slightly smaller than the periods for p-polarized incident light at the corresponding wavelength. This observation is consistent with the model illustrated in Fig. 2.43 and with transitions of type (a) in Fig. 2.41 but not with type (b) transitions. In transitions of type (b) the periodicity is fully determined by the transition of a QWS through the Fermi energy. For the two other types a different periodicity can result because of the selection rules for optical dipole transitions. Different 3d states act as intermediate/final state or initial state, respectively.

At about 5 ML Cu thickness an additional peak occurs for s-polarized incident light which is not (or almost not) visible for p-polarized incident light. This could be again just due to a sign change of the SH amplitude as discussed above for p-polarized light at $\lambda = 920$ nm. When growing Co on the Cu(001) substrate the SH intensity increases immediately with Cu coverage for p-polarized incident light, while for s-polarized incident light the SH intensity runs through a deep minimum before it increases again (see Fig. 2.25). This behaviour suggests that the SH light amplitude from a Co film has a sign opposite to that of the Cu substrate for s-polarized incident light. Therefore, a sign change is expected when growing Cu on top of the Co film, even without the presence of QWS. This can be observed in Fig. 2.45. At about 2 ML Cu thickness the intensity drops to nearly zero for all wavelengths. A sign change of the $E(2\omega)$ amplitude would be consistent with the measured intensity data. At $\lambda = 740$ nm there could be another sign change of $E(2\omega)$. However, for all other wavelengths and thicknesses the SH intensity is well above the background level at least for one magnetization direction.

The complications arising from the possible sign changes and deep minima in

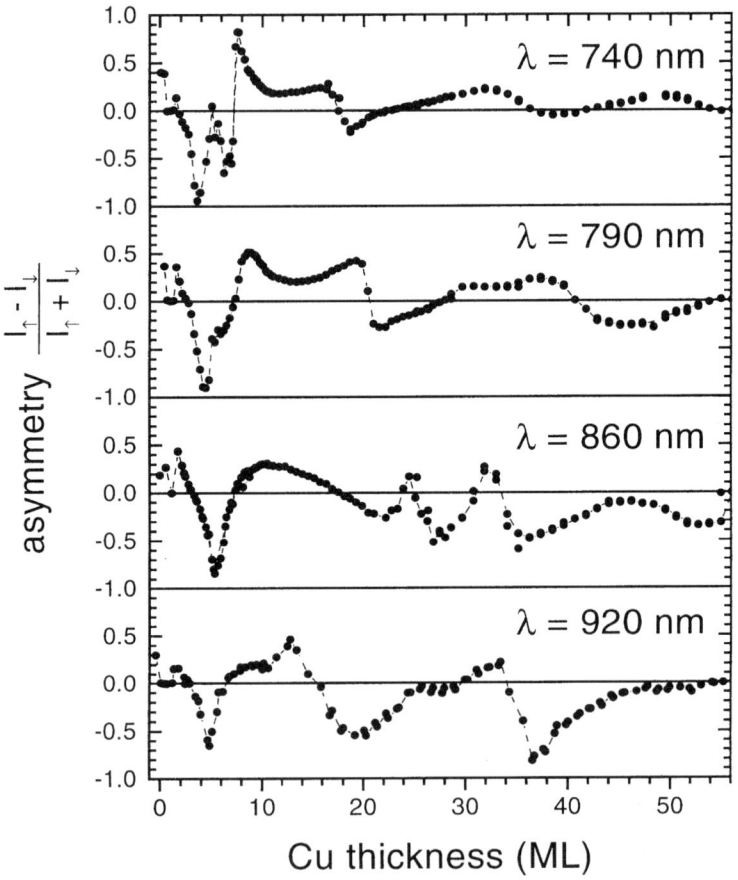

Fig. 2.46. Asymmetries A_p calculated from the measured data in Fig. 2.44.

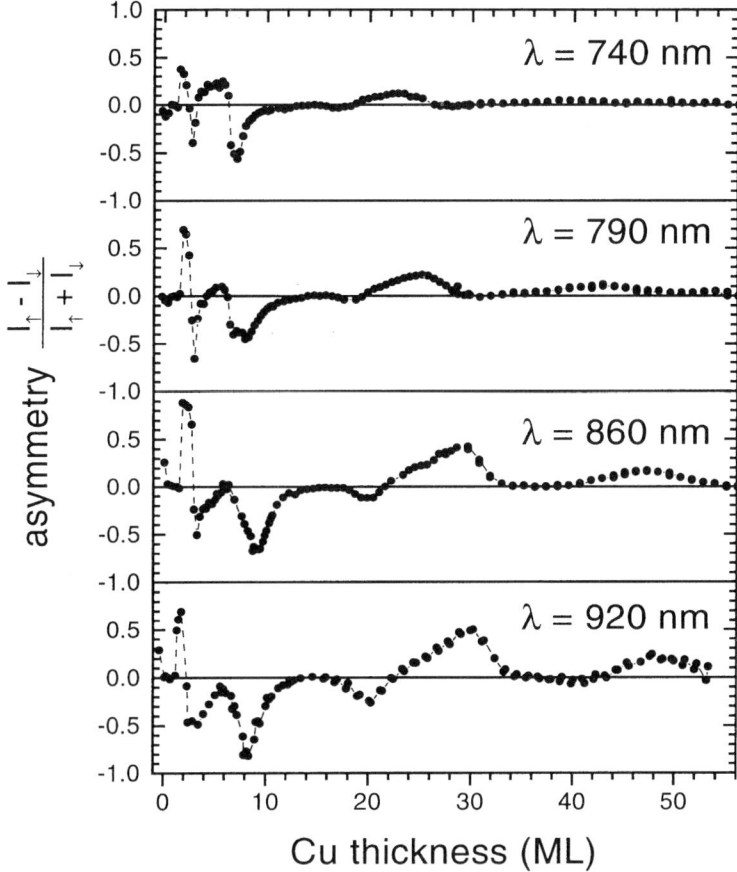

Fig. 2.47. Asymmetries A_s calculated from the measured data in Fig. 2.45.

the average SH intensity make the interpretation of the asymmetry curves difficult, which are shown in Fig. 2.46 for p-polarized incident light and Fig. 2.47 for s-polarized incident light. Several features in the asymmetry curves support the consideration of sign changes in the $E(2\omega)$ amplitude. At certain thicknesses the asymmetry increases, then rapidly changes sign, and increases again. This behaviour is expected from eqn (2.5.20) for a sign change in $\chi^{(2)}_{nm}$ and a constant $\chi^{(2)}_m$. The fast and strong oscillations in the low thickness range around 5 ML and the sign change at about 18 ML at $\lambda = 740$ nm to 35 ML at $\lambda = 920$ nm are probably caused by such an effect. This implies that in the average SH intensity minima and maxima are interchanged at larger Cu thicknesses and the longer of the two periods discussed above (12 ML instead of 9 ML at 740 nm) applies for the measurements with p-polarized incident light.

In conclusion, the quantum-well states induce periodic oscillations in the 'nonmagnetic' and in the magnetization-induced SHG as well. Because the amplitudes of these oscillations are much larger than the SH amplitude generated by the Cu surface, apparently shorter periods in the average SH intensity and a quite irregular behaviour in the asymmetry are found. The transverse Kerr geometry has been chosen for the above investigations because it shows the largest effects. However, it is a disadvantage of this geometry that the magnetization-induced SHG cannot be separated uniquely from the 'nonmagnetic' part because both $\chi_{\text{nm}}^{(2)}$ and $\chi_{\text{m}}^{(2)}$ produce p-polarized SH light. This problem is absent if the longitudinal geometry is chosen, because there $\chi_{\text{nm}}^{(2)}$ and $\chi_{\text{m}}^{(2)}$ generate SH light with orthogonal polarization as discussed in section 2.5.4. First experimental results in this geometry are presented in the next subsection.

2.7.4 MSHG from the Cu/Co/Cu(001) sandwich systems in the longitudinal Kerr geometry

Figure 2.48 shows the SH intensity from a Cu/10 ML Co/Cu(001) sandwich as a function of the Cu cover-layer in the longitudinal geometry at $\lambda = 790$ nm. Preparation conditions were similar to those of the measurements in the preceding subsection. Only the Cu cover layer was grown at a lower temperature

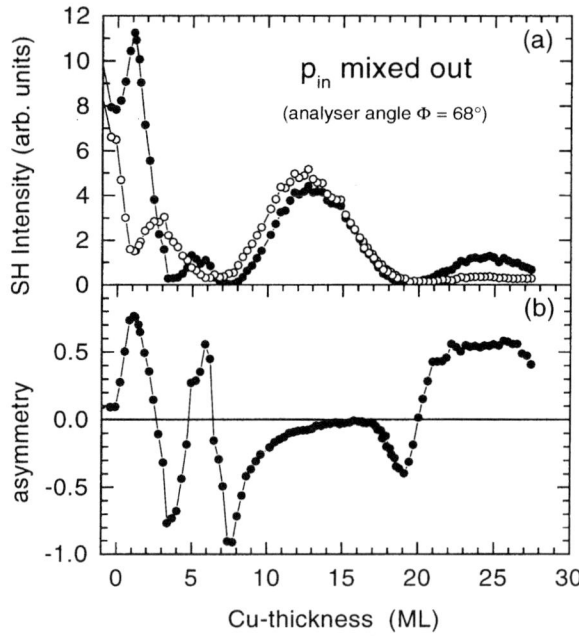

Fig. 2.48. SH intensity from a Cu/10 ML Co/Cu(001) sandwich as a function of the Cu cover-layer in the longitudinal geometry at $\lambda = 790$ nm and p-polarized incident light. Incident angle is $\theta_i = 39°$. In (a) the SH intensity measured at a fixed analyser angle $\Phi = 68°$ is shown and in (b) the calculated asymmetry $A_p(68°)$.

of 150 K. The incident light was p-polarized and the SH light was detected at an analyser angle of 68°, i.e. 22° from s-polarization. The incident angle was 39°. In (a) the intensity of the SH intensity for the magnetization in two opposite directions is shown and in (b) calculated asymmetry. The measurements were performed with saturated magnetization. Therefore the range of zero asymmetry below 1 ML Cu coverage caused by the flip of the easy axis is not observed. The average SH intensity shows a very similar behaviour to that of the corresponding measurements in transverse geometry, as expected. The 'nonmagnetic' tensor elements $\chi_{nm}^{(2)}$ are identical in the two Kerr geometries. As discussed in section 2.6.2 the Kerr angle for the uncovered Co film is small because of the near cancellation of magnetization-induced contributions from the Cu substrate/Co interface and the Co surface. Therefore only an asymmetry of about 0.1 results at the chosen analyser setting of 68°. With increasing Cu cover layer strong and fast oscillations are observed initially and an irregular behaviour at larger Cu thickness follows. While there are differences in the details the overall course of the curves resembles that of the corresponding curve for the transverse geometry in Fig. 2.45.

In this longitudinal geometry, however, it is possible to separate SH contributions from $\chi_m^{(2)}$ and $\chi_{nm}^{(2)}$ by measuring the pure s-polarized SH intensity (polarizer angle $\Phi = 90°$) and the p-polarized SH intensity ($\Phi = 0°$). This is shown in Fig. 2.49(b) and (c). For comparison the curve measured at $\Phi = 68°$ shown in Fig. 2.48(a) is reproduced in Fig. 2.49(a). The SH intensity curve for p-polarized incident light, which is proportional to $|\chi_{nm}^{(2)}|^2$, shows deep minima at about 4 and 7 ML Cu thickness where the rapid sign changes of the asymmetry shown in Fig. 2.48 occur. Also the sign change of the asymmetry at about 20 ML coincides with a minimum in the p-polarized SH intensity. (In Fig. 2.49 the sign change occurs somewhat earlier at 18 ML Cu thickness, which however is not significantly different because the thickness uncertainty in these measurements on wedges amounts to about 10%.) The sign change at 2 to 3 ML does not coincide with a minimum in the p-polarized SH intensity and therefore the magnetization-induced part $\chi_m^{(2)}$ must change sign there. In the s-polarized SH intensity (Fig.2.49(c)), which is proportional to $|\chi_m^{(2)}|^2$, a pronounced minimum is observed at that Cu thickness. The fact that the minimum intensity is not zero is caused by small alignment errors in the polarization of the incident light, azimuthal angle, etc. Remember that for mixed incident polarization also s-polarized SH light is generated by 'nonmagnetic' tensor elements. Because the s-polarized SH intensity is more than 30 times smaller than the p-polarized SH intensity at that thickness, even small misalignments cause a relatively large 'nonmagnetic' s-polarized SH light amplitude. Because the 'nonmagnetic' SH intensity becomes smaller at larger thickness the s-polarized 'nonmagnetic' background becomes smaller too. An oscillation period cannot be derived unambiguously from the present data. However, the variations with Cu thickness in $\chi_m^{(2)}$ are much less pronounced than in $\chi_{nm}^{(2)}$. The only exception is the strong increase of $|\chi_m^{(2)}|^2$ at 1 ML Cu coverage by more than a factor of 30 followed by a sign reversal of $\chi_m^{(2)}$ at 2 to 3 ML. A similar, but less strong dependence on low

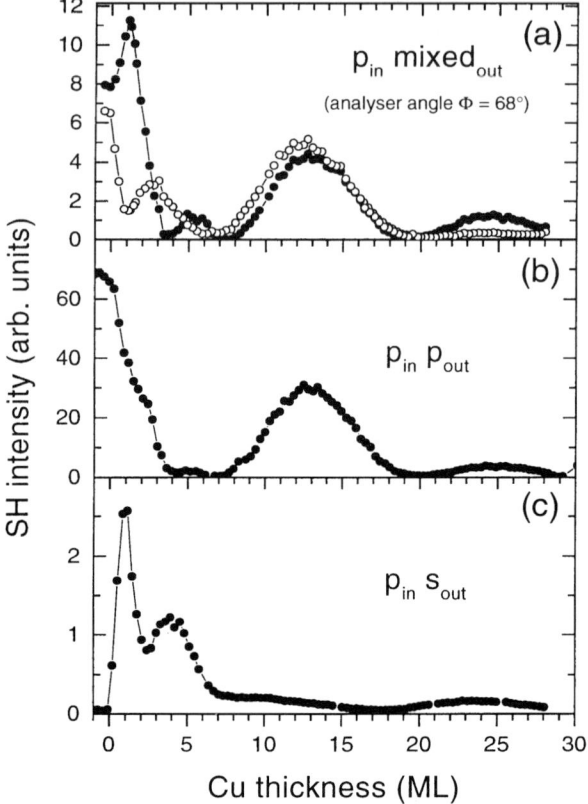

Fig. 2.49. Comparison of the p-polarized (b) and s-polarized (c) SH intensity from a Cu/10 ML Co/Cu(001) sandwich as a function of the Cu cover layer for p-polarized incident light at $\lambda = 790$ nm. For comparison the SH intensity at an analyser angle of $\Phi = 68°$ from Fig. 2.48(a) is shown in (a). The intensities shown in (b) and (c) are the average of the SH intensities for the magnetization in opposite directions.

Cu coverage has been observed in the linear Kerr signal and in the magnetic anisotropy[229,230] In ref. 229 an oscillating behaviour in the Kerr signal was found also for larger Cu thickness. A period of about 8 ML was observed in the linear Kerr signal at $\lambda = 633$ nm.

2.8 Summary and outlook

In this chapter the application of second harmonic generation from well characterized surfaces and interfaces and its dependence on the magnetization has been discussed. In general very large Kerr angles and Kerr asymmetries compared to the linear case have been observed. For MSHG from Fe(110) asymmetries up to $A = 97\%$ in the transverse Kerr geometry and Kerr angles up to

$\Phi_K^{(2,L)} = 86°$ in the longitudinal Kerr geometry have been observed, for example. The reason for the occurrence of such large magneto-optical effects lies partially in the fact that MSHG is described by a higher-rank tensor than the linear Kerr effect. In the case of MSHG, therefore, geometries can be found where a large 'magnetic' tensor element and only a small 'nonmagnetic tensor element contribute to the SH signal. This is not possible in the linear case because the amplitude of the reflected light is always much larger than the light component caused by the magnetization. Nevertheless, even in cases where quite high 'nonmagnetic' SH light amplitude contributes a very large magnetization-induced SH signal is observed. For p-polarized light incident at an angle of about 45° asymmetries larger the 60% were observed for clean Fe(110) in the transverse Kerr geometry.

The MSHG signal does not depend only on the ratio of 'magnetic' and 'nonmagnetic' tensor elements of $\chi^{(2)}$ but also quite strongly on the linear optical constants, i.e. the refractive index at the fundamental and SH wavelength, which in turn depend on the chosen geometry, leading to a reduction or enhancement of the nonlinear Kerr signal depending on the chosen geometry. This is analogous to the linear Kerr effect, where, for example, a large refractive index causes the Kerr effect to be much larger in the polar than in the longitudinal geometry, in general.

Extreme surface sensitivity has been proven by the thickness dependence of MSHG from thin Co and Fe films on Cu(001). The SHG signal and the nonlinear Kerr signal saturate nearly at that point where the surface is completely covered with the ferromagnetic material, implying an effective thickness in which the SH light is generated of the order of 1 to 2 ML only. The dependence of the MSH signal on adsorbates suggests that in many cases the contributions from the surface of the ultrathin film and the buried interfaces are of comparable size.

Establishing a quantitative relation between the MSH signal and the magnetization for a given geometry, wavelength, etc., is complicated by the fact that this signal depends strongly on electronic changes at the surface. For example, in the transverse geometry with p-polarized incident light a pronounced maximum around 4 ML in the Kerr asymmetry was found for Co films on Cu(001) as a function of the film thickness, while a first-principles calculation revealed only minor variations in the magnetic moments at that thickness range. These variations became even more dramatic in sandwich systems like Cu/Co/Cu(100) with strong variations of the electronic structure at the interfaces due to the appearance of quantum-well states. Oscillatory-like behaviour was found in the 'nonmagnetic' part of SHG and the magnetization-induced part was disturbed to such an extent that the variations in the measured asymmetry were mainly caused by the 'nonmagnetic' SHG.

However, these effects are not a drawback of MSHG alone. They appear in all optical methods as in the linear Kerr effect, although less pronounced. The advantages of MOKE and MSHG are fast and easy application of the method compared to real magnetometry in the case of ultrathin films. Estimates of magnetic moments from nonlinear as well as linear Kerr measurements can be made only on a qualitative basis without a detailed theoretical calculation. Nevertheless valuable information can be obtained from MSHG as was shown on the example of fcc Fe films on Cu(001). The constant asymmetry in the two structural phases I and II directly implies (nearly) constant magnetization at the surface/interface in the two phases.

MSHG is usually dominated by the dipolar term described by $\chi^{(2,D)}$, especially for ultrathin films. Nevertheless, significant contributions from the higher-order bulk allowed terms $\chi^{(2,Q)}$ may be present in the SHG signal as demonstrated by the anisotropic (M)SHG response from Co films on Cu(001). Especially in geometries probing small tensor elements $\chi_{ijk}^{(2,D)}$ they become more pronounced. In the case of ultrathin films on Cu(001) they resulted mainly from the substrate, so that the 'magnetic' SHG still results only from the surface or the interfaces. However, in cases of a ferromagnetic substrate or thicker ferromagnetic films these bulk contributions have to be considered.

A field less investigated up to now is to use MSHG as a very sensitive tool for determining magnetic anisotropies from the measurements of hysteresis loops. While for thicker films or systems with perpendicular anisotropy not much can be gained compared to linear MOKE, MSHG may become superior in the very low thickness range because the MSH signal is almost independent of thickness. Just as an illustration MSHG hysteresis loops as a function of the azimuthal angle are shown in Fig. 2.50. The asymmetry of the SH intensity from a 10 ML Co film on Cu(001) measured in the transverse Kerr geometry is shown as a function of the azimuth direction. The sample is rotated while the optical plane and the axis of the external magnetic field are kept fixed. Full hysteresis loops are shown for selected azimuthal angles. The uniaxial anisotropy causes the two-looped hysteresis curves in the intermediate axis (azimuth 45°) and zero remanence.[231] Along the easy-axis (azimuth 135°) a single square loop is observed. From the separation of the two loops in the intermediate axis the uniaxial anisotropy can be calculated.[232] This is not affected by the azimuthal anisotropy of SHG. On the other hand the variation of the asymmetry at saturation magnetization (open symbols in Fig. 2.50) does *not* directly exhibit the azimuthal dependence of the saturation magnetization because the 'nonmagnetic' SHG shows an azimuthal anisotropy as well. Therefore all the complications of the interpretations of the MSH signal discussed above are absent in those investigations restricted to the shape of the hysteresis curve which do not depend on a linear correlation of the magnitude of the MSH signal and the magnetization. On the other hand, if large electronic changes can be ruled out, like in temperature-dependent measurements, the nonlinear Kerr angle or the asymmetry is expected to be a very good relative measure of the magnetization, so that a quantitative analysis is possible directly without the aid of theoretical calculations.

Another application would be nonlinear Kerr microscopy as already demonstrated for bulk MSHG.[233] Even for surface MSHG sensitivity would be suffi-

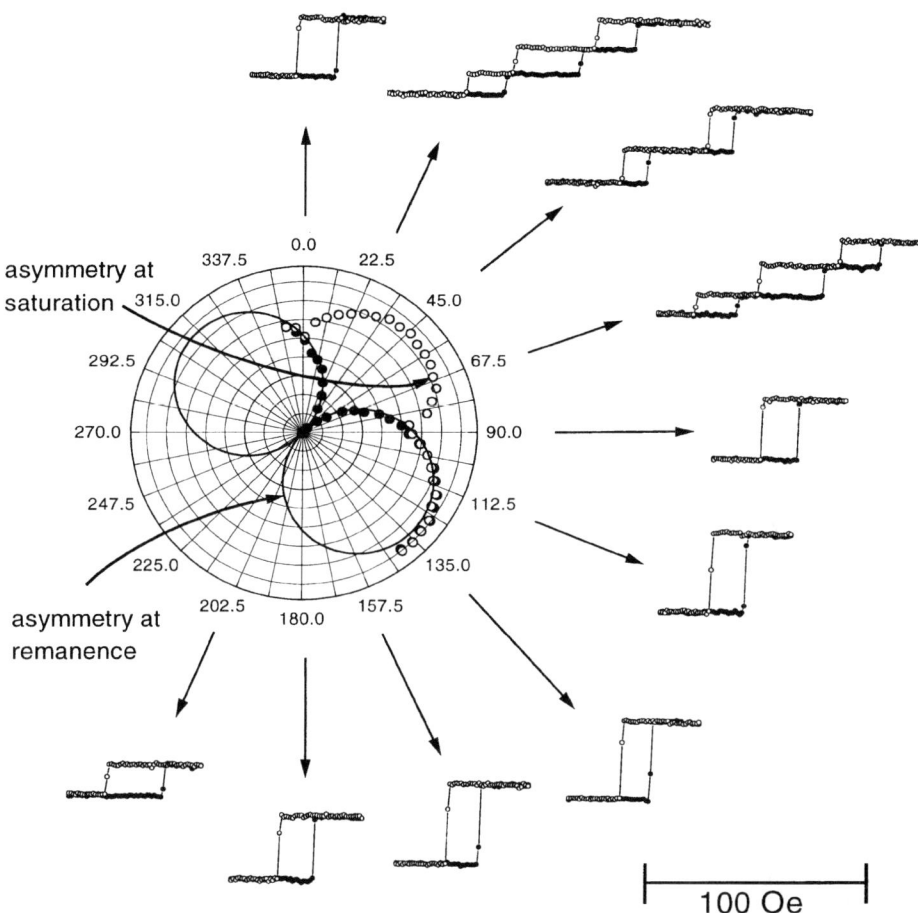

Fig. 2.50. Azimuthal anisotropy of MSHG from a 10 ML Co film on a slightly miscut Cu(001) sample. The crystal is rotated while the optical plane and the axis of the magnetic field remains fixed. Solid symbols, MSH asymmetry measured in remanence; open symbols, MSH asymmetry measured in saturation. The solid curve is a cosine function fitted to the solid symbols. Transverse Kerr geometry, incident angle $\theta_i = 38°$, $\lambda = 790$ nm. Hysteresis curves for selected azimuthal angles are shown at the circumference.

cient if one uses either an amplified Ti:sapphire system and a CCD camera or alternatively using a tightly focused spot for scanning. With the setup used in the experiments described in this chapter a minimum focus width of 25 µm of the fundamental beam could be achieved, limiting the possible resolution to about 12 µm. For Co films on Cu(100) the acquisition time for a single image point would be of the order of 10 ms to discriminate between two magnetization directions. An area of 1 mm² could be scanned within a few minutes. This is still much slower than what can be achieved with linear Kerr microscopy of thicker films. However, MSHG microscopy might become competitive in the monolayer or even submonolayer coverage regime.

Acknowledgements

The author would like to thank Q. Jin, H. Regensburger, and M. Straub for making available experimental results prior to publication. The authors wishes to express his gratitude to W. Hübner and Th. Rasing for stimulating discussions and especially J. Kirschner for his continuous support. This work was financially supported in part by the EC through grant ERB-EMRX-CT96-0015.

References

1. R.-P. Pan, H. D. Wei, and Y. R. Shen, *Phys. Rev.* B **39**, 1229 (1989).
2. W. Hübner and K.-H. Bennemann, *Phys. Rev.* B **40**, 5973 (1989).
3. E. R. Moog and S. D. Bader, *Superlattices Microstructures* **1**, 543 (1985).
4. S. D. Bader, E. R. Moog, and P. Grünberg, *J. Magn. Magnet. Mat.* **53**, L295 (1986).
5. J. Shen, R. Skomski, M. Klaua, H. Jenniches, S. Manoharan, and J. Kirschner, *Phys. Rev.* B **56**, 2340 (1997).
6. J. Zak, E. R. Moog, C. Liu, and S. D. Bader, *Phys. Rev.* B **43**, 6423 (1991).
7. G. Pénissard, P. Meyer, J. Ferré, and D. Renard, *J. Magn. Magnet. Mat.* **146**, 55 (1995).
8. L. M. Falicov, D. T. Pierce, S. D. Bader, R. Gronsky, H. J. Hopster, D. N. Lambeth, S. S. P. Parkin, G. Prinz, M. Salamon, I. K. Schuller, and R. H. Victoria, *J. Mater. Res.* **5**, 190 (1990).
9. J. Reif, J. C. Zink, C.-M. Schneider, and J. Kirschner, *Phys. Rev. Lett.* **67**, 2878 (1991).
10. J. Reif, C. Rau, and E. Matthias, *Phys. Rev. Lett.* **71**, 1931 (1993).
11. O. A. Aktsipetrov, O. V. Braginski, and D. A. Esikovi, *Sov. J. Quantum Electron.* **20**, 259 (1990).
12. S. B. Borisov and I. L. Lybchanksy, *Opt. Spectrosc. (USSR)* **61**, 801 (1986).
13. M. Fiebig, D. Frölich, B. B. Krichevtsov, and R. V. Pisarev, *Phys. Rev. Lett.* **73**, 2127 (1994).
14. M. Fiebig, D. Fröhlich, and H.-J. Thiele, *Phys. Rev.* B **54**, R12681 (1996).
15. V. N. Muthukumar, R. Valenti, and C. Gros, *Phys. Rev. Lett.* **75**, 2766 (1995).
16. V. N. Muthukumar, R. Valenti, and C. Gros, *Phys. Rev.* B **54**, 433 (1996).
17. R. V. Pisarev, V. V. Pavlov, A. Kirilyuk, and Th. Rasing, *J. Magn. Soc. Japan* **20**, 23 (1996).
18. V. V. Pavlov, R. V. Pisarev, A. Kirilyuk, and Th. Rasing, *Phys. Rev. Lett.* **78**, 2004 (1997).
19. U. Pustogowa, W. Hübner, and K. H. Bennemann, *Phys. Rev.* B **48**, 8607 (1993).
20. U. Pustogowa, W. Hübner, and K. H. Bennemann, *Appl. Phys.* A **59**, 611 (1994).
21. W. Hübner and K. H. Bennemann, *Phys. Rev.* B **52**, 13411 (1995).
22. M. G. Koerkamp and Th. Rasing, *J. Magn. Magnet. Mat.* **156**, 213 (1995).

23. T. M. Crawford, C. T. Rogers, T. K. Silva, and Y. K. Kim, *Appl. Phys. Lett.* **68**, 1573 (1996).
24. U. Pustogowa, W. Hübner, and K. H. Bennemann, *Surf. Sci.* **307–309**, 1129 (1994).
25. P. A. Franken, R. Hill, C. W. Peters, and G. Weinreich, *Phys. Rev. Lett.* **7**, 118 (1961).
26. F. Brown, R. E. Parks, and A. M. Sleeper, *Phys. Rev. Lett.* **14**, 1029 (1965).
27. S. Janz and H. M. van Driel, *Int. J. Nonlinear Opt. Phys.* **2**, 1–42 (1993).
28. J. W. Sipe and G. I. Stegmann, "Nonlinear optical response of metal surfaces," in *Surface Polaritons*, ed. by V. M. Agranovich and D. L. Mills (North-Holland, Amsterdam, New York, Oxford, 1982), Chap. 15, p. 661–702.
29. Y. R. Shen, "Application of optical second-harmonic generation to surface science" in *Chemistry and Structure at Interfaces: New Laser and Optical Techniques*, ed. by R. B. Hall and A. B. Eillis (VCH Publishers, Deerfield Beach, FL, 1986), Chap. 4, pp. 151–196.
30. G. L. Richmond, J. M. Robinson, and V. L. Shannon, *Prog. Surface Sci.* **28**, 1–70 (1988).
31. S. S. Jha, *Phys. Rev.* **140**, A2020 (1965).
32. J. Rudnick and E. A. Stern, *Phys. Rev.* B **4**, 4274 (1971).
33. M. Corvi and W. L. Schaich, *Phys. Rev.* B **33**, 3688 (1986).
34. M. G. Weber and A. Liebsch, *Phys. Rev.* B **36**, 6411 (1987).
35. A. Chizmeshya and E. Zaremba, *Phys. Rev.* B **37**, 2805 (1988).
36. A. Liebsch, *Phys. Rev.* B **40**, 3421 (1989).
37. A. Liebsch and W. I. Schaich, *Phys. Rev.* B **40**, 5401 (1989).
38. J. C. Quail and H. J. Simon, *Phys. Rev.* B **31**, 4900 (1985).
39. R. Murphy, M. Yeganeh, K. J. Song, and E. W. Plummer, *Phys. Rev. Lett.* **63**, 318 (1989).
40. D. Guidotti, T. A. Driscoll, and H. J. Gerritsen, *Solid State Commun.* **46**, 337 (1983).
41. T. A. Driscoll and D. Guidotti, *Phys. Rev.* B **28**, 1171 (1983).
42. J. A. Litwin, J. E. Sipe, and H. M. van Driel, *Phys. Rev.* B **31**, 5543 (1985).
43. H. W. K. Tom and G. D. Aumiller, *Phys. Rev.* B **33**, 8818 (1986).
44. K. Pederson and O. Keller, *J. Opt. Spectr. Am.* B **6**, 2412 (1989).
45. S. Janz, K. Pedersen, and H. M. van Driel, *Phys. Rev.* B **44**, 3943 (1991).
46. S. Janz, D. J. Bottomley, H. M. van Driel, and R. S. Timsit, *Phys. Rev. Lett.* **66**, 1201 (1991).
47. H. Ishida and A. Liebsch, *Phys. Rev.* B **50**, 4834 (1994).
48. Z. C. Ying, J. Wang, G. Andronica, J.-Q. Yao, and E. W. Plummer, *J. Vac. Sci. Technol.* A **11**, 2255 (1993).
49. The results of Janz *et al.* can probably be explained by a resonance effect (Th. Rasing, private communication).
50. J. Bloch, G. Lübke, S. Janz, and H. M. van Driel, *Phys. Rev.* B **45**, 12011 (1992).
51. R. J. M. Anderson and J. C. Hamilton, *Phys. Rev.* B **38**, 8451 (1988).
52. R.A. Bradley, R. Georgiadis, S. D. Kevan, and G. L. Richmond, *J. Vac Sic. Technol.* A **10**, 2996 (1992).
53. C. M. Li, L. E. Urbach, and H. L. Dai, *Phys. Rev.* B **49**, 2104 (1994).
54. E. K. L. Wong and G. L. Richmond, *J. Chem. Phys.* **99**, 5500 (1993).
55. E. K. L. Wong, K. A. Friedrich, and G. L. Richmond, *Chem. Phys. Lett.* **195**, 628 (1992).
56. R. Georgiadis and G. L. Richmond, *J. Chem. Phys.* **95**, 2895 (1991).
57. A. Friedrich, C. Shannon, and B. Pettinger, *Surf. Sci.* **251/252**, 587 (1991).

58. V. L. S. Daniel, A. Koos, and G. L. Richmond, *J. Phys. Chem.* **94**, 2091 (1990).
59. D. A. Koos and G. L. Richmond, *J. Phys. Chem.* **96**, 3770 (1992).
60. M. A. Hoffbauer, V. J. McVeigh, and M. J. Zuerlein, *J. Vac. Sci. Technol.* B **10**, 268 (1992).
61. J. Woll, G. Meister, U. Barjenbruch, and A. Goldmann, *Appl. Phys.* A **60**, 173 (95).
62. D. A. Koos, V. L. Shannon, and G. L. Richmond, *Phys. Rev.* B **47**, 4730 (1993).
63. L. E. Urbach, K. L. Percival, J. M. Hicks, E. W. Plummer, and H.-L. Dai, *Phys. Rev.* B **45** 3769 (1992).
64. S. Reiff and J. H. Block, *Surf. Sci.* **345**, 281 (1996).
65. G. Lüpke, G. Marowsky, R. Steinhoff, A. Friedrich, B. Pettinger, and D. M. Kolb, *Phys. Rev.* B **41**, 6913 (1990).
66. R. Vollmer, M. Straub, and J. Kirschner, *Surf. Sci.* **352–354**, 684 (1996).
67. A. V. Petukhov and A. Liebsch, *Surf. Sci.* **320**, L51 (1994).
68. A. V. Petukhov and A. Liebsch, *Surf. Sci.* **331–333**, 1335 (1995).
69. A. V. Petukhov and A. Liebsch, *Surf. Sci.* **334**, 195 (1995).
70. C. Jakobsen, D. Podenas, and K. Pedersen, *Surf. Sci.* **321**, 1 (1994).
71. A. V. Petukhov and A. Liebsch, *Surf. Sci.* **294**, 381 (1993).
72. H. W. K. Tom, C. M. Mate, X. D. Zhu, J. E. Crowell, T. F. Heinz, G. A. Somorjai, and Y. R. Shen, *Phys. Rev. Lett.* **52**, 348 (1984).
73. X. D. Zhu, Y. R. Shen, and R. Carr, *Surf. Sci.* **163**, 114 (1985).
74. S. G. Grubb, A. M. DeSantolo, and R. B. Hall, *J. Phys. Chem.* **92**, 1419 (1988).
75. D. Heskett, K. J. Song, A. Burns, E. W. Plummer, and H. L. Dai, *J. Chem. Phys.* **85**, 7490 (1986).
76. X. D. Zhu, Th. Rasing, and Y. R. Shen, *Chem. Phys. Lett.* **155**, 459 (1989).
77. S. Janz, K. Pedersen, and H. M. van Driel, *J. Vac. Sci. Technol.* A **9**, 1509 (1991).
78. J. Bloch, D. J. Bottomley, S. Janz, and H. M. van Driel, *Surf. Sci.* **257**, 328 (1991).
79. J. T. Yates, P. A. Thiel, and H. Weinberg, *Surf. Sci.* **82**, 45 (1979).
80. R. Vollmer, M. Straub, and J. Kirschner, *Surf. Sci.* **352**, 937 (1996).
81. X. D. Zhu, Th. Rasing, and Y. R. Shen, *Phys. Rev. Lett.* **61**, 2883 (1988).
82. D. Heskett, L. E. Urbach, K. J. Song, E. W. Plummer, and H. L. Dai, *Surf. Sci.* **197**, 225 (1988).
83. J. M. Chen, J. R. Bower, C. S. Wang, and C. H. Lee, *Opt. Commun.* **9**, 132 (1973).
84. H. W. K. Tom, C. M. Mate, X. D. Zhu, J. E. Crowell, Y. R. Shen, and G. A. Somorjai, *Surf. Sci.* **172**, 466 (1986).
85. K. J. Song, D. Heskett, H. L. Dai, A. Liebsch, and E. W. Plummer, *Phys. Rev. Lett.* **61**, 1380 (1988).
86. J. M. Hicks, L.E. Urbach, E. W. Plummer, and H.-L. Dai, *Phys. Rev. Lett.* **61**, 2588 (1988).
87. K. Giesen, F. Hage, H. J. Riess, W. Steinmann, R. Haight, R. Beigang, R. Dreyfus, P. Avouris, and F. J. Himpsel, *Phys. Scr.* **35**, 578 (1987).
88. G. Petrocelli, S. Martellucci, and R. Francini, *Appl. Phys.* A **56**, 263 (1993).
89. P. N. Argyres, *Phys. Rev.* **97**, 334 (1955).
90. K. H. J. Buschow, "Magneto-optical properties of alloys and intermetallic compounds," in *Ferromagnetic Materials*, ed. by E. P. Wohfarth and K. H. J. Buschow (North-Holland, Amsterdam, 1988), Vol. 4, p. 139.
91. J. Schoenes, "Magneto-optical properties of metals, alloys and compounds," in *Electronic and Magnetic Properties of Metals and Ceramics*, part I, ed. by R. W. Cahn, P. Haasen, and E. J. Kramer (VCH, Weinheim, 1992), Vol. 3A of *Material Science and Technology*, Chap. 2, pp. 147–255.

REFERENCES

92. S. D. Bader, *J. Magn. Magnet. Mat.* **100**, 440 (1991).
93. P. M. Oppeneer, J. Sticht, and F. Herman, *J. Magn. Soc. Japan* **15**, 73 (1991).
94. S. V. Halilov and R. Feder, *Solid State Commun.* **88**, 749 (1993).
95. G. Y. Guo and H. Ebert, *Phys. Rev.* B **50**, 10377 (1994).
96. T. Gasche, M. S. S. Brooks, and B. Johansson, *Phys. Rev.* B **53**, 296 (1996).
97. J. Kerr, *Philos. Mag.* **3**, 339 (1877).
98. G. Metzger, A. Pluvinage, and R. Tourguet, *Ann. Phys. (Paris)* **10**, 5 (1965).
99. M. J. Freiser, *IEEE Trans. Mag.* **MAG-4**, 152 (1968).
100. J. F. Dillon, "Magneto-optical properties of magnetic crystals," in *Magnetic Properties of Materials*, ed. by J. Smith (McGraw-Hill, New York, 1971), p. 149.
101. L. D. Landau and E. M. Lifschitz, *Elektrodynamik der Kontinua*, Vol. 8 of *Lehrbuch der Theoretischen Physik* (Akademie-Verlag, Berlin, 5th edition, 1985).
102. J. Zak, E. R. Moog, C. Liu, and S. D. Bader, *J. Appl. Phys.* **68**, 4203 (1990).
103. U. Tiwari, R. Ghosh, and P. Sen, *Phys. Rev.* B **49**, 2159 (1994).
104. C.-Y. You and S.-C. Shin, *Appl. Phys. Lett.* **69**, 1315 (1996).
105. R. Gamble and P. Lissberger, *J. Opt. Spectr. Am.* **5**, 1522 (1988).
106. M. Mansuripur, *J. Appl. Phys.* **67**, 6466 (90).
107. J. Zak, E. R. Moog, C. Liu, and S. D. Bader, *J. Magn. Magnet. Mat.* **89**, 107 (1990).
108. P. Yeh, *J. Opt. Spectr. Am.* **69**, 742 (1979).
109. D. R. Lide and H. P. R. Frederikse (eds), *CRC Handbook of Chemistry and Physics*, Vol. 8 (CRC Press, 74th edition, 1993).
110. K. Hellwege and O. Madelung (eds), *Group III: Crystal and Solid State Physics*, Vol. 19a of *Landolt-Börnstein* (Springer-Verlag, Berlin).
111. Y. R. Shen, "Surface probed by nonlinear optics," in *Surface Science: The First Thirty Years*, ed. by C. B. Duke (North-Holland, Amsterdam, 1994).
112. E. Ghahramani, D. J. Moss, and J. E. Sipe, *Phys. Rev. Lett.* **64**, 2815 (1990).
113. A. S. Arrott, "Epitaxial growth of metallic structures," in *Ultrathin Magnetic Structures*, ed. by J. A. C. Bland and B. Heinrich (Springer-Verlag, Berlin, 1994), Vol. 1, Chap. 5.1, pp. 177–215.
114. A. K. Schmid and J. Kirschner, *Ultramicroscopy* **42–44**, 483 (1992).
115. H.A. Wierenga, M. W. J. Prins, D. L. Abraham, and Th. Rasing, *Phys. Rev.* B **50**, 1282 (1994).
116. G. L. Eesley, *Phys. Rev. Lett.* **51**, 2140 (1983).
117. G. L. Eesley, *Phys. Rev.* B **33**, 2144 (1986).
118. P. B. Corkum, F. Brunel, N. K. Sherman, and T. Srinivasan-Rao, *Phys. Rev. Lett.* **61**, 2886 (1988).
119. J. F. Ready, *Effects of High-Power Laser Radiation* (Academic Press, New York, 1971).
120. H. A. Wierenga, W. de Jong, M. W. Prins, Th. Rasing, R. Vollmer, A. Kirilyuk, H. Schwabe, and J. Kirschner, *Phys. Rev. Lett.* **74**, 1462 (1995).
121. M. Straub, R. Vollmer, and J. Kirschner, *Phys. Rev. Lett.* **77**, 743 (1996).
122. Y. R. Shen, *The Principles of Nonlinear Optics* (Wiley, New York, 1984).
123. P. Guyot-Sionnest, W. Chen, and Y. R. Shen, *Phys. Rev.* B **33**, 8254 (1986).
124. T. F. Heinz, "Second-order nonlinear optical effects at surfaces and interfaces," in *Nonlinear Surface Electromagnetic Phenomena*, ed. by H.-E. Ponath and G. I. Stegman, Elsevier Science, B.V., 1991), Chap. 5, pp. 353–416.
125. J. E. Sipe, D. J. Moss, and H. M. van Driel, *Phys. Rev.* B **35**, 1129 (1987).
126. P. Guyot-Sionnest and Y. R. Shen, *Phys. Rev.* B **38**, 7985 (1988).
127. V. Mizrahi and J. Sipe, *J. Opt. Spectr. Am.* **5**, 660 (1988).

128. B. Koopmans, A. M. Janner, H. A. Wierenga, Th. Rasing, G. A. Sawatzky, and F. van der Woude, *Appl. Phys.* A **60**, 103 (1995).
129. U. Pustogowa, W. Hübner, and K. H. Bennemann, *Phys. Rev.* B **49**, 10031 (1994).
130. K. Böhmer, J. Hohlfeld, and E. Matthias, *Appl. Phys.* A **60**, 203 (1995).
131. W. Hübner, K. H. Bennemann, and K. Böhmer, *Phys. Rev.* B **50**, 17597 (1994).
132. J. C. Hamilton, R. J. M. Anderson, and R. L. Williams, *J. Vac. Sci. Technol.* B **7**, 1208 (1989).
133. W. Hübner, *Phys. Rev.* B **42**, 11553 (1990).
134. J. Kirschner, *Surf. Sci.* **138**, 191 (1984).
135. B. Koopmans, M. G. Koerkamp, Th. Rasing, and H. van den Berg, *Phys Rev. Lett.* **74**, 3692 (1995).
136. M. G. Koerkamp and Th. Rasing, *Surf. Sci.* **352–354**, 933 (1996).
137. R. Vollmer, M. Straub, and J. Kirschner, *J. Magn. Soc. Japan* **20**, 29 (1996).
138. R. Vollmer, unpublished.
139. C. M. Schneider, A. K. Schmidt, P. Schuster, H. P. Oepen, and J. Kirschner, "Influence of growth and structure on the magnetism of epitaxial cobalt films on Cu(001)," in *Magnetism and Structure in Systems of Reduced Dimensions* (Plenum, New York, 1993), p. 453.
140. J. R. Cerdá, P. L. de Andres, A. Cebollada, R. Miranda, E. Navas, P. Schuster, C. M. Schneider, and J. Kirschner, *J. Condens. Matter* **5**, 2055 (1993).
141. M. T. Kief and J. W. F. Egelhoff, *Phys. Rev.* B **47**, 10785 (1993).
142. G. L. Nyberg, M. T. Kief, and J. W. F. Egelhoff, *Phys. Rev.* B **48**, 14509.
143. J. J. de Miguel, A. Cebollada, J. M. Gallego, S. Ferrer, R. Miranda, C. M. Schneider, P. Bressler, J. Garbe, K. Bethke, and J. Kirschner, *Surf. Sci.* **211/212**, 732 (1989).
144. C. M. Schneider, P. Bressler, P. Schuster, J. Kirschner, J. J. de Miguel, and R. Miranda, *Phys. Rev. Lett.* **64**, 1059 (1990).
145. B. Heinrich, J. F. Cochran, M. Kowalewski, J. Kirschner, Z. Celinski, A. S. Arott, and K. Myrtle, *Phys. Rev.* B **44**, 9348 (1991).
146. P. Krams, F. Lauks, R. L. Stamps, B. Hillebrands, and G. Günterrodt, *Phys. Rev. Lett.* **69**, 3674 (1992).
147. H. P. Oepen, C. M. Schneider, D. S. Chuang, A. A. Ballentine, and R. C. O'Handley, *J. Appl. Phys.* **73**, 6186 (1993).
148. H. P. Oepen, A. Berger, C. M. Schneider, Th. Reul, and J. Kirschner, *J. Magn. Magnet. Mat.* **121**, 490 (1993).
149. H. A. Wierenga, W. de Jong, M. W. J. Prins, Th. Rasing, R. Vollmer, A. Kirilyuk, H. Schwabe, and H. Kirschner, *Surf. Sci.* **331–333**, 1294 (1995).
150. C. S. Wang, B. M. Klein, and H. Krakauer, *Phys. Rev. Lett.* **54**, 1852 (1985).
151. V. L. Moruzzi, P. M. Marcus, K. Schwarz, and P. Mohn, *Phys. Rev.* B **34**, 1784 (1986).
152. F. J. Pinski, J. Staunton, B. L. Gyorffy, D. D. Johnson, and G. M. Stocks, *Phys. Rev.Lett.* **56**, 2096 (1986).
153. V. L. Moruzzi, P. M. Marcus, and J. Kübler, *Phys. Rev.* B **39**, 6957 (1989).
154. J. Thomassen, F. May, M. Wuttig, and H. Ibach, *Phys. Rev. Lett.* **69**, 3831 (1992).
155. J. Giergiel, J. Kirschner, J. Landgraf, J. Shen, and J. Woltersdorf, *Surf. Sci.* **310**, 1 (1994).
156. T. Kraft, P. M. Marcus, and M. Scheffler, *Phys. Rev.* B **49**, 11511 (1994).
157. D. Li, M. Freitag, J. Person, Z. Q. Qiu, and S. D. Bader, *Phys. Rev. Lett.* **72**, 3112 (1994).

158. S. Müller, P. Bayer, C. Reischl, K. Heinz, B. Feldmann, H. Zillgen, and M. Wuttig, *Phys. Rev. Lett.* **74**, 765 (1995).
159. S. Müller, P. Bayer, A. Kinne, P. Schmailzl, and K. Heinz, *Surf. Sci.* **322**, 21 (1995).
160. D. J. Keavney, D. F. Storm, J. W. Freeland, I. L. Grigorov, and J. C. Walker, *Phys. Rev. Lett.* **74**, 4531 (1995).
161. T. Detzel, M. Donath, N. Memmel, and V. Dose, *J. Magn. Magnet. Mat.* **152**, 287 (1996).
162. U. Gradmann and H. O. Isbert, *J. Magn. Magnet. Mat.* **15–18**, 1109 (1980).
163. F. Baudelet, M.-T. Lin, W. Kuch, K. Meinel, C. M. Schneider, and J. Kirschner, *Phys. Rev. B* **51**, 12563 (1995).
164. M.-T. Lin, J. Shen, W. Kuch, H. Jenniches, M. Klaua, C. M. Schneider, and J. Kirschner, *Phys. Rev. B* **55**, 5886 (1997).
165. A. Kirilyuk, J. Giergiel, M. Straub, and J. Kirschner, *Phys. Rev. B* **54**, 1050 (1996).
166. Y.-M. Zhou, W.-Q. Zhang, L.-Q. Zhong, and D.-S. Wang, *J. Magn. Magnet. Mat.* **145**, L273 (1995).
167. R. D. Ellerbrock, A. Fuest, A. Schatz, W. Keunne, and R. A. Brand, *Phys. Rev. Lett.* **61**, 475 (1995).
168. W. A. A. Macedo and W. Keunne, *Phys. Rev. Lett.* **61**, 475 (1988).
169. Y. Tsunoda, N. Kinitomi, and R. W. Nicklow, *J. Phys. F* **17**, 2247 (1987).
170. W. Keunne, A. Schatz, R. D. Ellerbrock, A. Fuest, K. Wilmers, and R. A. Brand, *J. Appl. Phys.* **79**, 4265 (1996).
171. T. Detzel, M. Vonbank, M. Donath, and V. Dose, *J. Magn. Magnet. Mat.* **147**, L1 (1995).
172. C. Egawa, E. M. McCash, and R. F. Willis, *Surf. Sci.* **215**, L271 (1989).
173. J. Tomassen, B. Feldmann, and M. Wuttig, *Surf. Sci.* **264**, 406 (1992).
174. P. Xhonneux and E. Courtens, *Phys. Rev. B* **46**, 556 (1992).
175. H. Landskron, G. Schmidt, K. Heinz, K. Müller, C. Stuhlmann, U. Beckers, M. Wuttig, and H. Ibach, *Surf. Sci.* **256**, 115 (1991).
176. P. Bayer, S. Müller, P. Schmailzl, and K. Heinz, *Phys. Rev. B* **48**, 17611 (1993).
177. J. V. Barth and D. E. Fowler, *Phys. Rev. B* **52**, 1528 (1995).
178. T. F. Heinz, M. M. T. Loy, and W. A. Thompson, *Phys. Rev. Lett.* **54**, 63 (1985).
179. J. E. Sipe, V. Mizrahi, and G. I. Stegeman, *Phys. Rev. B* **35**, 9091 (1987).
180. P. Guyot-Sionnest and Y. R. Shen, *Phys. Rev. B* **35**, 4420 (1987).
181. T. Miyano, Y. Sakisaka, T. Komeda, and M. Onichi, *Surf. Sci.* **169**, 197 (1986).
182. G. Pirug, G. Brodén, and H. P. Bonzel, *Surf. Sci.* **94**, 323 (1980).
183. A. Wight, N. G. Condon, F. M. Leibsle, and G. Worthy, *Surf. Sci.* **331–333**, 133 (1995).
184. F. K. Schulte, *Surf. Sci.* **55**, 427 (1976).
185. P. J. Feibelmann, *Phys. Rev. B* **27**, 1991 (1983).
186. R. C. Jaklevic, J. Lambe, M. Mikker, and W. C. Vassel, *Phys. Rev. Lett.* **26**, 88 (1971).
187. R. C. Jaklevic and J. Lambe, *Phys. Rev. B* **12**, 4146 (1975).
188. M. Jalochowski and E. Bauer, *Phys. Rev. B* **38**, 5272 (1988).
189. P. Grünberg, R. Schreiber, Y. Pang, M. B. Brodsky, and H. Sowers, *Phys. Rev. Lett.* **57**, 2442 (1986).
190. M. N. Baibich, J. M. Broto, A. Fert, F. N. V. Dau, F. Petroff, P. Etienne, G. Creuzet, A. Friedrich, and J. Chazelas, *Phys. Rev. Lett.* **61**, 2472 (1988).
191. G. Binash, P. Grünberg, F. Saurenbach, and W. Zinn, *Phys. Rev. B* **39**, 4828 (1989).
192. S. S. P. Parkin, N. More, and K. P. Roche, *Phys. Rev. Lett.* **64**, 2304 (1990).

193. S. S. P. Parkin, *Phys. Rev. Lett.* **67**, 3598 (1991).
194. M. A. Ruderman and C. Kittel, *Phys. Rev.* **96**, 99 (1954).
195. Y. Yafet, *Phys. Rev.* B **36**, 3948 (1987).
196. P. Bruno and C. Chappert, *Phys. Rev. Lett.* **67**, 1602 (1991).
197. P. Bruno and C. Chappert, *Phys. Rev.* B **46**, 261 (1992).
198. D. M. Edwards, J. Mathon, R. B. Muniz, and M. S. Phan, *Phys. Rev. Lett*.**67**, 493 (1991).
199. J. E. Ortega, F. J. Himpsel, G. J. Mankey, and R.F. Willis, *Phys. Rev.* B **47**, 1540 (1993).
200. M.D. Stiles, *Phys. Rev.* B **48**, 7238 (1993).
201. M. van Schilfgaarde and W. A. Harrison, *Phys. Rev. Lett.* **71**, 3870 (1993).
202. P. Bruno, *J. Magn. Magnet. Mat.* **121**, 248 (1993).
203. W. R. Bennett, W. Schwarzacher, and J. W. F. Egelhoff, *Phy. Rev. Lett.* **65**, 3169 (1990).
204. A. Cebollada, R. Miranda, C. M. Schneider, P. Schuster, and J. Kirschner, *J. Magn. Magnet. Mat.* **102**, 25 (1991).
205. M. T. Johnson, S. T. Purcell, N. W. E. McGee, R. Coehoorn, J. aan de Stegge, and W. Hoving, *Phys. Rev. Lett.* **68**, 2688 (1992).
206. Z. Q. Qui, J. Pearson, and S. D. Bader, *Phys. Rev.* B **46**, 8659 (1992).
207. J. E. Ortega and F. J. Himpsel, *Phys. Rev. Lett.* **69**, 844 (1992).
208. P. Segovia, E. G. Michel, and J. E. Ortega, *Phys. Rev. Lett.* **77**, 3455 (1996).
209. C. Carbone, E. Vescuvo, O. Rader, W. Gudat, and W. Eberhardt, *Phys. Rev. Lett.* **71**, 2805 (1993).
210. G. K. Y. Chang and P. Johnson, *Phys. Rev. Lett.* **71**, 2801 (1993).
211. C. Carbone, E. Vescovo, R. Kläsges, W. Eberhardt, O. Rader, and W. Gudat, *J. Appl. Phys.* **76**, 6966 (1994).
212. C. Carbone, E. Vescovo, R. Kläsges, D. D. Sarma, and W. Eberhardt, *J. Magn. Magnet. Mat.* **156**, 259 (1996).
213. L. Nordström, P. Lang, R. Zeller, and P. H. Dederichs, *Europhys. Lett.* **29**, 395 (1995).
214. P. Lang, L. Nordström, K. Wildberger, R. Zeller, P. H. Dederichs, and T. Hoshino, *Phys. Rev.* B **53**, 9092 (1996).
215. P. van Geldern, S. Crampin, and J. E. Inglesfield, *Phys. Rev.* B **53**, 9115 (1996).
216. Y. Suzuki, T. Katayama, S. Yoshida, K. Tanaka, and K. Sato, *Phys. Rev. Lett.* **68**, 3355 (1992).
217. Y. Suzuki, T. Katayama, A. Thiaville, K. Sato, and S. Yoshida, *J. Magn. Magnet. Mat.* **121**, 539 (1993).
218. W. Geerts, Y. Suzuki, T. Katayama, K. Tanaka, K. Ando, and S. Yoshida, *Phys. Rev.* B **50**, 12581 (1994).
219. R. Mégy, A. Bonouh, Y. Suzuki, P. Beauvillain, P. Bruno, B. Lecuyer, and P. Veillet, *Phys. Rev.* B **51**, 5586 (1995).
220. P. Bruno, *Phys. Rev.* B **52**, 411 (1995).
221. Y. Suzuki and P. Bruno, *J. Magn. Magnet. Mat.* **144**, 651 (1995).
222. P. Bruno, Y. Suzuki, and C. Chappert, *Phys. Rev.* B **53**, 9214 (1996).
223. T. Katayama, Y. Suzuki, M. Hayashi, and A. Thiaville, *J. Magn. Magnet. Mat.* **126**, 527 (1993).
224. H. A. Wierenga, M. W. J. Prins, and Th. Rasing, *Physica* B **204**, 281 (1995).
225. A. Kirilyuk, M. G. Koerkamp, Th. Rasing, R. Mégy, and P. Beauvillain, *J. Magn. Soc. Japan* **20**, 361 (1996).

226. T. Luce, W. Hübner, and K. H. Bennemann, *Phys. Rev. Lett.* **77**, 2810 (1996).
227. C. L. Fu and A. J. Freeman, *Phys. Rev.*B **38**, 3016 (1987).
228. A. Berger, U. Linke, and H. P. Oepen, *Phys. Rev. Lett.* **68**, 839 (1992).
229. W. Weber, A. Bischof, R. Allenspach, C. Würsch, C. H. Back, and D. Prescia, *Phys. Rev. Lett.* **76**, 3424 (1996).
230. W. Weber, C. H. Back, U. Ramsperger, A. Vaterlaus, and R. Allenspach, *Phys. Rev.* B **52**, R14400 (1995).
231. P. Krams, B. Hillebrands, G. Günterodt, and H. P. Oepen, *Phys. Rev.* B **49**, 3633 (1994).
232. W. Weber, C. H. Back, A. Bischof, C. Würch, and R. Allenspach, *Phys. Rev. Lett.* **76**, 1940 (1996).
233. A. Kirilyuk, V. Kirilyuk, Th. Rasing, V. V. Pavlov, and R. V. Pisarev, *J. Magn. Soc. Japan* **20**, 129 (1996).

3
NONLINEAR MAGNETO-OPTICAL STUDIES OF ULTRATHIN FILMS AND MULTILAYERS

Th. Rasing
Research Institute for Materials, Nijmegen, The Netherlands

3.1 Introduction

The strong activity and success in magnetic thin-film research is based on the current developments in surface and material science, enabling the manipulation and control of thin-film structures on the atomic level and the parallel development of many spin-appended derivations of surface science techniques.[1] Some of the most exciting recent discoveries, like the giant magnetoresistance (GMR) and the oscillatory exchange coupling, are related to the properties of multilayers of alternating ferromagnetic and paramagnetic layers. As the interfaces between these layers appear to play an essential role for these phenomena and consequently for the device properties based on them, a detailed study of the magnetic interface properties is needed.

There exist several techniques (e.g. spin-polarized photoemission spectroscopy,[2] spin-polarized electron energy loss spectroscopy,[3] and spin-polarized low-energy electron diffraction[4]) to study the magnetic properties of clean surfaces. Unfortunately, (polarized) electrons are difficult to use for studying buried interfaces owing to their short mean free path. Since interfaces between thin metallic films are accessible by light, an optical technique would have significant advantage. The magneto-optical Kerr effect (MOKE) is well known and frequently used, for both fundamental studies as well as applications. This linear optical technique is based on the changes in the linear susceptibility as a function of the applied magnetic field. As this results in a (small) rotation of the polarization of light travelling through a magnetic material, it yields a probe for the bulk magnetization.[5] Though very sensitive and even applicable to monolayers, MOKE is not interface-specific.

Second harmonic generation (SHG) is a nonlinear optical technique that derives its interface sensitivity from the breaking of symmetry at boundaries between centrosymmetric media.[6-9] Though SHG was one of the first nonlinear optical effects discovered after the invention of the laser,[10] it took another two decades before the development of SHG as a tool for surface and interface science started.

INTRODUCTION

Though possible surface effects were already discussed in one of the early papers of Bloembergen,[11] at that time they were considered as minor, small contributions. The development of surface science and the pioneering work of the Shen group in Berkeley has changed this picture drastically. Shen's book on nonlinear optics gives a good introduction to this field.[6] In the past decade, SHG has been proven to be an extremely versatile surface and interface probe. Applications range from the study of the structure and symmetry of semiconductor and metal surfaces and interfaces,[8,12,13] molecular adsorbates, *in situ* studies of molecular adsorption, desorption and diffusion,[14] electric field dynamics at metal/semiconductor interfaces[15] and the electrochemistry at electrolytic interfaces.[16] As an optical, interface-sensitive technique, SHG is particularly attractive for the study of buried interfaces. The question arises whether SHG would also be sensitive to the magnetic properties of surfaces and interfaces, i.e. whether there would be a nonlinear, surface- and interface-sensitive equivalent of the MOKE technique.

The origin of MOKE lies in the spin–orbit coupling that acts like a magnetic field on the current induced by the electromagnetic field of the incident light.[5] This should also hold for the nonlinear contributions of the induced current, which are the origin of SHG, leading to a magetization-induced second harmonic generation (MSHG), or indeed a nonlinear magneto-optical Kerr effect (NOMOKE or NOLIMOKE). Based on symmetry arguments, Ru-Pin Pan *et al.* indeed showed that the presence of a magnetization would lead to new, nonzero *surface* contributions to the nonlinear optical response.[17] At almost the same time, Hübner and Bennemann independently calculated the nonlinear magneto-optical spectrum of Ni, based on a spin-dependent band-structure calculation.[18] They showed that this should indeed lead to observable effects, with magnetic contributions to the nonlinear tensor coefficients of more than 10%. The first experimental evidence for an MSHG effect was given by Reif *et al.* for a clean Fe(110) surface,[19] whereas Spierings *et al.* showed the first MSHG results from buried Co/Au interfaces.[20]

A strong demonstration of the surface and interface sensitivity of the MSHG technique was given by Wierenga *et al.*[21] by an *in situ* MSHG and MOKE study of the Co/Cu(100) system. These studies also showed that the nonlinear magneto-optical effects are quite sizeable: magnetic contrasts between signals with opposite magnetization directions of over 50% were observed. This indicates that the magnetization-induced tensor elements are of the same order of magnitude as the nonmagnetic ones. This is in strong contrast to the linear optical response, where the magnetization induces very small off-diagonal tensor elements. So far those MSHG measurements were based on magnetization-induced intensity changes (or in other words the transverse NOMOKE, as will be shown later). An obvious extension would be to look at the nonlinear equivalent of the MOKE rotation, i.e. the NOMOKE rotation of the SHG polarization. Pustogowa indeed predicted the appearance of large enhancements of the nonlinear Kerr angle, and showed this to be directly connected to the differences in the wave equations for the nonlinear and linear case.[22] (Actually, as we will discuss below, they showed that these differences lead to a *suppression* of the linear Kerr rotation, due to bulk transitions.) Experimentally, the first observation of a NOMOKE rotation was made by Aktsipetrov *et al.*, who measured a nonlinear Kerr rotation between 1° and 4° for a $Y_{2.5}Bi_{0.5}Fe_5O_{12}$ garnet film,[23] and Reif *et al.*[24] who found a nonlinear rotation of 14° for the Heussler alloy. Even larger effects were subsequently observed in experiments on thin Fe/Cr films and on single-crystalline Fe whiskers, showing enhancements of the nonlinear versus the linear MOKE rotation of over three orders of magnitude.[25] It appeared that not only the differences in the wave equations played a role, as pointed out by Pustogowa *et al.*,[22] but also that the differences in the symmetry properties between the nonlinear and linear optical response tensors can lead to additional enhancements of several orders of magnitude.

A very unusual kind of space and time symmetry breaking may occur in magnetic materials due to spin ordering in centrosymmetric crystals. It may happen that the spins in the magnetic unit cell are ordered in such a way that they break the inversion symmetry, while the crystallographic structure remains centrosymmetric. As a consequence, a nonlinear optical susceptibility arises that is sensitive to the spin orientation. This new type of susceptibility was recently observed in antiferromagnetic chromium oxide Cr_2O_3, opening new possibilities to study the electronic structure of magnetic materials.[26] These different symmetry properties also lead to novel possibilities of nonlinear magneto-optical imaging, as was recently demonstrated by some interesting results on antiferromagnetic materials[27] and on magnetic garnet films.[28] Also, it was found that the appearance of confined electronic (quantum-well) states in ultrathin films, which are thought to be responsible for the antiferromagnetic coupling, can have an enormous effect on the nonlinear optical response of such systems.[29] MSHG therefore seems to be a particularly powerful tool to study these confined states, especially when it can be done spectroscopically. Finally, the use of femtosecond laser sources also opens the way to study the spin dynamics on a subpicosecond timescale, a virtually unexplored area so far.[30]

In this chapter, we want to discuss the various aspects of nonlinear magneto-optics and its potential for magnetic thin-film research and applications. After an introduction to surface second harmonic generation and a brief discussion of the linear MOKE technique, the effects of the presence of a magnetization on second harmonic generation will be discussed, from both a phenomenological and a microscopic point of view. We will treat the various configurations of the nonlinear Kerr effect (longitudinal, transverse and polar), and compare them with their linear equivalents. The combination of large probing depth and interface specificity of this nonlinear optical technique offers interesting possibilities for the study of multilayers. However, this requires a detailed understanding and analysis of the nonlinear optical response of such metallic multilayers. This will therefore be treated in some detail. After a discussion of the experimental aspects, we will discuss various experimental case studies, addressing both the surface and interface sensitivity as well as the effects of multiple interface contributions to MSHG. The various possibilities of the large nonlinear magneto-optical Kerr rotations and the apparent extreme sensitivity of

(M)SHG to confined quantum-well states in very thin metallic films will be discussed in subsequent chapters. Bulk MSHG effects in antiferromagnetic structures as well as in magnetic garnets will be treated after that. Finally, we will discuss the new possibilities that nonlinear magneto-optical imaging offers and illustrate this with various domain and domain-wall studies in antiferromagnetic and magnetic garnet structures. A discussion of the future prospects will end this chapter.

3.2 Surface second harmonic generation

SHG arises from the nonlinear polarization $P(2\omega)$ induced by an incident laser field $E(\omega)$. This polarization can be written as an expression in $E(\omega)$:

$$P(2\omega) = \chi^{(2)}E(\omega)E(\omega) + \chi^{(Q)}E(\omega)\nabla E(\omega) + \cdots . \qquad (3.2.1)$$

The lowest-order term in eqn (3.2.1) describes an electric dipole source. Symmetry considerations show that this contribution is zero in a centrosymmetric medium, thus limiting electric dipole radiation to the interfaces where inversion symmetry is broken (see Fig. 3.1). The bulk second harmonic can now be described in terms of the much smaller electric quadrupole-like contributions (second term in eqn (3.2.1)). However, because of the large volume difference between interface and bulk, this does not necessarily mean that the total bulk second harmonic signal is negligible. Interface sensitivity needs to be verified for any given system. In particular, for insulating materials, and for semiconductors using below-bandgap excitation, these bulk effects can be large and sometimes even dominating. The very effective screening in metals usually limits the symmetry breaking to the first atomic layers[31] though sizeable bulk contributions have been observed as well.[32] Experimentally, surface sensitivity can often be demonstrated by *in situ* surface modification, like oxidation or CO adsorption. A complete theoretical treatment of the various contributions to surface SHG is given by Guyot-Sionnest *et al.*[33] There are various reviews that discuss the application of SHG to surface science,[34] semiconductor surfaces[12] or metals.[13] A more recent review, including the latest developments, can be found in the introduction of Chapter 2 by R. Vollmer in this book.

As will be shown in sec. 3.5, in the case of magnetic surfaces or interfaces, the

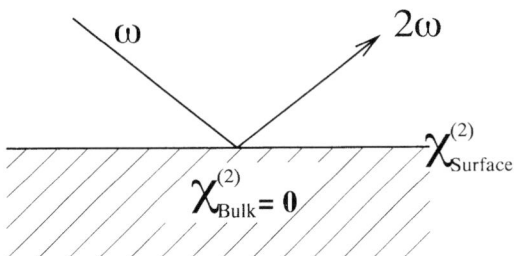

Fig. 3.1. A schematic view of second harmonic generation.

presence of the magnetization will not affect the inversion symmetry but will introduce extra nonzero surface tensor elements. Those will change sign when the direction of M is reversed, which leads to a nonlinear polarizability that will depend on M:

$$P(2\omega, \pm M) = (\chi^{(2),+} \pm \chi^{(2),-})E(\omega)E(\omega)$$
$$+ (\chi^{(Q),+} \pm \chi^{(Q),-})E(\omega)\nabla E(\omega) + \cdots . \quad (3.2.2)$$

The presence of these new tensor components that are odd in the magnetization will lead to the phenomenon of magnetization-induced second harmonic generation. Because M is an axial vector, its orientation is conserved under inversion, i.e. M does not break the bulk inversion symmetry. This means that the presence of a magnetization does not induce electric dipole contributions in the bulk and that the surface sensitivity is conserved. However, the time reversal symmetry breaking will induce new nonzero tensor elements for both $\chi^{(2)}$ and $\chi^{(Q)}$, which will be derived in sec. 3.5.

3.3 SHG from multilayers

To find the nonlinear optical response of a metallic multilayer we must know the nonlinear source polarization at all the interfaces, and calculate the generated SH fields. Nonlinear optical generation from simple multilayer systems has been treated by Dick et al. for layers of organic molecules,[35] by Yeganeh et al. for semiconductors[36] and by Koopmans et al. for thin C_{60} films.[37] Their results were analyzed either by simultaneously solving the boundary conditions at all interfaces, or by describing the multiple reflections of both the fundamental and the SH beams in terms of reflection and transmission coefficients. A well known concept in optics is to treat the boundary conditions at an interface, and the propagation of light through a homogeneous slab, in terms of matrices on the basis of forward- and backward-travelling waves. The matrices relate the field components on both sides of the interface and the layer respectively. Describing the full multilayer is thus reduced to a simple matrix multiplication (see for example[38]). Although this approach is completely analogous to solving simultaneously the boundary conditions at all interfaces, it has the strong advantage of being much more flexible when it comes to solving systems with different numbers of layers.

For our MSHG calculations we assume that the system contains only electric dipole sources of SH radiation at the interfaces. Because of the smallness of the surface nonlinear optical response, we can totally ignore the depletion of the incoming beam by the SHG. The model is in fact related to work by Sipe.[39] He developed a Green function formalism for calculating fields generated by sources in the presence of a multilayer geometry. We use the nonlinear boundary conditions as derived by Heinz,[40] giving the discontinuity of the second harmonic fields at the nonlinear interface (i.e. at the source). Our treatment of the emitted fields is similar to his approach. After introducing infinitesimal vacuum sheets between all the layers, we come to an equivalent description of the propagation of both the fundamental and the second harmonic, which is very convenient for converting the model into a computer program.

The influence of the total multilayer on the transmission of the fundamental wave can be described for p-polarized light as:

$$\begin{pmatrix} E^+_{\omega,p,0} \\ E^-_{\omega,p,0} \end{pmatrix} = \prod_{j=1}^{m} T_j^{\omega,p} \begin{pmatrix} E^+_{\omega,p,m+1} \\ 0 \end{pmatrix}, \quad (3.3.1)$$

where the system contains m layers and $E^+_{\omega,p,0}$ and $E^-_{\omega,p,0}$ represent respectively the forward and backward fields at the front side and $E^+_{\omega,p,m+1}$ the forward field at the back side of the sample (notice that there can be only a forward field at the back side of the sample). $T_j^{\omega,p}$ is obtained from the equation

$$T_j^{\omega,p} = M_{0,j}^{\omega,p} \otimes C_j^{\omega} \otimes M_{j,0}^{\omega,p}, \quad (3.3.2)$$

where $M_{0,j}^{\omega,p}$ describes the interface between media 0 and j (the index 0 refers to a vacuum sheet), and C_j^{ω} describes layer j:

$$M_{j-1,j}^{\omega,p} = \frac{1}{f_{j-1,j-1}^{\omega,+}} \begin{pmatrix} f_{j-1,j}^{\omega,+} & -f_{j-1,j}^{\omega,-} \\ -f_{j-1,j}^{\omega,-} & f_{j-1,j}^{\omega,+} \end{pmatrix},$$

$$f_{j',j}^{\omega,\pm} = \hat{n}_{j'}(\omega)\cos\hat{\theta}_{\omega,j} \pm \hat{n}_j(\omega)\cos\hat{\theta}_{\omega,j'},$$

$$C_j^{\omega} = \begin{pmatrix} e^{-i\hat{k}_{j,z}(\omega)t_j} & 0 \\ 0 & e^{i\hat{k}_{j,z}(\omega)t_j} \end{pmatrix}, \quad (3.3.3)$$

$$\hat{k}_{j,z}(\omega) = \frac{\hat{n}_j(\omega)\omega}{c}\cos\hat{\theta}_{\omega,j},$$

$$\hat{n}_j(\omega)\sin\hat{\theta}_{\omega,j} = \sin\theta_{\omega,0},$$

where $\theta_{\omega,0}$ is the (real) angle of incidence in the vacuum. With eqn (3.3.1) we relate the forward and backward fields at all interfaces in the multilayer to the incoming fundamental field. Using the ideas of Bloembergen and Pershan[11] we introduce an infinitesimal nonlinear sheet of thickness $2\delta_2$, inside an infinitesimal vacuum sheet of thickness $2\delta_1$ (see Fig. 3.2).

This sheet has a refractive index $\hat{n}_j^i$, the refractive index of the interface. If we take the x, z plane as the plane of incidence ($k_y = 0$), we find that the fundamental fields induce a 2ω polarization in the sheet:

$$P^{sh}_{2\omega,k}(\mathbf{r}_j, t) = P^{sh}_{2\omega,k,j} e^{i[2k_x(\omega)x - 2\omega t]}, \quad (3.3.4)$$

where $\mathbf{r}_j$ is lying in the plane of interface j, and

$$P^{sh}_{2\omega,k,j} = \chi^{(2)}_{klm,j}(2\omega)\mathscr{E}_{\omega,l,j}\mathscr{E}_{\omega,m,j}, \quad (3.3.5)$$

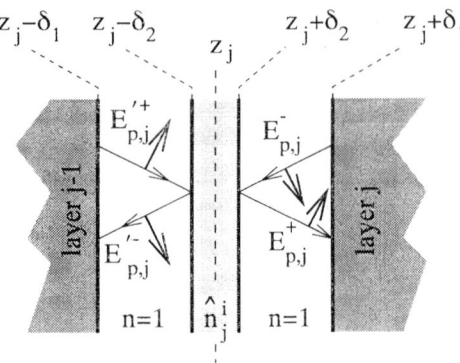

Fig. 3.2. Infinitesimal nonlinear sheet of thickness $2\delta_2$ inside infinitesimal vacuum sheet of thickness $2\delta_1$, at interface j at $z = z_j$. $E'^{\pm}_{p,j-1}$ represents the p-polarized second harmonic fields in the vacuum sheet with $z_j - \delta_1 < z < z_j - \delta_2$ ('just left of the nonlinear sheet') and $E^{\pm}_{p,j-1}$ the p-polarized SH fields in the vacuum sheet with $z_j + \delta_2 < z < z_j + \delta_1$ ('just right of the nonlinear sheet'); $\hat{n}^i_j$ is the refractive index of the interface. The label 2ω has been omitted for clarity.

with $k, l, m = x, y, z$ and where $\chi^{(2)}_{klm,j}(2\omega)$ represents the nonlinear susceptibility tensor elements of interface j. $\mathscr{E}_{\omega,j}$ is the total fundamental field in the nonlinear sheet and autosummation is implied. The total fundamental field in the nonlinear sheet is in general a sum of forward and backward fields, which are derived from the fields in the vacuum sheet:

$$\begin{pmatrix} \mathscr{E}^+_{\omega,p,j} \\ \mathscr{E}^-_{\omega,p,j} \end{pmatrix} = \frac{1}{u^{\omega,+}_j} \begin{pmatrix} u^{\omega,+}_j & -u^{\omega,-}_j \\ -u^{\omega,-}_j & u^{\omega,+}_j \end{pmatrix} \begin{pmatrix} E'^+_{\omega,p,j} \\ E'^-_{\omega,p,j} \end{pmatrix}, \quad (3.3.6)$$

with $u^{\omega,\pm}_j = \hat{n}^i_j(\omega)\cos\theta_{\omega,0} \pm \cos\hat{\theta}^i_{\omega,j}$ is the complex angle of propagation in the nonlinear sheet, and p indicates the polarization. The total fundamental field in the nonlinear sheet j follows from the equations:

$$\mathscr{E}_{\omega,x,j} = (-\mathscr{E}^+_{\omega,p,j} + \mathscr{E}^-_{\omega,p,j})\cos\hat{\theta}^i_{\omega,j},$$
$$\mathscr{E}_{\omega,z,j} = (\mathscr{E}^+_{\omega,p,j} + \mathscr{E}^-_{\omega,p,j})\sin\hat{\theta}^i_{\omega,j}. \quad (3.3.7)$$

Substituting these equations into eqn (3.3.5) gives the total polarization of the sheet. The presence of the source $P^{sh}_{2\omega,j}$ causes a discontinuity of the SH fields at interface j. This discontinuity is derived from the boundary conditions, accounting for the presence of a (nonlinear) source polarization.[6,40] The result is

$$\Delta E_{2\omega,x,j} = -\frac{ik_x(2\omega)}{[\hat{n}^i_j(2\omega)]^2} 4\pi P^{sh}_{2\omega,z,j},$$
$$\Delta H_{2\omega,y,j} = \frac{i2\omega}{c} 4\pi P^{sh}_{2\omega,x,j}. \quad (3.3.8)$$

SHG FROM MULTILAYERS

These equations express the difference between the complex amplitudes of the field components in the vacuum sheets on both sides of nonlinear sheet j:

$$\Delta E_{2\omega,x,j} = E_{2\omega,x,j} - E'_{2\omega,x,j},$$
$$\Delta H_{2\omega,y,j} = H_{2\omega,y,j} - H'_{2\omega,y,j}, \quad (3.3.9)$$

where $E'_{2\omega,j}$ and $H'_{2\omega,j}$ are the electric and magnetic field respectively in the vacuum sheet with $z_j - \delta_1 < z < z_j - \delta_2$, and $E_{2\omega,j}$ and $H_{2\omega,j}$ are the fields in the vacuum sheet with $z_j + \delta_2 < z < z_j + \delta_1$ (see Fig. 3.2). Separating forward and backward field components gives

$$\Delta E_{2\omega,x,j} = -\cos\theta_{\omega,0}(E^+_{2\omega,p,j} - E^-_{2\omega,p,j}) + \cos\theta_{\omega,0}(E'^+_{2\omega,p,j} - E'^-_{2\omega,p,j}),$$
$$\Delta H_{2\omega,y,j} = -(E^+_{2\omega,p,j} + E^-_{2\omega,p,j}) + (E'^+_{2\omega,p,j} - E'^-_{2\omega,p,j}). \quad (3.3.10)$$

Here we have used the nonlinear equivalent of Snell's law.[6] Because there is no dispersion in vacuum, we obtain $\theta_{2\omega,0} = \theta_{\omega,0}$. After introducing

$$\kappa_{2\omega,p,j} \equiv \frac{E^+_{2\omega,p,j} - E^-_{2\omega,p,j}}{E^+_{2\omega,p,j} + E^-_{2\omega,p,j}}, \quad (3.3.11)$$

and an analogous definition for $\kappa'_{2\omega,p,j}$, we derive

$$E'^+_{2\omega,p,j} + E'^-_{2\omega,p,j} = \frac{\Delta E_{2\omega,x,j} - \kappa_{2\omega,p,j}\cos\theta_{\omega,0} \cdot \Delta H_{2\omega,y,j}}{(\kappa'_{2\omega,p,j} - \kappa_{2\omega,p,j})\cos\theta_{\omega,0}} \quad (3.3.12)$$

The values of κ are easily derived from eqns (3.3.1), (3.3.2) and (3.3.3) (now for 2ω), once we divide the multilayer into two parts at interface j, and realize that the system has only outgoing SH fields at both ends. For the second harmonic wave outgoing at the back side, we find

$$\begin{pmatrix} E^+_{2\omega,p,j} \\ E^-_{2\omega,p,j} \end{pmatrix} = \prod_{q=j}^{m} T_q^{2\omega,p} \begin{pmatrix} E^+_{2\omega,p,m+1} \\ 0 \end{pmatrix}, \quad (3.3.13)$$

with $T_q^{2\omega,p}$ as in eqn (3.3.2). From this equation we directly calculate the value of $\kappa_{2\omega,p,j}$. The value of $\kappa'_{2\omega,p,j}$ is obtained after studying the SH wave outgoing at the front side. In the latter case the multilayer is treated in the backward direction, which implies exchanging $E^+_{2\omega,p,j}$ and $E^-_{2\omega,p,j}$ and replacing z by $-z$ in $C_j^{2\omega}$. This analysis also gives the ratio

$$\frac{E_{2\omega,p,0}}{E'^+_{2\omega,p,j} + E'^-_{2\omega,p,j}}. \quad (3.3.14)$$

As eqn (3.3.12) gives this denominator of the ratio, we can calculate $E^-_{2\omega,p,0}$, the outgoing SH field at the front side, which is exactly the field that we detect in a standard SH reflection geometry. Similar expressions can of course be derived for the s-polarized fields.

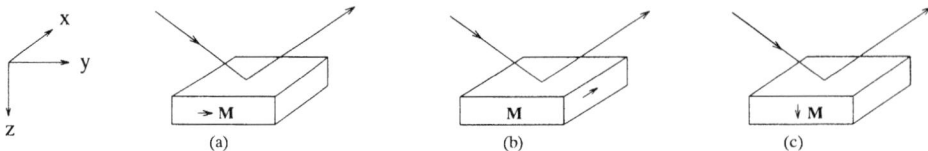

Fig. 3.3. Kerr effect configurations: (a) longitudinal, (b) transverse and (c) polar.

3.4 The magneto-optical Kerr effect

The presence of a time reversal symmetry breaking magnetization M introduces nonreciprocal optical effects, like polarization rotation. For metallic, reflecting media, these effects are summarized under the title magneto-optical Kerr effects. (For transparent media this leads to the so-called Faraday effects.) One distinguishes three different configurations: the polar, longitudinal and transverse Kerr effect (see Fig. 3.3), because the relative orientation of M with respect to the polarization as well as the optical plane of the light plays an important role.

Though the origins of NOMOKE and MOKE are the same (the spin–orbit coupling), it will be shown that there are some very basic differences between the two effects, leading to enormous enhancements of the NOMOKE versus the MOKE rotation. For that reason, we will first briefly describe the well known linear Kerr rotation, before introducing its nonlinear equivalent.

3.4.1 The linear magneto-optical Kerr effect (MOKE)

In a medium the E field should satisfy the wave equation

$$\nabla \times (\nabla \times E(r,t)) + \frac{\varepsilon(r,t)}{c^2} \frac{\partial^2}{\partial t^2} E(r,t) = 0, \qquad (3.4.1)$$

in which ε is the dielectric tensor of the medium. In the Fourier domain and with

$$k = \frac{\omega}{c} n, \qquad (3.4.2)$$

in which k is the wavevector and n the refractive index vector of the medium (defined as $n = n\hat{k}$ with n the refractive index of the medium and $\hat{k}$ a unit vector pointing in the direction of k), eqn (3.4.1) becomes

$$(nE(k,\omega))n - n^2 E(k,\omega) + \varepsilon(k,\omega) E(k,\omega) = 0. \qquad (3.4.3)$$

Rewriting this equation we get

$$(\varepsilon(k,\omega) + nn) E(k,\omega) = n^2 E(k,\omega). \qquad (3.4.4)$$

This is an eigenvalue equation with eigenvalues equal to n^2. For a nontrivial solution for E it is necessary that the determinant of $(\varepsilon(k,\omega) + nn - n^2)$ vanishes, which then determines the eigenvalues n^2.

The effect of a magnetization is included in the dielectric tensor. In a coordinate system fixed as in Fig. 3.3 and for an arbitrary direction of the magnetization vector M, the frequency-dependent linear dielectric tensor takes the following form:[41,42]

$$\varepsilon(\omega) = \begin{pmatrix} \varepsilon_0(\omega) & \varepsilon_1(\omega)m_z & -\varepsilon_1(\omega)m_y \\ -\varepsilon_1(\omega)m_z & \varepsilon_0(\omega) & \varepsilon_1(\omega)m_x \\ \varepsilon_1(\omega)m_y & -\varepsilon_1(\omega)m_x & \varepsilon_0(\omega) \end{pmatrix}, \quad (3.4.5)$$

in which m_x, m_y and m_z are the direction cosines of M. The diagonal components $\varepsilon_0(\omega)$ have purely nonmagnetic character (are symmetric under magnetization reversal) and the off-diagonal components $\varepsilon_1(\omega)$ are magnetic contributions (antisymmetric under magnetization reversal). From now on we will drop the frequency dependence of $\varepsilon_0(\omega)$ and $\varepsilon_1(\omega)$ for notation simplicity.

Using eqn (3.4.5) we can rewrite eqn (3.4.4) as

$$\begin{pmatrix} \varepsilon_0 + n_x^2 - n^2 & \varepsilon_1 m_z + n_x n_y & -\varepsilon_1 m_y + n_x n_z \\ -\varepsilon_1 m_z + n_x n_y & \varepsilon_0 + n_y^2 - n^2 & \varepsilon_1 m_x + n_y n_z \\ \varepsilon_1 m_y + n_x n_z & -\varepsilon_1 m_x + n_y n_z & \varepsilon_0 + n_z^2 - n^2 \end{pmatrix} \begin{pmatrix} E_x \\ E_y \\ E_z \end{pmatrix} = 0. \quad (3.4.6)$$

To simplify this equation we express n in terms of the angle of refraction θ_t. For this we use the coordinate system of Fig. 3.3:

$$n = n(0, \sin\theta_t, \cos\theta_t) \quad (3.4.7)$$

With this n eqn (3.4.6) becomes

$$\begin{pmatrix} \varepsilon_0 - n^2 & \varepsilon_1 m_z & -\varepsilon_1 m_y \\ -\varepsilon_1 m_z & \varepsilon_0 - n^2 \cos^2\theta_t & \varepsilon_1 m_x + n^2 \sin\theta_t \cos\theta_t \\ \varepsilon_1 m_y & -\varepsilon_1 m_x + n^2 \sin\theta_t \cos\theta_t & \varepsilon_0 - n^2 \sin\theta_t \cos\theta_t \end{pmatrix} \begin{pmatrix} E_x \\ E_y \\ E_z \end{pmatrix} = 0.$$

$$(3.4.8)$$

Now it is fairly easy to solve this equation for the three different Kerr effect configurations (see Fig. 3.3).

Polar Kerr effect

In the polar Kerr effect $m_x = m_y = 0$ and $m_z = 1$ (see Fig. 3.3). Solving eqn (3.4.8) and retaining only terms linear in ε_1 we then obtain as eigenvalues the complex refractive indices:

$$n_\pm^2 = \varepsilon_0 \pm i\varepsilon_1 \cos\theta_t, \quad (3.4.9)$$

and the eigenvectors are $\boldsymbol{E}^{\pm}(\omega) = \boldsymbol{E}^{\pm} e^{i\omega[t - (1/c)n_{\pm} \cdot r]}$, which are right- and left-circularly polarized waves. The $+$ and $-$ refer to right- and left-circularly polarized light, respectively. Thus we see that magnetism causes different refractive indices for right- and left-circularly polarized light.

The complex Kerr angle is defined as

$$\tan \Psi_K = \Phi_K + i\varepsilon_K = i \frac{E_r^+ - E_r^-}{E_r^+ + E_r^-}, \qquad (3.4.10)$$

where Φ_K is the Kerr angle, ε_K is the Kerr ellipticity, and $E_r^\pm$ the reflected field amplitudes. To express the latter in the incident field amplitude $E_i^\pm$, the Fresnel formulae should be used.

Longitudinal Kerr effect

Now $m_x = m_z = 0$ and $m_y = 1$ (see Fig. 3.3). Again retaining only terms linear in ε_1 we find that the eigenvectors of eqn (3.4.8) are also right- and left-circularly polarized waves with eigenvalues

$$n_\pm^2 = \varepsilon_0 \pm i\varepsilon_1 \sin \theta_t. \qquad (3.4.11)$$

Transverse Kerr effect

Now $m_y = m_z = 0$ and $m_x = 1$ (see Fig. 3.3). Solving eqn (3.4.8) again we obtain the eigenvalues

$$n_1^2 = \varepsilon_0, \qquad n_2^2 = \varepsilon_0 + \frac{\varepsilon_1^2}{\varepsilon_0}. \qquad (3.4.12)$$

So to first order in ε_1 the two eigenvalues are the same. This means that there is no difference in refractive indices for right- and left-circularly polarized light to first order in ε_1, and it follows that there is no transverse Kerr rotation or ellipticity to first order in ε_1. However, the *intensity* of the reflected light does depend on the magnetization.[43]

p-polarized incident light

In this case the incident electric field $\boldsymbol{E}_i$ is parallel to the plane of incidence and the Fresnel formula that relates $E_r^\pm$ to $E_i^\pm$ is

$$\frac{E_r^\pm}{E_i^\pm} = \frac{n_\pm^2 \cos \theta_i - \sqrt{n_\pm^2 - \sin^2 \theta_i}}{n_\pm^2 \cos \theta_i + \sqrt{n_\pm^2 - \sin^2 \theta_i}}, \qquad (3.4.13)$$

in which θ_i is the angle of incidence. For linearly polarized incident light $E_i^+ = E_i^-$ and so we can derive expressions for the complex linear Kerr angles for the polar and longitudinal cases. Using eqns (3.4.9), (3.4.10), (3.4.11) and

(3.4.13) and Snell's law $\sin\theta_r = \sin\theta_i/n$ we obtain the following expressions for the polar and longitudinal case, respectively:

$$(\tan\Psi_K^p)^{pol} = -\frac{1}{\sqrt{1+\chi_0}}\frac{\chi_1}{\chi_0}\cos\theta_i\frac{\cos(2\theta_i)+\chi_0}{\cos(2\theta_i)+\chi_0\cos^2\theta_i}, \qquad (3.4.14)$$

$$(\tan\Psi_K^p)^{long} = -\frac{1}{\sqrt{1+\chi_0}}\frac{\chi_1}{\chi_0}\frac{\sin\theta_i\cos\theta_i}{\sqrt{\cos^2\theta_i+\chi_0}}\frac{\cos(2\theta_i)+\chi_0}{\cos(2\theta_i)+\chi_0\cos^2\theta_i}. \qquad (3.4.15)$$

As before only terms linear in ε_1 are retained and we replaced ε_0 and ε_1 by their appropriate susceptibilities ($\varepsilon_0 = 1 + \chi_0$ and $\varepsilon_1 = \chi_1$).†

s-polarized incident light

Now the incident electric field E is perpendicular to the plane of incidence and we have to use the Fresnel formula

$$\frac{E_r^\pm}{E_i^\pm} = \frac{\cos\theta_i - \sqrt{n_\pm^2 - \sin^2\theta_i}}{\cos\theta_i + \sqrt{n_\pm^2 - \sin^2\theta_i}}. \qquad (3.4.16)$$

With $E_i^+ = E_i^-$ and using eqns (3.4.9), (3.4.10), (3.4.11) and (3.4.16) and Snell's law $\sin\theta_t = \sin\theta_i/n$ we obtain to first order in χ_1:

$$(\tan\Psi_K^s)^{pol} = -\frac{1}{\sqrt{1+\chi_0}}\frac{\chi_1}{\chi_0}\cos\theta_i, \qquad (3.4.17)$$

$$(\tan\Psi_K^s)^{long} = -\frac{1}{\sqrt{1+\chi_0}}\frac{\chi_1}{\chi_0}\frac{\sin\theta_i\cos\theta_i}{\sqrt{\cos^2\theta_i+\chi_0}}. \qquad (3.4.18)$$

3.5 Magnetization-induced second harmonic generation

Similar to the linear case, the presence of a magnetization M will affect the nonlinear optical susceptibility tensor $\chi^{(2)}$.

A quantum-mechanical derivation of the optical second-order nonlinear susceptibility can be found in Shen's book on nonlinear optics.[6] For the second-order dipole tensor elements one finds for the off-resonance case

$$\chi_{jkl}^{(2)}(2\omega) = -\frac{e^3}{\hbar^2}\sum_q\sum_{g,n,n'}\left(\frac{(r_j)_{gn}(r_k)_{nn'}(r_l)_{n'g}}{(2\omega - \omega_{ng})(\omega - \omega_{n'g})} + \text{five similar terms}\right)f_g(q), \qquad (3.5.1)$$

† In ref. 22 the prefactor $1/\sqrt{1+\chi_0}$ in eqn (3.4.15) is missing, probably because the factor $1/n$ ($\simeq 1/\sqrt{\varepsilon_0} = 1/\sqrt{1+\chi_0}$) in $\sin\theta_t = \sin\theta_i/n$ has been ignored in ref. 44.

where the summation runs over all possible states g, n, n' and wavevector q, $\hbar\omega_{ng} = E_n - E_g$, and $f_g(q)$ is the Fermi distribution function for the state $|g, q\rangle$.

There exist two ways in which the magnetization could influence the susceptibility:

1. The energy eigenvalues shift, which would result in different values of ω_{ng} for different spin states.
2. The electronic eigenfunctions change, because they are perturbed, which would lead to differences in the transition matrix elements $(r_i)_{gn}$ for the different spin states.

Both are a result of the combination of spin–orbit coupling and exchange interaction. Kittel has shown that the second effect has the largest influence on the optical susceptibility.[5]

Spin–orbit coupling changes the orbital momentum p of the electrons, and thus the polarization of the generated light, in much the same way as an external magnetic field would do,[18] so

$$-i\hbar\dot{p} = [H_{SO}, p] = [h(r)L \cdot S, p] = -h(r)p \times S. \quad (3.5.2)$$

In nonferromagnetic materials both spin states are equally occupied so no net effect occurs. However, in ferromagnetic materials, exchange interaction splits the band structure and spin states are not equally occupied, causing a net rotation of the polarization.[18]

Some of the first theoretical treatments of the effect of a magnetization on the nonlinear optical properties already appeared more than a decade ago for systems without inversion symmetry.[45] In that case, the presence of a magnetization lowers the symmetry of the crystal and gives rise to new nonzero tensor elements. Here we are concerned with the response of metallic thin films, where the bulk possesses inversion symmetry. For these systems Ru-Pin Pan et al.[17] showed how the presence of a magnetization leads to new nonzero tensor elements. In the following we will first show how these can be derived from simple symmetry arguments, and how they lead to nonlinear equivalents of the various MOKE effects.

Further details on the derivations of $\chi(M)$ can be found in Chapters 2 and 5 by Vollmer and by Hübner. One finds that the odd terms $\chi^{(2),-}$ are linearly dependent on the magnetization M, whereas the lowest-order effect of M on $\chi^{(2),+}$ is of second order.

So, just as in the linear case, the nonlinear Kerr effect is directly proportional to the magnetization. Combined with the interface sensitivity, this means that the nonlinear Kerr effect *directly probes interface magnetism*.

3.5.1 The nonlinear magneto-optical tensor components

As already mentioned, for materials that possess inversion symmetry, the nonlinear susceptibility tensor $\chi^{(2)}$ vanishes for the bulk and only gives a nonzero surface contribution due to the symmetry breaking at the surface. The presence of a magnetization M does not break the inversion symmetry, since M is an axial vector that conserves its orientation under inversion. However, M does lower the symmetry of the surface, introducing extra nonzero tensor elements.[17] These elements can be derived in the following way. Consider the (100) surface of a cubic crystal, and the longitudinal configuration, i.e. the magnetization parallel to the surface and in the plane of incidence (see Fig. 3.3).

In the absence of M, the (100) surface has 4mm symmetry. In the presence of a magnetization the symmetry is reduced. With $M = (M, 0, 0)$, the symmetry group is C_{4v}, i.e. the only nontrivial symmetry element not involving time inversion, i.e. reversal of the magnetization, is the reflection of the yz plane m_x. The combined reflection on the xz plane and reversal of the magnetization m_y^- is a symmetry transformation of the *magnetic* point group too. The symmetry operations then become

$$m_x^+ \circ \chi_{ijk}^{(2)}(M) = \chi_{ijk}^{(2)}(M),$$
$$m_y^- \circ \chi_{ijk}^{(2)}(M) = \chi_{ijk}^{(2)}(M).$$
(3.5.3)

For example, this gives

$$m_y^- \circ \chi_{zyy}^{(2)}(M,0,0) = \chi_{zyy}^{(2)}(M,0,0),$$
$$\Leftrightarrow + \chi_{zyy}^{(2)}(-M,0,0) = \chi_{zyy}^{(2)}(M,0,0),$$
(3.5.4)

$$m_y^- \circ \chi_{yyy}^{(2)}(M,0,0) = \chi_{yyy}^{(2)}(M,0,0),$$
$$\Leftrightarrow - \chi_{yyy}^{(2)}(-M,0,0) = \chi_{yyy}^{(2)}(M,0,0).$$
(3.5.5)

As a result, we can separate the nonlinear susceptibility into two sets of tensor elements: those which are odd (χ^-) and even (χ^+) in the magnetization. These tensor elements fulfil the relation:

$$\chi_{ijk}^{(2),\pm}(-M) = \pm \chi_{ijk}^{(2),\pm}(M).$$
(3.5.6)

The tensor elements for the transverse configuration can be derived in an analogous way. In this case, $M = (0, M, 0)$ and the corresponding symmetry operations are

$$m_x^- \circ \chi_{ijk}^{(2)}(0,M,0) = \chi_{ijk}^{(2)}(0,M,0),$$
$$m_y^+ \circ \chi_{ijk}^{(2)}(0,M,0) = \chi_{ijk}^{(2)}(0,M,0).$$
(3.5.7)

Finally we consider the case of a polar Kerr effect, i.e. with the magnetization normal to the surface plane: $M = (0, 0, M)$. For this case, the symmetry operations are

$$m_x^- \circ \chi_{ijk}^{(2)}(0,0,M) = \chi_{ijk}^{(2)}(0,0,M),$$
$$m_y^+ \circ \chi_{ijk}^{(2)}(0,0,M) = \chi_{ijk}^{(2)}(0,0,M),$$
$$R_{z,90°}^+ \circ \chi_{ijk}^{(2)}(0,0,M) = \chi_{ijk}^{(2)}(0,0,M).$$
(3.5.8)

Table 3.1 Nonzero tensor elements of the nonlinear susceptibility tensor $\chi^{(2)}(M)$ for the (001) cubic surface, with the magnetization parallel to the x axis (longitudinal configuration), y axis (transverse configuration) and z axis (polar configuration). For simplicity of notation the tensor components are indicated by their indices only

	Even in M	Odd in M
Longitudinal $M \parallel x$	$yzy = yyz$ $xzx = xxz$ zzz zyy zxx	$xyx = xxy$ $zyz = zzy$ yzz yyy yxx
Transverse $M \parallel x$	$xxz = xzx$ $yyz = yzy$ zxx zyy zzz	$yxy = yyx$ $zxz = zzx$ xxx xyy xzz
Polar $M \parallel z$	$xxz = xzx = yyz = yzy$ $zxx = zyy$ zzz	$xyz = xzy = -yxz = -yzx$

Table 3.1 gives the so-derived odd and even surface tensor components for all the Kerr configurations.[46] From Table 3.1 the effect of the magnetization on the SHG response can qualitatively be understood directly. Take for example an s-polarized input beam for the case of a longitudinal configuration. In that situation, there is only one even term (χ_{zyy}) and one odd term (χ_{yyy}) that will contribute to the MSHG response. Because the even term will give rise to a p-polarized output and the odd term will give an s-polarized SHG signal, the total output polarization can be varied by changing the direction of the magnetization. This will lead to a nonlinear Kerr rotation, similar as for the linear case. However, as we will see later, the effects can be much larger than in the linear case. One of the reasons is that the odd component is of the same order of magnitude as the even (there are also other, intrinsic, differences between the two cases that will be discussed in the next section).

For the transverse geometry, it follows from Table 3.1 that the generated MSHG signal will be p-polarized, for both s- and p-polarized input beams. Again take for example an s-polarized input. The MSHG signal will then get the form

$$I_{2\omega}(\pm M) \approx |\alpha \chi_{zyy}^{(2)} \pm \beta \chi_{xyy}^{(2)}|^2, \tag{3.5.9}$$

where the constants α and β include the Fresnel factors for the incoming and generated optical beams. This means that the presence of a magnetization will only affect the output intensity, not its polarization. So, similarly to the linear case, there is no nonlinear Kerr rotation in the transverse geometry. However, in contrast to the linear case, the magnetization-induced intensity changes can be very large (for the same reason as there are large nonlinear rotations). Coupled with the simplicity of the experimental configurations, this makes the transverse geometry very attractive. One can then define a magnetization contrast ρ as

$$\rho = \frac{I_{2\omega}(+M) - I_{2\omega}(-M)}{I_{2\omega}(+M) + I_{2\omega}(-M)}. \tag{3.5.10}$$

For the quadrupolar bulk term $\chi^{(Q)}$ of eqn (3.2.1), a similar analysis can be done, as was shown by Wierenga.[47] For $M = (0, M, 0)$ they can be derived for the 4mm symmetry in the following way:

$$\mathcal{M}_x^- \circ \chi_{jklm}^{(Q)}(My) = \chi_{jklm}^{(Q)}(My), \tag{3.5.11}$$

$$\mathcal{M}_y^+ \circ \chi_{jklm}^{(Q)}(My) = \chi_{jklm}^{(Q)}(My), \tag{3.5.12}$$

$$\mathcal{M}_z^- \circ \chi_{jklm}^{(Q)}(My) = \chi_{jklm}^{(Q)}(My), \tag{3.5.13}$$

$$\mathcal{R}_{y,90°}^+ \circ \chi_{jklm}^{(Q)}(My) = \chi_{jklm}^{(Q)}(My). \tag{3.5.14}$$

For the isotropic system with $M = My$ we have an identical set of symmetry operations, with one exception: $\mathcal{R}_{y,\alpha}^+ \circ \chi_{jklm}^{(Q)}(My) = \chi_{jklm}^{(Q)}(My)$ is a symmetry operation for all α. This leads to one extra relation between the even tensor elements, which is also found for the $\chi^{(Q)}$ of a nonmagnetized isotropic material:[48,49]

$$\chi_{xxxx}^{(Q)}(My) = \chi_{xxzz}^{(Q)}(My) + \chi_{xzzx}^{(Q)}(My) + \chi_{xzxz}^{(Q)}(My). \tag{3.5.15}$$

We also obtain one extra relation between the odd tensor elements†:

$$\chi_{xzzz}^{(Q)}(My) = \chi_{xxxz}^{(Q)}(My) + \chi_{xzxx}^{(Q)}(My) + \chi_{xxzx}^{(Q)}(My). \tag{3.5.16}$$

The elements of $\chi^{(Q)}(Mx)$ are obtained by exchanging y and x in Table 3.2 and the elements of $\chi^{(Q)}(Mz)$ are obtained by exchanging y and z in the same table.

Table 3.2 gives the corresponding bulk odd and even tensor components for all the Kerr configurations. As pointed out by Ru-Pin Pan et al.,[17] the tensor components of Table 3.1 fulfil another important relationship, which can be derived from the breaking of the time reversal symmetry due to a magnetic field. Including the explicit time and phase dependence, the induced polarization can be written as

$$P_{2\omega}^{(2)}(r,t) = \{\chi^+(M) + \chi^-(M)\} E_\omega(r,t) E_\omega(r,t). \tag{3.5.17}$$

† These relations can of course be obtained in a full analysis of the influence of a rotation over an arbitrary angle α. However, it is more convenient to do a first order-of-magnitude analysis after rotating over an infinitesimal angle δ.

Table 3.2 Nonzero tensor elements of the nonlinear bulk susceptibility tensor $\chi^{(Q)}(M)$ for a magnetized cubic crystal. For simplicity of notation the tensor components are indicated by their indices only

	Even in M	Odd in M
Longitudinal $M \parallel x$	$yyyy = zzzz$, $yyxx = zzxx$	$yyyz = -zzzy$, $yyzy = -zzyz$
	$yxxy = zxxz$, $yyzz = zzyy$	$yzzz = -zyyy$, $xyzx = -xzyx$
	$yzzy = zyyz$, $yxyx = xzxz$	$xxzy = -xxyz$, $yxzx = -zxyx$
	$yzyz = zyzy$, $xyyx = xzzx$	$yxxz = -zxxy$, $yzxx = -zyxx$
	$xxyy = xxzz$, $xyxy = xzxz$	$xyxz = -xzxy$, $yzyy = -zyzz$
Transverse $M \parallel y$	$xxxx = zzzz$, $xxyy = zzyy$	$xxxz = -zzzx$, $xxzx = -zzxz$
	$xyyx = zyyz$, $xxzz = zzxx$	$xzzz = -zxxx$, $yxzy = -yzxy$
	$xzzx = zxxz$, $xyxy = zyzy$	$yyzx = -yyxz$, $xyzy = -zyxy$
	$xzxz = zxzx$, $yxxy = yzzy$	$xyyz = -zyyx$, $xzyy = -zxyy$
	$yyxx = yyzz$, $yxyx = yzyz$	$yxyz = -yzyx$, $xzxx = -zxzz$
	$yyyy$	
Polar $M \parallel z$	$xxxx = yyyy$, $xxzz = yyzz$	$xxxy = -yyyx$, $xxyx = -yyxy$
	$xzzx = yzzy$, $xxyy = yyxx$	$xyyy = -yxxx$, $zxyz = -zyxz$
	$xyyx = yxxy$, $xzcz = yzyz$	$zzyx = -zzxy$, $xzyz = -yzxz$
	$xyxy = yxyx$, $zxxz = zyyz$	$xzzy = -yzzx$, $xyzz = -yxzz$
	$zzxx = zzyy$, $zxzx = zyzy$	$zxzy = -zyzx$, $xyxx = -yxyy$
	$zzzz$	

In the absence of dissipation (far away from resonances), time reversal implies

$$P_{2\omega}(r,t) \Rightarrow P_{2\omega}^*(r,t),$$
$$E_\omega(r,t) \Rightarrow E_\omega^*(r,t), \quad (3.5.18)$$
$$M \Rightarrow -M,$$

leading to

$$P_{2\omega}^*(r,t) = \{\chi^+(M) - \chi^-(M)\}E_\omega^*(r,t)E_\omega^*(r,t). \quad (3.5.19)$$

Adding eqn (3.5.17) and eqn (3.5.19) yields

$$P_{2\omega}(r,t) + \text{c.c.} = \chi^+(M)\{E_\omega(r,t)E_\omega(r,t) + \text{c.c.}\}$$
$$+ \chi^-(M)\{E_\omega(r,t)E_\omega(r,t) - \text{c.c.}\}. \quad (3.5.20)$$

The left-hand side of eqn (3.5.20) is real, and so must be the right-hand side, for all r and t. From this it immediately follows that, in the absence of dissipation, $\chi^+(M)$ is real and $\chi^-(M)$ is purely imaginary. The same is true for the quadrupole terms of Table 3.2. This result also follows from a microscopic treatment, as was recently shown by Hübner and Bennemann.[50]

3.5.2 The nonlinear magneto-optical Kerr effect

Strictly speaking, a nonlinear magneto-optical Kerr effect (NOMOKE) or rotation is a misleading expression, as it suggests a nonlinear dependence on the magnetization of the Kerr rotation. However, quantum-mechanical derivations show that $\chi^{(2)}(M)$ depends *linearly* on the magnetization in a similar way as the linear magneto-optical response (see also sec. 3.5.3 and Chapter 5 by Hübner). Because the nonlinear Kerr angle $\Phi_K^{(2)}$ is related to a ratio of magnetic and nonmagnetic $\chi^{(2)}$ tensor components (see below), $\Phi_K^{(2)}$ will depend linearly on M (just as the linear Kerr rotation $\Phi_K^{(1)}$). For that reason, it would be more correct to speak of a Kerr rotation of the nonlinear optical response.

However, in accordance with published literature so far, we will use the expression NOMOKE for its simplicity.

The nonlinear optical effects are caused by the nonlinear surface polarization $P(2\omega)$ that acts as a source term in the wave equation. The latter can now be written as

$$\nabla \times \nabla \times E(2\omega) + \frac{\varepsilon(2\omega)}{c^2}\frac{\delta^2}{\delta t^2}E(2\omega) = -\frac{1}{c_0 c^2}\frac{\delta^2}{\delta t^2}P(2\omega). \quad (3.5.21)$$

The solution of this wave equation included the solution of the homogeneous equation plus one particular solution of the inhomogeneous equation. Similarly to the linear case, the homogeneous part yields complex refractive indices for the transmitted second harmonic wave:

$$N_t^2 = \varepsilon_0(2\omega) \pm i\varepsilon_1(2\omega)\sin\theta_{2t}, \quad (3.5.22)$$

where θ_{2t} refers to the refraction angle of the 2ω beam. The inhomogeneous part has solutions that no longer relate to indices of refraction but directly to the surface susceptibility $\chi^{(2)}$.

For the calculation of the Kerr rotation the amplitude of the reflected electric field should be known. For this we have to make use of the Fresnel formulae for the nonlinear case. These formulae can be derived by using the continuity of the electric field parallel to the surface and of the magnetic field (p-polarized incident light is assumed).[6,11] This leads to the following two equations:

$$-\cos\theta_i E_r(2\omega) = \cos\theta_t E_t(2\omega) + \frac{\cos\theta_s N_s^2 \sin\theta_s}{\varepsilon_0(2\omega)[\varepsilon_0(2\omega) - N_s^2]}P(2\omega), \quad (3.5.23)$$

$$k_r E_r = k_t E_t + \frac{k_s \sin\theta_s}{\varepsilon_0(2\omega) - N_t^2}P(2\omega), \quad (3.5.24)$$

where E_r, k_r and E_t, k_t are the electric field wavevector of the reflected and transmitted waves and θ_t and θ_i are the refraction and incident angle (see Fig. 3.4). N_s is the refractive index of the source term:

$$N_s^2 = \varepsilon_0(\omega) \pm i\varepsilon_1(\omega)\sin\theta_s. \quad (3.5.25)$$

The angle θ_s determines the orientation of the source wavevector k_s, and is equal to the refraction angle of the fundamental ray. The direction of the source

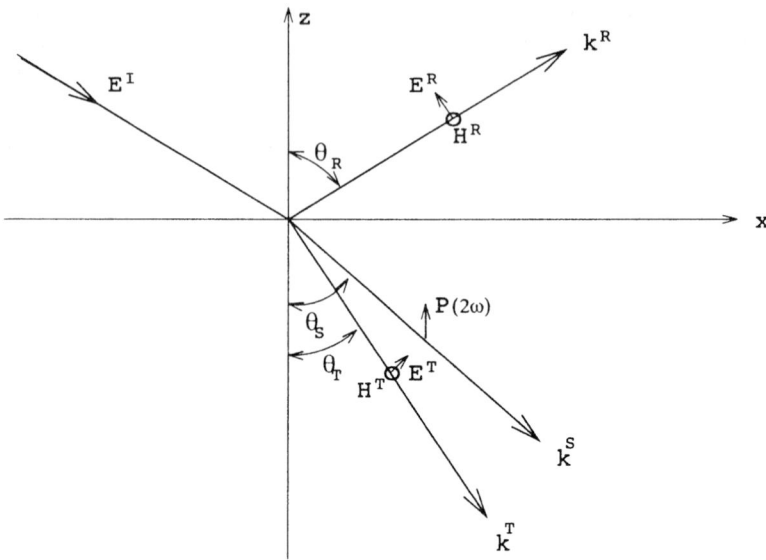

Fig. 3.4. The harmonic wave at the boundary of a nonlinear medium, with the electric field vector in the plane of reflection.

term $P(2\omega)$ is chosen perpendicular to the surface. The amplitude of the reflected electric field follows from the combination of these two equations:

$$E_r^\pm(2\omega) = \frac{1}{\varepsilon_0 c^2} \frac{P(2\omega)\sin\theta_s}{k_t\cos\theta_i + k_r\cos\theta_t} \frac{\varepsilon_0(2\omega)k_s\cos\theta_t - N_s^2\sin\theta_s}{\varepsilon_0(2\omega)[\varepsilon_0(2\omega) - N_s^2]}. \quad (3.5.26)$$

Magnetism is included in the indices of refraction (eqn (3.5.25)). With this formula for E_r, eqn (3.4.10) will lead to a longitudinal nonlinear Kerr angle of:[22]

$$\tan\Psi_K^{(2)} = i\left(\frac{\chi^{(2),-}}{\chi^{(2),+}} + (\text{h.o.})\right), \quad (3.5.27)$$

in which $\chi^{(2),+}$ and $\chi^{(2),-}$ are the nonmagnetic and magnetic contributions to $\chi^{(2)}$.

The big difference between the expressions for $\Psi_K^{(2)}$ and $\Psi_K^{(1)}$ is the factor $1/\sqrt{(1+\chi_0)(\cos^2\theta_i + \chi_0)}$ in eqn (3.4.18), which reduces the value for $\Psi_K^{(1)}$ in eqn (3.4.18). Therefore it is clear that the nonlinear Kerr rotation for all θ_i is always enhanced by a factor of $\sim[1+\chi_0(\omega)]$. For example, for Fe at $\theta_i = 45°$ and $\lambda = 830$ nm, this gives a reduction factor 0.15 for $\Phi_K^{(1)}$.

Above this effect, in the nonlinear case one can select large magnetic $\chi^{(2)}$ contributions by a proper choice of input polarization and angle of incidence. This is in contrast with the linear case where one always deals with small off-diagonal magnetic terms ($\chi^{(1),-}(\omega)$) relative to large diagonal nonmagnetic terms ($\chi^{(1),+}(\omega)$).

This can easily be demonstrated with a simple example. Consider the case of a longitudinal configuration, with $M \parallel x$ axis. Inspection of Table 3.1 shows that, for a nonmagnetic surface, the only nonzero tensor component that will be excited for an s-input polarization is $\chi^{(2)}_{zyy}$, which will lead to a purely p-polarized output. However, in the case of a magnetic surface, also the odd tensor component $\chi^{(2)}_{yyy}$ will be nonzero, leading to a turning of the SHG polarization away from the p direction. In the limit of a very small angle of incidence, the z component of the reflected SHG signal will be vanishingly small. This means that, in the limit of near-normal incidence, the SHG signal from the magnetic surface will be s-polarized, instead of p-polarized for the nonmagnetic case, i.e. a nonlinear Kerr rotation of 90°. A similar enhancement of the magnetic (off-diagonal) term relative to the nonmagnetic (diagonal) one is however not possible for the linear case. The same configuration only leads to a small linear Kerr rotation, given by $\Phi^{(1)}_K = \arctan(\chi^{(1)}_{yx}/\chi^{(1)}_{yy})$, because $\chi^{(1)}_{yx} \ll \chi^{(1)}_{yy}$.

For a p-polarized input and for the same configuration and angle of incidence, the nonlinear Kerr rotation will generally be much smaller. This is due to the fact that, for this polarization, there are three nonzero even tensor elements $\chi_{zzz}, \chi_{xzx}, \chi_{zxx}$, and two nonzero odd tensor components χ_{yzz}, χ_{yxx}. In this situation an increasing contribution from one tensor component is often (partly) compensated by a decrease in another term, resulting in less pronounced changes in the MSHG response and in smaller nonlinear Kerr rotations.

The large nonlinear Kerr effects can also be used to determine rather easily the easy axis of magnetization. This is particularly useful to distinguish between in-plane and out-of-plane magnetization (as was also pointed out by Hübner and Bennemann[50]). Again, consider the case of near-normal incidence.

For an s-polarized input beam, only the longitudinal geometry gives rise to (a rather strong) s-polarized MSHG output. If there is only a p-polarized output, one should look at the p-polarized input. In the polar geometry there will be an s-polarized component, whereas the transverse geometry only gives rise to p-polarized MSHG.

3.5.3 Estimates for $\chi^{(2)}(M)$

Some theoretical values for $\chi^{(2),-}$ have been estimated for nickel and iron. Ru-Pin Pan et al. made a rather crude approximation for Ni(001) by assuming that all odd and that all even tensor elements are equal, and obtained[17,51]

$$|\omega^2 \chi^{(2),-}|_{Ni} \sim 10^{12} \text{ V}^{-1}\text{s}^{-2}\text{ m} \sim 10^{-16} \text{ esu}. \qquad (3.5.28)$$

The same surface has been more thoroughly analyzed by Hübner et al. Including the complete band structure and dipole transition matrix elements they obtained[52,53]

$$|\omega^2 \chi^{(2)}_{xzz}|_{Ni} \sim 10^{17} \text{ V}^{-1}\text{s}^{-2}\text{ m}. \qquad (3.5.29)$$

Pustogowa et al. used a similar approach for the surface of iron and found[54,55]

$$|\omega^2 \chi^{(2)}_{xzz}|_{Fe} \sim 10^{16} \text{ V}^{-1} \text{s}^{-2} \text{ m}. \quad (3.5.30)$$

All these values are well above the detection limit of SHG, that is 10^9–10^{10} V^{-1}s^{-2} m or 10^{-19}–10^{-18} esu,[56] implying that magnetization-induced tensor elements should be observable.

Furthermore, the estimated ratios of odd and even elements are high enough to induce observable changes of the SH intensity on inverting the magnetization. For nickel both Ru-Pin Pan et al.[17] and Hübner[53] estimated

$$|\chi^{(2),-}/\chi^{(2),+}|_{Ni} = 0.07, \quad (3.5.31)$$

and for iron Pustogowa et al. deduced[54,55]

$$|\chi^{(2),-}/\chi^{(2),+}|_{Fe} = 0.18. \quad (3.5.32)$$

Reif et al. deduced the latter ratio by simply scaling the nickel ratio of eqn (3..31) by the ratio of the bulk magnetic moments for iron and nickel, $\mu_{Fe}/\mu_{Ni} = 3.8$, and obtained[19]

$$|\chi^{(2),-}/\chi^{(2),+}|_{Fe} = 0.27. \quad (3.5.33)$$

This value is in close agreement with the result of a simple analysis of their experiments on the Fe(110) surface, and also close to the value given by Pustogowa et al. (see eqn (3.5.32)). Scaling by the surface magnetic moments instead of the bulk magnetic moments does not significantly alter the estimate.[57] Using a similar approach for cobalt gives, with $\mu_{Co}/\mu_{Ni} = 3.1$,[57]

$$|\chi^{(2),-}/\chi^{(2),+}|_{Co} = 0.22. \quad (3.5.34)$$

Experimentally (see sec. 3.7.3) we find $|\chi^{(2),-}/\chi^{(2),+}|_{Co} = 0.6$ for an excitation wavelength of 800 nm (the estimates were done for 532 nm). For the Co/Cu interface we find a very similar ratio $|\chi^{(2),-}/\chi^{(2),+}|_{Co/Cu} = 0.89$.

All these esimates have been made for a particular frequency. In a number of recent papers, Hübner et al. present *ab initio* calculations of the linear and nonlinear magneto-optical spectra for Ni and Fe.[18,54] The nonlinear spectra show a much richer structure, which is connected with the differences in selection rules and the surface sensitivity of MSHG. Spectroscopic MSHG experiments therefore will be an excellent and sensitive way to probe magnetic surface states. So far, this has been a rather unexplored area of research.

3.6 Instrumentation

The study of nonlinear optical effects requires excitation by strong electromagnetic fields. Nonlinear optics therefore has developed strongly since the discovery of the laser, the more because the first laser happened to be a pulsed one. Considering the fact that, for surface nonlinear optics, input intensities are larger than 1 MW cm^{-2}, it is obvious that almost any material, and in particular a thin metallic film, would be immediately destroyed if it were subjected to such a laser power in a continuous mode. Instead, pulsed lasers are used, through which one can reach such peak powers, whereas the average power is usually far below 1 W.

The second harmonic signal (in photons per pulse) in reflection from a surface is given by[6]

$$S(2\omega) = \frac{32\pi^3\omega}{\hbar c^3 \varepsilon(\omega)\varepsilon(2\omega)^{1/2}} |\chi_s^{(2)}|^2 I^2(\omega) AT,$$

for an input laser pulse with intensity $I(\omega)$, cross-section A and pulse width T. Here $\chi_s^{(2)}$ indicates the effective surface nonlinear susceptibility, including the Fresnel factors for the transmission and reflection of the fundamental and second harmonic fields at the interface. For a metal surface, $\chi_s^{(2)}$ is of the order of 10^{-15} esu. A pulsed Nd:YAG laser at 1.06 μm with $I(\omega) = 10$ MW cm^{-2}, $A = 0.1$ cm^2 and $T = 10$ ns would yield 10^2 photons/pulse, or 3000 photons/second for a typical pulse repetition rate of 30 Hz. Including a typical overall detection efficiency of ~5% (resulting from the several colour filters, polarizers, monochromator and quantum efficiency of the photomultiplier), this yields a signal of 150 photon counts/second. This is a rather low, but easily detectable signal with modern (gated) photon counting techniques. Note that this means that the nonlinear yield $I(2\omega)/I(\omega) \approx 10^{-15}$. This requires an extremely strong reduction of the fundamental laser intensity on the detection side, which is obtained by using several colour filters and a monochromator. Through peak powers of 100 MW are readily available with standard YAG lasers, such powers cannot be used as they would damage the sample under investigation.

The recent developments in high-repetition-rate, mode-locked lasers have changed and dramatically improved the experimental possibilities. A mode-locked Ti:sapphire laser, with an average output power of 1 W, delivers pulses of about 100 fs at a repetition rate of 82 MHz. For a typical experimental condition (see the examples in secs 3.7–3.12), one uses an average input power of 100 mW focused on an area of 10^{-4} cm^2. This corresponds to a peak intensity of $I = 1.2 \times 10^8$ W cm^{-2}, or a laser fuence of 12 μJ cm^{-2}. For the same susceptibility as above ($\chi_s^{(2)} = 10^{-15}$ esu) this yields only 1.4×10^{-4} photons per pulse. However, due to the high repetition rate this corresponds to 1.2×10^4 photons per second, i.e. a *100 times increase* of the signal.

Figure 3.5 shows a schematic experimental setup for the MSHG experiments, with a Ti:sapphire laser for the incident fundamental radiation. The mode-locked Ti:sapphire laser is pumped by an argon ion laser. The output has a Gaussian mode and is continuously tunable from 720 nm to 1100 nm using two or three mirror sets. At 835 nm the laser has an average output power of approximately 1.3 W. The laser fuence is regulated by a set of neutral density filters. The polarization of the incoming fundamental beam is adjustable with a Soleil–Babinet compensator, which we use as a tunable half-wave plate. A lens

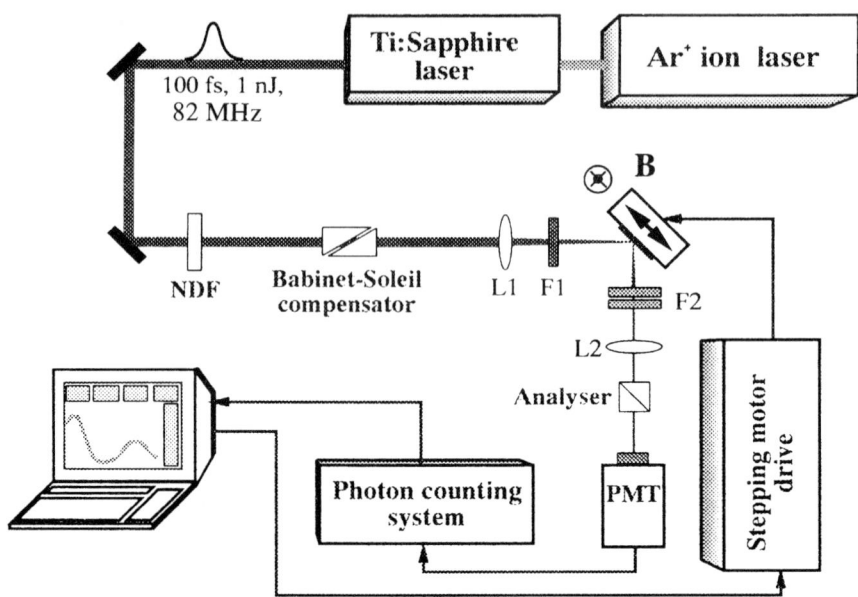

Fig. 3.5. Schematic representation of an MSHG setup: NDF, neutral density filter; L, lens; F, colour filter; B, electromagnet; PMT, photomultiplier tube.

is used to focus the beam on the sample to a spot diameter of about 100 μm. Before reaching the sample, the light is passed through a colour filter (CF1). It removes all light with frequencies below 570 nm, thus eliminating any possible signals at the second harmonic wavelength that have not been generated at the sample. The sample can be mounted between the poles of an electromagnet. For measurements on wedge-type multilayer samples, the sample is mounted on a computer-controlled micrometer stage. The latter allows the sample to be translated parallel to its surface without changing the alignment, thus facilitating thickness-dependent measurements in a very controlled and reliable way (see sec. 3.9). For anisotropy measurements, the sample can be mounted on a computer-controlled rotation stage. This sample holder is such that MSHG signals in transmission can also be measured. This technique was used for instance for studies on transparent garnet films (see sec. 3.11). The magnetization M is either parallel to the surface and in the plane of incidence (longitudinal configuration) or parallel to the surface and perpendicular to the plane of incidence (tranverse configuration). In the case of a perpendicular magnetization, the setup can also be modified to a polar geometry. Because of the enormous difference between the intensity of the reflected fundamental and the generated second harmonic, and the fact that in general the beams are not spatially separated, proper filtering is of vital importance. Therefore, two colour filters (CF2) are used to filter out the fundamental light. A spatial separation between fundamental and second harmonic beams can also be accomplished by making use of an Amice prism. A UV-transparent lens (L2) focuses the second harmonic light into the photomultiplier tube (PMT: Thorn EMI, 9789B). To

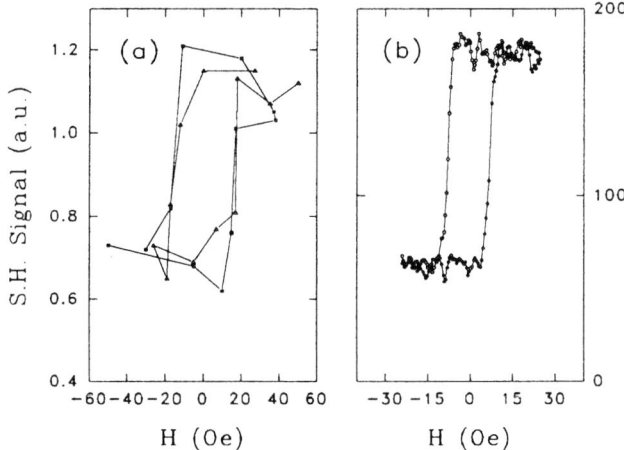

Fig. 3.6. MSHG hysteresis: (a) from a Co–Au film as measured with a Nd:YAG laser, and (b) from a Co–Cu film as measured with a Ti:sapphire laser.

improve the signal-to-noise ratio, a chopper in combination with a lock-in amplifier is used. In the longitudinal configuration an analyzer (GT) is used to measure the plane of polarization of the second harmonic signal. In the transverse configuration the signal is always p-polarized (see sec. 3.5), and thus no analyzer is needed. For small signals, a (gated) photon counting system can be used.

The advantage of a high-repetition-rate, mode-locked laser versus a low-repetition Q-switched laser is clearly demonstrated in Fig. 3.6. In Fig. 3.6(a), an MSHG hysterisis curve is plotted, measured on a 500 Å Co film, covered by 50 Å Au using the frequency-doubled output of a Nd:YAG laser at 532 nm as the fundamental incident radiation.[20] The data taking for this figure took about 2 h in total. In contrast, the hysteresis curve in Fig. 3.6(b) was measured within 5 min using a Ti:sapphire laser.[21] A further advantage of the use of a Ti:sapphire laser is its high stability and the fact that it can also easily be used for the linear MOKE experiments as well. In addition, this type of laser is easily tunable. Though the standard tunability range of a pulsed Ti:sapphire laser is limited to the range of about 720–1100 nm, this tuning range can be easily extended using different nonlinear optical techniques, like parametric oscillators and amplifiers.[6]

In order to measure the (relative) phase of the various nonlinear optical tensor components, a reference source on a computer-controlled micrometer stage can be inserted just before the sample (see Fig. 3.7). The relative phase of this reference signal with respect to that of the sample can be varied by translating the reference parallel to the incident beam, thus introducing an extra optical phase delay

$$\delta\phi = \left(\frac{4\pi \Delta n}{\lambda}\right) d.$$

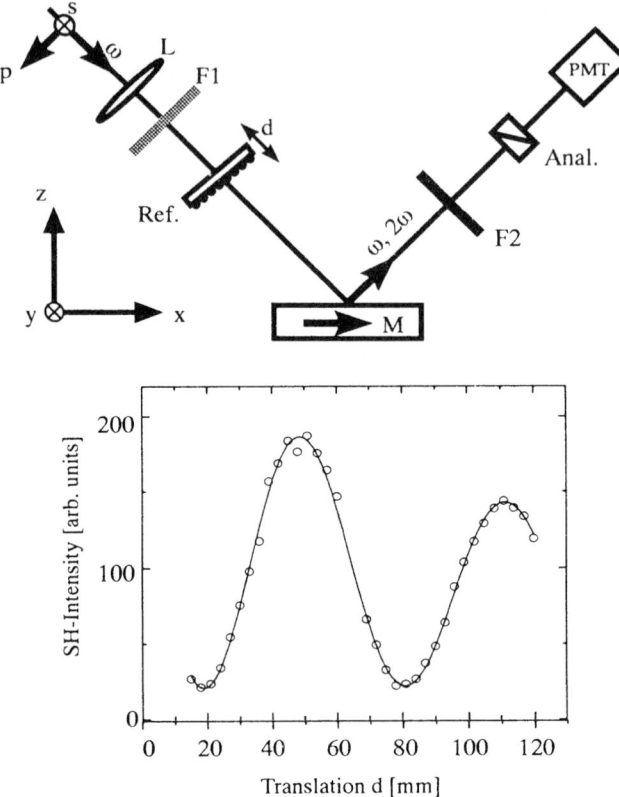

Fig. 3.7. Top: experimental geometry for phase measurement. A reference sample is placed between colour filter F1 and the sample. Below: an actual example of such phase measurement is shown.

This way of measuring the phase relies on the fact that the ambient is (slightly) dispersive. For a fundamental wavelength of 800 nm, $\Delta n = n(2\omega) - n(\omega) \approx 6.5 \times 10^{-6}$. This means that for a π phase shift, the reference source should be translated over $d = 4\Delta n/\lambda \simeq 32$ mm. As a reference source a glass slide covered on one side by a thin poled polymer film with a high $\chi^{(2)}$ is used. This has the advantage of avoiding complications due to biaxiality and the corresponding polarization rotation of standard nonlinear optical crystals as a reference. The reference intensity can be tuned by changing the angle of incidence on the glass slide.

3.7 Experimental case studies

Because all nonlinear magneto-optical effects find their origin in the effects of time reversal symmetry breaking on the nonlinear optical susceptibility tensor, we will start this section with a direct experimental demonstration of this effect.

Though studies of clean surfaces in an ultrahigh-vacuum environment are of interest and importance for the development of MSHG, one of the most attractive features of a nonlinear optical technique is the possibility to probe buried interfaces of multilayer systems. We have therefore devoted a substantial effort to the study of such buried magnetic interfaces. However, to prove experimentally the surface sensitivity of MSHG, *in situ* studies of clean surfaces with a possibility of surface modification are indispensable. The first of such experiments was reported by Reif *et al.*, who showed an MSHG effect on Fe(001) of about 25%, which disappeared in time due to surface contamination.[19] The first results on buried magnetic interfaces were obtained for Co/Au films.[20] Interface sensitivity was demonstrated here by varying the number of interfaces. Though both these types of early experiments were very important and stimulated a lot of theoretical and experimental work, clear demonstration of sensitivity for surface magnetism was obtained by simultaneously performed *in situ* MOKE and MSHG studies on fcc Co films on Cu(001) in UHV.[21,58,59] Comparison of MSHG with linear MOKE results in combination with CO and O_2 dosage experiments proved the surface sensitivity of MSHG. It also showed some very interesting results for extremely thin films (below 5 ML). After this, we will proceed with discussing the Cu/Co/Cu(001) multilayer system. Here, both the Co and the Cu overlayer thickness will be varied, which will give rise to very different results. In particular we will see how the Cu overlayer thickness variation on a 10 ML Co film will give rise to an extremely strong oscillatory-like behaviour of the MSHG response. This appears to be the first indication of the extreme sensitivity of MSHG for the presence of quantum-well states in thin metallic films. (More detailed studies of these phenomena will follow at the end of this section.) In the following sections nonlinear magneto-optical Kerr rotation experiments will be discussed, on a number of Fe and Co systems. Enormous enhancements (up to a factor of 10^3) of the nonlinear Kerr rotation relative to the linear equivalent are observed and can be explained by the fundamental differences between the two techniques as described in section 3.5.[25,60] After this, we will briefly discuss some first results on polycrystalline Ni surfaces and some very interesting applications of MSHG studies of magnetic quantum-well systems. It will appear that, in the presence of quantum wells, the latter totally dominate the nonlinear optical response. The reason for this extreme sensitivity appears to originate in the spatial distribution of the electronic density of states of these quantum-well states, though a real quantitative understanding is still lacking. Finally we will discuss very recent MSHG studies on antiferromagnetic and garnet films, and demonstrate the novel possibilities of nonlinear magneto-optical imaging in these materials.

3.7.1 Demonstration of even / odd tensor components

In sec. 3.5, we have shown that the presence of ***M*** leads to a magnetization-dependent nonlinear optical susceptibility tensor, $\chi^{(2)}$, with elements being either purely symmetric or antisymmetric with respect to ***M***. That is, upon reversal of the direction of M the $\chi^{(2)}$ components either remain unchanged or they undergo a π phase shift, respectively. The consequences of this phase shift are observed in the already mentioned nonlinear magneto-optical effects, which will be discussed further on. Here the existence of purely symmetric and antisymmetric susceptibility tensor elements will be demonstrated directly. This direct manifestation of the effects of time reversal symmetry breaking can be observed by measuring both the phase and the intensity changes of the second harmonic radiation in selected polarization combinations upon sign reversal of the magnetization M.

In the case of an 'isotropic' interface with $C_{\infty,v}$ symmetry, defined by the x, y plane with x being in the plane of incidence and with a magnetization parallel to x ('longitudinal configuration', see Fig. 3.7), the nonlinear susceptibility tensor is given by (see Table 3.1):

$$\chi_{ijk}^{(2)}(M) = \begin{pmatrix} 0 & 0 & 0 & 0 & \chi_{xzx}^{(\text{even})} & \chi_{xyx}^{(\text{odd})} \\ \chi_{yxx}^{(\text{odd})} & \chi_{yyy}^{(\text{odd})} & \chi_{yzz}^{(\text{odd})} & \chi_{yzy}^{(\text{even})} & 0 & 0 \\ \chi_{zxx}^{(\text{even})} & \chi_{zyy}^{(\text{even})} & \chi_{zzz}^{(\text{even})} & \chi_{zyz}^{(\text{odd})} & 0 & 0 \end{pmatrix}. \quad (3.7.1)$$

This magnetization-dependent susceptibility tensor consists of components that are either even (symmetric), $\chi_{ijk}^{(\text{even})}(M)$, or odd (antisymmetric), $\chi_{ijk}^{(\text{odd})}(M)$, in magnetization, i.e. upon sign reversal of the magnetization the odd components change sign whereas the even components remain unchanged.

In this longitudinal configuration one can find polarization combinations of the incoming fundamental beam and the generated second harmonic beam with the underlying tensor components being either purely even or purely odd (see Table 3.3). In the even configurations, s_{in}–p_{out} and p_{in}–p_{out}, both the phase and the magnitude of the generated second harmonic radiation remain unchanged upon sign reversal of the magnetization, whereas in the odd combinations, s_{in}–s_{out} and p_{in}–s_{out}, the SH radiation undergoes a π phase shift. In most experiments this phase information is lost, since only intensities are measured. Therefore, MSHG effects can only be observed in a more indirect way through

Table 3.3 Comparison of theory and experiment: phase difference $\Delta\varphi$ and relative magnetic effect ρ in the MSHG signal upon sign reversal of magnetization in selected polarization combinations

Polarization	$\chi^{(2)}$ components	Symmetry	$\Delta\varphi$ (deg) Exp.	$\Delta\varphi$ (deg) Theor.	ρ (%) Exp.	ρ (%) Theor.
s_{in}–p_{out}	$\chi_{zyy}^{(\text{even})}$	even	0 ± 5	0	≤ 1	0
s_{in}–s_{out}	$\chi_{yyy}^{(\text{odd})}$	odd	179 ± 5	180	≤ 1	0
p_{in}–p_{out}	$\chi_{zxx}^{(\text{even})}, \chi_{xzx}^{(\text{even})}, \chi_{zzz}^{(\text{even})}$	even	2 ± 5	0	≤ 1	0
p_{in}–s_{out}	$\chi_{yxx}^{(\text{odd})}, \chi_{yzz}^{(\text{odd})}$	odd	185 ± 5	180	≤ 1	0
s_{in}–$45°_{out}$	$\chi_{yyy}^{(\text{odd})}, \chi_{zyy}^{(\text{even})}$	mixed	30 ± 5		≈ 50	

interference effects between odd and even terms, i.e. in a polarization combination corresponding to a sum of even and odd tensor components, as in the case of s_{in}–$45°_{out}$. However, one can directly measure phase shifts of the SH signal by making use of an external SHG source with an adjustable relative phase.[61–63]

To demonstrate this phase shift, we studied MSHG in the reflection geometry on a magnetic multilayer system, consisting of a Co film (thickness 50 nm), covered with a 1.5 nm Rh film deposited by d.c. magnetron and RF diode sputtering, respectively (for further details on the sample, see ref. 64).

For the SHG experiments we used our standard Ti:sapphire laser setup, now including the components for the phase measurements (see Fig. 3.7).

The SH signal of the sample is superimposed by that of the reference SHG source, and interference is observed by measuring the total SH signal upon variation of the optical phase delay φ between these two SH sources. By translating the reference source along the path of the beam by d, we introduce an extra optical phase delay $\delta\varphi = (4\pi \Delta n/\lambda)d$, resulting in a cosine-like interferogram of the total SH signal:

$$I_{2\omega,tot}(d) = I_{2\omega,s} + I_{2\omega,r} + 2\sqrt{I_{2\omega,s}I_{2\omega,r}} \cos(\delta\varphi + \varphi), \quad (3.7.2)$$

where $I_{2\omega,s}$ and $I_{2\omega,r}$ indicate the SH signal generated by the sample and the reference, respectively. Δn is the dispersion of the ambient air and λ is the fundamental wavelength. Since the fundamental beam was focused on the sample, the interference signal depends on the position d with respect to the focus position d_0 as $I_{2\omega,r}(d) \propto 1/(d-d_0)^2$, and therefore the interference pattern does not have a perfect cosine form,[62,65] but is given by the expression

$$I_{2\omega,tot}(d) = I_{2\omega,s} + I_{2\omega,r}(d) + 2\alpha\sqrt{I_{2\omega,s}I_{2\omega,r}} \cos[(4\pi \Delta n/\lambda)d + \varphi], \quad (3.7.3)$$

with the coherence parameter α, describing the spatial and temporal coherence of the laser,[66] and the phase φ, being the sum of the phase shifts introduced by the sample and the reference. For a fundamental radiation of 840 nm, the dispersion of air at room temperature and normal atmospheric conditions is[67] $\Delta n = n(840 \text{ nm}) - n(420 \text{ nm}) \approx 6.5 \times 10^{-6}$, resulting in a periodicity of the interferogram of approximately 65 mm.

The interference patterns obtained in the odd p_{in}–s_{out} polarization combination upon sign reversal of the magnetic field are shown in Fig. 3.8(a). The data can be perfectly fitted by eqn (3.7.3), resulting in a phase difference $\Delta\varphi = \varphi(M^+) - \varphi(M^-) = (185 \pm 5)°$. This value as well as the obtained relative magnetic effect in the SH intensity of $\rho \leq 1\%$ is in excellent agreement with the theory, which predicts a π shift of the harmonic radiation. Figure 3.8(b) depicts the results of a phase measurements in the even s_{in}–p_{out} combination. The phase difference was $\Delta\varphi = (0 \pm 5)°$, and the relative magnetic effect was $\rho \leq 1\%$, both again in good agreement with the phenomenological model. Table 3.3 shows the experimental results for all selected polarization combinations.[64]

Though the present analysis was done for the electric dipolar response terms $\chi_{ijk}^{(2)}$, it can be simply shown that this odd/even division is a general rule,

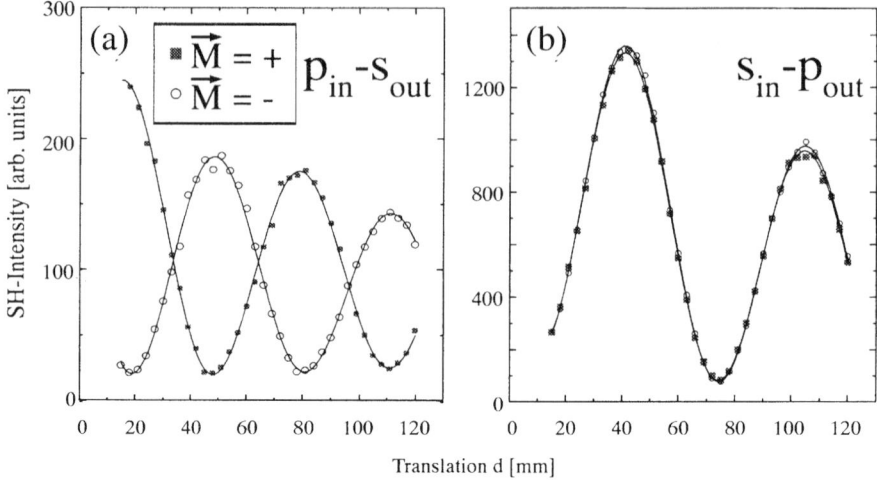

Fig. 3.8. (a) Interference patterns for the SHG for opposite magnetizations of the sample in the odd p_{in}-s_{out} configuration and best fits to eqn (3.7.3). The phase difference was found to be $\Delta\varphi = (185 \pm 5)°$. (b) The same for the even s_{in}-p_{out} configuration. The phase difference was found to be $\Delta\varphi = (0 \pm 5)°$.

irrespective of the mechanism for MSHG (see sec. 3.5) as long as the surface contains one mirror plane. That means it also holds for the electric quadrupolar or magnetic dipolar sources. Moreover, any possible linear magneto-optical effects (Kerr rotation) on the incident and outgoing waves do not affect the phase shifts considered here. The latter results in a mixture of s and p polarization of the fundamental wave. For example, the s-polarized incoming wave excites the y component of E in the magnetized medium, which is even in the longitudinal (M_x) and polar (M_z) magnetization, and x and z components, which are odd in $M_\parallel$ (the component in the mirror plane). For a p-polarized input beam one finds that $E_y(\omega)$ is odd in $M_\parallel$ while $E_x(\omega)$ and $E_z(\omega)$ are even. That means that the contribution of $P_y(2\omega)$ to the s-polarized MSHG is even in $M_\parallel$ while its contribution to the p component is odd, and so on. Analysing the MSHG response including linear optical effects we again find that the s MSHG is purely odd with respect to the inversion of $M_\parallel$, if the fundamental beam is purely p- or s-polarized, whereas the p component of the MSHG response is purely even in $M_\parallel$ under the same conditions.

3.7.2 MSHG as a probe of surface magnetism

Fe(110)

One of the most studied magnetic materials, Fe, has also been subject to several experimental and theoretical studies of nonlinear magneto-optics. The first experimental evidence of the influence of surface magnetism on second harmonic generation were reported on Fe(110) by Reif et al.[19]

The experiment was done on a clean Fe(110) surface in UHV, in the transverse geometry. For the excitation, the frequency-doubled output of a Nd:YAG laser at 532 nm was used, which provided 6 ns pulses at a repetition rate of 20 Hz. A clear difference in the SHG response for magnetization up and down could be observed that disappeared with time.

From an analysis of the data an initial magnetic effect of $\chi^{(2),-}/\chi^{(2),+} \approx 0.25$ was found. The decay of this ratio (with a time constant $\tau \approx 30$ min) was attributed to the disappearance of the surface magnetic moment due to the adsorption of background CO gas (the base pressure was $\approx 10^{-7}$ Pa).

Though these results also raised some questions (e.g. as to why the nonmagnetic contribution did not seem to change due to CO adsorption, in contrast to other observations[47]), they were very important as they demonstrated not only the magnetic effects in SHG, but also the potential of MSHG to probe (changes in) surface magnetism.

Recently, Vollmer *et al.* performed a very detailed study of Fe(110), using a mode-locked Ti:sapphire laser.[68] In the wavelength range of 740–840 nm they found that the surface sensitivity for MSHG depended strongly on the input polarization used for the experiment. For more details, see Chapter 2 by Vollmer in this book.

For the p-polarized input a sizeable bulk contribution was observed. Interestingly, our recent experiments with p-polarized excitation on an oxidized Fe(100) single-crystal surface in air also showed a strong effect of $\rho(pp) \approx 40\%$.[60] The latter thus may include bulk effects as well. For very clean surfaces, Vollmer *et al.* observed a nonlinear Kerr rotation close to 90°, which decreased strongly due to very small amounts of oxygen ($\leq 1\%$).[68] This strong decrease of $\Phi_K^{(2)}$ seems to be in contradiction with the observation of very large rotations, even for oxygen-covered surfaces (see sec. 8.1). Those results were obtained on a (100) surface and near-normal incidence. As Fe also forms magnetic oxides, this interesting system clearly requires more study (see for example, ref. 69).

Wierenga *et al.* studied the MSHG response for thin Fe films, epitaxially grown on a Cu(001) single-crystal substrate. Using the 800 nm output of the Ti:sapphire laser, a magnetic effect $\rho(pp) \approx 10\%$ was observed for a 20 ML bcc Fe film.[47] From a very crude analysis (ignoring phase factors), this would correspond to a ratio $\chi^{(2),-}/\chi^{(2),+} \approx 5\%$. Here one has to realize that this value described the effective result from two contributing interfaces (the Fe/vacuum and the Fe/Cu interface), which partly compensate each other. For a 2 ML fcc Fe film on Cu(001), it was found that $\rho(pp) = \rho(sp) = 0$. This is consistent with the prediction that one can make based on the tensor elements of Table 3.1. For a polar configuration, a p-polarized input only gives an s-polarized SHG, so $I_{pp}(2\omega) = \rho(pp) = 0$. In the case of an s-polarized output, only the odd (magnetic) component χ_{xyz} is excited. Apart from the fact that its absolute value was small, the absence of an even element prevents the possibility of nonlinear interference, and thus $\rho(sp) = 0$. Recently, Straub *et al.* have measured the nonlinear polar Kerr rotation of a 2 ML Fe film. At an angle of incidence of 38° they found a nonlinear polar Kerr rotation of $4° \pm 1°$.[70]

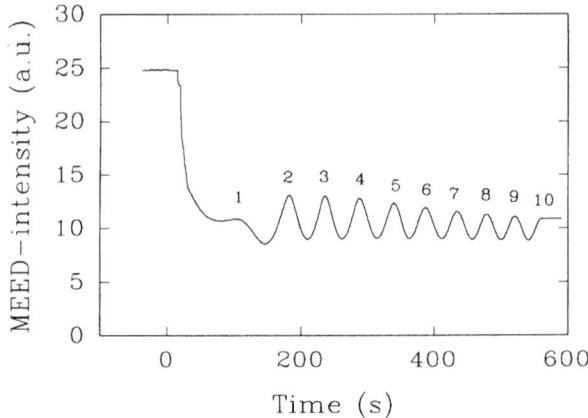

Fig. 3.9. This typical MEED pattern was taken while growing 10 ML of cobalt on Cu(001). A peak in the MEED intensity corresponds to a full monolayer.

Co / Cu(001)

Because the Co/Cu(001) system is one of the most thoroughly studied and well known examples of epitaxially grown thin magnetic films, it can serve as a perfect example to demonstrate the merits and limitations of a new technique such as MSHG. The studies on the Co/Cu(100) and Cu/Co/Cu(100) systems that will be described in the following sections were performed in a close collaboration of the Kirschner group in Halle and the Rasing group in Nijmegen. The actual experiments were done in Nijmegen, using the molecular beam epitaxy (MBE) growth chamber of the Halle group, which was shipped to Nijmegen for this purpose. MSHG in combination with MOKE was used to study thin Co films of thicknesses between 1 and 20 ML grown on a Cu(001) substrate.[21] The samples were prepared and studied in a UHV system with a base pressure of 5×10^{-11} torr. Details of the sample preparation can be found in ref. 21. Epitaxial growth was verified for every film by monitoring the (0,0) medium-energy electron diffraction (MEED) spot intensity while depositing (see Fig. 3.9). After preparation the film quality was checked by Auger electron spectroscopy (AES); all contaminations were below 1 at%, except carbon, which was typically 2–3%.

The easy axis of the film is parallel to the (110) direction. For the MSHG experiments the transverse Kerr geometry was chosen, i.e. the direction of the magnetization was perpendicular to the optical plane. The MOKE measurements were done in the longitudinal configuration. For the SHG experiments we used the output of the Ti:sapphire laser at 800 nm. The pulse fluence of the incoming beam was about 16 μJ cm^{-2}. At an angle of incidence of 35°, we have studied the p_{in}–p_{out} polarization combination (i.e. both fundamental and second harmonic beams are polarized in the plane of incidence).

Figure 3.10 shows the relative magnetic effect ρ and the MOKE amplitude

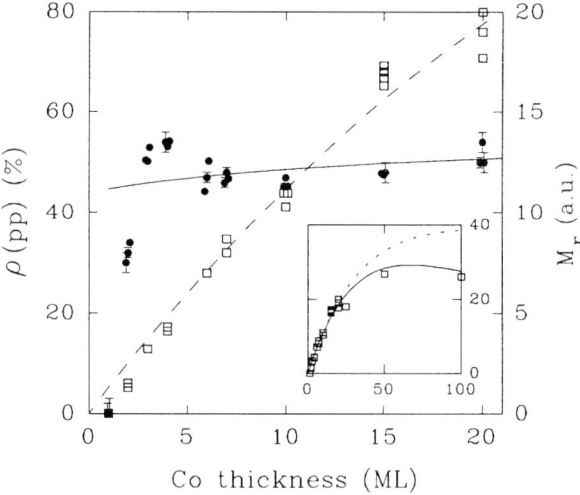

Fig. 3.10. Co film thickness dependence of the relative nonlinear magnetic effect $\rho(pp)$ and the MOKE amplitude M_r in Co/Cu(001): dots/solid line, experimental $\rho(pp)$ data and theoretical fit; open symbols/dashed line, MOKE data and fit. The inset shows the MOKE data and fits for thicker Co films: dashed line, Lambert–Beer law; solid line, multiple reflection model.

M_r as a function of the Co thickness for Co on Cu(001). The difference between the two sets of data is striking: M_r increases almost linearly with Co thickness, whereas ρ hardly changes after 3 ML of Co. The origin of this different behaviour lies in the probing depth of the two techniques. MOKE is a bulk probe, and the total Kerr rotation is nearly proportional to the amount of material, taking absorption losses into account. On the other hand MSHG is an interface-sensitive probe that is independent of the bulk film thickness. The deviation of the linear thickness dependence of M_r is a result of the absorption in the thin films. Usually this is taken into account by a Lambert–Beer type of analysis. A more accurate approach is to use a multiple reflection model, similar to Moog et al.[71] and Lissberger et al.[72] For thicknesses below 20 ML, both approaches give an equally good fit. Above 20 ML, the effects of multiple reflections start to play a noticeable role, as can be seen in the inset of Fig. 3.10. For the calculation, we used the bulk indices of refraction as listed in ref. 38. From the close agreement between experiment and calculation, we can conclude that the MOKE results can accurately be described by bulk refractive indices for Co thicknesses above 3 ML.

To analyse the MSHG results one should realize that there are two interfaces that contribute to the SH signal: the Co/Cu and the vacuum/Co interface. To determine the relative strength of the SH signal from these two interfaces we measured the SH signal from a Co film on Cu(001) as a function of carbon monoxide exposure, as gas adsorption is known to reduce strongly the SHG from metal surfaces.[9,13]

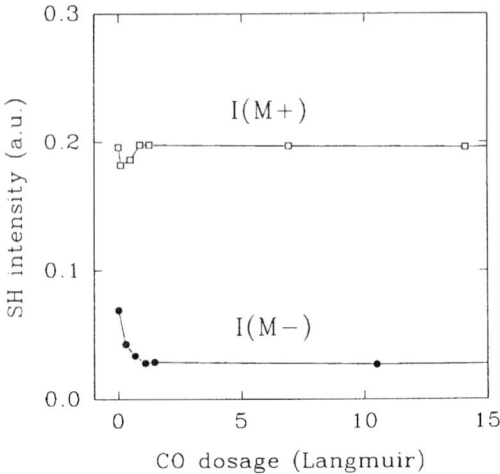

Fig. 3.11. The pp SH intensities of a 7 ML Co film on Cu(001) for positive and negative magnetic saturation as a function of CO dosage. The signals at 15 L CO remain constant until at least 40 L.

Figure 3.11 shows the pp SH intensities of a 7 ML Co film on Cu(001) for positive and negative magnetization as a function of CO dosage. We observe that the signals change until a dosage of 1 Langmuir (1 L = 10^{-6} torr s), whereafter they become constant until at least 40 L. The original value of $\rho \sim 45\%$ has increased to $\rho \sim 70\%$. Comparable effects have been observed on adsorbing O_2 and for different Co film thicknesses. We have observed that a dosage of a few Langmuirs of O_2 to magnetized Ni(110) and Fe(110) crystals reduces the pp SH signal, generated from a 532 nm Nd:YAG beam, by up to a factor of 20 depending on initial cleanliness. We therefore assume that the CO eliminates all SH contributions by the Co/vacuum interface.† The total response can then be calculated using the multiple reflection model of sec. 3.3 that includes the relevant nonlinear tensor elements and the boundary conditions for nonlinear sources at the interfaces.[73] The line in Fig. 3.10 shows the result of such a fit, which is seen to give a good description of the experimental results above 5 ML. For the analysis we only used one odd and one even term for each interface, like was done for the Co/Au example. This is of course a simplification, as Table 3.1 shows that for the pp configuration, there are three even and three odd terms. Including all these terms however gives too many fitting parameters, without yielding any more physical insight. Since for a fixed angle of incidence these various components can be scaled to each other, using only one even and one odd component still gives a realistic description.

For the Co/Cu interface we find the following ratio for these components:

$$\chi_{Co/Cu}^{(2),-}/\chi_{Co/Cu}^{(2),+} = 0.89.$$

† However, one should be careful with generalizing the effects of adsorption; see Chapter 2 by Vollmer.

We see that the magnetic contributions are of the same order of magnitude as the nonmagnetic ones. This accounts for the large relative effects ($\rho > 40\%$). Comparing the even and odd elements at the vacuum/Co and Co/Cu interfaces gives

$$\chi^{(2),-}_{\text{vac/Co}}/\chi^{(2),-}_{\text{Co/Cu}} = -1.4,$$

and

$$\chi^{(2),+}_{\text{vac/Co}}/\chi^{(2),+}_{\text{Co/Cu}} = -1.9.$$

So also for the vacuum/Co and Co/Cu interfaces, the amplitudes of both even and odd tensor elements are comparable. These ratios are found for any combination of one even and one odd tensor element. Recently, these experiments were repeated with a new system.[59] The results are in perfect agreement with those reported here, except for the fact that the background CO contamination was somewhat lower. This resulted in a larger Co/vacuum contribution and thus in a smaller total magnetic effect, in accordance with the analysis given here that shows that the two interface contributions are out of phase. The results described above show that above 5 ML both the MOKE and MSHG response can be accurately described by a multiple reflection model, using bulk indices of refraciton. The strong thickness dependence of the MSHG results below 5 ML cannot of course be described by such an approach, as it takes at least 2 ML to define an interface.

The microscopic origin of those results can be understood from electronic structure calculations, using a linear Korringa–Kohn–Rostoker (LKKR) method. With this method one calculates the magnetic moment per Co layer as a function of the Co layer thickness. Preliminary results indicate that the magnetic moments of such a Co film indeed change for the first few monolayers.[74] If we make the very bold assumption that the multiple reflection model works even between 1 and 5 ML and using the same bulk optical constants as before, we can then calculate the MSHG response from our Co/Cu(001) system, using the calculated magnetic moment distribution and assuming $\chi^{(2)}(M)$ to be proportional to the magnetic moment in the layers. The results of such a calculation are shown in Fig. 3.12. Despite the crude approximations, Fig. 3.12 shows a reasonable qualitative agreement with the experimental results plotted in Fig. 3.10, indicating that MSHG can indeed be used to probe the magnetic moments at magnetic interfaces, in accordance with the theoretical calculations of secs 3.3 and 3.5. On the other hand, the maximum of the theoretical ρ is at 2 ML, whereas the experiments show this to be 4 ML. This actually happens to coincide exactly with a recent observation in a MOKE study by Weber et al.[75] They found an oscillation in the magnetic anisotropy in ultrathin Co films grown on Cu(001), which were attributed to the periodic variations of the film morphology. Bloemen et al.[76] observed an oscillatory coupling between two Cu layers, separated by a Co wedge. The oscillations were attributed to the appearance of confined quantum-well states (QWS). Spin-dependent photoemission studies by Clemens et al. indeed support this interpretation.[77] Thus, an alternative explanation of the maximum at 4 ML would be the presence of QWS in the thin Co film. This

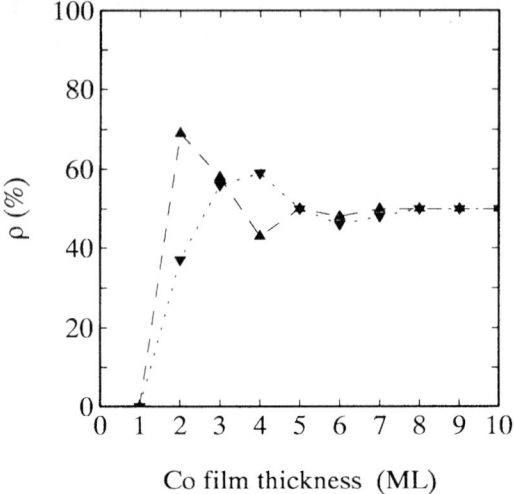

Fig. 3.12. Comparison between measured (▼) and calculated (▲) relative magnetic effect ρ for Co/Cu(001) as a function of Co thickness.

dilemma may be solved by doing spectroscopic MSHG. Some preliminary spectroscopic results for one Co thickness only are plotted in Fig. 3.13 for the tunability range of the Ti:sapphire laser.

Figure 3.13 shows monotonic decrease for $\rho(pp)$ between 2.9 and 3.5 eV, and a (smaller) increase for $\rho(sp)$. In a recent spectroscopic MOKE study, Suzuki *et*

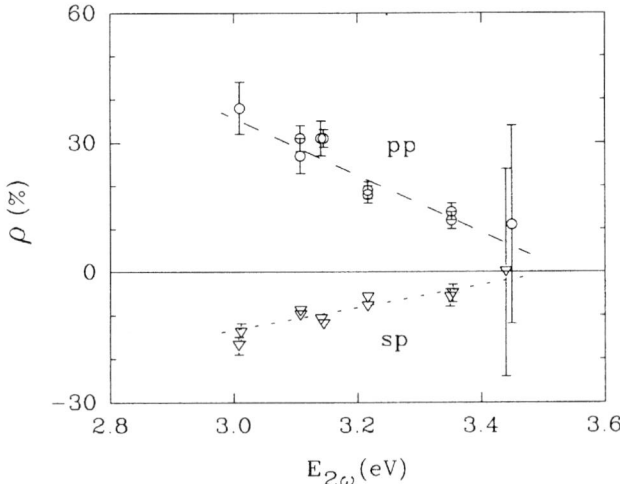

Fig. 3.13. The relative magnetic effect versus energy of the second harmonic photons in transverse MSHG on 10 ML Co/Cu(001). The dashed lines are guides to the eye.

al.[78] report a similar monotonic behaviour for the linear Kerr angle and ellipticity of fcc Co in this energy range. So a first conclusion would be that ρ changes because of the change of the odd tensor components in this frequency range. However, for a more quantitative analysis, more data and theory will be required.

3.7.3 MSHG studies of magnetic multilayers

Au/Co/Au

Some of the pilot experiments demonstrating the feasibility of MSHG were performed on Co/Au multilayers.[58] In contrast to the Cu/Co/Cu systems, which will be discussed later, Co/Au multilayers can be studied *ex situ* because of the protective inert Au layer. In that sense, they more resemble systems of practical interest. In the following we will discuss some of the earliest results on Co/Au multilayers, in which interface sensitivity was demonstrated by varying the number of interfaces.

The samples consisted of thin films of Co and Au, evaporated at a rate of about 2 Å s^{-1}, while the substrate was kept at room temperature (for details see ref. 58). Four systems were studied: sample A has one Co/Au interface, glass + 500 Å Co + 50 Å Au; sample B has two Co/Au interfaces, glass + 50 Å Au + 50 Å Co + 50 Å Au; sample C has three Co/Au interfaces, glass + 50 Å Co + 50 Å Au + 50 Å Co + 50 Å Au; the fourth sample is a 1500 Å thick Au film that served as a reference. Although the films are polycrystalline, they are isotropic on the scale of the laser beam diameter ($\sim 60 \text{ mm}^2$). For the experiments the 532 nm output of a frequency-doubled, Q-switched seeded Nd:YAG laser with 8 ns pulse width was used. The pulse intensity was kept below 7 mJ cm^{-2}. The samples were mounted between the poles of an electromagnet. The applied magnetic field was in the plane of the sample and was perpendicular to the plane of incidence (transverse geometry).

Figure 3.14 presents the results of experiments on the pure Au film and samples A, B and C at near-normal incidence and pp-like polarization. For each sample the second harmonic intensities for upward ($M\uparrow$) and downward ($M\downarrow$) saturation was measured. As is to be expected, the signal from the Au film is not dependent on the magnetization. The inversion of the $M\uparrow$ and $M\downarrow$ levels between samples A and B and between samples B and C is most striking. This is an extremely strong indication that the interfaces play a significant role.

Figure 3.15 shows a typical example of the samples under study. The complex coefficients $\hat{\alpha}_n$ that describe the local electromagnetic fields are calculated using

$$\hat{\alpha}_n = \frac{\hat{A}_n(\omega)^2}{\hat{S}_n(2\omega)}, \quad (3.7.4)$$

where $\hat{A}_n(\omega)$ is the ratio of amplitudes of the fundamental field at interface n

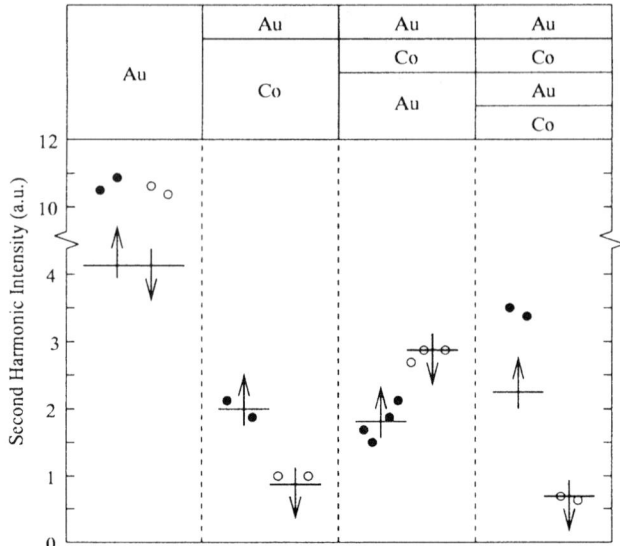

Fig. 3.14. The SH signals from the pure Au film and samples A (one Co/Au interface), B (two Co/Au interfaces) and C (three Co/Au interfaces). $M\uparrow$, film saturated upwards; $M\downarrow$, film saturated downwards.

and the incoming field outside the sample, and $\hat{S}_n(2\omega)$ is the ratio of the generated second harmonic field at interface n and the outgoing field. Both $\hat{A}_n(\omega)$ and $\hat{S}_n(2\omega)$ are calculated using the multiple reflection theory of sec. 3.

For the air/Au interface there ought to be no second harmonic generation for normal incidence and pp-like polarization, since χ_{xxx} is zero for an isotropic

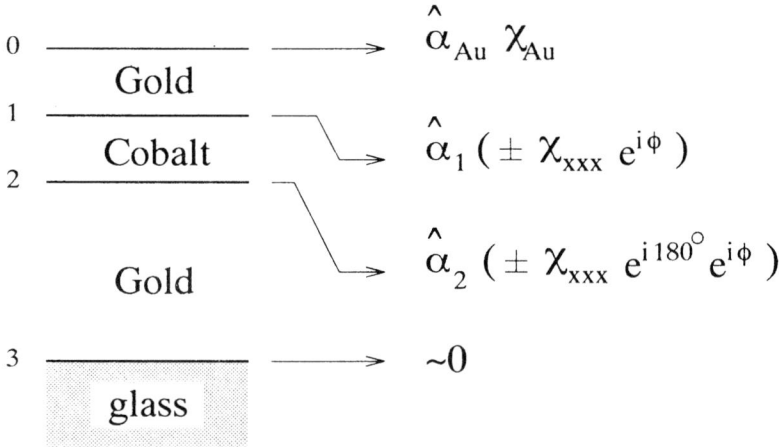

Fig. 3.15. A schematic example of the samples under study. The relative contributions of the interfaces are indicated.

nonmagnetic interface. However, the experiments on the pure Au film show significant SH signal in this configuration. This can be explained by the experimental limitations allowing a minimum angle of incidence of 4°. To first-order approximation we excite χ_{xzx}, χ_{xxz} and χ_{xzx}; however, the results will be analysed as though χ_{zxx} were the strongest tensor element. (Note that, for their calculations of the nonlinear Kerr rotation, Hübner and Bennemann assume that $\chi_{zxx} < \chi_{xzx}$.) Similar results can be obtained from any other combination.

Sample B contains two interfaces between Co and Au, which only differ by a mirror plane parallel to the interface. It was shown that the tensor elements of such interfaces have opposite phase. This relation reduces the number of free parameters in the systems to three: an even tensor element at the air/Au interface, $\chi^{(2)}_{zxx,\text{top}}$; an odd tensor element at each Au/Co interface, $\chi^{(2)}_{xxx,\text{bur}}(M)$; and ϕ, the phase difference between these two tensor elements. The values of these parameters are found by analysing the system with the model of sec. 3.3. The contributions from the Co/glass and the Au/glass interfaces are negligible for samples A and B, because of the thick metal layer between the air/Au and metal/glass interfaces. Furthermore we have measured the contributions of a comparable Co/quartz interface by exciting from the quartz side. From this it could be concluded that the contributions of the Co/glass interface in sample C are negligible.

The total second harmonic signal of the whole system is now described by

$$I(\pm M) = \left| \hat{\alpha}_{\text{Au}} \chi_{\text{Au}} \pm \sum_n (-)^{n-1} \hat{\alpha}_n \chi_{xxx}(M) e^{i\phi} \right|^2, \qquad (3.7.5)$$

where ϕ is the phase difference between χ_{Au} and $\chi_{xxx}(M)$. The '+' sign refers to $M\uparrow$, the '−' sign to $M\downarrow$.

The values of $\hat{\alpha}_{\text{Au}}$ and $\hat{\alpha}_t = \sum_n (-)^{n-1} \hat{\alpha}_n$ for all samples are listed in Table 3.4. The indices of refraction were obtained from Johnson and Christy.[79] We found that the calculated values for $\hat{\alpha}_{\text{Au}}$ and $\hat{\alpha}_t$ are not critically dependent on the exact thicknesses of the layers. Changing the size of both Co layers in sample C by as much as 10% caused variations of $\hat{\alpha}$ of only a few per cent.

The results of the calculations for samples A, B and C are indicated by the solid lines in Fig. 3.14. The values for the fitting parameters are: $\chi^{(2)}_{xxx,\text{bur}}(M)/\chi^{(2)}_{zxx,\text{top}} = 0.27$ and $\phi = 40°$ ($\chi^{(2)}_{zxx,\text{top}}$ is used as a scaling parameter).

Table 3.4 The $\hat{\alpha}$ values as calculated from multiple reflection theory. The phase of $\hat{\alpha}_{\text{Au}}$ was chosen to be zero

	$\hat{\alpha}_{\text{Au}}$	$\hat{\alpha}_t$
Pure Au	0.543	
Sample A	0.223	$0.142 e^{-i21°}$
Sample B	0.370	$0.112 e^{+i18°}$
Sample C	0.257	$0.123 e^{-i32°}$

The model clearly describes the inversion of $I(M\uparrow)$ and $I(M\downarrow)$ between samples A and B, and between samples B and C, and it does explain that sample C gives a larger magnetic field-induced SH effect than sample A. Although surprising and counterintuitive, as C involves deeper interfaces, it is purely the result of multiple reflections in these multilayer systems. The calculated values of the pure Au film and sample C are in better agreement with experiment than in the analysis of ref. 58.

In an earlier publication we found $\phi = 88°$, and we argued that this value was to be expected, because the time-inversion properties of $\chi(M)$ would predict a phase difference of $\phi = 90°$ (see sec. 3.5.1). However, this argument is only valid if dissipation is negligible. A brief look at the dielectric constants shows that this is probably not the case, as the energy of the second harmonic photons is near the interband transition of Au. This implies that the phase difference between odd and even elements is no longer *a priori* determined, and a deviation from 90° seems likely.

Obviously, our approach of using only one even and one odd nonlinear tensor component is a crude approximation. Hübner and Bennemann's results indicate that even contributions of relatively small tensor elements can be of major importance for the proper description of the nonlinear magnetic effect.[50] Nevertheless, the agreement with the experimental results is quite good. This is due to the fact that, for a fixed angle of incidence (as in the experiments described above), the contributions of the various $\chi^{(2)}$ components scale in the same way. The most important conclusion of these multilayer experiments is the fact that, for the analysis of such a multilayer system, a proper description of the linear and nonlinear electromagnetic response of the complete system, including multiple reflections, is an essential ingredient.

Cu / Co / Cu

An important application for the acclaimed *interface* specificity of MSHG will be the study of magnetic multilayers. For *in situ* multilayer studies, Cu/Co/Cu(001) structures will be used as our model system.[80] MSHG and MOKE experiments were done on a 10 ML Cu/Co/Cu(001) system, with varying Co thicknesses. The samples were prepared in the same way as before (see sec. 3.7.2). The Cu overlayers were grown at 1 ML/minute, while keeping the substrate at 70°C. For the MSHG and MOKE experiments we used our standard Ti:sapphire laser system.

Figure 3.16 shows the experimental Co thickness dependence of ρ for Cu/Co/Cu(001) plus the theoretical prediction based on the results of Fig. 3.10. For these calculations, the solutions for the Co/Cu interface were used directly in the Cu/Co/Cu(001) trilayers to calculate $\rho(pp)$ of these systems. The only fitting parameter was the nonmagnetic nonlinear contribution of the Cu/vacuum interface, which was not present before. The agreement between experimental and predicted dependence is quite good. More importantly, Fig. 3.16 indicates that such a multiple reflection model is *required* in order to understand the

EXPERIMENTAL CASE STUDIES 171

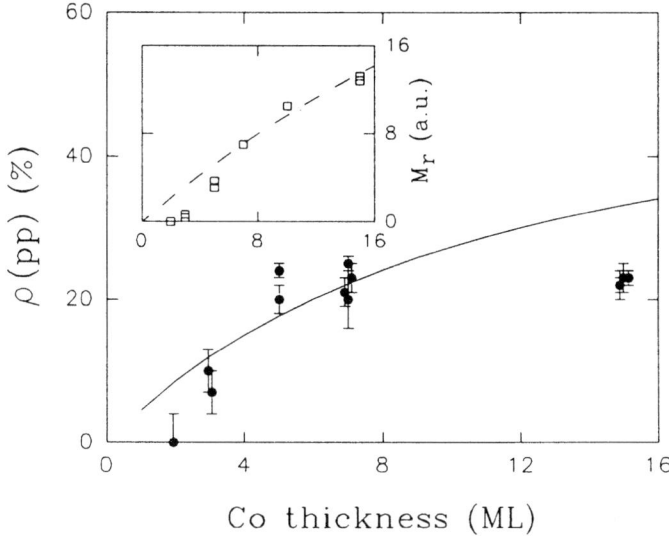

Fig. 3.16. Calculated and experimental Co film thickness dependence of $\rho(pp)$ in Cu/Co/Cu(001); dots, experimental data; solid line, calculation based on fitting parameters of Fig. 3.10. The inset shows the experimental and calculated MOKE amplitude.

MSHG response of such a multilayer at all. The Cu/Co/Cu trilayer is very symmetric and because of mirror symmetry $\chi^{(2)}_{Cu/Co} = -\chi^{(2)}_{Co/Cu}$. Thus, without taking the multiple reflections into account, which yield different local fields at the two interfaces, their total response and thus ρ would have been zero. (Here we ignore possible differences that may exist between the absolute values of the two interface contributions, due to differences in the local structure.) In Fig. 3.16 also the MOKE data for this system are plotted, with a theoretical fit including multiple reflections directly calculated from the Co/Cu results, using no adjustable parameters.

These results show that in principle we understand the MSHG response of a multilayer system, in which several interfaces contribute to the nonlinear optical signal. The next step will be to increase the number of layers up towards systems of practical interest, consisting of 30 to 50 layers. Another interesting option is whether it would be possible to 'tune' to a specific interface, using either wavelength or angle of incidence as controlling parameter. For the bulk response using MOKE, Penissard et al. have demonstrated this beautifully. By changing the wavelength, they could select a specific Co layer of a Co/Au multilayer sample.[81] In principle this must also be possible for the interface contributions, but so far this is an unexplored area.

3.7.4 Other MSHG results

In a number of recent papers, Crawford et al. have demonstrated the possibilities of MSHG for the study of the technologically important NiFe alloy. Thin

NiFe films on an Al_2O_3 substrate show strong MSHG effects in all the difference geometries (polar, longitudinal and transverse) that are three orders of magnitude stronger than their linear equivalents.[82,83] The thickness dependence could be modelled very well with the multiple reflection theory as discussed in sec. 3.3. Subsequently, MSHG was used to study the surface oxidation and annealing of NiFe films.[84] The MSHG response of Cu/NiFe multilayers and the effects of Co addition on the MSHG response were shown to correlate well with the observed GMR effects in these spin valve structures.[85] Wierenga et al. have observed a 10% transverse MSHG effect from the ferrimagnetic material $Mn_{0.6}Zn_{0.35}Fe_{2.05}O_4$.[69] These results show the possibilities of MSHG to study technologically interesting materials.

3.8 Nonlinear Kerr rotations

3.8.1 NOMOKE rotation from Fe

The very first experiments on the nonlinear Kerr rotation from thin, centrosymmetric magnetic films were stimulated by the prediction of large enhancements of $\Phi_K^{(2)}$ by Pustogowa and Hübner.[22] To enable direct comparison with their theoretical results, experiments were performed on Fe thin films and single crystals. All experiments were done *ex situ*, and therefore required protective cover layers. The first sample consisted of a thin Fe film (thickness 2 nm), covered with a 2 nm Cr film deposited by RF diode and d.c. magnetron sputtering, respectively (see ref. 25). As a substrate we used a (100) silicon wafer, with a thermal oxide layer of about 525 nm. For the second harmonic experiments we used the 770 nm output of the Ti:sapphire laser. The pulse width was 70 fs and the input power was 100 mW focused on a spot diameter of 100 μm. The experiments were done in the longitudinal configuration, i.e. the magnetization M was in the plane of the sample and in the optical plane of incidence (see inset of Fig. 3.17). Figure 3.17 shows the polarization dependence of the SH signal for an s-polarized input at an angle of incidence of 45° and for M along $\hat{x}$ and $-\hat{x}$ respectively. The difference between the two minima of the curves corresponds to $2\Phi_K^{(2)} \approx 34°$, i.e. a nonlinear Kerr rotation $\Phi_K^{(2)}$ of 17°.

In this geometry we measured a linear Kerr angle of 0.03°, using the same Ti:sapphire input beam. These results correspond to an enhancement of almost *three orders* of magnitude for the nonlinear Kerr rotation! The small linear rotation can be compared with the bulk Fe value of 0.1°.

These observations can be understood by an analysis based on the theoretical considerations of sec. 3.5 and including the multiple reflections as discussed in sec. 3.3.

The nonlinear Kerr angle can be expressed in the s and p components of the reflected SH field, denoted by $E_s(2\omega)$ and $E_p(2\omega)$, respectively. The relevant tensor components and their appearance for s- and p-polarized input

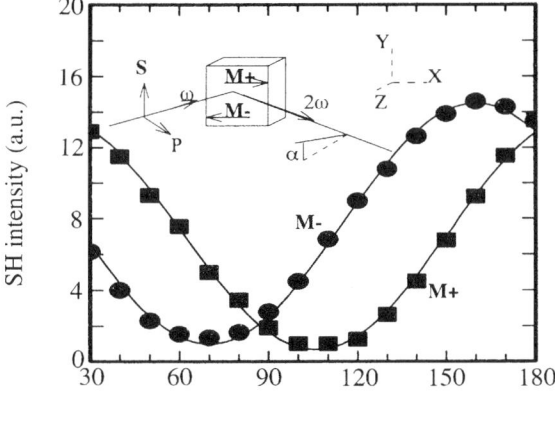

Fig. 3.17. Output polarization dependence of SHG reflection from an Fe/Cr interface, for s-polarized input: squares, $M \parallel \hat{x}$; filled dots, $M \parallel -\hat{x}$. The inset shows the experimental configuration.

have been repeated in Table 3.5. Defining $R \equiv \mathrm{Re}[E_s(2\omega)/E_p(2\omega)]$, $I \equiv \mathrm{Im}[E_s(2\omega)/E_p(2\omega)]$ and $A^2 = R^2 + I^2$, we get

$$\Phi_K^{(2)} = \tfrac{1}{2} \arctan[2R/(1-A^2)] + \phi_0, \qquad (3.8.1)$$

with $\phi_0 = 0$ for $A^2 \leq 1$, $\phi_0 = 90°$ for $A^2 > 1$ and $R \geq 0$, and $\phi_0 = -90°$ for $A^2 > 1$ and $R < 0$. Equation (3.8.1) is completely analogous to the earlier expression (3.5.27), as derived by Pustogowa et al.[22] It is easily verified that in the limit $A \ll 1$ eqn (3.8.1) reduces to $\Phi_K^{(2)} = R$. However, the nonlinear case generally is far from this limit, since $\Phi_K^{(2)}$ can become as large as $90°$.

Inspection of Table 3.5 shows that the s-input configuration is particularly simple, with only one even, χ_{zyy}^+, and one odd, χ_{yyy}^-, contributing element per interface. For normal incidence, only the χ_{yyy}^- contribution survives. From

Table 3.5 The nonzero elements of the SH susceptibility tensor for an isotropic surface in the longitudinal configuration ($M \parallel \hat{x}$). The two columns list the elements that are even and odd in the magnetization, respectively. The occurrence of the elements in the p- and s-input configuration is indicated within brackets

χ^+		χ^-	
zxx	(p)	yxx	(p)
zyy	(s)	yyy	(s)
zzz	(p)	yzz	(p)
xzx = xxz	(p)	zyz = zzy	
yzy = yyz		xyx = xxy	

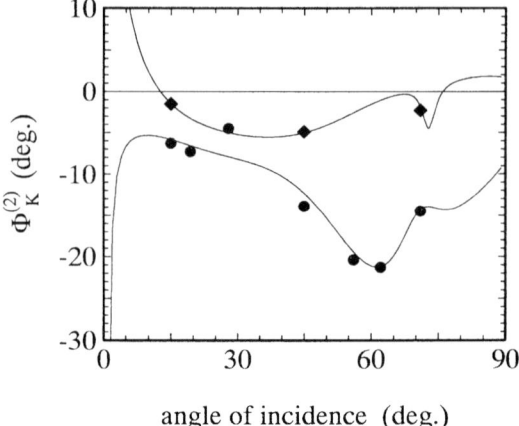

Fig. 3.18. Nonlinear Kerr rotation $\Phi_K^{(2)}$ for an Fe/Cr multilayer as a function of the angle of incidence: dots, s-input polarization; diamonds, p-input polarization. The curves are theoretical fits.

$E_p(2\omega) = 0$ one then finds $\Phi_K^{(2)} = \pm 90°$. Moving away from normal incidence, the ratio $|E_s(2\omega)/E_p(2\omega)|$, and as a consequence $|\Phi_K^{(2)}|$ decreases so that $\Phi_K^{(2)}$ is tunable over a wide range while scanning the angle of incidence.

A similar argument for a large $\Phi_K^{(2)}$ close to normal incidence is found for a p-polarized input configuration. Table 3.5 shows that, close to normal incidence, the dominating even tensor components have one z index, and thus vanish at normal incidence, whereas the dominant odd tensor element, χ_{yxx}^-, is finite at normal incidence.

Figure 3.18 shows the observed nonlinear Kerr rotation for both s- and p-polarized input, as a function of the angle of incidence. The solid curves in Fig. 3.18 are theoretical fits, based on a multiple reflection model, with the unknown interface tensor elements of Table 3.5 as parameters and the following assumptions. Only the Fe is expected to contribute to the magnetic (odd) nonlinear susceptibility, because the Cr film is antiferromagnetic. Furthermore, from experiments on Cr films we know that the SHG response from Cr is very weak, so that the major contribution to the nonmagnetic nonlinear susceptibility is also expected to originate from the Fe film. Therefore we assigned effective SHG susceptibilities to the Fe/Cr interface layer. We verified that the actual position within the Cr/Fe top layer did not significantly change the results of our fits. In addition to these SHG sources at the top layer, we incorporated a nonmagnetic SHG source at the Si/silicon oxide interface. The bulk optical constants of the metals and silicon were obtained from refs 79 and 86 respectively.

Because of the limited number of parameters involved in the s-polarization configuration (χ_{yyy}^- and χ_{zyy}^+ at the top layer, and χ_{zyy}^+ at the Si/silicon oxide interface), one finds a unique fit to these experimental data points. The fit in Fig. 3.18 includes a relative maximum of $|\Phi_K^{(2)}|$ near $\theta_i \approx 65°$ that is due to an enhancement effect through multiple reflections in the thick silicon oxide layer. Similar, but smaller, enhancement factors due to a substrate are also known for the linear Kerr angle.[87]

For the p-polarization case, several combinations of tensor elements give satisfying fits. Figure 3.18 gives one such solution, obtained by choosing fixed values for the relative phase factors and fitting the absolute values of the tensor components.

These results show once more that, for a proper understanding and analysis of the nonlinear optical response, the total electromagnetic response, linear and nonlinear including multiple reflections, has to be considered. The importance of such substrate effects can also be illustrated in the following way. From our thin-film analysis we can simply calculate the response of an Fe surface, by letting the thickness of the Fe film go to infinity. Similarly, we can vary the thickness of the silicon oxide layer for the thin-film structure.

Figure 3.19 shows the results of such simulations. For the thin oxide layer, the enhancement effect around $\theta_i = 60°$ has totally vanished, which demonstrates the importance of the multiple reflections and of the local field effects and thus the role of the substrate. Without going through a complete calculation, one can simply get an intuitive idea about this role. Let us compare two systems: (i) the surface of a bulk magnetic medium with refractive index $n = n_B$, and (ii) a thin film of the same material on a substrate with $n = n_S$ and neglecting the SHG contribution from the film/substrate interface. In the limits of vanishing film

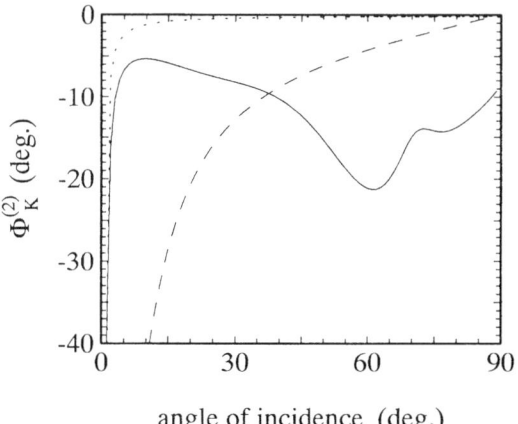

Fig. 3.19. Nonlinear Kerr rotation $\Phi_K^{(2)}$ for an Fe/Cr multilayer and a bulk Fe surface for s-polarized input as a function of the angle of incidence. The bold solid line is the theoretical fit of Fig. 3.18. The dashed line is a simulation for $\Phi_K^{(2)}$ of a clean Fe surface, based on the multilayer fit. The dotted line is a simulation for $\Phi_K^{(2)}$ for an Fe/Cr multilayer on a silicon wafer with a 10 nm oxide.

thickness and $\sin\theta \ll n$ and using the standard Fresnel reflection theory, one can easily show that for s-polarized input the ratio of the reflected SHG fields due to the odd (E_s) and even (E_p) terms is given by

$$\frac{E_s(2\omega)}{E_p(2\omega)} \propto \frac{1+R_s(2\omega)}{1+R_p(2\omega)} = \frac{1}{n(2\omega)}, \qquad (3.8.2)$$

with n equal to n_B or n_S. Similar relations hold for p-polarized input. Equation (3.8.2) indicates that the MSHG effects will be more pronounced for a thin film on a low-refractive-index substrate ($n_S \ll n_B$) than for a bulk material.

For the bulk surface, Fig. 3.19 predicts a smooth variation of $\Phi_K^{(2)}$ as a function of the angle of incidence. In accordance with eqn (3.8.1), the $\Phi_K^{(2)}$ can be tuned at will between 0° and 90°, by varying the angle of incidence. This is a direct result of the fact that, near normal incidence, the even contribution $\chi_{zyy}^{(2),+}$ vanishes, whereas the odd magnetic term $\chi_{yyy}^{(2),-}$ gives a finite contribution.

The prediction of this large tunable nonlinear Kerr rotation was confirmed by experiments on single-crystalline Fe whiskers.[88] The (100) surfaces of the whiskers were capped by MBE-grown Au and Cr layers, while some experiments were also performed on uncapped, oxidized Fe whiskers. The experiments were done using the 833 nm output of the Ti:sapphire laser. Figure 3.20 shows the measured angular dependence of the nonlinear Kerr rotation in the longitudinal configuration and s-polarized excitation. In this figure the experimental results can be seen for an Fe sample with a Cr top layer, an Fe sample with a Au top layer and for an uncapped oxidized Fe sample. We find a maximum Kerr angle of 80° at an angle of incidence of 6°. This corresponds with an enhancement of more than a factor 10^3, compared with the value for the linear Kerr rotation of 0.03° as obtained in the same experimental setup, using the fundamental incident beam.

The solid line in Fig. 3.20 is a theoretical fit for the Fe/Cr sample based on eqn (3.8.1) and a model for SHG from interfaces according to sec. 3.3. The experimental results are in quite good agreement with the predicted behaviour from Fig. 3.19. The inset in Fig. 3.20 shows the experimental results for the longitudinal configuration with p-polarized incident light. In Table 3.5 it can be seen that now we are dealing with two odd tensor components, both of which give rise to s-polarized SHG, and three even components, which produce p-polarized SHG. Therefore it is to be expected that for the same angles of incidence the ratio $|E_s(2\omega)/E_p(2\omega)|$ and thus $\Phi_K^{(2)}$ will be smaller compared to the s-input configuration. Near normal incidence the influence of the components χ_{ijk} with a 'z' for i, j or k will again be small. Since all even tensor components have at least one z index, the p-polarized SHG will then be small. The s-polarized SHG, which is now dominated by χ_{yxx}^{-}, is finite. The result is that for small angles of incidence the Kerr rotation will again increase. This is clearly confirmed by our experiments and in good agreement with the theoretical θ_i dependence of recent calculations of Hübner et al.[89]

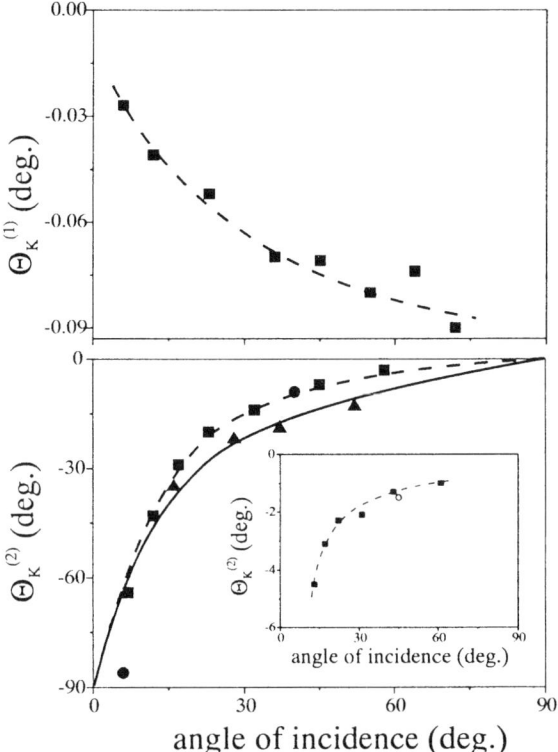

Fig. 3.20. Linear and nonlinear Kerr rotation for an Fe surface for s-polarized input in the longitudinal configuration as a function of the angle of incidence. Triangles, Fe/Cr; dots, Fe/Au; squares, uncapped Fe. The solid line is a theoretical fit for the Fe/Cr sample. The dashed line is the theoretical prediction for a clean Fe surface from ref. 25. The inset shows $\Phi_K^{(2)}$ for p-polarized input. The open circle is the calculation from ref. 22.

At an angle of incidence of 45° we find $\Phi_K^{(2)} = 1.2°$, in excellent agreement with the theoretical prediction of 1.4° from Pustogowa and Hübner.[22]

In the longitudinal configuration we also measured the wavelength dependence of both $\Phi_K^{(1)}$ and $\Phi_K^{(2)}$ at an angle of incidence of 45° (see Fig. 3.21). The average value for $\Phi_K^{(1)}$ is 0.062°, in excellent agreement with the bulk Fe value of 0.061° of ref. 90. Figure 3.21 shows that in the tuning range of the Ti:sapphire laser the Kerr rotations are constant, in agreement with the theoretical prediction of Pustogowa et al.[22] From their calculations it follows that for $\Phi_K^{(2)}$ a strong wavelength dependence is expected above 900 nm and below 600 nm. Future experiments in these wavelength ranges should be done.

3.8.2 NOMOKE rotation from Ni

Though the first theoretical calculations for magnetic $\chi^{(2)}$ components were

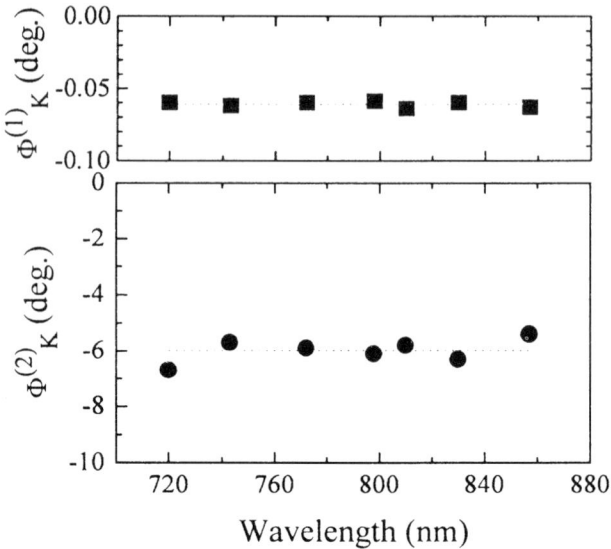

Fig. 3.21. Experimental wavelength dependence of the nonlinear (dots) and linear (squares) Kerr angle for the s-input longitudinal configuration at an incident angle of 45°.

done for Ni,[18] so far there has been little experimental MSHG work on this material. Recently, Böhmer et al. published the results of an MSHG study of polycrystalline nickel surfaces.[91] The experiments were done in air, using picosecond laser pulses at 600 nm for the excitation. In the longitudinal configuration, they observed a nonlinear Kerr rotation of 4°. In the transverse configuration they found anisotropies of 13% and 19%, for s → p and p → p polarization combinations respectively, by measuring the azimuthal magnetization dependence. From this azimuthal dependence, the authors also concluded that the (antiferromagnetic) nickel oxide layer does not affect the results. From a comparison of their measured MOKE and MSHG hysteresis loops, there may be an indication of a slightly different behaviour for the two cases. Though the accuracy was not good enough to draw definite conclusions from these results, this is an interesting point for future studies. Especially for the surface of a bulk crystal, the surface domain structures may be different from that of the bulk, due to differences in required magnetic energy to order these domains. Finally it should be mentioned that preliminary MSHG studies on a clean Ni surface by Wierenga et al. were unsuccessful.[47] In contrast to the results discussed above, these experiments were performed using the frequency-doubled output of a Nd:YAG laser at 532 nm as the fundamental excitation. Comparing the two wavelengths used (600 and 532 nm respectively) with the theoretical work of Hübner et al.,[18] one finds that the magnetic component $\chi^{(2)}_{xzz}$ shows a maximum near 600 nm, and a zero near 532 nm. As this may be accidental, a more complete spectroscopic MSHG study is clearly desirable.

3.9 MSHG studies of quantum-well states

3.9.1 Quantum-well states in thin metallic films

One of the exciting recent discoveries in magnetic multilayer systems is the observation of the oscillatory exchange coupling between two ferromagnetic films as a function of a nonmagnetic spacer layer. The subsequent discovery of the existence of quantum-well states in such very thin metallic films, which can be responsible for this observed coupling, has only increased the interest in these systems.

In contrast to semiconductor multilayers, where quantum wells are the result of differences in the bandgaps of alternating materials, for metallic thin films, quantum wells arise as a result of the differences in the density of states (DOS) between neighbouring layers. In a simple picture, these differences in the density of states give rise to a potential step at the interfaces. Electrons impinging on such an interface will be partly reflected, and depending on the thickness of the film, a standing electronic wave can be obtained, i.e. the system acts as a kind of Fabry–Perot interferometer for electrons.[92] This is in principle true for any combination of metallic films, or for a metallic film on a nonmetallic substrate.[93] In the case of a magnetic/nonmagnetic interface, the differences in the DOS, and thus the potential step, will be spin-dependent, due to the spin dependence of the band structure of the magnetic film. This will lead to a spin-dependent scattering at the interfaces, and thus to a spin-dependent quantization condition. As a result, even in a nonmagnetic spacer layer, the quantum-well states may become spin polarized.

Consider the situation of a thin paramagnetic film (P) of thickness d, sandwiched between two ferromagnetic (F) layers (see Fig. 3.22). A conduction electron propagating inside the P layer will be partly reflected at the P/F interface due to the differences in the electronic potentials. Because of the spin polarization of the F material, this potential difference and thus the reflection coefficient will be spin-dependent: $r = r^{\downarrow(\uparrow)}$. For an incident electron of wavevector $\mathbf{k}^i = (\mathbf{k}_\|, \mathbf{k}_\perp)$, the reflected wavevector is $\mathbf{k}^r = (\mathbf{k}_\|, -\mathbf{k}_\perp)$. An electron wave that has made a round trip in the spacer, i.e. that is reflected at interfaces 1 and 2, will experience a phase shift

$$\Delta\phi = 2k_\perp d + \phi_1 + \phi_2, \tag{3.9.1}$$

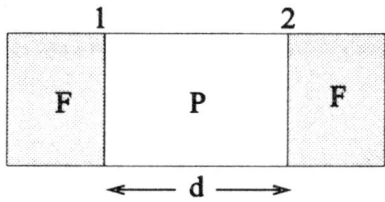

Fig. 3.22. Schematic picture of two ferromagnetic (F) films separated by a paramagnetic (P) spacer layer of thickness d.

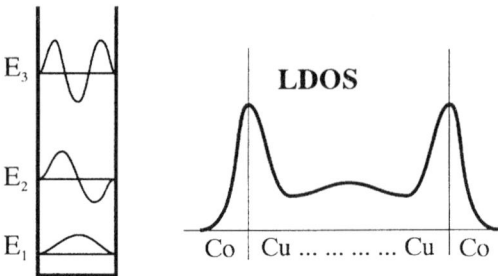

Fig. 3.23. Schematic picture of quantum-well states in a confined potential (left) and local density of states for a Co/10 ML Cu/Co quantum-well system, showing strong localization at the interfaces (right). Adapted from ref. 95.

where ϕ_i indicates the phase shift upon reflection at interface i. Constructive interference then leads to the condition for standing electron waves

$$\Delta\phi = 2n\pi, \qquad \text{integer } n. \tag{3.9.2}$$

This gives for the electronic wavefunctions inside the P spacer

$$\Psi(r) \sim e^{i(k_\parallel \cdot r_\parallel + k_\perp z)}(1 + |r_1 r_2| e^{i(2k_\perp d + \phi_1 + \phi_2)}). \tag{3.9.3}$$

The corresponding ($k_\parallel$ projected) density of states is

$$D(k_\parallel, E) \sim \frac{dk_\perp}{dE} \int_0^d dz \, |\Psi|^2, \tag{3.9.4}$$

with the interference part

$$\Delta D \sim 2d \frac{dk_\perp}{dE} |r_1 r_2| \cos(2k_\perp d + \phi_1 + \phi_2)$$

$$\sim \mathrm{Im}\left\{ i2d \frac{dk_\perp}{dE} r_q r_2 \, e^{i2k_\perp d} \right\}. \tag{3.9.5}$$

The simple picture (Fig. 3.23) shows that, as a function of the interlayer thickness, peaks appear in the density of states: the quantum-well states. These states have been observed by photoemission experiments.[94]

The same states are expected also to be responsible for the observed oscillatory coupling. This can be seen if the spin dependence of the reflection coefficient is taken into account. Summing over the spin and $k_\parallel$, one obtains for the ferromagnetic and antiferromagnetic configurations

$$\Delta D_\mathrm{F}(E) \sim \mathrm{Im} \int dk_\parallel \, 2id \frac{dk_\perp}{dE} (r_1^\uparrow r_2^\uparrow + r_1^\downarrow r_2^\downarrow) e^{izk_\perp d}, \tag{3.9.6}$$

$$\Delta D_\mathrm{AF}(E) \sim \mathrm{Im} \int dk_\parallel \, 2id \frac{dk_\perp}{dE} (r_1^\uparrow r_2^\downarrow + r_1^\downarrow r_2^\uparrow) e^{izk_\perp d}. \tag{3.9.7}$$

The interlayer exchange coupling is proportional to the difference in energy between the two configurations:

$$E_F - E_{AF} = \int_{-\infty}^{E_F} dE\, (E - E_F)\{\Delta_F(E) - \Delta D_{AF}(E)\}$$

$$\sim -\mathrm{Im} \int dk_\parallel \int_{-\infty}^{E_F} dE\, 4\Delta r_1 \Delta r_2\, e^{i2k_\perp d}. \qquad (3.9.8)$$

This shows how the period of the coupling depends on the wavevector $k_\perp$ of the spacer, as well as on the differences in the reflection coefficient at the interfaces for spin up and down. The oscillatory density of states is expected also to be responsible for the observed oscillatory behaviour of the nonlinear magneto-optical response because of the strong interface amplitude of the local density of states.[95]

3.9.2 Cu / Co / Cu(001)

A good candidate to observe these QWS is the Co/Cu/Co system. The DOS for Cu is very similar to the minority-spin DOS of Co. However, the majority DOS of Co is very different (Fig. 3.24). Therefore, the minority spin will hardly 'see' the interface whereas the majority spins will be more strongly scattered, resulting in a spin-dependent confinement in the Cu layer. As such confinement will lead to an oscillating density of states, optical transitions and in particular the MSHG response should also be affected by the presence of such

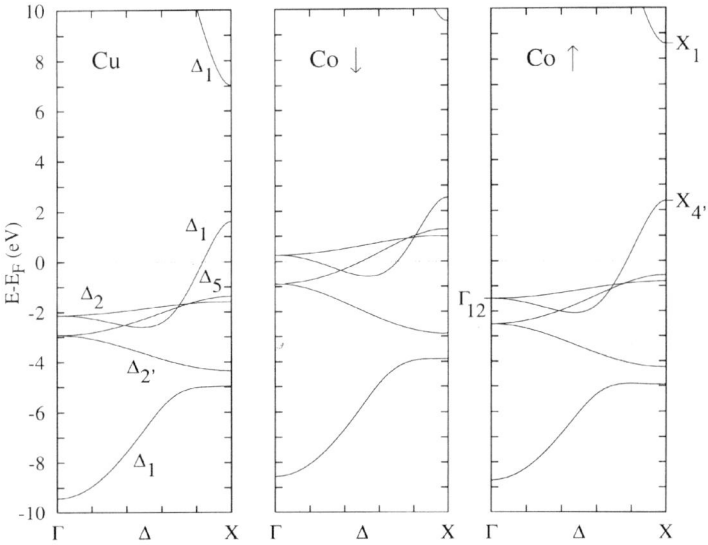

Fig. 3.24. Band structure along the $\Gamma-\Delta-X$ line of Cu and fcc Co; the zero of energy is taken at the Fermi level. From ref. 96.

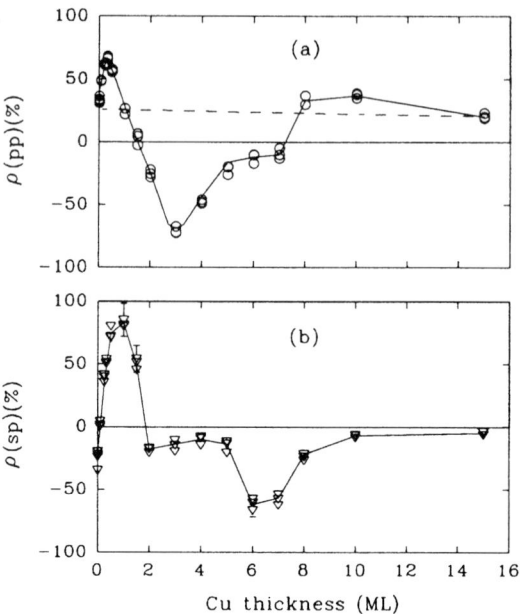

Fig. 3.25. The relative magnetic effects ρ(pp) and ρ(sp) as a function of Cu coverage on a 10 ML Co film on Cu(001): (a) circles, ρ(pp); dashed line, result of model calculation including optical interference but neglecting quantum-well oscillations; (b) triangles, ρ(sp). The solid lines are guides to the eye.

quantum-well states. To investigate this hypothesis further, we have studied the Cu/Co(001) system, where quantum-well states have been clearly identified.[94,97,98] For the experiments we used 10 ML Co grown on Cu(001) as a substrate.

The SHG intensity as a function of the Cu overlayer thickness showed an oscillatory-like behaviour with a period of about 10 ML. This effect becomes even more dramatic if we study the nonlinear magneto-optical response. In the present setup the MSHG experiments were done in the transverse geometry. Therefore, no polarization rotation is expected, but instead the intensity should vary with changing direction of M.

Figure 3.25 shows the MSHG results for this Cu/Co/Cu system, as a function of Cu coverage. We observe strong oscillatory-like variations in the relative magnetic effect ρ(pp) with amplitudes up to $\rho = 75\%$. From the simultaneously measured MOKE amplitude we conclude that, above 2 ML of Cu, the MOKE results are consistent with a normal multiple reflection analysis, using bulk refractive indices and a *constant* Kerr rotation. In the MSHG data of Fig. 3.25, two interfering periods may be distinguished: one of about 5 ML modulated by a substructure of 2–3 ML. Interestingly, the 5 ML periodicity has been observed by photoemission,[94,97,98] whereas such a short-period oscillation has been observed by Johnson *et al.*[99] in MOKE experiments on the Co/Cu/Co system.

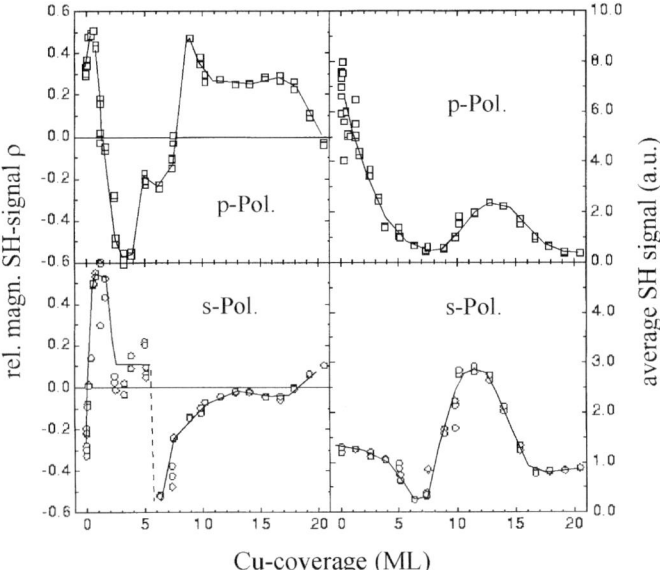

Fig. 3.26. Relative magnetic effect $\rho(sp)$ and $\rho(pp)$ and total SHG response (right) as a function of Cu coverage on a 10 ML Co film on Cu(001).

The extreme sensitivity of MSHG for these oscillations may be understood from its surface and interface specificity, in combination with results of recent calculations that indicate that for very thin films these quantum-well states primarily affect the density of states at the interfaces [100].

Figure 3.26 shows the relative and total SHG response for the same system (but from a different experimental run in Halle).[59] It is interesting to see the similarities and differences in the observed periods in Figs 3.25 and 3.26. The transverse magnetic signal ρ shows almost the same short-period oscillations, whereas the total SHG intensity oscillates with a period of about 10 ML. This apparent discrepancy might be due to some coincidental cancellation effects in the total response. On the other hand, the fast oscillations might be artificially induced by sign changes in the nonmagnetic part of the response.[101] Unfortunately, the complexity of the present experimental configuration does not allow a rigorous analysis of these data. The experiments were done for the p_{in}-p_{out} polarization combination. That means that for each (magnetic) interface, there are three even and three odd nonzero tensor components. Including the various phase factors yields far too many variables to make a meaningful fit. However, here we can at least use a simplified approach as was done for the Co/Cu(001) case, to prove that the observed effects cannot be due to changes in the magnetic moments.

Assuming a bulk-like electronic structure of the Cu films, the Cu thickness dependence of $\rho(pp)$ can be calculated from our multiple reflection model using the values of the various tensor elements as derived from our experiments. The dashed line in Fig. 3.25(a) shows the result of such a calculation, clearly indicating that classical interference by no means explains the observed strong oscillations. The calculations of the local magnetic moments (as done for Co/Cu(001)) showed a similar negative result: the induced moments in Cu and their variation with thickness are absolutely too small to be able to explain even qualitatively the observed oscillations.[74] The conclusion must be that the observed oscillatory response is due to the effect that the quantum-well states have on the nonmagnetic $\chi_{Cu}^{(2)}$ response. Considering that only a small fraction of the electronic states are really confined, the observed size of the MSHG effects due to these QWS is huge.

A more detailed discussion about the possible origin of the various observed periods can be found in the next section.

3.9.3 Au / Co / Au(111)

Recent MOKE experiments on Au(111)/Co(0001) systems showed an oscillatory dependence of the polar Kerr angle as a function of Au overlayer thickness.[102] The observed period of about 7 ML could however not be related to any spanning vector of the constructed Fermi surface. In addition, there appeared to be a slight wavelength dependence of the observed period. However, these MOKE studies are complicated by the fact that the QWS-induced oscillations in the Kerr angle are a very small fraction of the already low MOKE signal (observed oscillation amplitude in ref. 102 was about 0.001°, on a Kerr angle of 0.01). In this part we will discuss the unambiguous observation of QWS in Au(111) overlayers on Co(0001) by MSHG. The oscillations are found both in the SHG intensity and in the nonlinear magneto-optical Kerr effects measured as a function of the gold overlayer thickness. For magnetic measurements, we used both polar and transverse configurations, giving rise to polarization rotations and MSHG intensity changes, respectively. The spectral dependences of the observed signals show that interband transitions in the overlayer metal are responsible for the thickness-independent part of the second harmonic intensity, which is therefore reduced at longer wavelengths. In contrast, the oscillatory part hardly depends on the photon energy in the low-energy part of the spectrum.

Samples and experimental setup

The samples were step-shaped wedges of Au(111) grown on top of a thin (5–20 monolayers (ML)) Co(0001) film on a thick Au buffer layer on a float glass substrate. The cobalt films have also been grown as steps, with a few different thicknesses (see Fig. 3.27). Because of a strong interface-induced perpendicular magnetic anisotropy in this system,[103] we had a possibility to use either polar or transverse magneto-optical configurations depending on the cobalt thickness.

For the MSHG measurements, the output from the Ti:sapphire laser (100 fs pulse width, repetition rate of 82 MHz) was focused onto the sample, which

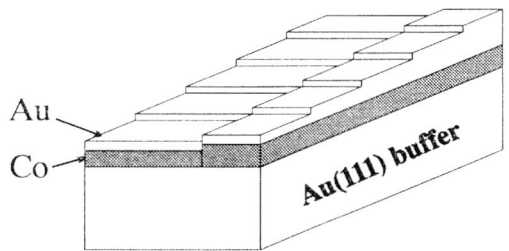

Fig. 3.27. Au(111)/Co double wedge sample structure.

could be moved with the help of a stepping motor in a magnetic field that was either in-plane or perpendicular to the sample. After proper filtering, the outcoming specular second harmonic (SH) light was detected with a photomultiplier. In the polar configuration, a Kerr rotation of the SH polarization was measured. In the transverse configuration for both p and s incoming light polarizations the SH output was always strictly p-polarized (i.e. in the plane of incidence). As a magnetic signal, we measured the normalized intensity difference for the magnetization up and down. We also measured the linear MOKE response using the fundamental beam.

Figure 3.28 shows the SH intensity measured for p- and s-polarization inputs at different wavelengths, displaying a strong oscillatory behaviour as a function of the overlayer thickness. Evidently, these oscillations are a dominating feature of the curves, especially at longer wavelengths. A similar oscillatory behaviour can be found for the MSHG response, both for the polarization rotation in the polar configuration and for the magnetization contrast in the transverse one (Fig. 3.29). All observed thickness dependences could be fitted with the formula

$$I_{\text{SH}} = A \exp\left(-\frac{d_{\text{Au}}}{\delta}\right) \sin\left(2\pi \frac{d_{\text{Au}}}{\Lambda} + \gamma\right) + B. \quad (3.9.9)$$

From Fig. 3.28 it can be seen that the phase of the oscillations (γ) is wavelength-dependent (p polarization) while the period Λ remains approximately constant. Figure 3.30 shows the period Λ for p and s incoming polarizations as a function of the second harmonic photon energy.

The observed period of about 14 ML is approximately twice the value obtained with the reported linear MOKE value of 7.7 ML.[102] This strong discrepancy between our results might be assigned to the different wavelength region used for the MOKE experiments (540–630 nm). To fill the gap, MOKE measurements were done using the same Ti:sapphire laser at $\lambda = 850$ nm. This resulted in a period of 7 ± 1 ML. That is, in the case of $\lambda_{\text{MOKE}} = \lambda_{\text{SH}}$, the total excitation energy was equal for the two experiments. We also investigated the situation $2\lambda_{\text{MOKE}} = \lambda_{\text{SH}}$, to study possible effects of initial and intermediate or intermediate and final states. However, again, the observed MSHG period was always *twice* as large as the period detected with MOKE. Thus, the difference in

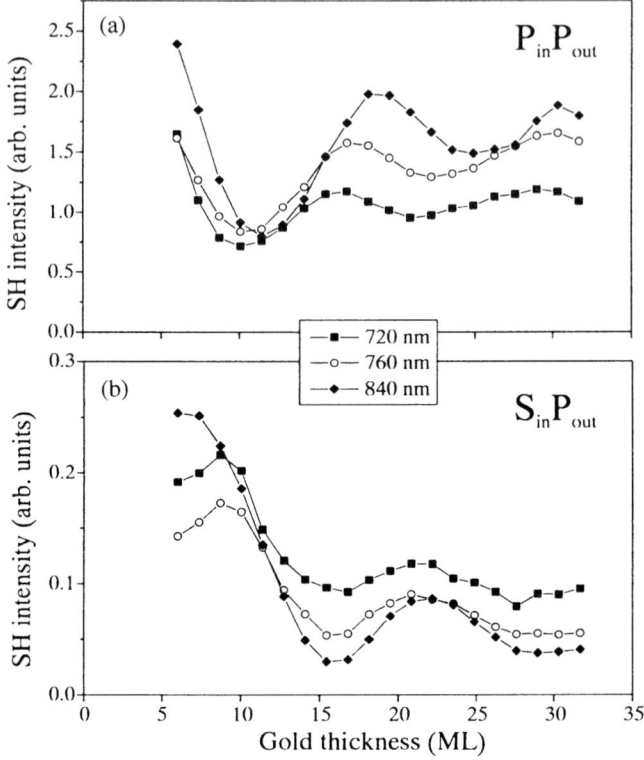

Fig. 3.28. (a) SH intensity (p_{in}–p_{out} polarization combination) as a function of the Au(111) overlayer thickness on a 20 ML thick Co film. (b) Same for s_{in}–p_{out}.

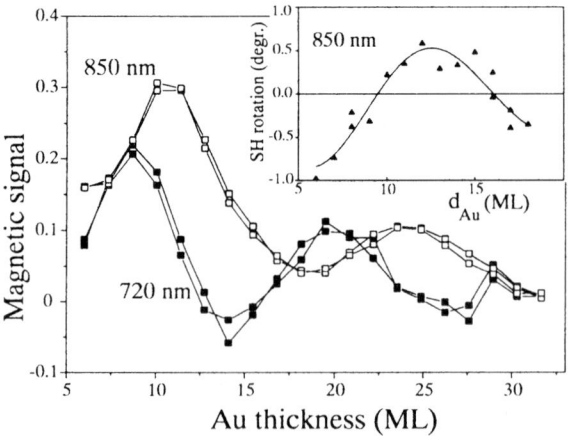

Fig. 3.29. Relative magnetic signal and SH polarization rotation (in inset) as a function of the gold thickness.

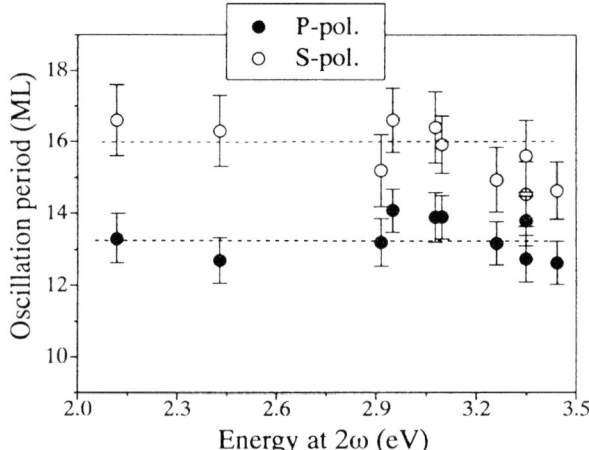

Fig. 3.30. The fitted values for the oscillation periods for s and p polarizations as a function of the second harmonic photon energy. Dashed lines are guides for the eye.

periods cannot be explained by the different total excitation energy ($\hbar\omega$ versus $2\hbar\omega$).

As has already been discussed, for a thin film with identical interfaces, the corresponding tensor elements on the opposite film interfaces are related to each other by a mirror symmetry and therefore differ only by a phase factor of 180°. Hence the resulting total SHG signal arises from the competition between the signals from the two film interfaces, which predominantly cancel each other out and only depends on the difference in their local fields. In other words, the nonlinear polarization $P(2\omega)$ is an *odd* function with respect to the film symmetry plane.

With this approach, the influence of QWS on MSHG would be largely cancelled out too because every QWS contributes symmetrically (via its local density of states) to the χ tensor elements of both interfaces. Even for the nonsymmetric geometry (like our case of Co/Au and Au/air interfaces) one may still argue that the corresponding electron wavefunctions are rather symmetric once they form a confined state.

Following these arguments, one may expect no QWS effects on the MSHG signal at all! But this is contrary to the experimental observations of a total domination of QWS on the SHG response. That we do observe a (rather strong) signal is partly related to the fact that the local electromagnetic fields at the two interfaces are different (this follows from Fresnel formulae) and partly from the (a)symmetry of the QWS wavefunctions. In a simple textbook picture the confined QWS have alternating odd and even character, i.e. the asymmetry of the QWS wavefunctions is repeated with the double period (see Fig. 3.23). This asymmetry can be expressed as a relative phase factor between the two QWS interface contributions (Co/Au and Au/air). Because the total SHG response results from a *coherent* superposition of these interface contributions, this total

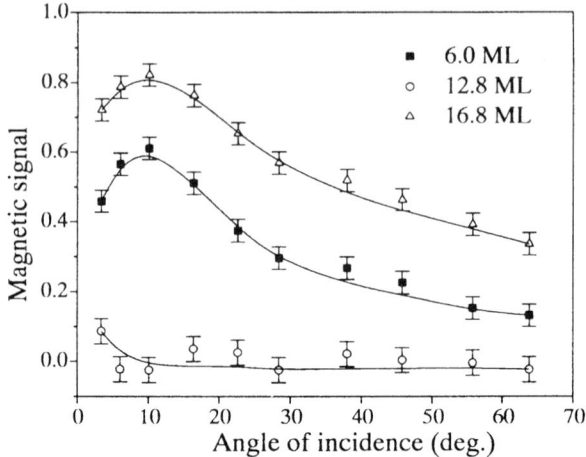

Fig. 3.31. Relative magnetic signal as a function of the angle of incidence for three different gold thicknesses (indicated in the figure) and for s_{in}–p_{out} polarization combination. Solid lines are theoretical fits taking into account only the interface contributions.

response will also display a periodic behaviour with the double period, despite the fact that the individual contributions oscillate with the single period. We should emphasize that it is the wavefunction phase at a given interface which plays a crucial role and, without interference, the effect would be unobservable. We also stress that the thickness dependence of the local fields can be neglected for these ultrathin films.

To test these ideas, we decomposed the total MSHG signal into the contributions from different interfaces. This is possible using a transfer matrix technique (as described in sec. 3.3) once enough independent experimental data are available. For this purpose, first the angle-of-incidence dependences of the MSHG signals were measured for each gold overlayer thickness value (see Fig. 3.31). Next, a fit of these data was performed, using the χ tensor elements as fitting parameters. For the fit to be unique, one has to use either s_{in}–p_{out} or q_{in}–s_{out} (q is the polarization between p and s) polarization combinations. In both cases, there is only one even (and one odd) χ component per (magnetic) interface.

The major part of the signals can be described very well with a model taking into account only the interface MSHG (see Fig. 3.31). Oscillatory behaviour of the signals naturally originates from the corresponding behaviour of the tensor components. Figure 3.32 shows the individual χ components of the two gold layer interfaces as a function of the film thickness, displaying an oscillatory behaviour with a mean period of around 6–8 ML, i.e. the same period as observed by MOKE. However, the resulting SHG intensity and magnetic signal perfectly fit the experimentally observed slowly oscillating behaviour with the double period. This means that, while the local density of states and hence the χ tensor at each interface show the standard QWS period, the resulting total

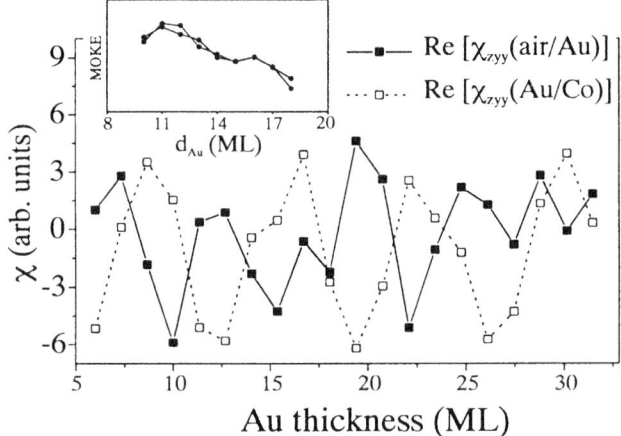

Fig. 3.32. Tensor components of different Au(111) film interfaces as a function of the film thickness. Inset shows the Kerr ellipticity as a function of gold thickness. The visible oscillations have a period of about 7 ML, which is roughly equal to that for the tensor components.

response includes also the phase between the corresponding elements and therefore allows a slower variation. This is possible, of course, only because the interfaces are not independent as soon as every QWS wavefunction is located at both interfaces simultaneously.

For the linear MOKE technique, the oscillatory part of the signal is provided by the narrow region along the Au/Co interface, where spin polarization of electrons is affected the most. Therefore it is related to the local density of states at this interface only, which oscillates with a single QWS period.

Figure 3.33 shows the dependence of the oscillation amplitude A relative to the background B as a function of the SH photon energy. Also plotted is the optical absorption curve for gold taken from ref. 104. While the A/B value for the s-polarized incoming light remains approximately constant (80–90%), for p polarization this ratio definitely shows a decrease towards higher energies. This increase actually coincides with the fundamental absorption edge in gold, which strongly suggests optical interband transitions as the origin for the thickness-independent part of the SH intensity.

In contrast we can suppose that the strong oscillatory behaviour of SHG at lower photon energies is not related to direct interband transitions. Therefore it is necessary to consider other possibilities, namely that either (i) the (spin-polarized) band structure of cobalt serves as a source for either initial or final states, or (ii) intraband transitions in gold are responsible for the effect. It has been shown, however, that in the case of transitions between spin-polarized states, the SH intensity should be considerably different for the opposite directions of the sample magnetization.[105] This does not appear to be the case for the considered samples, where the relative magnetic effect almost always

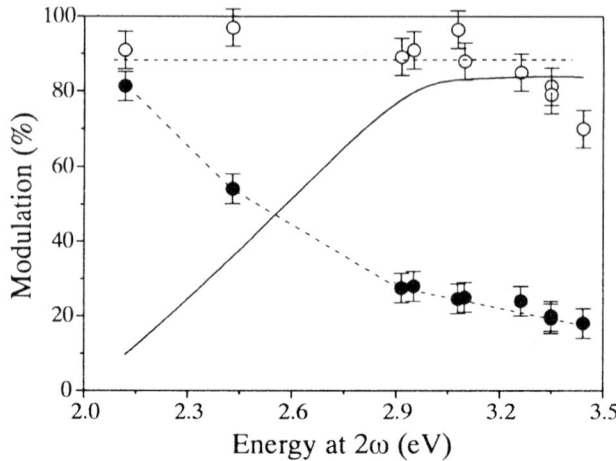

Fig. 3.33. Fitted values for the oscillation amplitude relative to the thickness-independent background as a function of the second harmonic photon energy for s (open circles) and p polarizations (closed circles). Solid line shows the absorption in gold. From ref. 104.

oscillates around zero or very close to it (Fig. 3.29). Hence the direct influence of the cobalt band structure can be ruled out.

As for the intraband transitions, they might be of much more importance in very thin films in comparison to the bulk metals. Indeed, for ultrathin films, the component of the wavevector perpendicular to the film plane becomes quantized. Therefore the requirement of its conservation is lifted, and transitions between the subsequent QWS become possible. Such transitions at low photon energy obviously only take place in the vicinity of the Fermi surface. This fact can naturally explain why the oscillation period does not considerably depend on the photon energy.

3.9.4 Cu / Co / Au(111)

The quantum confinement in ultrathin films depends sensitively on the band structure, in particular on the extremal k vectors.[92] Therefore it will be of interest also to look at Cu(111) oriented films, in comparison with Cu(001) that were already discussed in sec. 3.9.2. Also, the previous section showed the strong advantage of a wedge-shaped sample, which actually corresponds to a large set of samples grown simultaneously, so that all interfaces are grown under exactly the same conditions. This allows a much better comparison and quantitative analysis of the observed magneto-optical response.

The Cu samples were prepared in the same way as the Au discussed before (wedge-shaped Cu(111) on top of 5 ML Co(0001) on a thick Au buffer layer on a float glass substrate). The copper wedge was covered by 10 ML of gold for protection.

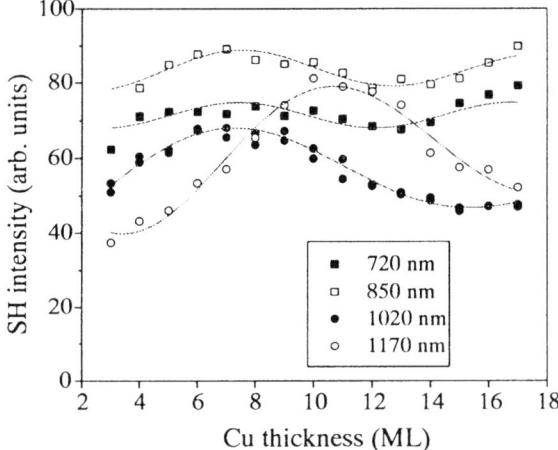

Fig. 3.34. Second harmonic intensity as a function of the overlayer thickness for different wavelengths (indicated in the figure) of incident (p-polarized light) for a Cu(111) wedge. Solid lines are theoretical fits to the data with eqn (3.9.9).

Figure 3.34 shows the SHG intensity for the Cu(111) overlayer, displaying a strong oscillatory behaviour as a function of the Cu thickness. Though qualitatively similar to the Au overlayer case, the frequency dependence is clearly different. Figure 3.35(a) shows the energy dependence of the oscillation amplitude A relative to the background B. Figure 3.35(b) shows the same for the period Λ. All these quantities A, B and Λ were obtained by fitting the observed thickness dependence to eqn (3.9.9) from sec. 3.9.3. Also plotted is the optical absorption curve for copper, taken from ref. 106.

For the Cu(111) overlayer, the amplitudes behave in a way very similar to the case of gold. However, there is one exception: the fundamental absorption edge

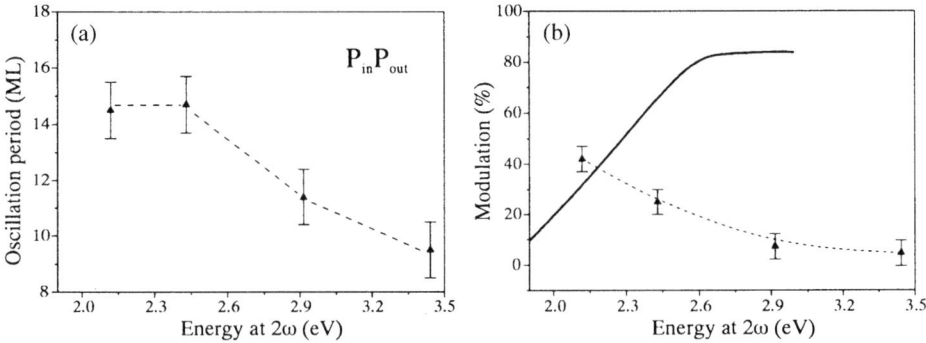

Fig. 3.35. The fitted values for (a) the oscillation periods and (b) the oscillation amplitude relative to the thickness-independent background as a function of the second harmonic photon energy. Solid line shows the absorption in Cu. Adapted from ref. 106.

for copper is shifted to lower energies (2.1 eV instead of 2.5 eV for gold). In other aspects, the band structures of Cu and Au are very similar. Therefore, the observed change in relative amplitude is just shifted to lower energies (Fig. 3.35(b)). In contrast to the case of gold, for copper there is also a considerable change in the oscillation period occurring around 2.7 eV. This does not coincide with the fundamental absorption edge, but can possibly be ascribed to the beginning of some type of optical interband transition at this energy.

Thus, apart from this energy-dependent period, the copper results are very similar to those of gold. The main difference was the observed amplitude of the intensity oscillations, which was much smaller than for Au(111). In analogy to Au, we would then expect also for Cu(111) the oscillation period to be half of that observed by MSHG. That means that the QWS oscillation in Cu(111) should have a period of 5 to 7 ML, depending on the energy. The recently detected linear MOKE oscillations on such samples[107] indeed showed a period of 7 ML. Hence, the period doubling is confirmed also for this system.

3.10 MSHG studies of antiferromagnetic structures

In this chapter we will show that also in antiferromagnetic crystals the interference between magnetically odd and even tensor elements will lead to observable nonlinear magneto-optical effects that are directly related to the magnetic ordering. In particular, this has recently led to the observation of the antiferromagnetic domain structure and the spin-flop domains in Cr_2O_3.

3.10.1 Cr_2O_3

As all recent MSHG work on Cr_2O_3 has been performed and published by M. Fiebig, D. Fröhlich and R. V. Pisarev, we will try to follow mostly their approach and notations (see also refs 26, 27, 108–110). In Cr_2O_3 there are two processes that contribute to the SHG signal. A magnetic dipole (MD) contribution

$$M(2\omega) \propto \chi_m^{(2)} : E(\omega)E(\omega),$$

induced by the electric field $E(\omega)$ of the incident light wave, is allowed for all temperatures, but does not couple to the magnetic ordering of the crystal. It is described by the time-variant nonlinear susceptibility χ_{ijk}^m (axial i tensor). Below the Néel temperature $T_N = 307.6$ K an electric dipole (ED) contribution

$$P(2\omega) \propto \chi_e^{(2)} : E(\omega)E(\omega)$$

becomes allowed because space and time reversal symmetry operations are simultaneously broken by the antiferromagnetic (AFM) order of spins. This contribution is described by the time-noninvariant nonlinear susceptibility χ_{ijk}^e (polar c tensor). Quadrupole contributions to the SHG signal are neglected since they have not been observed in Cr_2O_3 up to now.

Thus the electric dipole term $\chi_e^{(2)}$ is related to the magnetic ordering whereas the magnetic dipole term $\chi_m^{(2)}$ is not. Usually, MD transitions can be neglected in comparison with ED transitions. However, in this case the electric dipole transitions become allowed only below T_N due to the magnetic ordering of the spins. Thus these transitions are not very strong and their contributions to the nonlinear susceptibilities may be comparable to the magnetic dipole contributions. As will be shown below, this also implies that the magnetic effects in SHG from Cr_2O_3 do depend on specific resonance conditions.

For circularly polarized light incident along the 3-fold z axis of Cr_2O_3, one finds for the SHG intensity $I(2\omega)$:

$$I(2\omega) \sim \left[|\chi_m^{(2)}|^2 + |\chi_e^{(2)}|^2\right] \times \left(|E_+|^4 + |E_-|^4\right)$$
$$\sim 2\left[\chi_m^{(2)'} \chi_e^{(2)'}\right] \times \left(|E_+|^4 - |E_-|^4\right), \qquad (3.10.1)$$

where χ' and χ'' are the real and imaginary parts of the nonlinear susceptibility tensors, and $E_{+/-}$ refer to right-/left-handed circularly polarized fields. Owing to the presence of the interference term in eqn (3.10.1) the resultant SH intensity depends on the helicity of the pumping beam and the direction of the AFM order parameter l, which defines the sign of $\chi_e^{(2)}$. We can therefore write eqn (3.10.1) like

$$I_{\sigma,l}(2\omega) \propto [C + \text{sgn}(l)\text{sgn}(\sigma)\cdot\Delta]I_\sigma^2(\omega). \qquad (3.10.2)$$

C corresponds to the first two terms in eqn (3.10.1) whereas $\text{sgn}(l)\text{sgn}(\sigma)\cdot\Delta$ describes the interference. The sign of the interference contribution can be changed either by reversing the direction of the circular polarization σ or by the time reversal operation, which is equivalent to a reversal of the AFM vector l. The latter describes the relative orientation of the spins in the unit cell. In Cr_2O_3, the AFM spin ordering can be such that $l = 1$ or $l = -1$, leading to 180° domain structures. The two 180° domains should therefore reveal the same SH spectra but with a reversed dependence on the circular polarization.

For the experiments, a frequency-tripled Nd:YAG laser was used that pumps an optical parametric oscillator, to obtain the tuning range of interest. The Cr_2O_3 samples were grown by the Vernuil method and their size was about 10 mm².

Figure 3.36 shows the SHG spectrum of Cr_2O_3 for incoming light with two opposite circular polarizations. The interference of time-invariant and time-noninvariant contributions to the SHG leads to a pronounced polarization dependence. In agreement with eqn (3.10.2) the dependence can be reversed if the laser spot is moved to another domain. The polarization dependence vanishes at the Néel temperature.[26]

Concluding, it can be said that nonlinear magneto-optical spectroscopy yields unique opportunities to study the magnetic properties of antiferromagnetic materials. This also offers interesting opportunities to study so-called biased spin valve structures, where an AFM material is used to bias a coupled multilayer structure.

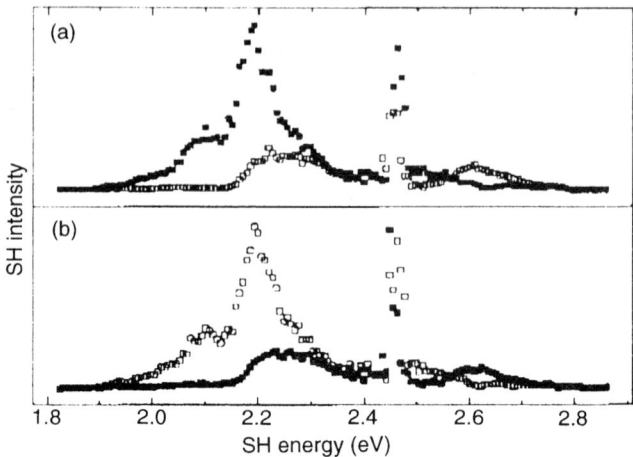

Fig. 3.36. SH spectrum of Cr_2O_3 in zero magnetic field at $T = 10$ K. Full (open) squares refer to right- (left-) circularly polarized light incident along the z axis. (a) and (b) correspond to two domains with opposite orientation of the AFM vector. From ref. 26.

3.10.2 Other AFM materials

Though most work up till now has been done on Cr_2O_3, there are a few MSHG studies of other AFM materials reported. Haematite (α-Fe_2O_3) has the same crystallographic structure as Cr_2O_3, but unlike the latter the inversion symmetry in haematite is not broken by the AFM ordering of the spins. A weak but magnetic SHG signal was observed below the so-called Morin temperature T_M (≈ 260 K).[111] In yttrium manganite $YMnO_3$, the same authors observed an MSHG signal below the Néel temperature ($T_N = 75$ K), which was purely magnetic, i.e. the signal was not sensitive for the change in helicity of the excitation light. This is in agreement with the crystal symmetry of $YMnO_3$. Magnetic contrast was obtained by a clever trick: instead of intrinsic interference, an external SHG reference signal (from quartz) was used to obtain magnetic contrast. This was also used to visualize the 180° domain structure in this crystal.[112]

In the AFM structure NiO, Janner et al. have observed a resonance SHG signal near 3.9 eV[113] for epitaxially grown thin films. No signal was observed for a pure single crystal, which may suggest a strain-induced SHG for the thin films.

3.11 MSHG studies of magnetic garnets

Magnetic garnets were studied intensely some decades ago, because of possible application for magnetic storage.[114–116] Therefore their crystallographic and magnetic structures are well known, which makes them interesting testing materials for a new technique like MSHG. In this section we will show how a simultaneous breaking of space and time reversal symmetry in thin garnet films leads to bulk crystallographic and magnetization-induced contributions to the nonlinear optical susceptibility, which gives rise to a new transverse magneto-optical effect linear in M. We will also show how the two different contributions can easily be separated based on their different symmetry properties.

3.11.1 Symmetry properties of thin garnet films

Bulk crystals of magnetic garnets like yttrium iron garnet $Y_3Fe_5O_{12}$ possess a crystallographic as well as a magnetic centrosymmetric structure.[114,115] Consequently, SHG is forbidden in the electric dipole approximation. However, thin films of magnetic garnets possess magnetic and magneto-optical properties different from those in bulk crystals.[116] It was shown that in thin magnetic garnet films the inversion symmetry may be broken due to a distortion of the crystal structure, because of a lattice mismatch between the magnetic film and the nonmagnetic substrate. The substrate orientation therefore is also very important to determine the film symmetry (see e.g. ref. 110). This is evidenced by the observation of SHG and electric-field-induced linear optical effects.[23,109,110] The symmetry analysis of the crystallographic SHG contributions was given by Pisarev et al.[110] Here we will give a short summary for the films of 100, 111, 210 and 110 symmetry that were studied in the experiments.

The point group of the films can be derived from the cubic garnet point group m3m, by retaining only those symmetry elements that leave the film invariant. Because all elements have to leave the surface normal invariant, the new point groups are polar and have no inversion symmetry.

(001) The 4-fold axis [001] is directed along the normal of the film; the mirror plane and the 2-fold axes that are perpendicular to [001] are lost and the point group of the film becomes 4mm (C_{4v}). SHG is allowed in this point group and the nonlinear polarization $P^{(2)}(2\omega)$ has the following components:

$$P_x^{(2)} = 2\chi_{xxz}E_xE_z,$$
$$P_y^{(2)} = 2\chi_{yyz}E_yE_z, \quad (3.11.1)$$
$$P_z^{(2)} = 2\chi_{zxx}(E_x^2 + E_y^2) + \chi_{zzz}E_z^2,$$

with $\chi_{xxz} = \chi_{yyz}$. From eqn (3.11.1) it follows that at normal incidence $P(2\omega) = 0$.

(111) The 3-fold axis is along the normal of the film and the relevant noncentrosymmetric point group is 3m (C_{3v}). The nonlinear polarization has the following form:

$$P_x^{(2)} = 2\chi_{xxz}E_xE_z + \chi_{xxx}(E_x^2 - E_y^2),$$
$$P_y^{(2)} = -2\chi_{xxx}E_xE_y + 2\chi_{yyz}E_yE_z, \quad (3.11.2)$$
$$P_z^{(2)} = 2\chi_{zxx}(E_x^2 + E_y^2) + \chi_{zzz}E_z^2,$$

where the y axis is chosen perpendicular to the mirror plane m.

(210) The only symmetry element for these films is the mirror plane m, which is perpendicular to the film surface and contains the axes [210] and [1$\bar{2}$0]. Thus the relevant point group is monoclinic m (C_{1h}) and the nonlinear polarization has the following form:

$$P_x^{(2)} = \chi_{xxx} E_x^2 + \chi_{xyy} E_y^2 + \chi_{xzz} E_z^2 + 2\chi_{xxz} E_x E_z,$$
$$P_y^{(2)} = 2\chi_{yxy} E_x E_y + 2\chi_{yzy} E_z E_y, \quad (3.11.3)$$
$$P_z^{(2)} = \chi_{zxx} E_x^2 + \chi_{zyy} E_y^2 + \chi_{zzz} E_z^2 + 2\chi_{xzx} E_x E_z.$$

(110) The 2-fold axis [110] is along the normal of the film surface and the point group of the film is mm2 (C_{2v}). The nonlinear polarization has the following form:

$$P_x^{(2)} = 2\chi_{xzx} E_x E_z,$$
$$P_y^{(2)} = 2\chi_{yyz} E_y E_z, \quad (3.11.4)$$
$$P_z^{(2)} = \chi_{zxx} E_x^2 + \chi_{zyy} E_y^2 + \chi_{zzz} E_z^2,$$

and, like for (100), $\boldsymbol{P}(2\omega) = 0$ at normal incidence.

3.11.2 Nonlinear magneto-optical response of thin garnet films

In the electric dipole approximation the nonlinear optical polarization $\boldsymbol{P}(2\omega)$ of a magnetic medium possessing a spontaneous magnetization $\boldsymbol{M}(0)$ can be written in the form

$$P^i(2\omega) = \chi_{ijk}^{(2)}(-2\omega, \omega, \omega) E_j(\omega) E_k(\omega)$$
$$+ i\chi_{ijkl}^{(3)}(-2\omega, \omega, \omega, 0) E_j(\omega) E_k(\omega) M_l(0), \quad (3.11.5)$$

where $E_j(\omega)$ and $E_k(\omega)$ are the incoming fundamental fields. The polar tensor components $\chi_{ijk}^{(2)}$ describing the crystallographic contribution have been given in eqns (3.11.1)–(3.11.4). An axial tensor $\chi_{ijkl}^{(3)}$ of rank 4 describes the magnetization-induced contribution. It is also allowed in noncentrosymmetric crystals. These two contributions to the nonlinear polarization $\boldsymbol{P}(2\omega)$ may coexist in the same medium. They are both spontaneous, and no external field is required to observe them in a single-domain state. Because they possess different transformation properties under the symmetry operations of the medium, they vary differently when the incident polarization $E_i(\omega)$ varies with respect to the crystal axes. This gives the possibility to separate these two different contributions. An alternative way to accomplish this is by studying their temperature behaviour: the magnetization-induced contribution should vanish at the transition from a magnetically ordered to a paramagnetic state.

Table 3.6 Basic parameters of the samples

Substrate orientation	Film symmetry	Film composition	Lattice parameter (Å)		
			Film	Substrate	Misfit
(001)	4mm	$(YbPr)_3(FeGa)_5O_{12}$	12.4140	12.3787	0.0353
(111)	3m	$(YLuBi)_3(FeGa)_5O_{12}$	12.3720	12.3794	−0.0074
(210)	m	$(YPrLuBi)_3(FeGa)_5O_{12}$	12.5276	12.4789	0.0487
(110)	mm2	$(YBi)_3(FeGa)_5O_{12}$	12.382	12.377	0.005

3.11.3 Experimental results and discussion

The magnetic films were grown by a liquid-phase epitaxial method. Four different types of magnetic garnet thin films with substrate orientations (001), (111), (210) and (110) have been studied. Samples of these types differed in film and substrate compositions (see Table 3.6). Thin wafers of gadolinium gallium garnet $Gd_3Ga_5O_{12}$ (GGG) and substituted GGG with a larger lattice parameter have been used as substrates. The largest mismatch between lattice parameters of film and substrate was in the case of the (210) film, and the smallest one in the (110) film. In the (111) film the mismatch was negative. The MSHG experiments were done with the Ti:sapphire setup at 841 nm, in the transmission geometry and with an applied magnetic field up to $H = 2.3$ kOe applied along the y axis in the plane of the films. In this way any influences of linear magneto-optical effects like Faraday rotation or magnetic circular dichroism are avoided.

At 841 nm (1.474 eV) the linear absorption coefficient of magnetic garnet films and bulk crystals is of the order of $\alpha = 10–20\,\mathrm{cm}^{-1}$,[114] but at the frequency of the second harmonic (2.948 eV), α is three orders of magnitude larger. Therefore, in transmission experiments the detected SHG signal can only escape from a backside layer with a thickness of about 1 μm. Under such circumstances the phase-matching conditions are unimportant.[6]

Rotating the sample by 360° around the z axis allows one to register the rotational anisotropy of the SHG signal with the magnetization being kept along the y axis and the incoming and outgoing linear light polarization being fixed along the x or y axis. As we will show below, such an approach allows unambiguous separation between crystallographic and magnetization-induced SHG signals due to their different transformation properties.

(111) Figure 3.37 shows the experimental results for the (111) oriented film for different polarization combinations and directions of magnetic field. In a demagnetized sample the rotational anisotropy is characterized by a 60° periodicity, as was reported previously.[23,110,117] However, in a magnetized sample the rotational anisotropy is characterized by a 120° periodicity. In the xx polarization geometry there is a strong magnetic effect that shifts the observed 3-fold symmetry by 60°, whereas in the yy geometry, no magnetic effect is observed.

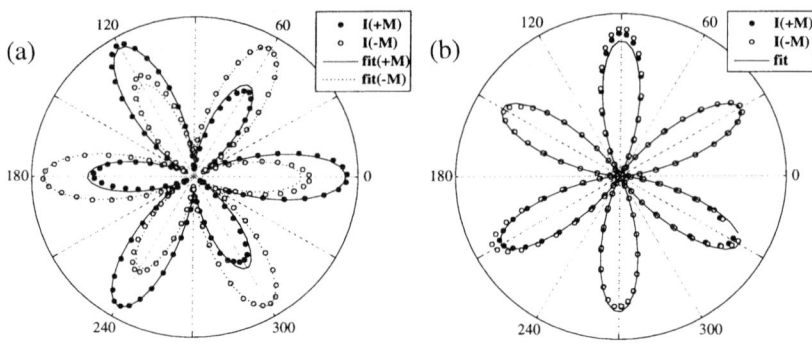

Fig. 3.37. The rotational anisotropy of the SHG intensity from a (111) film: (a) *xx* polarization combination, (b) *yy* polarization combination.

(111) films have the point group symmetry 3m and the SHG at normal incidence in transmission is allowed for the crystallographic and magnetization-induced parts of eqn (3.11.5). The crystallographic part has been discussed above.

The magnetization-induced part depends on the combination of incident–outgoing polarizations, as is shown in Table 3.7 for the case of the electric dipole approximation. Taking into account both crystallographic and magnetic contributions, we get for the SHG intensity

$$I_{xx}(2\omega, \phi) = E^4(A\cos^2 3\phi + BM^2 + 2CM\cos 3\phi),$$
$$I_{yy}(2\omega, \phi) = E^4 A \sin^2 3\phi,$$
(3.11.6)

where ϕ is the angle between the direction of magnetization M and the x axis in the (111) plane, A, B and C are combinations of the real and imaginary parts of the complex nonlinear susceptibility components $\chi^{(2)}_{ijk}$ and $\chi^{(3)}_{ijkl}$, and the indices xx and yy denote input–output polarizations of the light. In the case of xx polarization the SHG intensity comprises a pure crystallographic term $\approx A$, a magnetization-induced term $\approx BM^2$ and an interference term $\approx CM$. Owing to the interference term CM the 6-fold symmetry of the SHG intensity is transformed into a 3-fold symmetry for xx and yx polarizations, in accordance with

Table 3.7 Crystallographic and magnetization-induced susceptibilities for (111) films

Input–output polarizations	Crystallographic	Magnetization-induced
xx	$\chi^{(2)}_{xxx}$	$i\chi^{(3)}_{xxyy}$
yx	$-\chi^{(2)}_{xxx}$	$I\chi^{(3)}_{xxyy}/3$
xy	$\chi^{(2)}_{xxx}$	0
yy	$-\chi^{(2)}_{xxx}$	0

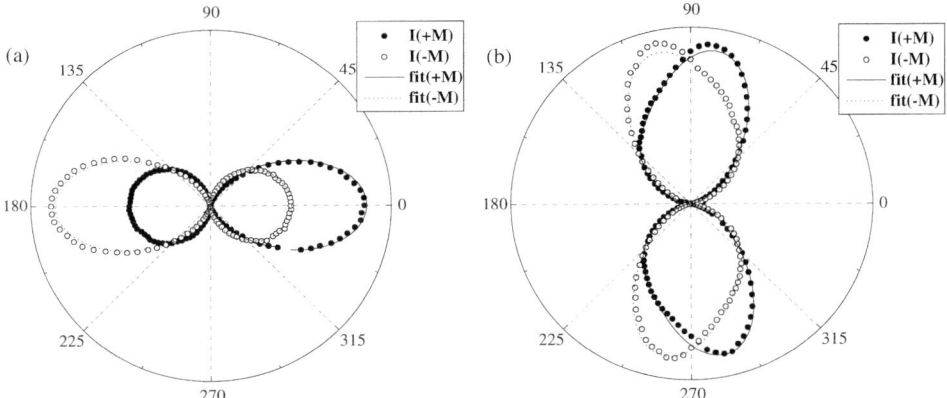

Fig. 3.38. The rotational anisotropy of the SHG intensity from a (210) film: (a) *xx* polarization combination, (b) *yy* polarization combination.

Fig. 3.37(a). The fit of the experimental data to eqns (3.11.6) is extremely good. Equations (3.11.6) also show the absence of a magnetic effect for the *yy* polarization, as was confirmed by the experiment (Fig. 3.37(b)). The same was true for the *xy* polarization. This is a direct confirmation of the electric dipole origin of the SHG for the crystallographic and magnetization-induced contributions.

(210) Figure 3.38 shows clear magnetic effects for both polarization combinations, though again the *xx* effects are much stronger.

(210) films have point group symmetry m. The crystallographic and the magnetization-induced contributions to the SHG are allowed at normal and oblique incidence and thus they can interfere in a magnetized sample. The magnetization-induced polarizations can be written as

$$P_{xx}(2\omega, \phi) = E^2 M \left[B \cos^4 \phi - A \sin^4 \phi + \tfrac{3}{4}(A - B)\sin^2 2\phi \right],$$
$$P_{yy}(2\omega, \phi) = E^2 M \sin 2\phi \, (A \cos^2 \phi + B \sin^2 \phi). \quad (3.11.7)$$

The solid lines in Fig. 3.38 are best-fit calculations using crystallographic and magnetization-induced (eqns (3.11.7)) contributions to the nonlinear polarization. The SHG intensity in films of this orientation was one to two orders of magnitude larger as compared to films of other orientations.

(001) Figure 3.39 shows the rotational anisotropy pattern for the (001) sample. The SHG vanished without application of the magnetic field. In a magnetized sample the SHG intensity has the same value for opposite orientations of magnetization.

(001) films have the point group 4mm, and at normal incidence the crystallographic SHG is symmetry forbidden, but the magnetization-induced contribution

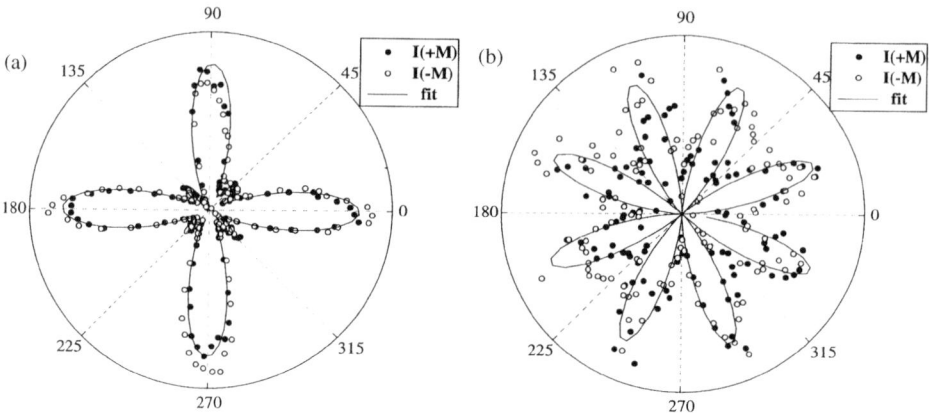

Fig. 3.39. The rotational anisotropy of the SHG intensity from a (001) film: (a) *xx* polarization combination, (b) *yy* polarization combination.

is allowed. This means that for these films, an *SHG signal appears by applying a magnetic field*. The intensity $I(2\omega)$ of the SHG should vary as

$$I_{xx}(2\omega, \phi) = E^4 M^2 (A - B\sin^2 2\phi + C\sin^4 2\phi),$$
$$I_{yy}(2\omega, \phi) = \tfrac{1}{4} E^4 M^2 C \sin^2 4\phi,$$

(3.11.8)

where ϕ is the angle between the direction of magnetization M in the (001) plane with respect to the (100) axis. A, B and C are combinations of the real and imaginary parts of nonlinear susceptibility coefficients χ_{ijkl}, as for example $A = ({\chi_{xxyy}^2}' + {\chi_{xxyy}^2}'')$. Interference terms are not present in films of this symmetry. In all cases and in accordance with eqns (3.11.8) the SHG intensity was not sensitive to the magnetization reversal by 180°, though the nonlinear polarization $P(2\omega)$ is itself proportional to a spontaneous magnetization M.

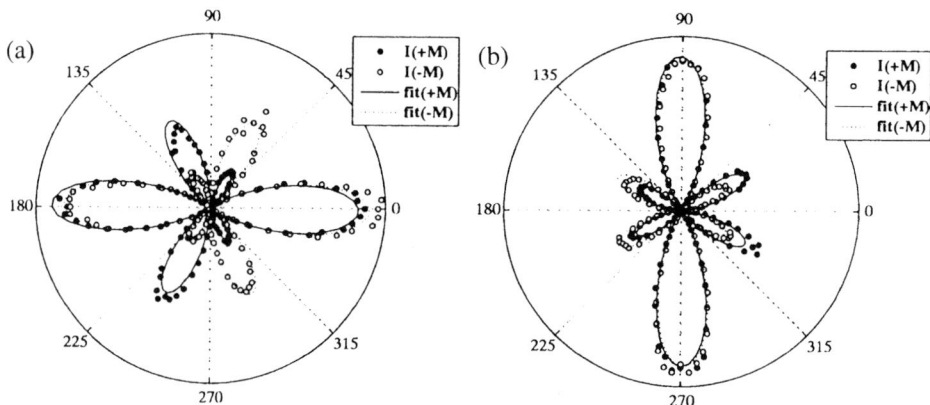

Fig. 3.40. The rotational anisotropy of the SHG intensity from a (110) film: (a) *xx* polarization combination, (b) *yy* polarization combination.

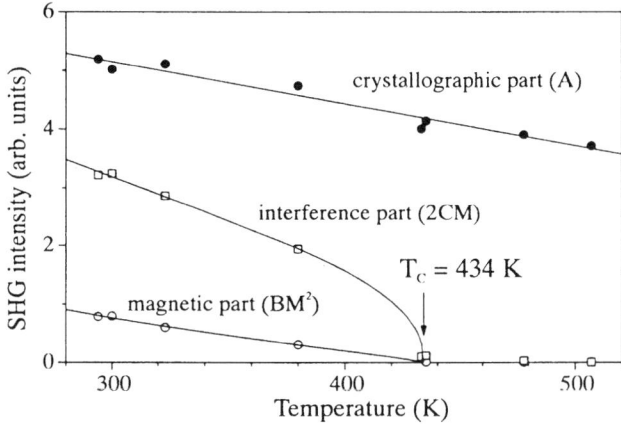

Fig. 3.41. Temperature variations of crystallographic, magnetization-induced and interference terms in the SHG intensity for the (111) film.

(110) Figure 3.40 shows the rotational anisotropy pattern in a (110) film.

(110) films have the point group symmetry mm2. We found that the magnetization-induced contributions to the nonlinear polarization are characterized by equations similar to those for (210) films (see eqns (3.11.7). Equations (3.11.7) describe the rotational anisotropy of the SHG in (110) films very well, though the rotational patterns in (210) and (110) films look different (see Figs 3.38 and 3.40).

In order to find another proof of the different origins of the two contributions to the SHG, the rotational anisotropy of the SHG has been studied as a function of temperature for several films. Figure 3.41 shows the results for the (111) film. For the temperature dependence of the interference term we found $CM \sim (1 - T/T_C)^B$ with $B = 0.6$. Whereas the crystallographic contribution decreased linearly with temperature, the magnetization-related contributions vanished at T_C.[118]

In a recent paper, Aktsipetrov et al.[119] report MSHG effects in iron garnet films in the Faraday configuration ($M \parallel k$). A Faraday rotation of the SHG waves is observed, where the effect for the fundamental wave is supposed to be negligible. For the (210) orientation, magnetization-induced changes in the amplitude and phase of the different Fourier components of the anisotropic signal are observed. These effects are attributed to magnetostriction in the substrate/magnetic film structure. From the much larger signals for (210) and (110) relative to (111) films, it was concluded that the orthorhombic magnetic anisotropy plays an important role in the MSHG conversion. However, no clear separation between crystallographic and magnetic contributions is given.

3.12 Nonlinear magneto-optical imaging

Magnetization-induced second harmonic generation can be shown to offer new possibilities for magnetic domain imaging. The use of an optical response that is governed by a higher-rank tensor offers sensitivity to additional combinations of magnetization directions and optical wavevector and polarization, which can be demonstrated beautifully in magnetic garnet films of different crystallographic orientations. A symmetry analysis of nonlinear magneto-optical imaging of magnetic domains and domain walls is first presented. Gradient terms are shown to give rise to the MSHG via spatial derivatives of the magnetization. The nonvanishing independent elements of the relevant tensors are derived for cubic media and different contributions to the MSHG image from domains and domain walls are analysed for thin magnetic films with different symmetry. It is shown that measurements of polarization properties of the MSHG response may yield information about the relative importance of different magnetization-induced contributions and also the type of domain walls. Experimental results for garnet films of odd/even symmetry will be shown, indeed proving the novel possibilities of nonlinear magneto-optical imaging. Finally, we will also show how the sensitivity of MSHG for the antiferromagnetic ordering can be used to visualize antiferromagnetic domain structure.

3.12.1 Domain and domain-wall contributions in nonlinear magneto-optical microscopy

For nonuniform magnetic media the nonlinear optical susceptibility tensor $\chi^{(2)}$ may be presented as a sum of different terms (here we omit the frequency arguments and skip the usual superscript (2) for the nonlinear susceptibility tensor $\chi_{ijk}^{(2)}(2\omega, \omega, \omega)$):

$$\chi_{ijk} = \chi_{ijk}^{[0]} + \chi_{ijkL}^{[1]} M_L + \chi_{ijkLM}^{[2]} M_L M_M$$
$$+ \chi_{ijklM}^{[3]} \nabla_l M_M + \chi_{ijkLmN}^{[4]} M_L \nabla_m M_N + \cdots, \qquad (3.12.1)$$

where $\chi^{[0]}$ is the nonmagnetic part of χ while $\chi^{[1]}$ and $\chi^{[2]}$ describe the effect of the *local* magnetic order. Capital letters are used to denote the indices of the axial magnetization vector $\boldsymbol{M}$. In eqn (3.12.1) we also introduce gradient terms $\chi^{[3]}$ and $\chi^{[4]}$, which are nonvanishing in the presence of a nonuniform magnetization. Similar gradient terms were introduced into the theory of linear-optical domain imaging.[120] We note that all tensors with an odd number of polar (small) indices vanish for centrosymmetric media. In that case only the gradient terms ($\propto \chi^{[3]}$ and $\chi^{[4]}$) contribute to the nonlinear source $\boldsymbol{P}(2\omega)$.

Below, we focus in particular on the nonlinear-optical properties of different thin magnetic garnet films. The theoretical consideration is, however, more general since it is based only on symmetry arguments and therefore can be applied to other magnetic systems with the same symmetry. The *bulk* of a perfect garnet crystal is cubic and centrosymmetric.[121] In thin films grown on an imperfectly matched substrate, however, the inversion symmetry is lifted via a distortion of the lattice (see sec. 3.11.1), so that all terms on the right-hand side of eqn (3.12.1) are symmetry-allowed. On the other hand, the lattice distortion is assumed to be small so that the lattice is close to the centrosymmetric arrangement in the perfect crystal. This assumption is essential for an experimental detection of the gradient effects on MSHG. One obviously expects that in most cases the gradient terms in eqn (3.12.1) are relatively small corrections to the leading nonmagnetic $\chi^{[0]}$ and local $\chi^{[1]}$ and $\chi^{[2]}$ magnetic terms. In a thin garnet film with a 'nearly centrosymmetric' lattice, the importance of these local terms may be reduced so that the relative weight of the gradient terms $\chi^{[3]}$ and $\chi^{[4]}$ is enhanced.

A coordinate system is introduced with the z axis being normal to the film, and the x and y axes lying in the film plane. We assume that there is at least one symmetry reflection plane normal to the (nonmagnetic) film which coincides with the $y = 0$ plane. The analysis is performed for films with an even-fold rotation symmetry as well as for those with an odd-fold one. For the thin garnet films, for example, those grown on (001) and (110) substrates have even-fold rotation symmetry (C_{4v} and C_{2v}, respectively) while those grown on (111) faces have the odd-fold rotation symmetry C_{3v}. The MSHG sources are calculated in two domains and in different parts of the domain wall between them. One part of the domain wall is assumed to be along the y axis (wall A) while another one along the x axis (wall B, see Fig. 3.42). Because of the growth-induced magnetic anisotropy, the easy axis is normal to the film plane so that within the domains the magnetization vector $\boldsymbol{M}$ is along z but antiparallel for two neighbouring

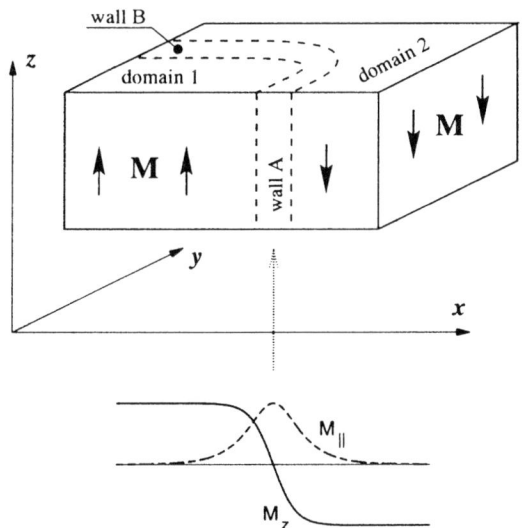

Fig. 3.42. Sketch of the domain-wall configuration studied in section 3.12.1. The $y = 0$ plane coincides with the mirror symmetry plane of the film lattice. The $x = 0$ plane is also a mirror symmetry plane for an even-fold rotation symmetry film, while $x \to -x$ is not a symmetry operation for an odd-fold rotation symmetry film. The lower panel shows the spatial variations of the normal ($\boldsymbol{M}_z$) and parallel ($\boldsymbol{M}_\parallel$) components of the magnetization. For the Bloch wall, $\boldsymbol{M}_\parallel$ is along y, whereas for the Néel wall, $\boldsymbol{M}_\parallel$ is along x.

domains. In the domain wall, however, a parallel component of M is present. In the case of a Bloch wall A, this component is along the y axis, whereas for a Néel wall A, it is along x (see Fig. 3.42). We also assume that the thickness of the domain wall is much smaller than the optical wavelength λ whereas the size of the domains is much larger than λ. We will analyse MSHG in transmission through the film. The fundamental beam is taken to be incident along the z axis and polarized along the x axis. The nonlinear polarization P along z does not radiate into the z direction and the components of P along x and y generate SHG light with two different polarizations. Because the (linear-optical) Faraday rotation of the fundamental and SHG polarizations within the thin magnetic film is small, it will be neglected.

Applying the symmetry operations, we find that for an even rotation symmetry film the nonmagnetic nonlinear polarization $\chi^{[0]}$ is purely polarized along z and therefore does not contribute to the MSHG image of either the domains or the domain walls. The contribution of the term linear in the magnetization ($\propto \chi^{[1]}$) of eqn (3.12.1) also vanishes in the domains but gives a finite contribution in the domain wall where a parallel magnetization is present. In the Bloch wall A, the nonlinear polarization can be written as

$$P_x^{[1]}(2\omega) = \chi_{xxxY}^{[1]} E_x^2(\omega) M_Y. \tag{3.12.2}$$

For the Néel wall A, on the other hand, the MSHG response arising from the second term of eqn (3.12.1) is polarized along the y axis.

In a similar way, one has no MSHG from the third term ($\propto \chi^{[2]}$) in the domains. The only nonvanishing contribution of, for example, the Bloch domain wall A and the Néel wall B is given by

$$P_y^{[2]}(2\omega) = \chi_{yxxYZ}^{[2]} E_x^2(\omega) M_Y M_Z. \tag{3.12.3}$$

We note, however, that $M_Y M_Z$ is an odd function of the coordinate along the wall normal. Since the 'brightness' of the wall image is proportional to the polarization $\langle P_y^{[2]}(2\omega) \rangle_{d_r \times d_r}$ that is averaged over the resolution length d_r of the optical objective, the nonlinear source (3.12.3) generated in a thin wall (much thinner than d_r) does not contribute to the image. Taking the next term of eqn (3.12.1) into account, one has a contribution to the MSHG in transmission at normal incidence:

$$P_y^{[3]}(2\omega) = \chi_{yxxxZ}^{[3]} E_x^2(\omega) \nabla_x M_Z \tag{3.12.4}$$

for both Bloch and Néel domain walls polarized along y (A walls). In B walls this contribution is polarized along x. The MSHG source (3.12.4) is therefore always polarized parallel to the wall and is independent of the type of the wall. The last term of eqn (3.12.1) can produce a contribution via terms containing $M_I \nabla_x M_I$ or $M_I \nabla_y M_I$ and therefore vanishes after integration over the domain wall. The results of the present analysis are summarized in Table 3.8. For the fundamental light polarized along the y axis we find exactly the same polarizations for the different contributions to the MSHG image.

Table 3.8 Polarization of the MSHG wave generated in an even rotation symmetry film via different terms $\chi^{[n]}$ of eqn (3.12.1). The fundamental beam is polarized along either x or y directions. The relationship to different components of the film magnetization is explicitly shown

Term n	Domain $M \parallel z$	Bloch wall A $M_\parallel \parallel y$	Bloch wall B $M_\perp \parallel x$	Néel wall A $M_\perp \parallel x$	Néel wall B $M_\parallel \parallel y$
0	—	—	—	—	—
1	—	$P_x \propto M_Y$	$P_y \propto M_X$	$P_y \propto M_X$	$P_x \propto M_Y$
2	—	—	—	—	—
3	—	$P_y \propto \nabla_x M_Z$	$P_x \propto \nabla_y M_Z$	$P_y \propto \nabla_x M_Z$	$P_x \propto \nabla_y M_Z$
4	—	—	—	—	—

Using the results from Table 3.8 one can simulate the expected domain images for Bloch and Néel walls A and B. The results are plotted in Fig. 3.43. The MSHG intensity along the y axis is recorded and the relevant elements of $\chi^{[1]}$ and $\chi^{[3]}$ are assumed to be of the same order of magnitude. As can be seen from Table 3.8, there is no SHG light generated within the domains in the even

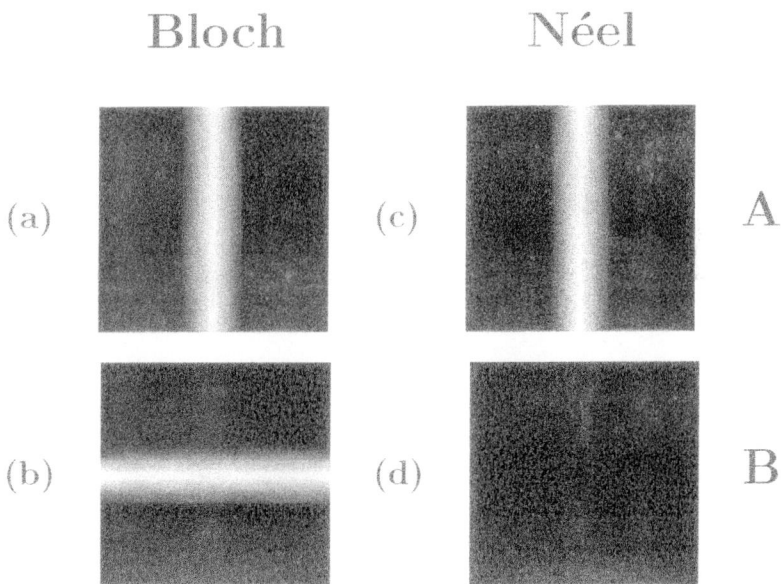

Fig. 3.43. Simulated images of fragments A (a, c) and B (b, d) of a Bloch (a, b) and Néel (c, d) wall in an even-fold rotation symmetry film. The MSHG response is polarized along the vertical y azimuth. The fundamental field is assumed to be polarized purely along the x (horizontal) or y azimuth. The thickness of the domain wall is much smaller than the resolution of the imaging objective. The brightness of the wall seen in (a) depends on the relative magnitude of the $\chi^{[3]}$ term, whereas that seen in (b) depends on the relative magnitude of the $\chi^{[1]}$ term.

Table 3.9 Polarization of MSHG generated in an odd rotation symmetry garnet film via different terms $\chi^{[n]}$ of eqn (3.12.1)

n	Domain	Bloch wall A	Bloch wall B	Néel wall A	Néel wall B
0	P_x	P_x	P_x	P_x	P_x
1	$P_y \alpha M_Z$	$P_x \alpha M_Y$	$P_y \alpha M_X$	$P_y \alpha M_X$	$P_x \alpha M_Y$
2	$P_x \alpha M_Z^2$	$P_x \alpha M_Y^2$	$P_x \alpha M_X^2$	$P_x \alpha M_X^2$	$P_x \alpha M_Y^2$
3	—	$P_y \alpha \nabla_x M_Z$	$P_x \alpha \nabla_y M_Z$	$P_y \alpha \nabla_x M_Z$	$P_x \alpha \nabla_y M_Z$
4	—	$P_y \alpha M_Y \nabla_x M_Z$ $+ M_Z \nabla_x M_Y$	$P_y \alpha M_X \nabla_y M_Z$ $+ M_Z \nabla_y M_X$	$P_x \alpha M_X \nabla_x M_Z$ $+ M_Z \nabla_x M_X$	$P_x \alpha M_Y \nabla_y M_Z$ $M_Z \nabla_y M_Y$

rotation symmetry film. Also, only y-polarized MSHG can be generated in the Néel wall if the fundamental beam is polarized along x or y axes. Therefore, if the MSHG signal is not purely polarized along y, the wall is of the Bloch type. Moreover, the relative weight of the $\chi^{[1]}$ and $\chi^{[3]}$ contributions can then be found from the polarization properties of the MSHG light generated by the Bloch domain wall.

A similar analysis can be given for odd-fold rotation symmetry films. Following the same arguments as were used above, one can find the symmetry-allowed contributions to the SHG wave generated in transmission by a fundamental wave at normal incidence. The results are collected in Table 3.9.

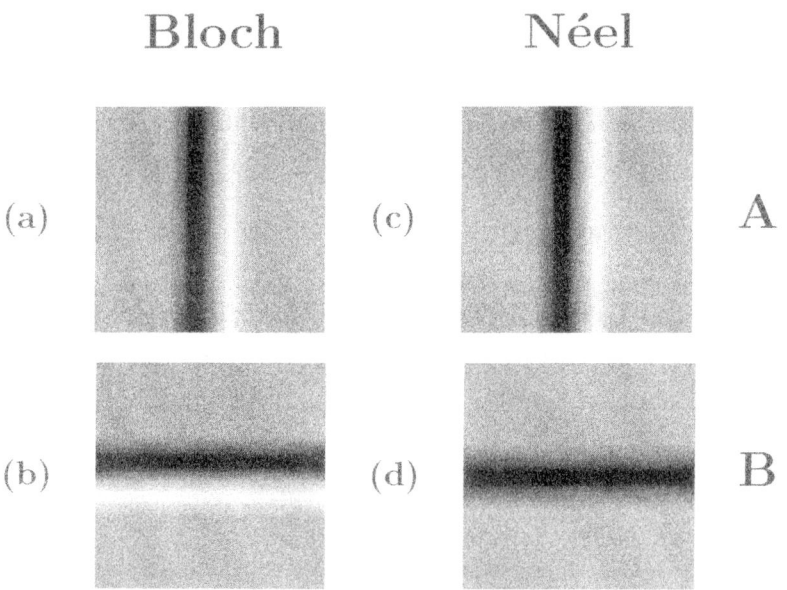

Fig. 3.44. Simulated images of fragments A (a, c) and B (b, d) of a Bloch (a, b) and Néel (c, d) wall in an odd-fold rotation symmetry film. The MSHG response is polarized along the vertical y azimuth. The fundamental field is assumed to be polarized purely along the x or y azimuth.

The lowering of the film symmetry is seen to result in a much larger number of nonvanishing contributions. The MSHG image of the magnetic domain structure is therefore expected to be more complex. One can nevertheless find some simple rules that may be used to analyse the image. For example, the polarization of the SHG response from the domains possesses the information about the relative importance of the local terms that are even ($\chi^{[0]}$ and $\chi^{[2]}$) or odd ($\chi^{[1]}$) in the magnetization. Then, the Néel wall B (parallel to the symmetry plane $y = 0$) should look like a 'dark' line if only the y-polarized MSHG is recorded. The 'brightness' of the Néel wall A (relative to the 'brightness' of the domains) then brings the information about the relative importance of the terms linear in the magnetization $\chi^{[1]}$ and $\chi^{[3]}$. Similarly, if the domain walls are of the Bloch type, the relative 'brightness' of the wall A in y-polarized MSHG light tells us the relative weight of the two gradient terms $\chi^{[3]}$ and $\chi^{[4]}$, etc. Figure 3.44 gives the simulated images for this situation.[122]

The analysis above shows that MSHG imaging can be used to study domains and domain walls in magnetic films. The nonlinear magneto-optical response is shown to give contributions that are proportional to the local magnetization and its spatial derivatives. The latter contributions are the only dipole-allowed source of MSHG in centrosymmetric media and may produce an important contribution to the MSHG image of domain walls in crystals where the inversion symmetry is lifted. MSHG imaging of the domain structure with the use of different polarizations may bring information about the type of the domain wall and the relative weight of different contributions to the MSHG response. In the next section, some first experimental results will be presented and discussed.

3.12.2 Nonlinear magneto-optical microscopy of domains in magnetic garnet films

To demonstrate the new possibilities of nonlinear magneto-optical domain imaging, we studied several garnet films as discussed in sec. 3.11. The reason is that these garnet films show a very strong MSHG response and they are also known to possess a very clear labyrinth-type domain structure, with typical domain sizes of a few micrometers. The latter makes optical imaging very attractive (see also refs 28, 123).

The experimental setup of our nonlinear magneto-optical microscope is schematically presented in Fig. 3.45. As a light source we used the Ti:sapphire

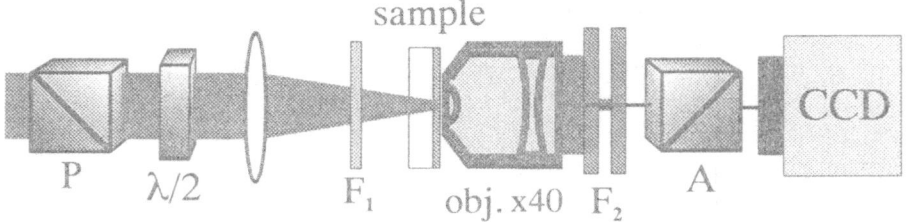

Fig. 3.45. Schematic nonlinear magneto-optical microscopy setup.

laser operating at the wavelength of 775 nm. A half-wavelength plate was used to rotate the linear polarization of the incoming light, which was focused on the sample into a spot of about 70 μm diameter. The average power of the pump beam on the sample was 100 mW, resulting in a peak power of nearly 4 GW cm^{-2}. We magnified the exposed area by a ×40 (NA = 0.65) objective in combination with an achromatic concave lens. After appropriate filtering the generated second harmonic intensity was imaged with a cooled CCD camera. The subtraction of the Gaussian-like background was applied afterwards to remove the spot-profile inhomogeneity in the image intensity due to the pump beam.

In accordance with the previous section, both even- and odd-fold rotation symmetry films were studied, that is (210) and (111) films. Faraday hysteresis measurements show that both films have a remanence magnetization, but only in the (111) film is the magnetization exactly perpendicular to the film surface. In the (210) sample the magnetization is tilted at an angle of about 18° with respect to the film normal. This value was derived from the difference between the saturation and the remanence magnetization, taking into account that at remanence the sample is in a single-domain state, which is supported by direct imaging.

The domain pattern of the demagnetized films was initially tested using our setup as a linear Faraday microscope. Figure 3.46(a) shows a typical labyrinth-type domain structure for the (210) film where the dark/light areas indicate 'up' and 'down' domains.

Next, second harmonic images of this very same domain structure were taken, for various values of the incoming linear polarization with respect to the crystal symmetry plane m (Fig. 3.46(b)–(f)). The SH images were recorded without analysing the outgoing light polarization. Remarkable changes in the magnetic contrast and in the SH intensity for the (210) garnet film were thus found. At 0°, i.e. the polarization parallel to m, the SHG domain pattern appears to be exactly the same as in the linear light. To follow all subsequent changes appearing in the magnetic structure, the domain walls in this image are marked with dashed lines (Fig. 3.46(b)). Rotating the incoming light polarization by 10°, a subdivision of the original domains was clearly observed (Fig. 3.46(c)). This subdivision is even more sharp at larger angles and at 35° the SH intensities in subdomains I and III become equal (Fig. 3.46(d)). At 90° and at 145° the magnetic structure looks very similar to the cases 0° and 35° respectively, but the contrast appears to be shifted by half a domain width (Fig. 3.46(e), (f)).

To analyse the observed images, one should recall that, in this configuration of normal incidence (and without polarization analysis), MSHG can only probe in-plane magnetization components, as follows from the MSHG selection rules from the previous section. There it was shown that for the even-fold symmetry films there is no contrast between up and down domains. This means that the different SH intensities observed in Fig. 3.46 correspond to different *in-plane* magnetizations. Therefore we can conclude that four domain types appear to exist in the (210) film ($M_I \neq M_{II} \neq M_{III} \neq M_{IV}$).

The recorded images can be analysed with the help of the rotational anisotropy measurements (see sec. 3.11), i.e. from the dependences of the SH intensity on

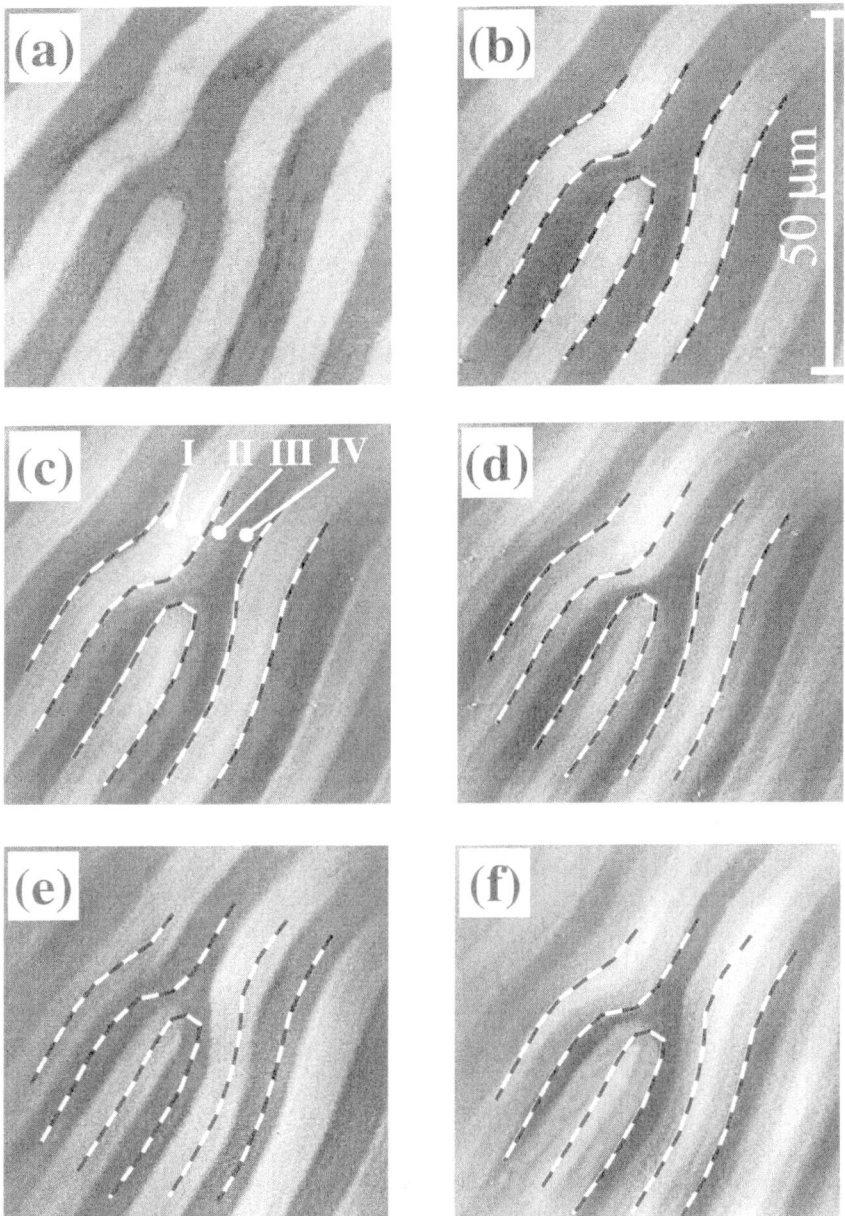

Fig. 3.46. (a) Linear and (b)–(f) second harmonic images of the magnetic domain structure in (210) oriented film. Input polarization was (b) 0°, (c) 10°, (d) 35°, (e) 90°, (f) 145° with respect to the symmetry plane m. Different subdomains (see Fig. 3.47) are indicated by I–IV. Dashed lines indicate the original position of domain walls (see text).

the azimuthal position of the sample with respect to the incoming light polarization and fixed (in-plane) magnetization. By a straightforward transformation, the

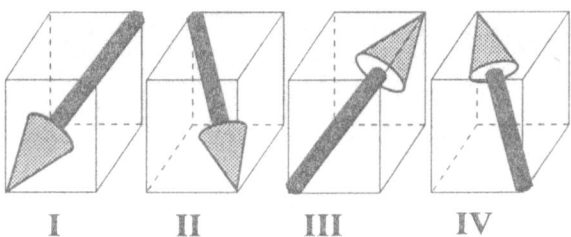

Fig. 3.47. Magnetization directions in four different domains in (210) film, as derived from the images of Fig. 3.46. Symmetry plane m coincides with the side-face plane.

MSHG dependence as a function of the incoming light polarization is obtained, with the in-plane M fixed in the sample. It is then possible to estimate the in-plane magnetization in every subdomain. In total, an appropriate model of the (210) domain structure can be derived (Fig. 3.47) with every 'up' and 'down' domain subdivided into two subdomains with different in-plane components. The in-plane magnetizations of the neighbouring domains are not collinear and the absolute value of the in-plane magnetization is the same in every subdomain. We should note that the observed image appeared to exist through the whole film, as was checked by changing the focusing depth, and can thus not be related to closure domains at the surface.

Thus, a nontrivial modulated domain structure is shown to exist in the (210) oriented garnet film, which only becomes distinguishable using combined linear and nonlinear magneto-optical microscopy. Attempts to observe domain walls in these films have not been successful so far.

Similar measurements on the (111) film showed no magnetic contrast, which means that no in-plane magnetization exists in the (111) sample. An analyser was then used to study the MSHG polarization rotation. To avoid any influence of the linear polarization rotation effect, a carefully crossed polarizer/analyser configuration was used. Surprisingly, the outcoming SH signal showed a strong contrast in this case (Fig. 3.48(b), though the linear picture showed no contrast between the domains at all (Fig. 3.48(a)). Hence, a *different Faraday rotation value* for the SH light is found in the different domains. Indeed, the resulting SH polarization is a vector sum of crystallographic and magnetic contributions, the latter having different signs for the two opposite domains. As soon as the polarization of the crystallographic part is not equal to that of the fundamental light (except for certain high-symmetry directions), the final SH polarization states are not symmetric with respect to the incoming light polarization (see insets in Fig. 3.48). This configuration clearly demonstrates the difference in the mechanism responsible for the magneto-optical contrast in the two cases, and may be used to make a correlation between the domain structure and crystallographic axes. The contrast disappears when the incoming light polarization coincides with one of the sample symmetry planes, in accordance with the symmetry analysis of sec. 3.12.1.

Figure 3.48(a) shows that, in the linear Faraday image, the domain walls show

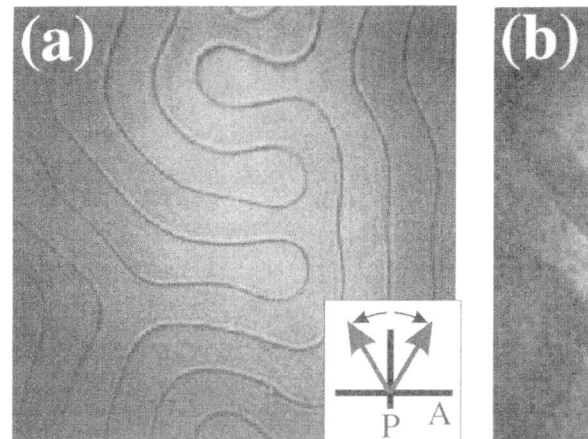 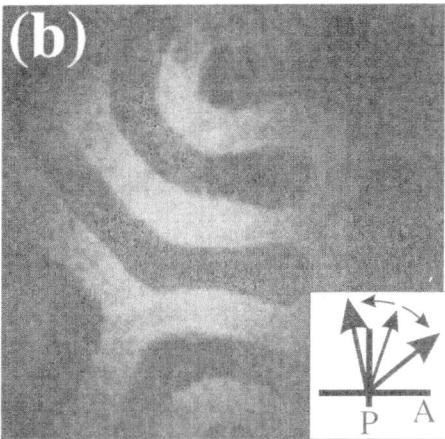

Fig. 3.48. (a) Linear and (b) second harmonic images of the magnetic domain structure in (111) oriented film obtained with crossed input and output polarizers.

up as dark lines (due to the scattering of the incident light on the walls). In the MSHG images domain walls similar to the predictions in Fig. 3.44 were sometimes observed. However, the signal-to-noise ratio at this stage was not good enough to make definite statements about the type of domain walls. This will require further study with an improved microscopy setup.

3.12.3 Nonlinear magneto-optical microscopy of antiferromagnetic domains in Cr_2O_3

In sec. 3.10 the sensitivity of MSHG for antiferromagnetic (AFM) ordering was already demonstrated. Here we will show how this sensitivity can be exploited to image antiferromagnetic domains, as was originally reported by Fiebig et al.[27]

The occurrence of AFM domains was predicted by Néel.[124] Owing to absence of a macroscopic magnetization, AFM domains are very hard to observe directly. Polarized light microscopy[125,126] and diffracted X-ray topography[127-129] are indirect techniques in the sense that they are sensitive not to the ordering of the spins, but to slight changes of the lattice associated with the magnetic ordering. Thus, these methods are suitable for the observation of orientation domains but they do not allow the distinction of 180° domains.[129] Neutron diffraction topograph is a technique that is suitable for the investigation of domains in antiferromagnets with any point symmetry,[130] since neutrons interact with spins and thus directly respond to the magnetic ordering. However, typical images of specimens with sizes of the order of several millimetres require exposure times of several hours with a resolution not better than 70 μm.[129]

In sec. 3.10, MSHG results on Cr_2O_3 were discussed. Cr_2O_3 belongs to the group of 180° antiferromagnets. Using resonant MSHG with circularly polarized light, domain observation appears to be quite straightforward.

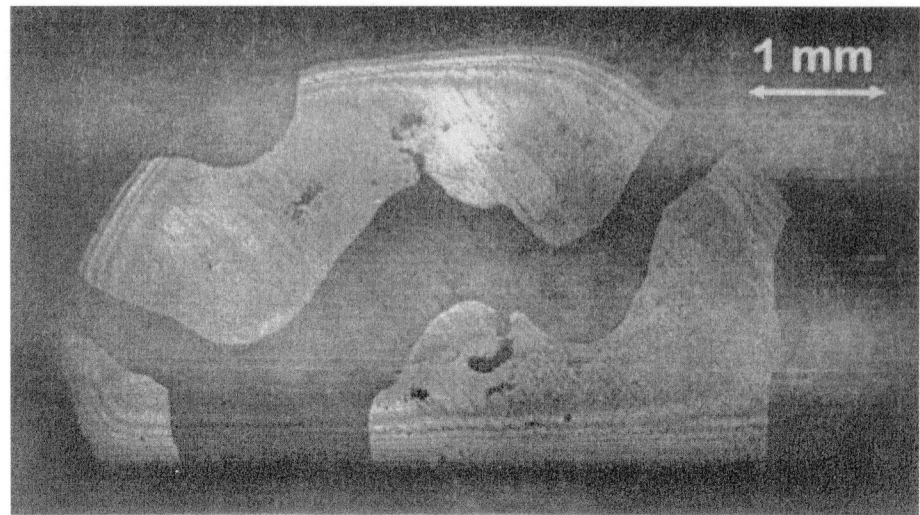

Fig. 3.49. SH image of Cr_2O_3 at $E_{SH} = 2.22$ eV with circularly polarized light. Sample size is 6×3 mm^2, thickness is 70 μm, $T = 2$ K, exposure time 3 min. The light/dark areas indicate different 180° domains. From ref. 27.

For the experiments a Nd:YAG pumped OPO system was used to generate the proper incident wavelength at 1.05 eV, which yields an SH signal at 2.1 eV that is strong for right- and weak for left-circularly polarized light.[27] The infrared beam from the OPO is enlarged to the size of the sample by means of a telescope. The helicity of the beam is set by use of a quarter-wave plate. SH light from the sample is projected on a liquid-nitrogen-cooled CCD camera. Scattered laser light is suppressed by use of optical filters. The X-ray-oriented bulk samples (size 10 mm^2 to 30 mm^2) are cut perpendicular to the optical axis from boules of Cr_2O_3 grown by the Verneuil method. Light is incident along the optical axis of the crystals. The spatial resolution is about 10 μm. Figure 3.49 was obtained at room temperature with right- and left-circularly polarized light, respectively. The exposure time for the sample of 6 mm $\times$ 3 mm was 3 min.

Figure 3.49 shows a clear domain structure in the Cr_2O_3 sample. By changing the polarization direction, the reversed contrast was observed, in accordance with the theoretical predictions of sec. 3.10. Heating above T_N made the domain structure disappear, only to reappear again after cooling down below T_N.

By applying a magnetic field along the z axis in the AFM phase, the Cr spins flop into the basal plane, maintaining their AFM order. The 3-fold rotation axis of the crystal is lost due to this transition. Consequently, three orientational domains (l_1, l_2, l_3) arise for either one of the two 180° domains (l^+, l^-). Thus, six different domains $(l_{1,2,3}^{+,-})$ are possible in the spin-flop phase. Despite a number of experimental investigations, the orientation of spins in the spin-flop phase of Cr_2O_3 is still an open problem.

Recently, Fiebig *et al.* have determined the domain structure in the spin-flop phase, making use of the two-photon selection rule for the point groups $\underline{2}/m\,(l\,\|\,y)$ and $2/\underline{m}\,(l\,\|\,x)$.[108] For these experiments, the Cr_2O_3 sample cut was perpendicular to the optical axis and mounted in a liquid-helium cryostat at $T = 1.8\,K$. A split coil magnet could deliver fields up to $6.6\,T$ along the optical axis. Three images with light polarized parallel to the respective 2-fold axes (y axes) of the crystal are sufficient for the identification of the six possible domains.

In conclusion, we have demonstrated in this section a new type of nonlinear magneto-optical microscopy which, in combination with standard linear microscopy, yields a wealth of additional and complementary information about magnetic domain structures. In particular, the higher-rank tensor and a polarization analysis of the incoming fundamental and/or second harmonic light gives information about the magnetic structure that cannot be obtained in a single configuration with linear microscopy. The sensitivity of MSHG to the breaking of crystallographic inversion symmetry also gives new possibilities for the observation of interface domain structures. This, together with the demonstrated sensitivity for antiferromagnetic domains, makes the further development of nonlinear magneto-optical microscopy very promising.

3.13 Summary and future prospects

In this chapter the newly developed nonlinear magneto-optical technique of magnetization-induced second harmonic generation was introduced and illustrated with various examples. MSHG combines interface specificity with large nonlinear magneto-optical effects, which makes it particularly attractive for the study of magnetic thin films and multilayers. Though the origin of the nonlinear magneto-optical effects is the same as for the linear case, namely the spin–orbit coupling, the effects can be quite different. This is partly related to the different (inhomogeneous) character of the wave equation. This leads to nonlinear Kerr rotations that are about one to two orders of magnitude larger than their linear equivalents. The differences in the symmetry properties of the nonlinear and linear optical tensors can be exploited to obtain further enhancements of up to three orders of magnitude. Yet, the interface sensitivity of SHG for centrosymmetric media is preserved. This interface sensitivity as well as the extreme sensitivity to quantum-well states in ultrathin metallic films has been discussed and demonstrated. The intrinsic differences between the linear and nonliner magneto-optical response can also lead to totally new effects, as was demonstrated for magnetic garnets and antiferromagnetic materials. This new area of nonlinear optics is not only interesting from a fundamental point of view, but also may be quite important for more applied studies. The first explored area so far is the application of nonlinear magneto-optics for the study of magnetic domain structures, showing that nonlinear imaging techniques can yield unique information regarding the domain structure in antiferromagnetic materials. For multilayered systems, MSHG could provide information about domain structures at deeper interfaces. Other future applications could be for *in situ* annealing studies of multilayer samples. The effects of sample preparation conditions, like sputtering pressure, substrate temperature, deposition speed, etc., on the interface magnetic properties can be probed by MSHG. In particular in combination with linear magneto-optics, this will give a powerful analysis tool for magnetic multilayer systems. Finally, the use of femtosecond pulsed lasers also opens the way for dynamic studies in the subpicosecond time range. This so far almost unexplored topic has become feasible by the availability of tunable, amplified femtosecond lasers. Nonlinear magneto-optics is expected to play an important role here.

Acknowledgements

I would like to thank all the (former and present) members of our research group in Nijmegen who have contributed to the work described here. In particular I want to mention H. A. Wierenga, R. Stolle, M. Groot Koerkamp, B. Koopmans, P. van Gelderen, F. Manders, W. de Jong, A. F. van Etteger, V. Kirilyuk, A. Petukhov, K. J. Veenstra and A. Kirilyuk. I thank D. Fröhlich for permission to use the figures for Cr_2O_3, R. Vollmer for his critical reading of the manuscript and W. Vink for all the assistance in preparing the manuscript. This work was part of the research programme of the Stichting voor Fundamenteel Onderzoek der Materie (FOM) and financially supported by the Nederlandse Organisatie voor Wetenschappelijk Onderzoek (NWO) and by European Research Programmes Brite Euram, Human Capital and Mobility (ERBCHBGCT930444, ERBCHRXCT940563, ERBCHB1CT951761, ERBFMB1CT960837) and the TMR network NOMOKE.

References

1. L. Falicov et al., *J. Mater. Res.* **5**, 1299 (1990) and references therein.
2. M. Campagna, D. T. Pierce, F. Meier, K. Sattler, and H. C. Siegmann, *Adv. Electr. Electr. Phys.* **41**, 113 (1976).
3. D. L. Abraham and H. H. Hopster, *Phys. Rev.* **B62**, 1157 (1989).
4. See for example: *Polarized Electrons in Surface Physics*, R. Feder, ed. (World Scientific, Singapore, 1985).
5. P. N. Argyres, *Phys. Rev.* **97**, 334 (1955).
6. Y. R. Shen, *The Principles of Nonlinear Optics* (Wiley, New York, 1984).
7. Th. Rasing, *Appl. Phys.* **A59**, 531 (1994).
8. T. F. Heinz, in *Nonlinear Surface Electromagnetic Phenomena*, H. E. Ponath and G. I. Stegeman, eds. (North Holland, Amsterdam, 1991), pp. 335.
9. G. L. Richmond, J. M. Robinson, and V. L. Shannon, *Prog. Surf. Sci.* **28**, 1 (1988).
10. P. A. Franken, A. E. Hill, C. W. Peters, and G. Weinreich, *Phys. Rev. Lett.* **7**, 118 (1961).

11. N. Bloembergen and P. S. Pershan, *Phys. Rev.* **128**, 606 (1962).
12. J. F. McGilp, *Prog. Surf. Sci.* **49**, 1 (1995).
13. S. Janz and H. M. van Driel, *Int. J. Nonlinear Opt. Phys.* **2**, 1 (1992).
14. X. D. Zhu, Th. Rasing, and Y. R. Shen, *Phys. Rev. Lett.* **61**, 2883 (1988); *Chem. Phys. Lett.* **155**, 459 (1989).
15. W. de Jong, A. F. van Etteger, C. A. van 't Hof, P. J. van Hall, and Th. Rasing, *Surf. Sci.* **331–333**, 1372 (1995).
16. O. A. Aktsipetrov and E. D. Mishina, *Sov. Phys. Dokl.* **29**, 37 (1984).
17. Ru-Pin Pan, H. D. Wei, and Y. R. Shen, *Phys. Rev.* **B39**, 1229 (1989); Ru-Pin Pan and Y. R. Shen, *Chin. J. Phys. (Tapei)*, **25**, 175 (1987).
18. W. Hübner and K. H. Bennemann, *Phys. Rev.* **B40**, 5973 (1989).
19. J. Reif, J. C. Zink, C. M. Schneider and J. Kirschner, *Phys. Rev. Lett.* **67**, 2878 (1991).
20. G. Spierings, V. Koutsos, H. A. Wierenga, M. W. J. Prins, D. Abraham, and Th. Rasing, *Surf. Sci.* **287**, 747 (1993); *J. Magn. Magn. Mat.* **121**, 109 (1993).
21. H. A. Wierenga, W. de Jong, M. W. J. Prins, Th. Rasing, R. Vollmer, A. Kirilyuk, H. Schwabe, and J. Kirschner, *Phys. Rev. Lett.* **74**, 1462 (1995).
22. U. Postogowa, W. Hübner, and K. H. Bennemann, *Phys. Rev.* **B49**, 10031 (1994); *Appl. Phys.* **A59**, 611 (1994).
23. O. A. Aktsipetrov, O. V. Braginskii, and D. A. Esikov, *Sov. J. Quantum Electron.*, **20**, 259 (1990).
24. J. Reif, C. Rau, and E. Matthias, *Phys. Rev. Lett.* **71**, 1931 (1993).
25. B. Koopmans, M. Groot Koerkamp, Th. Rasing, and H. v.d. Berg, *Phys. Rev. Lett.* **74**, 3692 (1995).
26. M. Fiebig, D. Fröhlich, B. B. Krichevtsov, and R. V. Pisarev, *Phys. Rev. Lett.* **73**, 2127 (1994).
27. M. Fiebig, D. Fröhlich, G. Sluyterman v.L., and R. V. Pisarev, *Appl. Phys. Lett.* **66**, 2906 (1995).
28. A. Kirilyuk, V. Kirilyuk, Th. Rasing, V. V. Pavlov, and R. V. Pisarev, *J. Magn. Soc. Jpn.* **20**, 361–364 (1996).
29. A. Kirilyuk, Th. Rasing, R. Mégy, and P. Beauvillain, *Phys. Rev. Lett.* **77**, 4608–4611 (1996).
30. E. Beaurepaire, J. C. Merle, A. Daunois, and J. Y. Bigot, *Phys. Rev. Lett.* **76**, 4250 (1996).
31. M. Weber and A. Liebsch, *Phys. Rev.* **B35**, 741 (1987); A. Liebsch, *Phys. Rev.* **B36**, 7378 (1987).
32. Recent work by R. Vollmer, M. Straub, and J. Kirschner, *Surf. Sci.* **352–354**, 684 (1996), shows that bulk contributions can be sizeable even for metallic surfaces, depending on the polarization and frequency of the incident fundamental radiation.
33. P. Guyot-Sionnest and Y. R. Shen, *Phys. Rev.* **B35**, 4420 (1987); P. Guyot-Sionnest, W. Chen, and Y. R. Shen, *Phys. Rev.* **B33**, 8254 (1986).
34. Y. R. Shen, 'Applicaton of optical second-harmonic generation to surface science', in *Chemistry and Structure at Interfaces: New Laser and Optical Techniques*, R. B. Hall and A. B. Ellis, eds. (VCH Publishers, Deerfield Beach, FL, 1986), Chap. 4, p. 151.
35. B. Dick, A. Gierulski, G. Marowsky, and G. A. Reider, *Appl. Phys.* **B38**, 107 (1985).
36. M. S. Yeganeh, J. Qi, J. P. Culver, A. G. Yodh, and M. C. Tamargo, *Phys. Rev.* **B46**, 1603 (1992).
37. B. Koopmans, A. Anema, H. T. Jonkman, G. A. Sawatzky, and F. van der Woude, *Phys. Rev.* **B48**, 2759 (1993).

38. E. D. Palik, ed. *Handbook of Optical Constants of Solids* (Academic Press, Orlando, FL, 1985).
39. J. E. Sipe, *J. Opt. Soc. Am.* **B4**, 481 (1987).
40. T. F. Heinz, *Nonlinear Optics of Surfaces and Adsorbates* (Thesis, Lawrence Berkeley Laboratory, Berkeley, USA, 1982).
41. R. P. Hunt, *J. Appl. Phys.* **38**, 1652 (1967).
42. C.-Y. You and S.-C. Shin, *Appl. Phys. Lett.* **69**, 1315 (1996).
43. M. J. Freiser, *IEEE Trans Magn.* **4**, 152 (1968).
44. U. Pustogowa, W. Hübner, and K. H. Bennemann, *Appl. Phys.* **A59**, 611 (1994).
45. N. N. Akkmediev, S. B. Borisov, A. K. Zvezdin, I. L. Lyubchanskii and Yu. V. Melikhov, *Sov. Phys. Sol. State* **27**, 650 (1985); S. Kielich and R. Zawadny, *Acta Physica Polonica* **A43**, 579 (1973); *Optica Acta* **20**, 867 (1973).
46. Note that Table 3.1 is slightly different from the results obtained by Ru-Pin Pan in ref. 17; this is probably due to the overlooking of additional symmetry elements of the 100 surface.
47. H. A. Wierenga, *Magnetization Induced Optical Second Harmonic Generation on Magnetic Multilayer* (PhD Thesis, University of Nijmegen, The Netherlands, 1995).
48. P. Guyot-Sionnest and Y. R. Shen, *Phys. Rev.* **B38**, 7985 (1988).
49. B. Koopmans, *Interface and Bulk Contributions in Optical Second-Harmonic Generation* (PhD Thesis, University of Groningen, The Netherlands, 1993).
50. W. Hübner and K. H. Bennemann, *Phys. Rev.* **B52**, 13411 (1995).
51. W. Hübner and K. H. Bennemann, *Vacuum* **41**, 514 (1990).
52. W. Hübner and K. H. Bennemann, *Surf. Sci.* **242**, 299 (1991).
53. W. Hübner, *Phys. Rev.* **B42**, 11553 (1990).
54. U. Pustogowa, W. Hübner, and K. H. Bennemann, *Phys. Rev.* **B48**, 8607 (1993).
55. U. Pustogowa, W. Hübner, and K. H. Bennemann, *Surf. Sci.* **307–309**, 1129 (1994).
56. Y. R. Shen, *Annu. Rev. Mater. Sci.* **16**, 69 (1986).
57. A. J. Freeman and C. L. Fu, *J. Appl. Phys.* **61**, 3356 (1987).
58. H. A. Wierenga, M. W. J. Prins, D. L. Abraham, and Th. Rasing, *Phys. Rev.* **B50**, 1282 (1994).
59. R. Vollmer, A. Kirilyuk, H. Schwabe, J. Kirschner, H. A. Wierenga, W. de Jong, and Th. Rasing, *J. Magn. Magn. Mat.* **148**, 295 (1995).
60. M. Groot Koerkamp and Th. Rasing, *J. Magn. Magn. Mat.* **156**, 213 (1996).
61. R. K. Chang, J. Ducuing, and N. Bloembergen, *Phys. Rev. Lett.* **15**, 6 (1995).
62. R. Stolle, G. Marowsky, E. Schwarzberg, and G. Berkovic, *Appl. Phys.* **B63**, 491 (1996).
63. R. Superfine, J. H. Huang, and Y. R. Shen, *Opt. Lett.* **15**, 1276 (1990).
64. R. Stolle, K. J. Veenstra, F. Manders, Th. Rasing, H. van den Berg, and N. Persat, *Phys. Rev.* **B55**, R4925 (1997).
65. K. Kemnitz *et al.*, *Chem. Phys. Lett.* **131**, 285 (1986).
66. G. Berkovic and E. Schwarzberg, *Appl. Phys.* **B53**, 333 (1991).
67. D. R. Lide (Editor in Chief), *CRC Handbook of Chemistry and Physics* (CRC Press, Boston, 1991).
68. R. Vollner, M. Straub, and J. Kirschner, *Surf. Sci.* **352**, 937 (1996).
69. Th. Rasing and H. A. Wierenga, *Ferroelectrics* **162**, 217 (1994).
70. M. Straub, R. Vollmer, and J. Kirschner, *Phys. Rev. Lett.* **77**, 743 (1996).
71. E. R. Moog, C. Liu, S. D. Bader, and J. Zak, *Phys. Rev.* **B 39**, 6949 (1989).
72. R. Gamble and P. H. Lissberger, *J. Opt. Soc. Am.* **a5**, 1533 (1988).
73. H. A. Wierenga, M. W. J. Prins, and Th. Rasing, *Physica* **B204**, 281 (1995).

REFERENCES

74. P. van Gelderen, S. Crampin, Th. Rasing, and J. Inglesfield, *Phys. Rev.* **B54**, R2343 (1996).
75. W. Weber, C. H. Back, A. Bischof, Ch. Würsch, and R. Allenspach, *Phys. Rev. Lett.* **76**, 1940 (1996).
76. P. J. H. Bloemen, M. T. Johnson, M. T. H. van de Vorst, R. Coehoorn, J. J. de Vries, R. Jungblut, J. aan de Stegge, A. Reinders, and W. J. M. de Jong, *Phys. Rev. Lett.* **72**, 764 (1994).
77. W. Clemens, T. Kachel, O. Rader, E. Vescovo, S. Blügel, C. Carbone, and W. Eberhardt, *Solid State Commun.* **81**, 739 (1992).
78. T. Suzuki, D. Weller, C. A. Chang, R. Savoy, T. Huang, B. A. Gurney, and V. Speriosu, *Appl. Phys. Lett.* **64**, 2736 (1994).
79. P. B. Johnson and R. W. Christy, *Phys. Rev.* **B9**, 5056 (1974); *Phys. Rev.* **B6**, 4370 (1972).
80. H. A. Wierenga, W. de Jong, M. W. J. Prins, Th. Rasing, R. Vollmer, A. Kirilyuk, H. Schwabe, and J. Kirschner, *Surf. Sci.* **331–333**, 1294 (1995).
81. G. Penissard, P. Meyer, J. Ferre, and D. Renard, *J. Magn. Magn. Mat.* **146**, 55 (1995).
82. T. M. Crawford, C. T. Rogers, T. J. Silva, and Y. K. Kim, *Appl. Phys. Lett.* **68**, 1573 (1996).
83. T. M. Crawford, C. T. Rogers, T. J. Silva, and Y. K. Kim, *IEEE Trans. Magn.* **32**, 4087 (1996).
84. T. J. Silva, T. M. Crawford, C. T. Rogers, and Y. K. Kim, *OSA Techn. Digest Series* **11**, 299 (1996).
85. T. M. Crawford, C. T. Rogers, T. J. Silva, and Y. K. Kim, *J. Appl. Phys.*, **8**, 4354 (1997).
86. D. E. Aspnes and A. A. Studna, *Phys. Rev.* **B27**, 985 (1983).
87. D. O. Smith, *J. Appl. Phys.* **36**, 1120 (1965).
88. M. Groot Koerkamp and Th. Rasing, *Surf. Sci.* **352–354**, 933 (1996).
89. W. Hübner and K. H. Bennemann, *Phys. Rev.* **B52**, 13411 (1995).
90. A. J. Kolk and M. Orlovic, *J. Appl. Phys.* **34**, 1060 (1963).
91. K. Böhmer, J. Hohlfeld, and E. Matthias, *Appl. Phys.* **A60**, 203 (1995).
92. P. Bruno, *J. Appl. Phys.* **76**, 6972 (1994).
93. O. Keller, A. Zayats, A. Liu, K. Pedersen, F. A. Pudomin, and E. A. Vinogradov, *Opt. Commun.* **115**, 137 (1995).
94. J. E. Ortega, F. J. Himpsel, G. J. Mankey, and R. F. Willis, *Phys. Rev.* **B47**, 1540 (1993); *Phys. Rev. Lett.* **69**, 844 (1992).
95. P. van Gelderen, S. Crampin, and J. Inglesfield, *Phys. Rev.* **B53**, 9115 (1996).
96. P. van Gelderen, *Spin-Polarized Quantum Well States in Interlayers and Overlayers* (*Electronic Structur Calculations*), University of Nijmegen (1995).
97. N. B. Brookes, Y. Chang, and P. D. Johnson, *Phys. Rev. Lett.* **67**, 354 91991).
98. C. Carbone, E. Vescovo, O. Rader, W. Gudat, and W. Eberhardt, *Phys. Rev. Lett.* **71**, 2805 (1993).
99. M. T. Johnson, S. T. Purcell, N. W. E. McGee, R. Coehoorn, J. aan de Stegge, and W. Hoving, *Phys. Rev. Lett.* **68**, 2688 (1992).
100. See for example, S. Krompiewski, F. Süss, B. Zellermann, and U. Kray, *J. Magn. Magn. Mat.* **148**, 198 (1995) and other contributions to the 14th Int. Coll. Magn. Films and Surf. and E-MRS Symp. on Magnetic Ultrathin films, Multilayers and Surfaces, September 1994, Düsseldorf, *J. Magn. Magn. Mat.* **148** (1995).
101. R. Vollmer *et al.*, to be published.

102. R. Mégy, A. Bounouh, Y. Suzuki, P. Beauvillain, P. Bruno, C. Chappert, B. Lecuyer, and P. Veillet, *Phys. Rev.* **B51**, 5586 (1995).
103. C. Chappert, K. Le Dang, P. Beauvillain, H. Hurdequint, and D. Renard, *Phys. Rev.* **B34**, 3192 (1986).
104. D. E. Aspnes, E. Kinsbron, and D. D. Bacon, *Phys. Rev.* **B21**, 3290 (1980).
105. T. A. Luce, W. Hübner, and K. H. Bennemann, *Phys. Rev. Lett.* **77**, 2810 (1996).
106. Landolt-Börnstein: *Eigenschaften der Materie in ihren Aggregatzustánden*, 8, Teil: *Optische Konstanten* (Springer-Verlag, Berlin, 1962).
107. A. Bounouh, C. Train, P. Beauvillain, P. Bruno, C. Chappert, R. Mégy, and P. Veillet, *J. Magn. Magn. Mat.* **165**, 484 (1997).
108. M. Fiebig, D. Fröhlich, and H. J. Thiele, *Phys. Rev.* **B54**, R12681 (1996).
109. B. B. Krichevtsov, V. V. Pavlov, and R. V. Pisarev, *Sov. Phys. Solid State* **31**, 1142 (1989).
110. R. V. Pisarev, B. B. Krichevtsov, V. N. Gridnev, V. P. Klin, D. Fröhlich, and Ch. Pahlke-Lerch, *J. Phys.* **C5**, 8621 (1993).
111. R. V. Pisarev, M. Fiebig, and D. Fröhlich, *Ferroelectrics* **204**, 1 (1997).
112. M. Fiebig, D. Fröhlich, and R. V. Pisarev, St. Leute, *Appl. Phys.* **B66**, 265 (1998).
113. A.-M. Janner, *Second-Harmonic Generation, a selective probe for excitons* (Ph.D Thesis, University of Groningen, The Netherlands, 1998).
114. G. Winkler, *Magnetic Garnets* (Vieweg, Braunschweig, 1981).
115. *Physics of Magnetic Garnets*, A. Paoletti, ed. (North Holland, Amsterdam, 1978).
116. *Magnetic Garnet Films*, special issue of *Thin Solid Films* **114** (1984).
117. G. Petrocelli, S. Martelucci, and M. Richetta, *Appl. Phys. Lett.* **63**, 3402 (1993).
118. V. V. Pavlov, R. V. Pisarev, A. Kirilyuk, and Th. Rasing, *Phys. Rev. Lett.* **78**, 2004 (1997).
119. O. A. Aktsipetrov, V. A. Aleshkevich, A. V. Melnikov, T. V. Misuryaev, T. V. Murzina, and V. V. Randoshkin, *J. Magn. Magn. Mat.* **165**, 421 (1997).
120. R. Schäfer and A. Hubert, *Phys. Stat. Sol.* (a) **118**, 271 (1990); V. Kambersky, *ibid*, **125**, K117 (1991).
121. A. Oleš, F. Kajzar, M. Kucab, and W. Sikora, *Magnetic Structures Determined by Neutron Diffraction* (Institute of Nuclear Techniques, Cracow, 1970).
122. A. V. Petukhov, I. L. Lyubchanskii, and Th. Rasing, *Phys. Rev.* **B** (accepted).
123. V. Kirilyuk, A. Kirilyuk, and Th. Rasing, *Appl. Phys. Lett.* **70**, 2306 (1997).
124. L. Néel, *Proceedings of the International Conference on Theoretical Physics*, Kyoto, Tokyo (1953), p. 701.
125. W. L. Roth, *J. Appl. Phys.* **31**, 2000 (1960).
126. H. Kondoh and T. Takeda, *J. Phys. Soc. Jpn.* **19**, 2041 (1964).
127. S. Saito, *J. Phys. Soc. Jpn.* **17**, 1287 (1962).
128. B. K. Tanner, M. Safa, D. Midgley, and J. Bordas, *J. Magn. Magn. Mater.* **1**, 337 (1976).
129. J. Baruchel, *Physica* **B192**, 79 (1993).
130. H. Alperin, *Proc. Int. Conf. Magn.* 128 (1973).

4

FEMTOSECOND TIME-RESOLVED LINEAR AND SECOND-ORDER REFLECTIVITY OF METALS

J. Hohlfeld, U. Conrad, J. G. Müller, S. S. Wellershoff and
E. Matthias
Freie Universität Berlin, Berlin, Germany

4.1 Introduction

Second harmonic generation (SHG) in reflection off surfaces is a background-free technique that offers several advantages, covered by an extensive literature that cannot all be cited here. We only want to recall some features of the technique and quote a few selected examples from the literature.

For centrosymmetric materials SHG is sensitive to surface properties like symmetry, electronic structure, roughness and adsorbates. The symmetry dependence is equivalent to a 2D crystallography and has been demonstrated for insulators,[1] semiconductors,[2,3] metals,[4] and monolayers.[5,6] The symmetry dependence also makes SHG perceptive to surface[7-10] and bulk magnetization.[11]

The sensitivity to electronic structure[12] allows identification of surface states[13] and dynamical screening at metal surfaces,[14] as well as monitoring of transient electron temperatures.[15] Also, surface plasma oscillations in small metal islands and clusters have been investigated by SHG.[16,17] Regarding surface morphology, the influence of surface steps was studied[18,19] and it was shown that roughness can cause a large enhancement of the SHG yield.[20] The literature on adsorbate studies with SHG is abundant and we only cite here surface diffusion of CO on Ni(111)[21] and two typical examples for oxygen on Cu(110)[22] and on Ag(110).[23] But perhaps the most exciting development is the application of SHG to the study of electronic structure of interfaces,[24] magnetism[25] in multilayer systems and the investigation of quantum wells.

These fascinating aspects of SHG are still more elating when taking into account that the availability of femtosecond laser pulses opens the possibility to carry out ultrafast time-resolved SHG experiments in pump–probe mode. This gives access to electron dynamics at surfaces, in multilayers, in adsorbates and in magnetic systems. Striking examples are ultrafast order–disorder transitions in semiconductors[2,26-28] and the observation of surface optical phonons.[29]

Another interesting aspect would be to uncover possible differences between electron dynamics at surfaces or interfaces and in the bulk by pump–probe

SHG. However, such type of investigation requires detailed knowledge about the influence of electron temperature T_e on the linear dielectric functions of the fundamental, $\varepsilon(\omega)$, and second harmonic light, $\varepsilon(2\omega)$, as well as on the nonlinear susceptibility $\chi^{(2)}$, in the formalism of Sipe et al.[30] Information about the electron temperature dependence of the dielectric functions can readily be obtained from comparison of transient linear reflectivity measurements with theory. However, the absence of quantitative theoretical predictions about $\chi^{(2)}(T_e)$ limits conclusions about electron temperature dynamics from transient SHG measurements.

In view of this situation, we decided to carry out a careful study of the electron temperature variation of the dielectric function for noble metals and then predict the time dependence of pump–probe SHG on the basis of Fresnel factors only.[30] Deviations of the measured SHG relaxation from these predictions are then interpreted as electron temperature dependence of the nonlinear susceptibility $\chi^{(2)}$. Accordingly, in sec. 4.2 we will describe measurements of transient linear reflectivities $R(t)$ of Cu, Ag and Au. Theoretical analysis of the data then provides information about the variation of R with electron temperature.

Section 4.3 is devoted to SHG experiments. The first subsection deals with the formalism for data analysis. In sec. 4.3.2 we will present pump–probe measurements on polycrystalline Cu, Ag and Au and analyse the second-order reflectivities on the basis of the electron temperature dependence of the dielectric functions $\varepsilon(\omega)$ and $\varepsilon(2\omega)$.

In sec. 4.3.3 we will present polarization-dependent SHG measurements on Al, Cu, Ti, Cr and Ni (demagnetized) in order to demonstrate the influence of electronic structure. In these experiments, the yield of s- and p-polarized second harmonic (SH) light is observed as a function of polarization of the fundamental light.

Section 4.3.4 deals with SHG measurements of magnetization dynamics in polycrystaline Ni. These experiments demonstrate that time-resolved SHG is the technique of choice for investigating ultrafast electron and spin relaxation processes because of the much greater sensitivity of SHG to electronic symmetry, compared to linear reflectivity. It is shown that for Ni the electron temperature is established in about 0.3 ps and that magnetization breakdown and recovery closely follow electron heating and relaxation.

4.2 Transient linear reflectivities

4.2.1 Dynamics of optically excited electrons

In this section we will introduce the electron distribution dynamics following optical excitations and its influence on the linear reflectivities of Cu, Ag and Au. It provides the general background for the interpretation of femtosecond time-resolved data on both linear and second-order reflectivities of these metals.

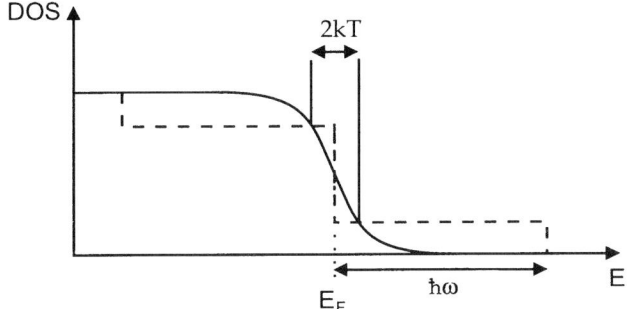

Fig. 4.1. Electron distribution in the conduction band generated initially by optical excitation with photon energy $\hbar\omega$ (dashed line) and the resulting electron distribution after thermalization (solid line). The size of the initial distribution is magnified about a factor of 4.

Electron distribution dynamics

The excited electron relaxation in noble metals and the resulting redistribution of the injected energy is governed by the degree of optical excitation, by electron–electron and electron–phonon scattering, by ballistic energy transport through nonequilibrium electrons, and finally by diffusive energy transport through thermalized electrons.

Optical excitation by a femtosecond pump pulse creates a non-Fermi distribution of the conduction electrons, which is centred around the Fermi level.[31] The width and height of this rectangular distribution, shown in Fig. 4.1, are given by twice the pump photon energy and the density of absorbed energy, respectively.

Immediately following optical excitation, the energy is redistributed among the electrons through inelastic electron–electron collisions with a typical rate of 10^{14} s^{-1}, leading to a hot Fermi distribution with temperature $T_e \gg T_0$ (T_0 is the temperature before irradiation). Related changes in electronic occupancy are much larger within the energy interval of $\pm kT_e$ around the Fermi level than the initial changes due to optical excitation, but the corresponding width of $2kT_e$ is always small compared to $2\hbar\omega$ (cf. Fig. 4.1). Competing with this thermalization process is ballistic energy transport by nonequilibrium electrons with approximately Fermi velocity. Advancing electron thermalization reduces the amount of nonequilibrium electrons and diminishes the importance of ballistic energy transport. Diffusive energy transport takes place by already thermalized electrons. Both mechanisms lead to an effective energy flow out of the initially excited volume.

During and following electron thermalization there is energy transfer from the electron gas to the lattice through electron–phonon scattering. This energy transfer causes the electron gas and lattice to reach a local equilibrium after a few picoseconds at $T_e = T_l$ (T_l denotes lattice temperature) $> T_0$.

The final stage of the equilibration process is thermal diffusion within the lattice. The lattice temperature recovers on a microsecond timescale to the starting temperature T_0 plus a small fluence-dependent rise ΔT. Because of the long timescale it will not be discussed here.

Regarding the thermalization of the electron gas, two different electron thermalization times are introduced in the literature. The first one is the total thermalization time, which refers to the establishment of a complete Fermi distribution. The second one refers to thermalization within a small energy range and is useful for pump–probe experiments. It gives the time required by the electron gas to reach the Fermi distribution within an energy interval around a specific energy determined, for example, by the probing photon. Since the electronic occupancy reaches the Fermi distribution much faster for large electron energy differences to the Fermi level compared to small ones, total thermalization times are always longer than or, at best, equal to thermalization times at a certain energy.

Experimental and theoretical investigations of electron thermalization demonstrated that the total thermalization time decreases with increasing absorbed pump fluence. For example, by time-resolved photoemission spectroscopy on thin gold films Fann et al.[31] determined total electron thermalization times of about 1300 fs and 700 fs for absorbed fluences of 120 μJ cm^{-2} and 300 μJ cm^{-2}, respectively. A somewhat contrasting behaviour is reported by Sun et al.,[32] who used transient thermoreflection spectroscopy to explore the dependence of electron thermalization in gold on the probing photon energy. They found almost constant electron thermalization times of about 500 fs for electron energy mismatches ≥ 0.3 eV with respect to the Fermi level in the fluence range 2.5–200 μJ cm^{-2}. This is about a factor of 2.6 shorter than the thermalization time determined within a ± 0.3 eV interval around the Fermi level and the corresponding total thermalization time reported by Fann et al.[31]

In our time-resolved measurements of second-order reflectivities, the absorbed fluence was about 1.5 mJ cm^{-2}. Extrapolating the fluence and energy dependence of electron thermalization times given above, these absorbed fluences result in total thermalization times of about 150 fs, which reduce to 60 fs at the probed electron energy. Since these times are comparable to the used probe pulse width of about 100 fs, it is justified to neglect the influence of nonequilibrium electrons on our second harmonic data and to interpret them in terms of the two-temperature model.

Two-temperature model

The two-temperature model (TTM) was developed by Anisimov et al.[33] It treats the electron gas and the phonon bath as two distinguishable coupled systems under the assumption that each of them stays in local thermal equilibrium. When the Debye temperature of the sample is smaller than the starting temperature T_0, the TTM describes the time variation of electron and lattice temperatures by the coupled equations

$$C_e(T_e)\frac{\partial T_e}{\partial t} = \frac{\partial}{\partial z}\left(K_e \frac{\partial T_e}{\partial z}\right) - g(T_e - T_l) + P(z,t), \quad (4.2.1)$$

$$C_l \frac{\partial T_l}{\partial t} = g(T_e - T_l), \quad (4.2.2)$$

$$P(z,t) = (1-R)I_0 \alpha \, e^{-\alpha|z|} e^{-(t/\tau)^2}. \quad (4.2.3)$$

Contributions of lattice diffusivity and temperature gradients parallel to the surface are neglected since they are small compared to the diffusivity of the electron gas and to the gradients perpendicular to the surface. The source term $P(z,t)$ depicts heating of the electrons by a pump pulse of intensity I_0 and duration τ. Absorption is determined by the extinction coefficient α and the linear reflectivity R. The electron thermal conductivity perpendicular to the surface is denoted by K_e. Energy transfer from the electrons to the lattice is governed by the temperature difference and the electron–phonon coupling constant g. Lattice and electronic heat capacities are given by C_l and C_e, respectively.

The heat capacity of the electron gas depends linearly on electron temperature T_e: $C_e(T_e) = A_e T_e$, where A_e is a constant.[34] According to Sommerfeld's model,[34] the electron thermal conductivity is $K \propto T_e/\nu$, where the collision frequency ν is the sum of electron–electron and electron–phonon collision frequencies, $\nu = \nu_{ee}(T_e) + \nu_{ep}(T_l)$.[35] Their dependence on temperature is approximately $\nu_{ee}(T_e) \propto T_e^2$ and $\nu_{ep}(T_l) \propto T_l$. At lower temperatures $\nu_{ee}(T_e) \ll \nu_{ep}(T_l)$, and it is not before $T_e \sim 8000$ K where ν_{ee} takes over. Note, however, that these considerations pertain only once the electron temperature is established.

Numerical solutions of the two-temperature model which demonstrate the different influence of electron–phonon coupling, thermal electron conductivity and pump-pulse intensity on the time evolution of electron and lattice temperatures at the surface are shown in Fig. 4.2.

The curves in Fig. 4.2 illustrate that the different effects of g and K_e on the relaxation allow an unambiguous determination of both quantities for a given intensity. When electron diffusion dominates the relaxation, T_e decays almost exponentially. For $K_e = 0$ on the other hand, a linear decrease of electron temperature is observed during the first several picoseconds. Neglecting electron diffusion, the TTM reduces for large electron temperatures ($T_e \gg T_l$) to a linear dependence on time (see Fig. 4.2(b))

$$T_e(t) = T_{e,\max} - (g/A_e)t. \quad (4.2.4)$$

Thus, any observation of a linearly decreasing electron temperature is a unique signature of absent electron diffusion. The same signature is expected when the temperature gradient vanishes, $\nabla_z T_e = 0$ (cf. eqn (4.2.4)).

Regarding the validity of the TTM it should be remembered that the model

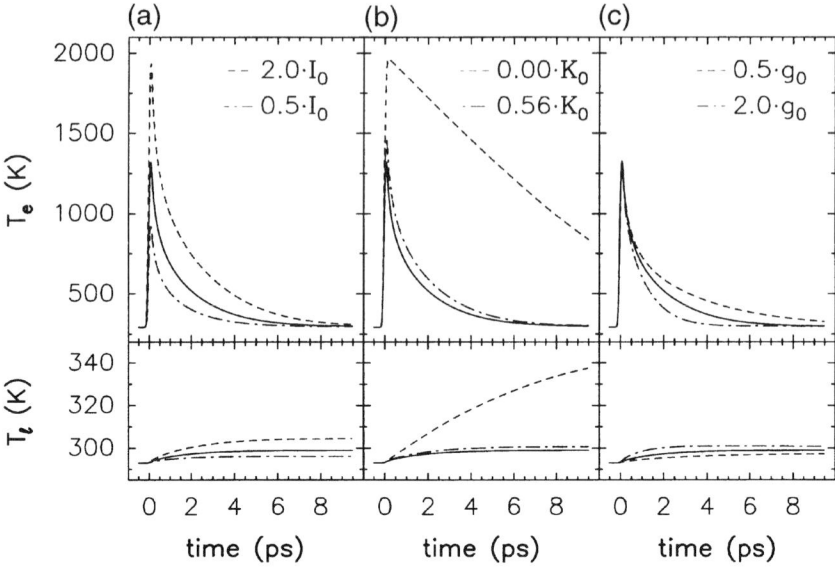

Fig. 4.2. Numerical results for transient electron (upper frames) and lattice (lower frames) temperatures at the surface for three different pump intensities I (a), electronic thermal conductivities K_e (b) and electron–phonon coupling constants g (c). Solid lines represent the reference values $I_0 = 1 \times 10^{15}$ W m^{-2}, $K_0 = 1$ W m^{-1} K^{-1} and $g_0 = 1 \times 10^{16}$ W m^{-3} K^{-1}.

assumes the existence of an electron temperature and therefore presupposes instantaneous electron thermalization. Consequently, its applicability depends on heating rate and time resolution. For high pump fluences the TTM can be utilized after a few hundred femtoseconds; for small fluences its use is not justified during much longer time intervals. In the latter case, either one has to extend the TTM by a rate equation which considers the time dependence of nonequilibrium electrons or, alternatively, one needs to solve the corresponding Boltzmann equation.[32,36] In any case, after electron thermalization is completed the TTM is valid and can be used to interpret transient reflectivity data at pump–probe delays greater than the electron thermalization time. Notice, however, that the strong dependence of electron temperature relaxation on the vertical distribution of T_e, which is created during the electron thermalization process, restricts the stringency of such interpretation to uniformly heated samples. This result will be of great importance for the interpretation of film-thickness-dependent measurements of transient linear reflectivities presented in sec. 4.2.4.

Since the measured time change of linear and nonlinear reflectivities results from the convolution of the time-dependent electron temperature and related variations of optical properties, any data analysis in terms of electron temperature dynamics requires knowledge about the variation of linear and nonlinear reflectivities with electron temperature. This point will be addressed in the following section.

4.2.2 Temperature dependence of linear reflectivities

Linear optical properties are described by the dielectric function $\varepsilon(\omega) = \varepsilon_1(\omega) + i\varepsilon_2(\omega)$ and most analyses of transient reflectivity measurements are based on the assumption of a linear variation of the reflectivity R with T_e:

$$R(T_e) = R_0 + \left[\frac{\partial R}{\partial \varepsilon_1}\frac{\partial \varepsilon_1}{\partial T_e} + \frac{\partial R}{\partial \varepsilon_2}\frac{\partial \varepsilon_2}{\partial T_e}\right]\Delta T_e. \qquad (4.2.5)$$

However, it has been demonstrated that the time dependences of transient linear reflectivity and electron temperature are not, in general, proportional to each other.[37] This observation enforces the use of a more sophisticated theoretical model to describe the variation of $\varepsilon(\omega)$ with electron temperature. Such a model has to consider the following temperature-related effects which may affect $\varepsilon(\omega)$:[38] *Broadening of the Fermi distribution* affects all interband transitions that originate or terminate at states near the Fermi level. *Volume thermal expansion* decreases the plasma frequency and causes shifts and warping of the electronic energy bands. In thin films, *thermal expansion* generates shear strain, which would not affect the free-electron gas but could split degenerate energy bands and distort the band structure. The *electron relaxation times* decrease with increasing phonon population. The corresponding *enhancement of electron–phonon interaction* also shifts and warps the energy bands.

The situation simplies in noble metals when photon energies near the interband transition threshold are used. In this case, the temperature dependence of $\varepsilon(\omega)$ is dominated by the broadening of the Fermi distribution. The resulting variations of linear reflectivity with electron temperature are to first order proportional to changes in electronic occupancy around the Fermi level but have opposite sign. This behaviour is illustrated in Fig. 4.3 and explained in the caption. In this context it is useful to define the energy mismatch ΔE between photon energy $\hbar\omega$ and interband transition threshold (ITT), $\Delta E = \hbar\omega - ITT$.

Our theoretical calculations of the electron temperature dependence of $\varepsilon(\omega)$ are based on models derived by Jah and Warke[39] and Rustagi.[40] In these models, $\varepsilon(\omega)$ is obtained solving the Liouville equation for the one-electron density matrix in first-order perturbation theory using the rapidly varying phase approximation.

In cubic crystals where transverse and longitudinal dielectric functions are identical, $\varepsilon(\omega)$ is found to be

$$\epsilon(\omega) = 1 - \frac{\omega_p^2}{\omega(\omega + i/\tau_c)} - \frac{4\pi e^2}{m^2\omega^2}$$
$$\times \sum_{kk'bb'} \frac{(\hbar\omega + i/\tau_{bb'})\langle b\mathbf{k}|p_i|b'\mathbf{k}'\rangle\langle b'\mathbf{k}'|p_i|b\mathbf{k}\rangle}{(E_{b'k'} - E_{bk})(E_{b'k'} - E_{bk} - \hbar\omega - i/\tau_{bb'})}$$
$$\times [f_0(E_{b'k'}) - f_0(E_{bk})]. \qquad (4.2.6)$$

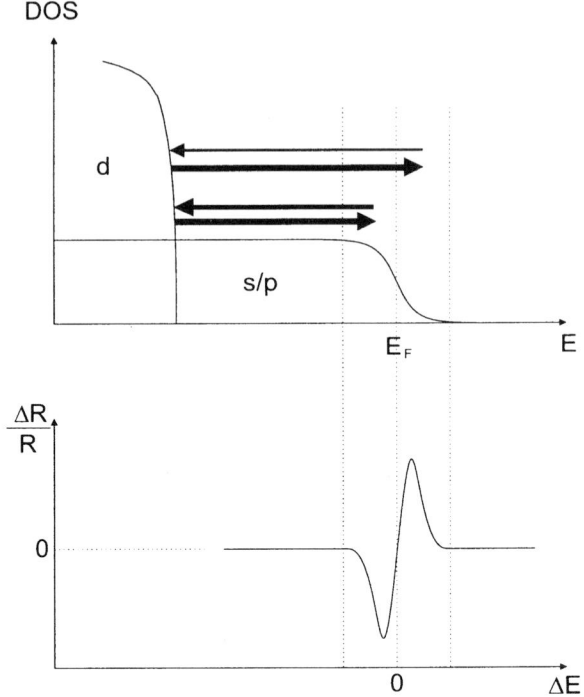

Fig. 4.3. Change in reflectivity versus energy mismatch ΔE (lower part). The upper part illustrates the cause: a decrease in electronic occupancy below the Fermi level increases the linear absorption for photon energies somewhat smaller than the interband transition threshold, thereby reducing the linear reflectivity; an increase of electronic occupancy above the Fermi level, on the other hand, reduces linear absorption and enhances the reflectivity.

Here, ω_p is the plasma frequency, $\tau_c = \tau_0[E_F/(\hbar\omega - E_F)]^2$ the energy-dependent relaxation time of the conduction electrons, $\langle b\mathbf{k}|p_i|b'\mathbf{k}'\rangle$ the transition matrix element for $b\mathbf{k} \to b'\mathbf{k}'$ transitions, $\tau_{bb'}$ the corresponding relaxation time, $E_{b\mathbf{k}}$ the energy of the state $|b\mathbf{k}\rangle$, and

$$f_0(E_{b\mathbf{k}}) = \frac{1}{\exp[E_{b\mathbf{k}} - E_F(T_e)/(kT_e)] + 1} \qquad (4.2.7)$$

the value of the Fermi distribution at this energy. The electron temperature dependence of the Fermi energy is given by:[41]

$$E_F(T_e) = E_{F0}\left[1 - \frac{\pi^2}{12}\left(\frac{kT_e}{E_{F0}}\right)^2\right], \qquad (4.2.8)$$

where E_{F0} denotes the Fermi energy at $T_e = 0$.

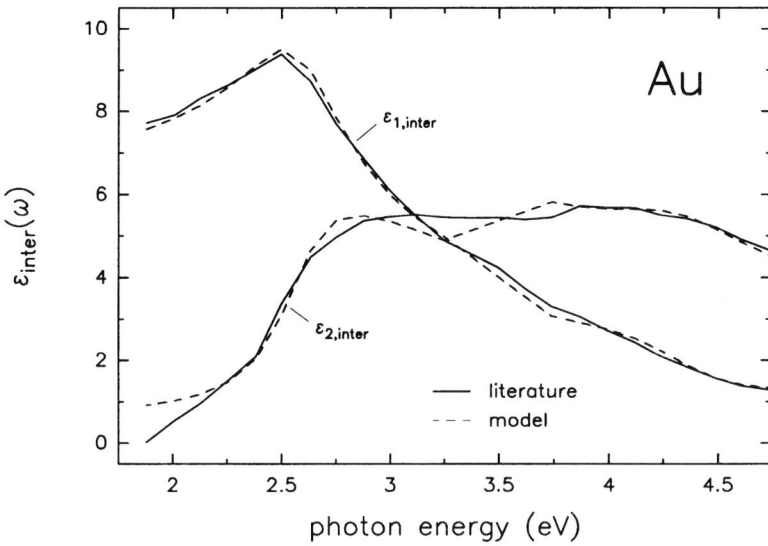

Fig. 4.4. Comparison of the calculated frequency dependence of the real (ε_1) and imaginary (ε_2) parts of the interband contribution of the dielectric function $\varepsilon_{\text{inter}}(\omega)$ with experimental data for Au, reported in the literature.[44]

The first two terms of eqn (4.2.6) describe the plasma or intraband contribution $\varepsilon_{\text{intra}}(\omega)$, the last one the interband part $\varepsilon_{\text{inter}}(\omega)$. In order to solve eqn (4.6) we used the electric dipole approximation and the approximation of electron-momentum-independent transition matrix elements, taking only prominent transitions into account. Furthermore, we assumed the involved energy bands to be either perfectly parabolic, $E_{b\mathbf{k}} = E_b + \hbar^2 k^2/(2m)$, or totally flat in the case of the d band, $E_{b\mathbf{k}} = E_b$. The approximation of electron-momentum-independent transition matrix elements can be expressed in the form:

$$\langle b\mathbf{k} | p_i | b'\mathbf{k}' \rangle^2 = mE_{b \to b'}/2, \qquad (4.2.9)$$

and the potential energy of the parabolic conduction bands E_c is related to the upper edge of the flat d bands, which is defined to be zero, $E_d = 0$.

With values for E_{F0} and ω_p taken from the literature,[42,43] we fitted the calculated frequency dependence of the dielectric function at a fixed temperature of 293 K to existing experimental data reported in ref. 44. As an example, the quality of a typical fit is shown for Au in Fig. 4.4

In this fit, we accounted for contributions from other transitions[38,45] by adding a constant value $\Delta \varepsilon_1$ to the real part of the dielectric function. The determined fitting parameters τ_0 as well as $E_{d \to c}$, τ_{dc} and E_c for the most important d band to conduction band transitions are listed in Table 4.1.

Once the frequency dependence of the linear dielectric function is known, its variation with electron temperature can be calculated, using eqn (4.2.6). Insert-

Table 4.1 Values of the Fermi energy at zero temperature E_{F0}, plasma frequency ω_p, electron collision times in the conduction band τ_0, energies related to the transition matrix elements as defined by eqn (4.2.9), lifetimes of excited d electrons and potential energies of the conduction band with respect to the upper limit of the d band. These numbers were the most important parameters when fitting the theoretical frequency dependence of ε to the literature data for Cu, Ag and Au

Metal	E_{F0} (eV)	$\hbar\omega_p$ (eV)	τ_0 (fs)	$E_{d\to c}$ (eV)	τ_{dc} (fs)	E_c (eV)
Cu	7.00	9.06	5.6	1.18	10.1	−4.79
Ag	5.48	9.14	30.9	2.10	6.3	−1.39
Au	5.51	8.90	7.7	1.53	4.3	−2.94

ing all fitting parameters, the dependence of linear reflectivity R and extinction coefficient α on ε and T_e can be expressed in the form:

$$R(T_e, \omega) = \frac{2A(T_e, \omega) + A^2(T_e, \omega) + \varepsilon_2^2(T_e, \omega) - (2A(T_e, \omega))^{3/2}}{2A(T_e, \omega) + A^2(T_e, \omega) + \varepsilon_2^2(T_e, \omega) + (2A(T_e, \omega))^{3/2}}, \quad (4.2.10)$$

$$\alpha(T_e, \omega) = \frac{2\omega}{c} \frac{\varepsilon_2(T_e, \omega)}{\sqrt{2A(T_e, \omega)}}, \quad (4.2.11)$$

where

$$A(T_e, \omega) = \varepsilon_1(T_e, \omega) + \sqrt{\varepsilon_1^2(T_e, \omega) + \varepsilon_2^2(T_e, \omega)}. \quad (4.2.12)$$

The change of linear reflectivity normalized to the reflectivity at room temperature, $[R(T_e) - R(293\,K)]/R(293\,K) = \Delta R/R$, is shown in Fig. 4.5 as a function of photon energy for Au at five different electron temperatures. These calculated changes of reflectivity around the interband transition threshold are in qualitative agreement with our simplified discussion in Fig. 4.1 and with corresponding measurements on gold reported by Schoenlein et al.[46]

The electron temperature dependence of R is also of interest for the analysis of time-resolved SHG measurements discussed in sec. 4.3.2. Therefore, using eqn (4.2.10) we calculated the change of reflectivity with electron temperature for Cu, Ag and Au at photon energies of 2 and 4 eV, corresponding to the fundamental and second harmonic in our SHG experiments. The results are displayed in Fig. 4.6. They represent typical examples of the dramatic variation with electron temperature that R can experience at different photon energies. Three main conclusions can be derived from the calculated curves:

1. The linear reflectivity depends nonlinearly on electron temperature within a wide range, in particular for small electron temperatures for which usually a linear dependence is assumed.
2. For a negative energy mismatch ΔE ($\hbar\omega = 2\,eV$ in the case of Cu and Au) the variation of R with electron temperature begins to saturate when the

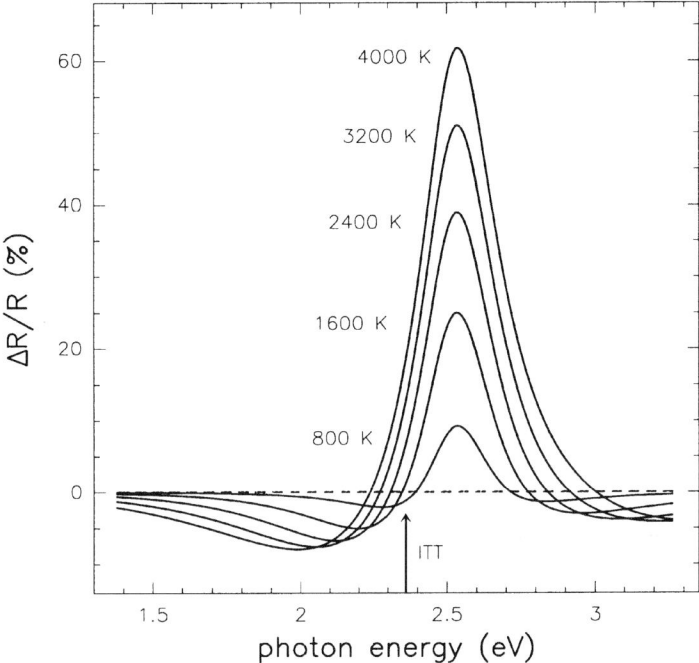

Fig. 4.5. Change of reflectivity as a function of photon energy, calculated for bulk Au for five electron temperatures and normalized to the reflectivity at room temperature. The arrow indicates the interband transition threshold of 2.38 eV. Notice the saturation with increasing temperature for photon energies around 2 eV.

thermal energy kT_e approaches ΔE. This behaviour may be attributed to a saturation of changes in electronic occupancy near the Fermi energy, i.e. at small values of ΔE. Notice, however, that the observed saturation behaviour is restricted to linear reflectivities at negative ΔE. At positive ΔE ($\hbar\omega = 4\,\text{eV}$ in the case of Ag) neither R nor the extinction coefficient α saturate at any ΔE. Hence, there is no unique relation between saturation of electronic occupancy and optical properties as, for example, claimed in ref. 46.

3. The normalized changes of R depend strongly on the absolute values of ε_1 and ε_2. The huge values of $\Delta R/R$ observed for Ag at 4 eV result from the fact that 4 eV is the 'apparent' plasma frequency of silver, i.e. $\varepsilon_1(4\,\text{eV}) \approx 0$. The constancy of $\Delta R/R$ at 2 eV, on the other hand, results from the fact that photons of 2 eV do not cause an interband transition.

The predicted saturation of $\Delta R/R$ for $kT_e \geq \Delta E$ in the case of Cu and Au is an important result for the interpretation of the time-resolved measurements of second-order reflectivities in sec. 4.3.2.

4.2.3 Experimental results

The discussion of transient linear reflectivity measurements will be grouped

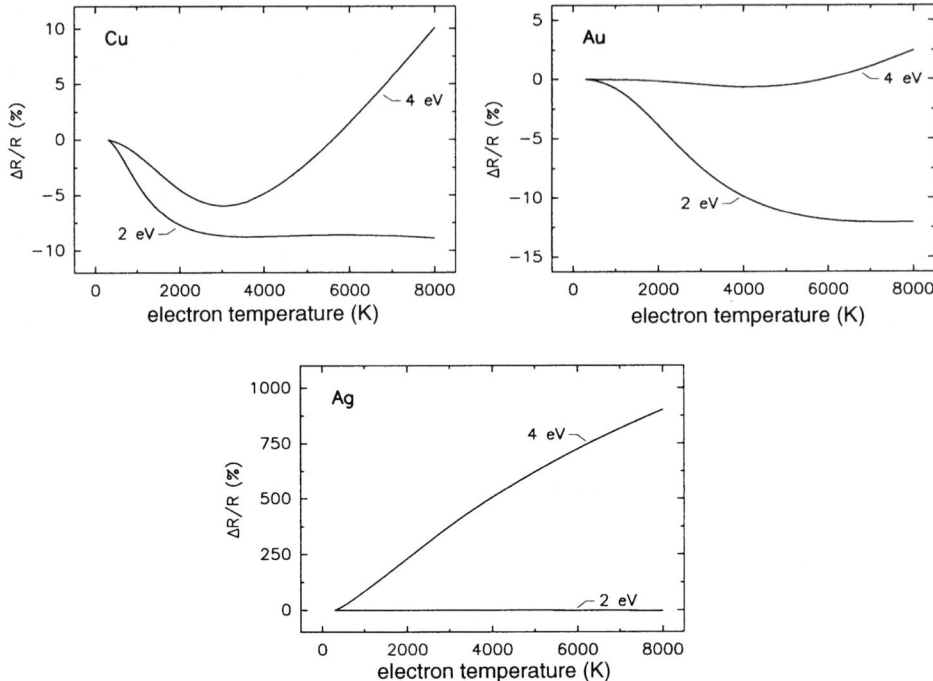

Fig. 4.6. Calculated change of reflectivity with electron temperature, normalized to $R(293\,K)$ at fixed photon energies of 2 and 4 eV for Cu, Ag and Au.

according to the purpose of our investigations. First, we will present results obtained with polycrystalline bulk samples of Cu, Ag and Au. They served as a test for our theoretical model of the electron temperature dynamics and the accompanying variations of the linear dielectric function. Secondly, we will show the time dependence of linear reflectivities of thin gold films in the thickness range from 10 to 500 nm, which establishes a mean free path of ballistic electrons of about 100 nm at pump fluences of $1\,mJ\,cm^{-2}$. Finally, we will present results on the pump intensity dependence of $\Delta R(t)$ for a 30 nm gold film which support the predicted saturation behaviour of R.

Experimental setup

The first pump–probe measurements of linear reflectivities, carried out on polycrystalline bulk samples of Cu, Ag and Au, were conducted at the 'Max-Born-Institut für nichtlineare Optik und Kurzzeitspektroskopie' in Berlin/Adlershof. In these experiments, probe-beam energies of 2 eV were used for Cu and Au, and 4 eV for Ag. A colliding-pulse mode-locked dye laser was employed to generate pulses of about 100 pJ/pulse and 65 fs duration at a photon energy of 2 eV with a repetition rate of 76 MHz. These pulses were amplified up to about 50 μJ in four dye amplification stages which were pumped by an excimer laser of 308 nm with a repetition rate of a few hertz. Owing to amplification, the pulses were broadened to about 100 fs. Pump and probe pulses were generated by a 1:1 beam splitter and focused to 500 and 200 μm spots at incident angles of 43° and 47°, respectively. For measurements of transient reflectivities at 2 eV, the probe pulses were attenuated by a factor of 5×10^3 using a neutral density filter (NG 10, Schott) of 1 mm thickness. Corresponding measurements at 4 eV were conducted by frequency doubling the probe pulses in a 0.3 mm thick BBO crystal and suppressing the residual fundamental by use of a 2 mm thick UG5 Schott filter. The probed area was almost homogeneously heated and additional heating due to probe pulses could be neglected.

A reference autocorrelation of the pulses was occasionally taken by placing a plane-parallel mirror, mounted on a translation stage, in front of the sample to direct the beams into a BBO crystal. This allowed the control of pulse length and calibration of zero delay between pump and probe pulses. Since the laser pulse energies showed strong fluctuations, a reference signal proportional to the probe pulse energy was simultaneously recorded and all transient reflectivity data were sorted according to distinct energy intervals defined by this reference signal. The sorting intervals were chosen to limit the probe pulse energy variations to 5%. All measurements were carried out on open samples at room temperature.

Comparison with model calculations

Transient linear reflectivities measured for Cu and Au at photon energies of 2 eV and for Ag at 4 eV are shown in Figs 4.7 and 4.8. They are compared to model calculations, indicated by solid lines.

No dependence of R on pump–probe delay was found for Ag at 2 eV and for Au at 4 eV, while Cu showed a similar effect at both photon energies. As predicted by theoretical calculations of $R(T_e)$ in Fig. 4.6, the change in reflectivity for Ag at 4 eV is an order of magnitude larger and opposite in sign compared to the effect in Cu or Au at 2 eV. The opposite sign can be understood by comparing the photon energy to the interband transition threshold. The rapid decline of reflectivity for Cu and Au is consistent with a negative energy mismatch ($\Delta E = -150$ meV for Cu and -380 meV for Au), monitoring the fast decrease of electronic occupancy below the Fermi level due to heating. For Ag, on the other hand, ΔE is positive but small ($+20$ meV), leading to a reduced absorption and consequently to an increased reflection. The one order of magnitude larger effect reflects the much smaller value of ε_1 for Ag compared to Cu and Au.

The good agreement between theoretical results and experimental data for all three metals justifies the use of the model outlined in sec. 4.2.2. This is quite satisfying, in particular when considering that all measurements were carried out in air. Especially for Cu there will be an oxide coating, and the small discrepancy between the data and the theoretical solid line which develops for Cu on a picosecond timescale (see Fig. 4.7) is most likely caused by it. This interpretation

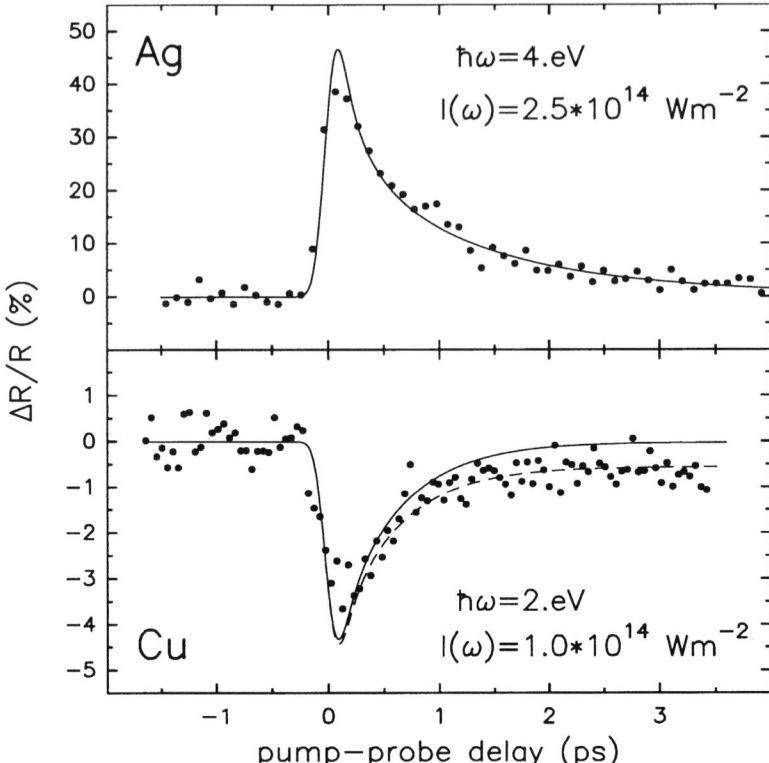

Fig. 4.7. Time dependence of relative linear reflectivities for Ag and Cu at photon energies of 4 eV and 2 eV, respectively. Solid lines represent best fits based on model calculations with electron–phonon coupling constant g and electronic heat conductivity K_e as free parameters (see Table 4.2).

is supported by the fact that this deviation between experiment and theory can be corrected for (dashed curve) by a term that is proportional to the lattice temperature increase. The reason for such correction is provided by the fact that the optical properties of semiconductors such as CuO and CuO_2 are sensitively affected by the lattice temperature.[47] We want to emphasize, however, that this correction does not affect the time dependence of the electron temperature.

The theoretical results were fitted to the data by using only two fitting parameters, the electronic thermal conductivity and the electron–phonon coupling constant. No further fitting parameters were necessary to describe the temperature dependence of the linear dielectric function. The best values of K_e and g are listed for all three metals in Table 4.2 together with values reported in the literature.

The values of K_e and g agree with literature data for all three metals. We checked the value of g obtained for Au by also measuring the transient reflectivity of a 20 nm thin gold film, for which electron diffusion is negligible.[32]

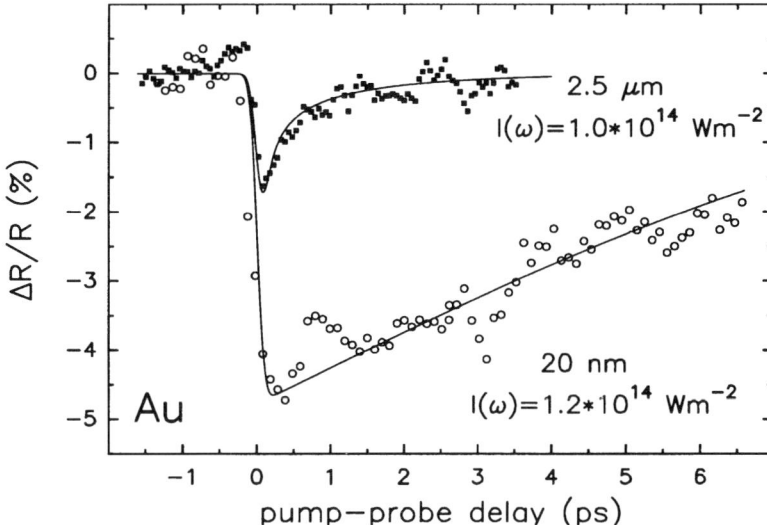

Fig. 4.8. Time dependence of relative linear reflectivities for a bulk sample (2.5 μm) of Au and a 20 nm thick gold film. The probing photon energy was 2 eV. Solid lines correspond to model calculations with $g = 1.1 \times 10^{16}$ W m^{-3} K^{-1} in both cases. Electron diffusion was taken into account for the bulk sample but neglected for the 20 nm film.

The data are also shown in Fig. 4.8. The linear decay of transient reflectivity and the considerably longer relaxation time testify to the absence of thermal diffusion. The maximum change of R is more than twice as large for the thin film compared to the one for a thick sample, while the ratio of the pump intensities used in both cases was only 1.2 : 1. This demonstrates the importance of energy transport by optically excited electrons during the first 100 fs after excitation by the pump pulse. For the 20 nm film, the solid line represents a calculation that omits electron diffusion but uses the same g as obtained for the 2.5 μm sample. The excellent agreement between theory and experiment provides confidence in the value of g given in Table 4.2.

4.2.4 Penetration depth of deposited energy

The rapid decay of the transient reflectivity for a 2.5 μm Au sample in Fig. 4.8

Table 4.2 Values of electronic thermal conductivity in units of W m^{-1} K^{-1} and electron–phonon coupling constants in units of 10^{15} W m^{-3} K^{-1} as determined from our experimental data (exp) and reported in the literature (lit)

Metal	K_{exp}	K_{lit}	g_{exp}	g_{lit}
Cu	385(60)	385[34]	47(24)	47–100[48–50]
Ag	418(65)	418[34]	11(3)	15–55[51–53]
Au	176(30)	317[34]	11(3)	10–40[31,32,48,53–57]

confirms that diffusive energy transport together with electron–phonon coupling causes an almost exponential relaxation behaviour of electron temperature. The linear decay in the 20 nm thin film results from electron–phonon coupling only, proving that there is no vertical temperature gradient. In addition, the much larger reflectivity change obtained for the 20 nm film compared to the 2.5 μm sample indicates a considerably higher density of deposited energy. These observations provide clear evidence that the energy deposition within the 2.5 μm sample is much deeper than 20 nm.

The penetration depth of deposited energy, λ_E, is governed by the optical penetration depth of the pump pulse and the subsequent energy transport by ballistic electrons and by electronic thermal diffusion. Thus, λ_E increases with time. Its time dependence is to first order for ballistic electrons $\lambda_E \propto t$, and for thermal diffusion $\lambda_E \propto \sqrt{t}$. The value of λ_E ($t = \tau$) for the duration of the pump pulse determines the maximum density of deposited energy as well as the peak electron and lattice temperatures at the surface. Hence, knowledge of $\lambda_E(\tau)$ is of great importance for many nonequilibrium processes at or near the surface, like electron and spin dynamics, ultrafast phase transitions and ablation thresholds, or adsorbate reactions.

A reduction of film thickness to values equal to or smaller than $\lambda_E(\tau)$ will not only confine the energy transport to shorter times but also cause an increase of deposited energy density. Thus, $\lambda_E(\tau)$ can be studied by measuring the film thickness dependence of transient linear reflectivities. Its maximum value corresponds to a film thickness for which a nearly exponential relaxation of transient reflectivity turns into a linear one and where the change of reflectivity begins to increase. The advantage of this technique is its simplicity and its applicability to films that are deposited on opaque or strongly absorbing substrates which inhibit measurements of transient reflectivity in the front-pump/back-probe configuration.[36,58] It also yields $\lambda_E(\tau)$ without much theoretical effort required by other techniques like, for example, the pulse length dependence of ablation thresholds.[59]

We investigated transient linear reflectivities of 13 commercial gold films (Berliner Institut für Optik) with thicknesses 10–100 nm (in steps of 10 nm) plus 200, 300 and 500 nm deposited from the same batch of gold onto optical-grade fused silica. No auxiliary Cr layer was used to improve adhesion. To avoid misleading results by deteriorating film properties, the film quality was controlled both by spectroscopic measurements of the static linear reflectivity and by the reproducibility of repeated recordings of the transient reflectivity. All experiments were carried out in air at room temperature. For the pump–probe measurements we used a commercial (Coherent) femtosecond laser system in our own laboratory. Wavelengths and widths of pump and probe pulses were 400 nm/200 fs and 500 nm/100 fs, respectively.

Static spectroscopy of the films with unpolarized light was done with a commercial spectrometer (Beckmann, model du 40). The results for reflectivity $R(\omega)$ and transmission $T(\omega)$ of seven different thicknesses are displayed in Fig. 4.9. With the exception of the 10 nm film, the data for all other films matched

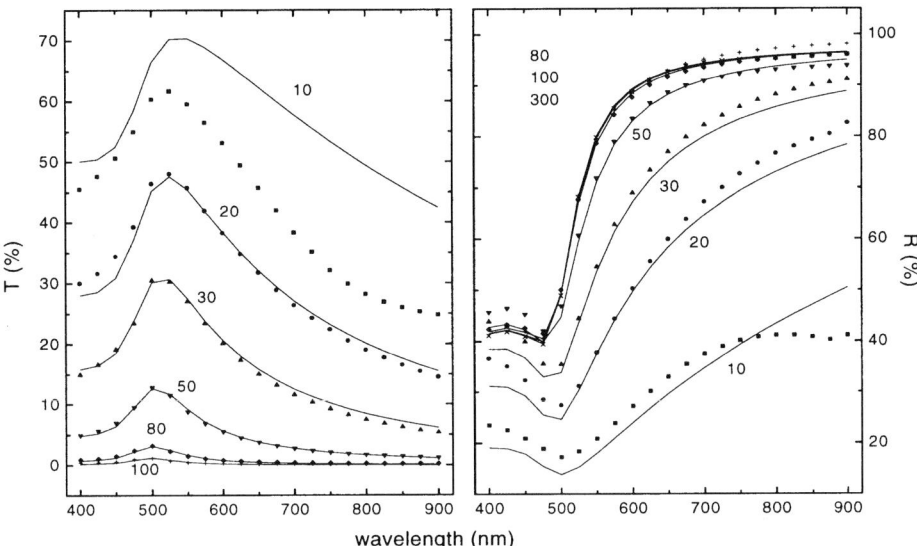

Fig. 4.9. Measured (dots) and calculated (lines) static reflectivities and transmissions for gold films of various thicknesses on fused silica as a function of wavelength.

well the theoretical predictions (solid lines) obtained by calculating the thickness dependence of $R(\omega)$ and $T(\omega)$ with the usual multiple reflection model,[60] using optical constants of the bulk. The excellent agreement between experiment and theory (within the experimental error of $\pm 10\%$) for film thicknesses down to 20 nm proves the optical constants to be indepenent of film thickness above 20 nm and testifies to a high film quality. The discrepancy for the 10 nm film was caused by the fact that it was no longer a closed film, as revealed by force microscopy inspection.

The second test of film quality was the reproducibility of transient reflectivity results within $\pm 5\%$ for repeated measurements at changing spot sites of the same film. Such tests were carried out on different days on all films and verified the high film homogeneity. This procedure ensured that the observed changes of transient linear reflectivities are exclusively caused by the film thickness itself, thus signalling the reduced energy transport with decreasing film thickness. Only for the 10 nm thick film did we observe deviations which originate from imperfections in film morphology.

The results of transient reflectivity measurements at 2.48 eV photon energy for 13 films of different thicknesses are displayed in Fig. 4.10(a). As indicated, we used both s- and p-polarized probe light but there is no qualitative difference between the two polarization directions. The two main features are (1) the dramatic increase of $\Delta R/R$ with decreasing film thickness and (2) the linear decay for film thicknesses ≤ 100 nm. These results are well reproduced by numerical solutions of the TTM—including multiple reflection optics in the source term—shown in Fig. 4.11. There is, however, a small deviation between

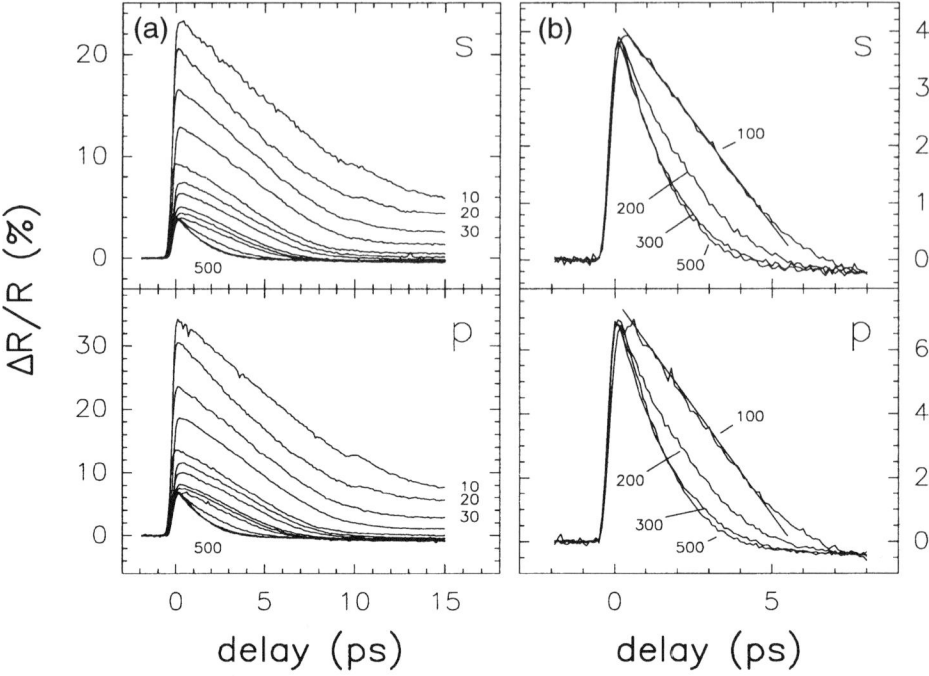

Fig. 4.10. (a) Transient linear reflectivities of gold films with various thicknesses, obtained with s- and p-polarized probe pulses of 500 nm wavelength and 100 fs duration. The film thickness ranged from 10 to 100 nm in steps of 10 nm plus 200, 300 and 500 nm, as indicated by the numbers. (b) Magnified display of transient reflectivities for film thicknesses ≥ 100 nm, illustrating the transition from linear decay into an almost exponential decay when passing from 100 nm to thicker films.

theory and experiment at short times for films ≤ 100 nm, which is more evident in Fig. 4.12 and will be discussed below. It indicates the effect of ballistic motion, which was not taken into account in the calculations.

The transition from linear to exponential decay becomes obvious in the magnified display in Fig. 4.10(b). While the relaxation of transient reflectivity for the 100 nm film is linear up to 4 ps (indicated by the straight line), it turns into an almost exponential decay with increasing film thickness. There is little further change when going from 300 nm to 500 nm films and we generally expect the transient reflectivities for thicker films to be identical to the 500 nm data. This change of decay pattern provides clear evidence for the dominance of diffusive energy transport for films thicker than 100 nm. The termination of the linear decay at 100 nm together with the large enhancement of the reflectivity change when going to thinner films prove that the initial energy deposition depth is λ_E (≈ 100 fs) = 100 nm. The mechanism for it is ballastic motion of hot electrons.[36] In comparison, the optical penetration depth of the 3.1 eV pump pulses is only 12 nm.

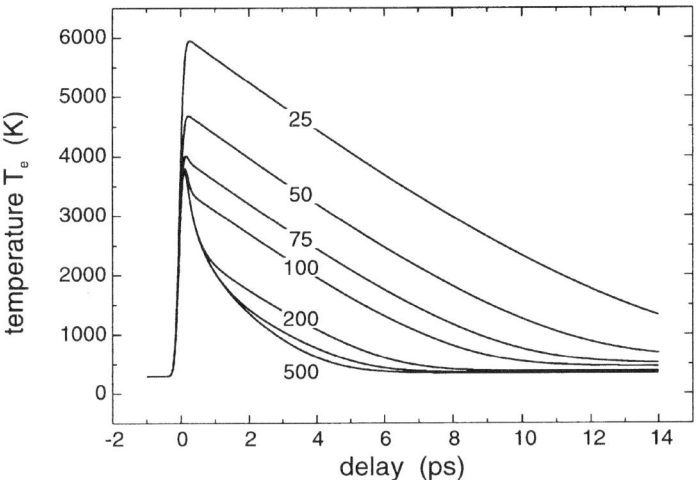

Fig. 4.11. Transient electron temperatures calculated for various film thicknesses using the TTM and multiple reflection theory. Film thicknesses in nm are indicated by numbers.

Further evidence for ballistic energy transport is presented in Fig. 4.12, where the transient reflectivities during the first 1.5 ps are compared to calculations for 80, 100 and 200 nm films. Good agreement is observed for all three film thicknesses at delays larger than 1.3 ps. For shorter times, however, the predicted electron temperatures always overshoot the measured ones and the

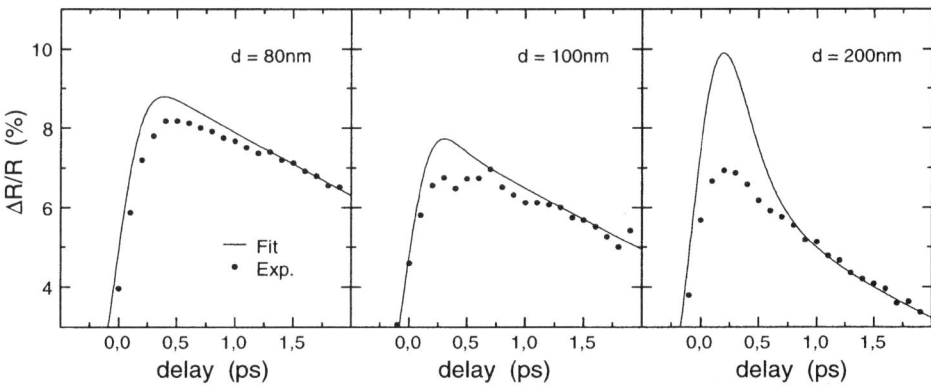

Fig. 4.12. Comparison of transient linear reflectivities at $\lambda = 500$ nm for film thicknesses of 80, 100 and 200 nm with TTM model calculations (solid lines) fitted to the data for $t > 1.3$ ps. The deviation increasing with film thickness for $t \leq 1.3$ ps and the good agreement for longer times demonstrate the importance of energy transport by nonthermalized electrons and point to an electron thermalization time of about 1.3 ps in agreement with refs 31 and 32.

deviation between experiment and theory increases with film thickness. In this context, it is important to remember that the two-temperature model neglects finite electron thermalization times and ballistic energy transport. Hence, the data in Figs 4.10 and 4.12 establish the importance of nonthermal energy transport. From the comparison in Fig. 4.12 we can also deduce an electron thermalization time of about 1.3 ps, in agreement with the literature.[31,32] We conclude that $\lambda_E = 100$ nm represents the mean free path of ballistic electrons, which corroborates results by Suárez et al.[36] who report that for thicknesses ≤ 100 nm electron transport across single-crystalline or polycrystalline gold films is 'mostly ballistic'.

Summarizing this section, we demonstrated that systematic film-thickness-dependent measurements of transient linear reflectivities yield information about the penetration depth of the deposited energy. For probing pulse lengths of 100 fs a value of $\lambda_E = 100$ nm was derived from conspicuous changes of the reflection dynamics and from comparison with theory.

4.2.5 Saturation of linear reflectivity

One major result of the theoretical model for the electron temperature dependence of linear reflectivities was the predicted saturation behaviour for photon energies somewhat smaller than the interband transition threshold (cf. Fig. 4.5). Although such saturation is implicitly mentioned in several papers,[32,46] it has not been experimentally demonstrated so far. In addition, the interpretation of

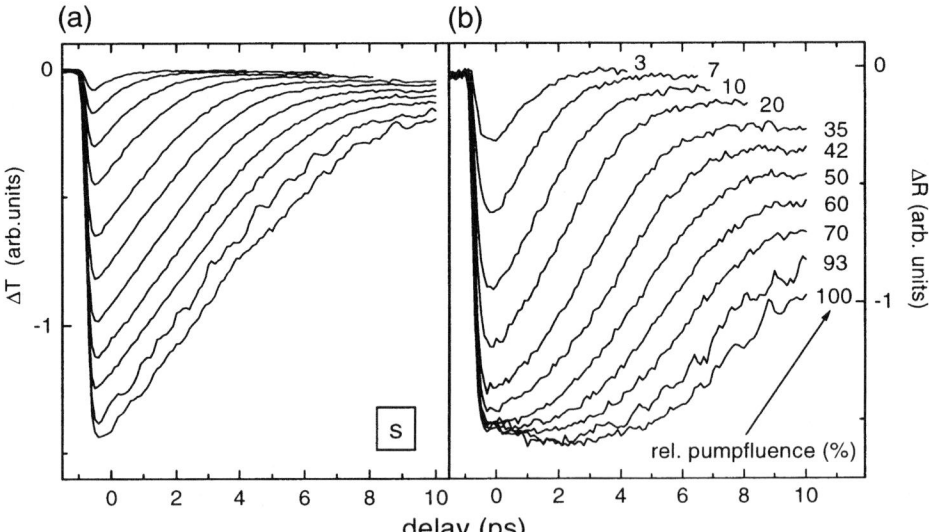

Fig. 4.13. Time dependence of (a) transmission and (b) reflectivity changes of a 30 nm gold film for s-polarized probe pulses at 546 nm and 11 different pump fluences, given in percentages of 180 μJ cm^{-2}.

transient second-order reflectivities presented in sec. 4.3.2 uses the conjecture that linear reflectivity and second-order susceptibility have similar temperature dependences. Also from that point of view saturation of reflectivity changes are of interest. For these reasons we decided to investigate the pump intensity dependence of transient linear reflectivities for gold to check the predicted saturation behaviour.

To reach the necessary electron temperatures we used a 30 nm film, taking advantage of the increase of absorbed energy density with decreasing film thickness, which was discussed in the last section. The wavelength for the pump light was 400 nm and for the probe beam 546 nm, the latter corresponding to an energy mismatch of $\Delta E = -110$ meV. The change of reflectivity as a function of delay for 11 different pump fluences is shown in Fig. 4.13(b). Since the largest pump intensity used in these measurements corresponds to the pump intensity applied for the film-thickness-dependent transient reflectivities in Fig. 4.10, it is evident that saturation and flattening of the transient reflectivities observed at pump–probe delays ≤ 2 ps for relative pump intensities $\geq 70\%$ are exclusively caused by a saturation of $R(T_e)$ and do not originate from dramatic changes of the electron temperature dynamics. This result is supported by simultaneously measured transient transmissivities, which are shown in Fig. 4.13(a) and which unambiguously prove a similar temperature dynamics for all pump intensities.

4.3 Second harmonic generation

Second-order reflectivity or second harmonic generation (SHG) in reflection is governed by a complex combination of linear and nonlinear optical properties of the material. Thus, its dependence on electron temperature is far more complicated than in the linear case. On the other hand, there are advantages which make it worth while to cope with the difficulties of interpreting SHG data. The sensitivity of SHG to transient electron temperatures is enhanced up to one order of magnitude compared to the linear reflectivity.[15,61] Furthermore, time-resolved pump–probe SHG offers the possibility to investigate electron dynamics at surfaces and interfaces.[62]

4.3.1 Formalism for data analysis

The physics of reflection SHG can be sketched as follows. An incident field $E(\omega)$ of fundamental light generates a local electric field $E_{\text{loc}}(\omega)$ in the sample, the strength of which is determined by the linear dielectric function $\varepsilon(\omega)$. This local field drives a second-order polarization $P(2\omega)$ which is given in the electric dipole approximation by

$$P(2\omega) = \chi^{(2)}(2\omega; \omega, \omega) : E_{\text{loc}}(\omega) \cdot E_{\text{loc}}(\omega). \qquad (4.3.1)$$

The second-order susceptibility $\chi^{(2)}(2\omega; \omega, \omega)$ describes the intrinsic ability of the sample to generate second harmonic radiation and acts as the source term for a frequency-doubled local electric field $E_{tot}(2\omega)$. The reflected second harmonic field $E(2\omega)$ is then governed by $\varepsilon(2\omega)$.

For centrosymmetric media such as polycrystalline noble metals, $\chi^{(2)}$ is surface- and interface-sensitive. The surface sensitivity of $\chi^{(2)}$ may be explained by von Neumann's principle,[63] which states that any theoretical description of physical properties of a sample has to obey at least the same symmetry rules as the described sample itself. Since $\chi^{(2)}$ is a third-rank tensor it is in centrosymmetric media restricted to a surface or interface region of atomic dimensions where inversion symmetry is broken.

According to a phenomenological model presented by Sipe et al.,[30] the different influences of $\varepsilon(\omega)$, $\varepsilon(2\omega)$ and $\chi^{(2)}(2\omega;\omega,\omega)$ on $E(2\omega)$ may be factorized in the following way:

$$E(2\omega) \propto F(2\omega) \cdot \chi^{(2)} \cdot f(\omega) \cdot |E(\omega)|^2. \qquad (4.3.2)$$

Here, the so-called Fresnel factors $f(\omega)$ and $F(2\omega)$ contain all optical information like refractive indices for both frequencies, angle of incidence and polarization for both fundamental and second harmonic radiation. Their explicit form is[30]

$$f(\omega) = \begin{pmatrix} f_c^2 t_p^2 \cos^2\phi \\ t_s^2 \sin^2\phi \\ f_s^2 t_p^2 \cos^2\phi \\ 2f_s t_p t_s \cos\phi \sin\phi \\ 2f_c f_s t_p^2 \cos^2\phi \\ 2f_c t_p t_s \cos\phi \sin\phi \end{pmatrix}, \quad F(2\omega) = \begin{pmatrix} A_p F_c \cos\Phi \\ A_s \sin\Phi \\ A_p N^2 F_s \cos\Phi \end{pmatrix}, \qquad (4.3.3)$$

with the abbreviations

$$f_s = \frac{\sin\Theta}{n(\omega)}, \quad f_c = \sqrt{1-f_s^2}, \quad t_p = \frac{2\cos\Theta}{n(\omega)\cos\Theta + f_c},$$

$$t_s = \frac{2\cos\Theta}{\cos\Theta + n(\omega)f_c}, \quad A_{p/s} = \frac{2\pi T_{p/s}}{\cos\Theta}, \quad N = n(2\omega). \qquad (4.3.4)$$

Lower-case or capital letters denote quantities of the fundamental or second harmonic, respectively, and equivalent expressions for $F_{s/c}$ and $T_{p/s}$ are used with N. The dielectric function and its temperature dependence discussed in sec. 3.2.2 enter eqns (4.3.2)–(4.3.4) through the indices of refraction $n = \sqrt{\varepsilon}$ for both frequencies. The angles ϕ and Φ determine the polarization of fundamental and second harmonic, respectively. Θ is the angle of incidence. The geometry of SHG experiments defining all angles is shown in Fig. 4.14.

It is important to notice that $f(\omega)$ and $F(2\omega)$ sample the material within the optical penetration depth for ω and 2ω, while $\chi^{(2)}$—in dipole approximation—is

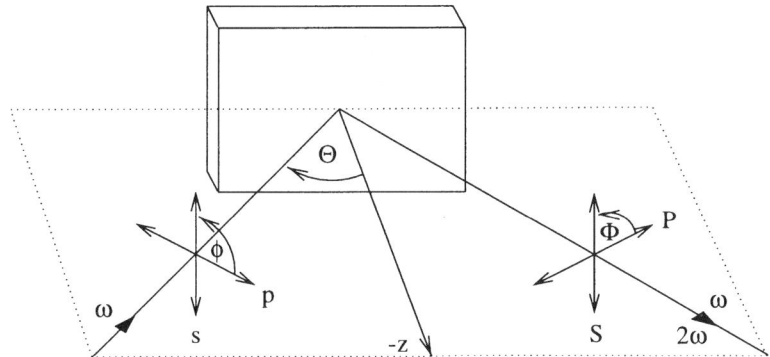

Fig. 4.14. Definition of angles used in eqns (4.14)–(4.16).

surface-sensitive for centrosymmetric media. Hence, in order to uncover possible differences in electron dynamics between bulk or surfaces and interfaces, it is necessary to isolate the electron temperature dependence of each individual factor. Unfortunately, knowledge of the electron temperature dependence of $\chi^{(2)}$ is scarce.[64,65] However, it can be seen from the general structure of $\chi^{(2)}$,

$$\chi_{ijk}(T_e) \propto \sum_{k,l,l',l''} \frac{\langle l, k |r_i| l', k \rangle \langle l', k |r_j| l'', k \rangle \langle l'', k |r_k| l, k \rangle}{E_{l'',k} - E_{l,k} - 2\hbar\omega + i\hbar\alpha}$$

$$\times \left[\frac{f(E_{l'',k}, T_e) - f(E_{l',k}, T_e)}{E_{l'',k} - E_{l',k} - \hbar\omega + i\hbar\alpha} - \frac{f(E_{l',k}, T_e) - f(E_{l,k}, T_e)}{E_{l',k} - E_{l,k} - \hbar\omega + i\hbar\alpha} \right], \quad (4.3.5)$$

that elevated electron temperatures can in principle affect the susceptibility (1) by altering the transition matrix elements, (2) by changing the denominators and (3) through the Fermi distributions. However, transition matrix elements and energy eigenvalues are not independent. Assuming that both remain constant, the variation of $\chi^{(2)}$ with electron temperature is contained in the Fermi distributions and can be estimated. For example, when photon energy and energy differences in one of the denominators are near resonance, an enlarged density of vacant intermediate states will enhance the SHG. As illustrated in Fig. 4.15, when the Fermi distribution is broadened by heating, interband transitions with negative energy mismatch will then cause a resonance enhancement of $\chi^{(2)}$. In the experiments described below, this will be the case at ω for Cu and Au and at 2ω for Ag. Consequently, when $\chi^{(2)}$ is dominated by the same electronic transitions as the linear dielectric function, we expect a qualitatively similar behaviour for $\chi^{(2)}$ and for R, except for the opposite sign.

In contrast to linear reflectivity, SHG originates from anharmonic electron motion. Hence only a small contribution to SHG stems from conduction electrons, while the major yield is determined by the structure of the intermediate states. For this reason $\Delta\chi^{(2)}$ should usually be considerably larger than ΔR such is the case for Cu and Au and fundamental photons of 2 eV. However, as

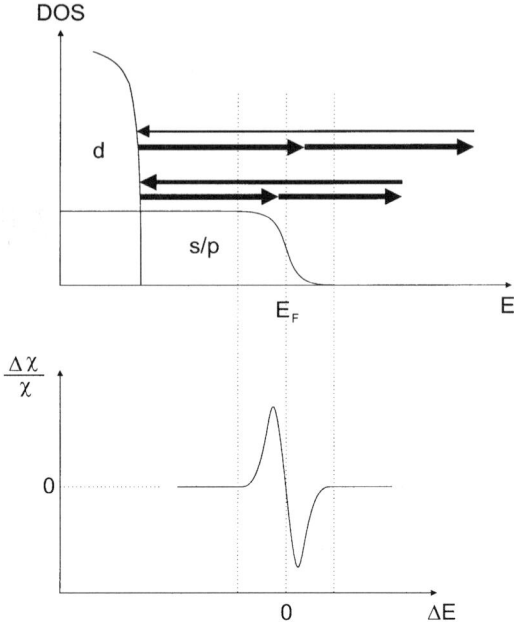

Fig. 4.15. Variation of $\chi^{(2)}$ with energy mismatch ΔE of the interband transition in noble metals. A decrease in electronic occupancy below the Fermi level causes a resonance enhancement of χ^2 for photon energies somewhat smaller than the interband transition threshold. The corresponding increase of electronic occupancy above the Fermi level, on the other hand, reduces the resonance enhancement and causes a reduction of χ^2. $\Delta E = \hbar\omega - ITT$.

already discussed in sec. 4.2.2 the relative changes of R are much larger for Ag at 4 eV, which is the apparent plasma frequency for Ag. For this reason the relative changes of $\chi^{(2)}$ and R are expected to be comparable in the case of Ag.

In the absence of firm theoretical predictions, the only way to determine the electron temperature dependence of $\chi^{(2)}$ is its extraction from experimental data. Since the factors $\chi^{(2)}$, $f(\omega)$ and $F(2\omega)$ contributing to SHG (cf. eqn (4.3.2)) are determined by sample symmetry and by polarization of both fundamental and second harmonic light, symmetry considerations and measurements of SH yield for different polarization combinations are a first step to reach this goal. Table 4.3 lists the tensor components of $\chi^{(2)}$ for the most common surface symmetries and polarization combinations, where 'mix' stands for 45°. Lower-case (capital) letters denote polarizations of the fundamental (second harmonic).

For isotropic surfaces of polycrystalline metals there are five nonvanishing tensor elements of $\chi^{(2)}$ but, since $x = y$, only three of them are independent. This reduces the problem to the elements χ_{zzz}, $\chi_{xzx} = \chi_{yzy}$ and $\chi_{zxx} = \chi_{zyy}$. To separate their different contributions to SHG one has to examine the polarization combinations $s \to P$, mix($\phi = 45°$) $\to S$ and $p \to P$ in Table 4.3. The corresponding fields are given by (cf. eqns (4.3.2) and (4.3.3))

Table 4.3 Non-vanishing tensor elements of $\chi^{(2)}$ that contribute to SHG for different surfaces and polarization combinations. Mix means 45° and lower-case (capital) letters denote polarizations of fundamental (second harmonic)

	s → S	s → P	p → S	p → P	mix → S	mix → P
(100) or isotropic	–	χ_{zyy}	–	χ_{zzz}, χ_{zxx} χ_{xzx}	χ_{yzy}	χ_{zzz}, χ_{zxx} χ_{xzx}, χ_{zyy}
(110)	–	χ_{zyy}	–	χ_{zzz}, χ_{zxx} χ_{xzx}	χ_{yzy}	χ_{zzz}, χ_{zxx} χ_{xzx}, χ_{zyy}
(111)	χ_{yyy}	χ_{zyy}	χ_{yxx}	χ_{zzz}, χ_{zxx} χ_{xzx}	χ_{yzy}	χ_{xzx}, χ_{zxx} χ_{xzx}, χ_{zyy}

s → P:
$$E_P(2\omega) \propto A_P \varepsilon(2\omega) F_s \chi_{zyy} t_s^2 |E_s(\omega)|^2, \quad (4.3.6)$$

mix → S:
$$E_S(2\omega) \propto A_S \chi_{yzy} f_s t_p t_s |E_{mix}(\omega)|^2, \quad (4.3.7)$$

p → P:
$$E_P(2\omega) \propto \left[A_P F_c \chi_{xzx} 2 f_c f_s t_p^2 \right. $$
$$\left. + A_P \varepsilon(2\omega) F_S \left(\chi_{zxx} f_c^2 t_p^2 + \chi_{zzz} f_s^2 t_p^2 \right) \right] |E_p(\omega)|^2. \quad (4.3.8)$$

Knowledge about the variation of $\varepsilon(\omega)$ and $\varepsilon(2\omega)$ with electron temperature provided in sec. 4.2.2 can now be used to predict the effect of transient electron temperatures on SHG caused entirely by the Fresnel factors. Assuming temperature-independent tensor elements, the temperature variation of the three terms entering p → P SHG is shown in Fig. 4.16 for an isotropic Cu surface.

Comparing these results with the electron temperature dependence of linear reflectivities (cf. Fig. 4.6) we notice that the relative changes of SH yield are about a factor of 5 larger. This is in accordance with observation by Sokolowski-Tinten et al.[66] who first reported the extreme sensitivity of SHG on ε.

One can predict the time relaxation of SHG (SH_p) on the presumption that only the Fresnel factors depend on electron temperature. Comparing this quantity with the measured (SH_m) time relaxation opens a way to uncover an eventual dependence of $|\chi_{ijk}|^2$ on electron temperature. If we denote the SHG yield for the probe pulse alone by SH_0, we can define the difference

$$\Delta SH = SH(T_e) - SH_0$$

between SHG with and without pump pulse. Assuming now that the SH yield is given by only one single tensor element χ_{ijk}, one can cast the relative change between predicted and measured SHG in the form

$$\frac{\Delta |\chi_{ijk}|^2}{|\chi_{ijk}|^2} = \frac{\Delta SH_m - \Delta SH_p}{SH_p(T_e)}. \quad (4.3.9)$$

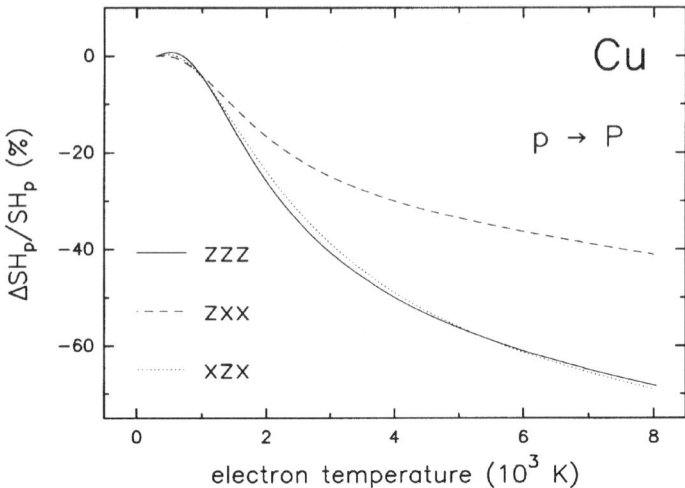

Fig. 4.16. Predicted relative change with electron temperature of the three terms in eqn (4.20) contributing to p → P polarized SHG assuming a temperature-independent $\chi^{(2)}$.

As can be seen from Table 4.3 the polarization combinations s → P and mix → S are sensitive to only one tensor element for isotropic surfaces. However, when just one polarization combination is measured, eqn (4.3.9) can only be used for the mix → S polarization combination because P-polarized SHG may be affected by nonlocal bulk contributions generating for isotropic surfaces a field (cf. eqn (4.3.10) in ref. 30)

$$E_P^{NL}(2\omega) \propto A_p F_S \gamma \left(E_s^2 + E_p^2 \right). \tag{4.3.10}$$

To find out whether or not there are nonlocal contributions to SHG and to determine the different magnitudes of χ_{zzz}, χ_{xzx} and χ_{zxx}—which all contribute to SHG in the p → P polarization combination (eqn (4.3.8))—we analysed the polarized SH yield from Cu, Ag and Au for different polarizations of the fundamental at 2 eV photon energy. Only P-polarized SH radiation varying with $\cos^4 \phi$ was detectable for all three metals. Since no S-polarized SH was found for arbitrary input polarization, we conclude that χ_{xzx} is vanishingly small (cf. eqn (4.19)). As discussed by Petrocelli et al.[67] it is reasonable to assume $\chi_{xzx} \approx \chi_{zxx}$ for polycrystalline metal surfaces. Hence, χ_{zxx} is also small and γ remains the only possible source of P-polarized SHG generated by s-polarized input fields. However, no SHG yield was detected for this s → P polarization combination and therefore γ is concluded to be negligible.

These findings are further supported by the different magnitudes of the Fresnel factors multiplied by the single elements χ_{ijk} and γ. Their relative values are listed in Table 4.4 for Au and demonstrate that the Fresnel factors connected to χ_{zzz} used for normalization are the smallest ones. Since the SHG yield was found to be dominant for the product of Fresnel factors times χ_{zzz}, the ratios of χ_{zxz}, χ_{zxx}, γ and χ_{zzz} are even smaller.

Table 4.4 Relative magnitude of Fresnel factors, M_{rel}, for Au connected to single tensor elements χ_{ijk} and to γ which may contribute to P-polarized SHG for p-polarized fundamental light

Tensor element	χ_{zzz}	χ_{xzx}	χ_{zxx}	γ
M_{rel}	1	12.7	5.8	3.8

Summarizing this discussion, the dipole-allowed tensor element χ_{zzz} describing the response of the near-surface region where symmetry is broken is the *only* source of SHG and relative changes of $|\chi_{zzz}|^2$ may be derived from time-resolved SH data using eqn (4.3.9).

4.3.2 Pump–probe SHG

In this section we present time-resolved second-order reflectivities measured on polycrystalline Cu, Ag and Au samples and derive the electron temperature dependence of the second-order susceptibilities. The results for the different metals are then compared to each other and discussed in terms of the conjecture that linear reflectivity and second-order susceptibility have a similar dependence on electron temperature.

Experimental setup

Measurements of transient second-order reflectivities were carried out with the same CPM laser system as used for transient linear reflectivities at 2 eV and 4 eV, but now similar pump and probe pulse energies (12 μJ to 16 μJ) and foci (≈ 250 μm) were used. This unusual choice of comparable pump and probe fluences, which led to strong heating by the probe pulses, was necessary to obtain a detectable SHG yield. The probe pulses monitored an almost homogeneously heated central area of the spatial temperature profile generated by the pump pulses, since the effective area probed by SHG corresponded to about half of the beam area. As in the linear case, we took care to ensure that the applied fluences did not induce damage.

Samples were placed inside a dark box with a 2 mm OG1 Schott filter at the entrance, thus cutting off all radiation at the SH frequency arising from optical elements along the beam path and scattered light from the excimer laser.

As illustrated in Fig. 4.17, pump and probe beams generated specular reflections of SH radiation with a coherent superposition of the two in the middle. By turning the sample, SHG could be observed in either one of the three outgoing directions. The SH radiation in this middle direction corresponds to the cross-correlation of pump and probe pulse and was used to calibrate zero delay and adjust a perfect spatial overlap between pump and probe foci. When measuring time-resolved SHG the specular reflection of the probe pulse was led into a monochromator with two 2 mm UG5 Schott filters and a polarizer in front of the entrance slit to cut out the fundamental and to select the desired polarization direction. All other signals were blocked by an aperture. The monochromator

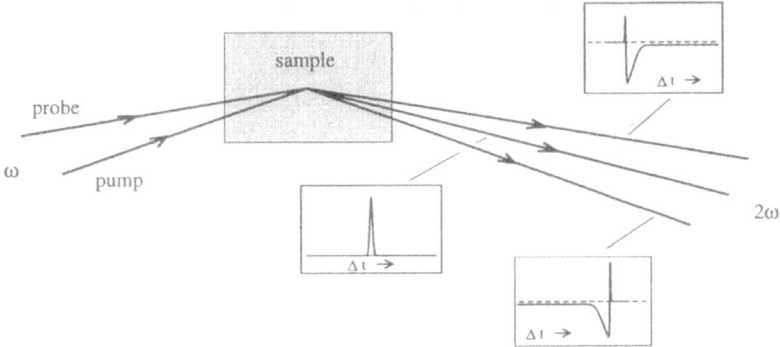

Fig. 4.17. Scheme of incident and emerging beam directions with observed shapes of the SHG signals in the insets.

led the light directly into a photomultiplier, whose output pulses were amplified by a boxcar integrator and then digitized and stored by a computer for each laser shot. Like for linear reflectivities, the recorded SHG signals were sorted according to a reference signal proportional to the probe pulse energy.

Results for Cu, Ag and Au

Relative changes of measured SH yield, normalized to the signal obtained in the absence of the pump beam, are shown in Figs 4.18–4.20 for Au, Cu and Ag. All

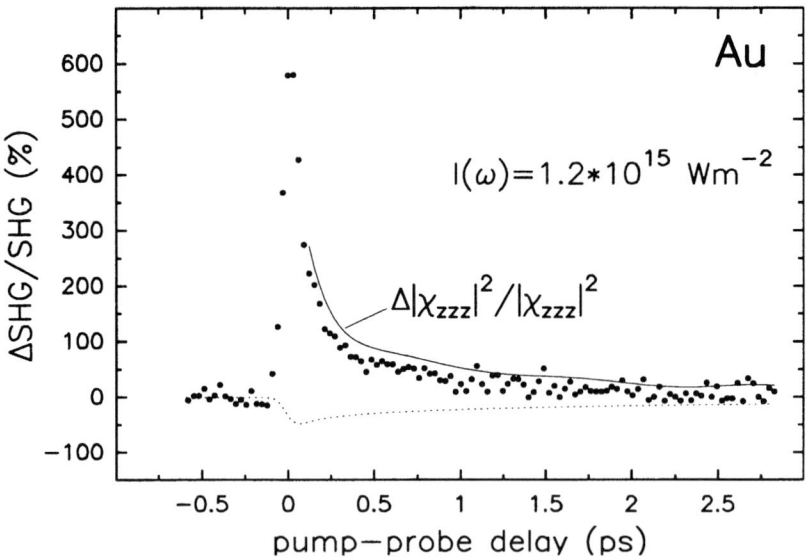

Fig. 4.18. Relative changes of SHG yield (dots) versus pump–probe delay for Au. The data have been normalized to the yield for probe pulses alone. Dotted lines indicate the expected electron temperature dependence of SHG from Fresnel factors only. Solid lines represent the difference between smoothed experimental data and dotted lines and represent the relative change of $|\chi_{zzz}|^2$.

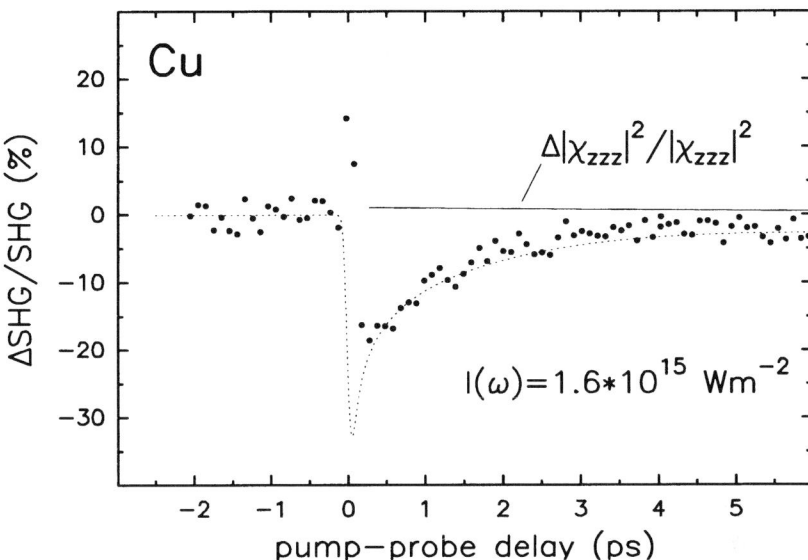

Fig. 4.19. Relative changes of SHG yield (dots) versus pump–probe delay for Cu. For the definition of solid and dotted lines see Fig. 4.18.

data were recorded in air at room temperature with the p → P polarization combination.

Each point displayed in these curves represents the average value of about

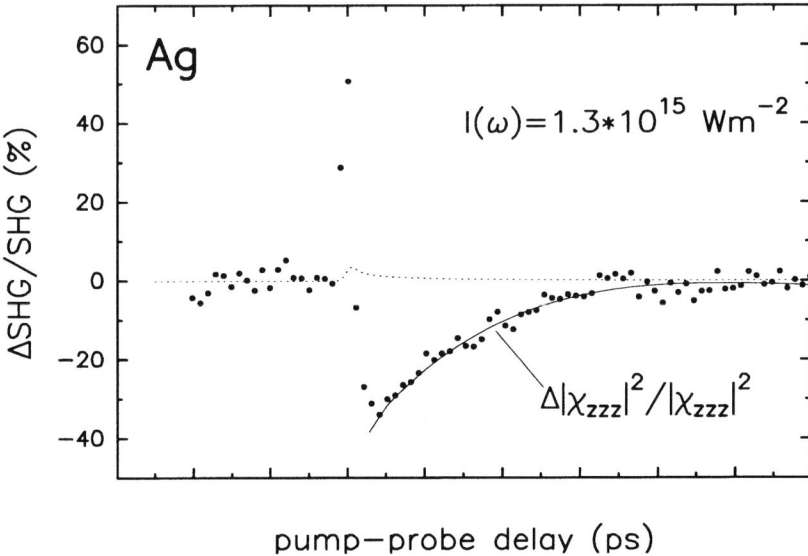

Fig. 4.20. Relative changes of SHG yield (dots) versus pump–probe delay for Ag. For the definition of solid and dotted lines see Fig. 4.18.

1000 probe pulses. The peaks around zero delay correspond to a coherent superposition between the p-polarized pump and probe beams and will not be discussed here. Dotted lines indicate predictions (cf. eqn (4.3.8)) for constant χ_{zzz} with the electron temperature exclusively contained in the Fresnel factors. Solid lines represent smoothed relative changes of $|\chi_{zzz}|^2$ as determined by eqn (4.3.9). For Cu, the excellent agreement between measured and predicted SH proves that $|\chi_{zzz}|^2$ is to first order independent of electron temperature within the explored temperature range. In contrast, for Ag and Au the strong deviation of dotted lines from the data shows that the temperature dependence of SHG is almost entirely contained in the nonlinear susceptibility.

Using the connection between electron temperature and timescale given by the TTM and the analysis of the linear reflectivities in sec. 4.2.1, we converted the timescale to temperature and deduced the relative change of $|\chi_{zzz}|^2$ with electron temperature. The results are plotted in Fig. 4.21. Again, it is obvious

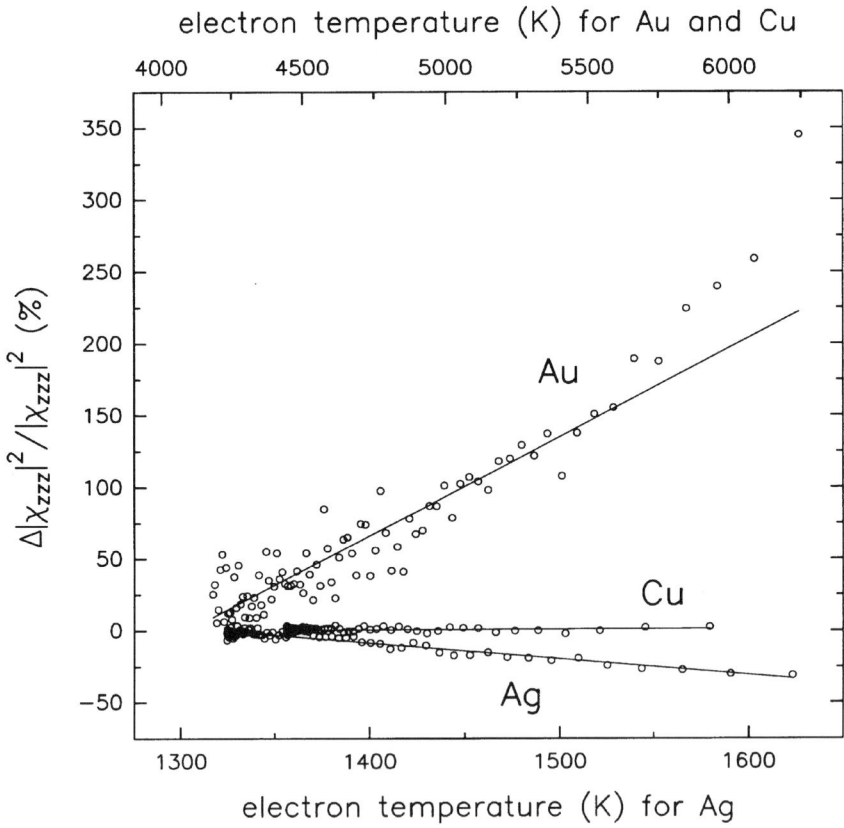

Fig. 4.21. Relative changes of $|\chi_{zzz}|^2$ with electron temperature for Cu, Ag and Au. These changes were derived from the time dependence of second-order susceptibilities using the relation between T_e and time delay provided by the analysis of transient linear reflectivities.

that $|\chi_{zzz}|^2$ does not depend on electron temperature for Cu. For both Ag and Au, a linear dependence of $\Delta|\chi_{zzz}|^2$ on T_e is found, but the corresponding slopes have different magnitudes and opposite signs.

Discussion of the temperature dependence of $\chi^{(2)}$

The purpose of the following comparative discussion of the different behaviour of $\chi^{(2)}$ for Cu, Ag and Au is to show that these observations are in qualitative agreement with the predictions of $\chi^2(T_e)$ outlined in sec. 4.3.1.

The most apparent difference between Cu and Au on the one hand and Ag on the other is the SHG resonance enhancement. It takes place in the first intermediate state at $\omega = 2\,\text{eV}$ for Cu and Au, and in the highest intermediate state at $2\omega = 4\,\text{eV}$ for Ag. As a consequence, for Ag we expected the relative changes of $R(4\,\text{eV})$ and $|\chi^{(2)}(4\,\text{eV}:2\,\text{eV},2\,\text{eV})|^2$ to be of comparable magnitude but opposite sign. Comparison of Figs 4.7 and 4.20 showed that this is true. This observation also provides evidence that any possible influence of surface contamination is of minor importance for Ag.

Another result is that temperature- and wavelength-dependent SHG measurements yield complementary information. Li et al.[68] reported that the frequency dependence of SHG on Ag(111) near the interband transition threshold is dominated by Fresnel factors, whereas we find for 2 eV fundamental photons the temperature dependence of SHG on polycrystalline Ag is mostly contained in $\chi^{(2)}$. This is no contradiction but demonstrates instead that a complete set of information is needed to pave the way for a unique interpretation of $\chi^{(2)}$ in terms of band structure.

At first glance it is astonishing that the relative changes of $|\chi_{zzz}|^2$ are negligible within the experimental errors for Cu, while for Au these changes are about two orders of magnitude larger than those of $R(\omega)$ (compare Figs 4.8 and 4.18). In both cases $\chi^{(2)}$ is resonantly enhanced in the first intermediate state at ω. We cannot exclude that the temperature dependence of SHG for Cu is affected by the oxide layer. However, as theoretically expected[47] and proven by the transient linear reflectivity of Cu in Fig. 4.7, the oxide layer should affect the measured time dependence of SHG predominantly for time delays $\geq 1\,\text{ps}$ and not so much the amplitude near zero delay.

Therefore, we propose that the different sizes of $\Delta|\chi_{zzz}|^2$ for Cu and Au originate from the unlike energy mismatch, ΔE, between the used photon energy and the interband transition thresholds, which is $\Delta E(\text{Au}) = -380\,\text{meV}$[46,49] and $\Delta E(\text{Cu}) = -150\,\text{meV}$.[70] Although peak electron temperatures of about 7000 K for Cu and 6700 K for Au were reached by simultaneous heating with pump and probe pulses, the average electron temperature generated by probe pulses alone was approximately 4300 K for Cu and 3400 K for Au, corresponding to 390 meV and 310 meV. Thus, the width of the Fermi distribution induced by only the probe pulses was much larger than ΔE for Cu but smaller than ΔE for Au. Consequently, monitored changes of electronic occupancy induced by only the probe pulses were saturated in the case of Cu, whereas this was not the case for Au. As a result, heating by pump pulses influenced the SHG only for Au, while for Cu the probe pulse SHG was unaffected by pump pulses. This qualitatively explains the nearly one order of magnitude larger SH yield at negative delay times for Cu compared to Au. We conclude that changes in $\chi^{(2)}$ saturate with electronic occupancy governed by T_e. This implies that the heating rate must be carefully considered in all pump–probe SHG experiments.

Finally, we want to examine the question whether the linear temperature dependence of $\chi^{(2)}$, displayed in Fig. 4.21, proves identical electron dynamics at the surface and in the bulk. In this context, we recall that the electron temperature dependence of $\chi^{(2)}$ was derived from time-resolved SHG by using the correlation between T_e and the decay time of transient linear reflectivities. Hence, this method actually presupposes identical electron dynamics at the surface (few uppermost monolayers) and in the bulk (optical penetration depth). That such is the case can nevertheless be justified by comparing the temperature dependences of $\Delta \chi^{(2)}$ in Fig. 4.21 with calculated temperature variations of ΔR (cf. Fig. 4.22).

The qualitative agreement between $-\Delta R/R$ in Fig. 4.22 and $\Delta |\chi_{zzz}|^2/|\chi_{zzz}|^2$ in Fig. 4.21 confirms our predictions about $|\chi_{zzz}(T_e)|^2$ and justifies the assumption that SHG can be viewed as a cascade of transitions where each can be affected by a change of the Fermi distribution due to heating. The deviation of

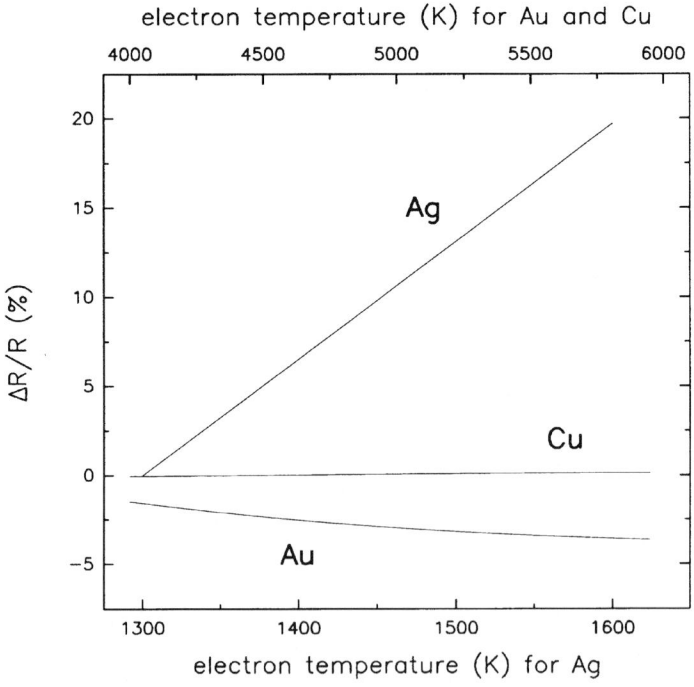

Fig. 4.22. Calculated normalized temperature variations of linear reflectivities for Cu and Au at 2 eV and for Ag at 4 eV. The temperature ranges correspond to those in Fig. 4.21.

the data from linearity, observed for Au at temperatures above 5500 K, could be attributed either to different electron dynamics at the surface and in the bulk during the first 200 fs or—more likely—to the onset of d → d → Fermi level transitions which do not contribute to R. Hence, at present it remains an open question whether we see an indication of different electron dynamics at the surface and in the bulk of polycrystalline Au.

4.3.3 Polarization-dependent SHG

All SHG measurements presented in sec. 4.3.2 were carried out with a fixed p → P polarization combination of fundamental and second harmonic light. In this section we present experimental evidence that even time-integrated SHG can provide qualitative information about the electronic structure of a material if the polarization is varied.[22,71,72] Since SHG hinges on the anharmonic motion of electrons it is intuitively clear that there should be a difference whether s or d electrons are excited and—pertaining to the geometry—that this should be seen in a different way, depending on the oscillation direction of the field vector with regard to the surface plane. To demonstrate this we will show a few examples of the variation of polarized SHG yield with the polarization of the fundamental light for polycrystalline Al, Cu, Cr, Ni and Ti.[73]

Measurements were carried out with 1 ps/600 nm pulses generated by a synchronously pumped dye laser amplified to about 20 μJ/pulse in three dye amplification stages, pumped by 8 ns/532 nm pulses from a Nd:YAG laser with a repetition rate of 10 Hz. Again, the experiments were performed in air at room temperature using an angle of incidence of 45°. Great care was taken not to damage the samples.

According to Table 4.3 and eqn (4.3.7) the polarization dependence of S-polarized SH yield should be material-independent for isotropic surfaces and is given by

$$I_S(2\omega, \phi) \propto |A_S \chi_{xzx} f_s t_p t_s I_0|^2 \cos^2 \phi \sin^2 \phi. \qquad (4.3.11)$$

As a representative example for s-band metals, the data obtained for polycrystalline Cu are shown in Fig. 4.23.

The dominant effect, caused by the different anharmonicity of the electron motion in and out of plane, is the one order of magnitude large P-polarized SH yield at $\phi = 0°$ compared to the maximum S-polarized yield at $\phi = 45/135°$. For the same reason, the S-polarized SHG has no unique relation to the electronic structure of the metal and consequently its variation with polarization of the fundamental exhibits the same pattern for all metals. Hence, no further data will be presented for S-polarized SHG. In contrast, the pattern for P-polarized SHG will change conspicuously when either s/p or d electrons dominate the SH yield. For buried d bands this causes a strong frequency dependence.[67] For 600 nm fundamental photons (2.07 eV) this behaviour will be demonstrated for Cu, Al, Cr and Ni.

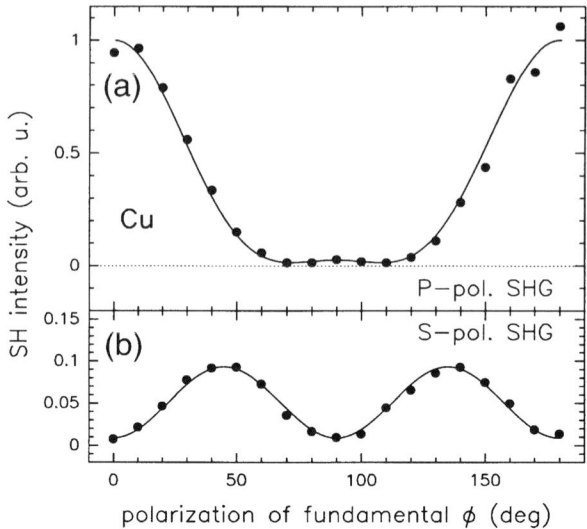

Fig. 4.23. Variation of (a) P- and (b) S-polarized SH yield with polarization of the 600 nm fundamental incident at 45° for polycrystalline Cu. The intensity scale is identical for both S and P polarization.

Figures 4.23 and 4.24 show that the ratios of P-polarized SH yield generated by p(0°, 180°)- and s(90°)-polarized fundamental are indeed strikingly different for Al and Cu on the one hand, and Cr and Ni on the other. The fact that almost no P-polarized SH is generated by s-polarized fundamental for Al and Cu is in accordance with the expected response of free electrons. Partially localized d electrons in Cr and Ni, to the contrary, exhibit a maximum in P-polarized SHG at $\phi = 90°$, i.e. for s-polarized fundamental. The polarization dependence of SHG as a fingerprint of the electronic structure at surfaces and interfaces was first discussed by Hübner et al.[72] By calculating the dipole transition matrix elements which contribute to $\chi^{(2)}$ (cf. eqn (4.3.5)) within a tight-binding-like model, they demonstrated that $I_{p \to p}(2\omega)$ is much larger than $I_{s \to p}(2\omega)$ whenever only s/p electrons are involved in the SHG process, but of comparable magnitude in case that the SH is generated by d electrons. Consequently, all alkali metals should behave like Al and all transition metals like Ni and Cr. The noble metals on the other hand should behave like Al for photon energies smaller than the interband transition threshold, but like Ni and Cr when $\hbar\omega \geq ITT$. Such behaviour was proved for polycrystalline Cu by Petrocelli et al.[67]

Although Hübner et al.[72] explain the experimental results for Cu, Al, Cr and Ni shown in Figs 4.23 and 4.24, the authors did not elaborate on the possibility that the mobility of d electrons may differ significantly even among the transition metals.

To demonstrate the importance of the d-electron mobility we confront the

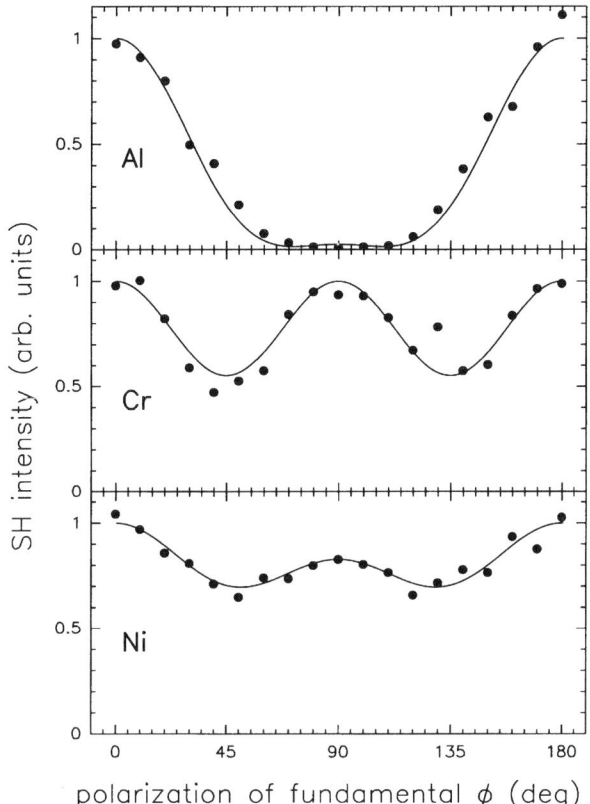

Fig. 4.24. Variation of P-polarized SH yield with polarization of the 600 nm fundamental light incident at 45° for polycrystalline Al, Cr and Ni. All curves were normalized to 1 at $\phi = 0°$.

data for Ni and Cr, where d-electron mobility is much smaller than the mobility of the s and p electrons, with those for Ti, shown in Fig. 4.25. From the data it is apparent that the s- and d-electron mobility is of the same order. The similarity of the experimental results obtained for Ti and Al is evidence that a large ratio of $I_{p \to p}(2\omega)/I_{s \to p}(2\omega)$ points to a high mobility of the electrons, notwithstanding whether they are s/p or d electrons. This is confirmed by the values of hopping integrals given in ref. 74, which also predicts the same behaviour for Sc. It would be of interest to check this conjecture by similar SHG measurements on this element.

4.3.4 Electron and magnetization dynamics of Ni

Dependence of SHG on magnetization

The sensitivity to electronic structure and symmetry also suggests that SHG will

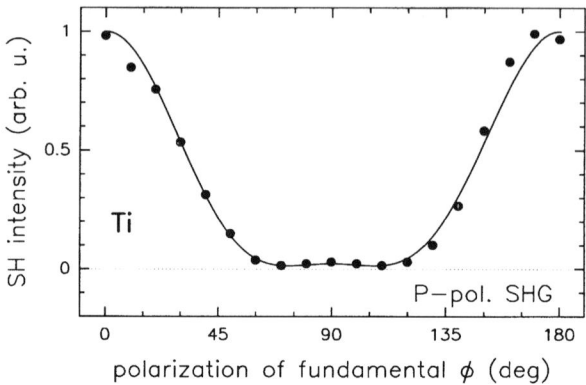

Fig. 4.25. Variation of P-polarized SH yield with polarization of 600 nm fundamental incident at 45° for polycrystalline Ti (film).

be able to recognize surface magnetization M. This was first pointed out by Pan et al.[7] who showed by symmetry considerations that for a magnetized reflective plane the second-order susceptibility tensor $\chi^{(2)}(M)$ can be divided into two terms of even and odd symmetry with respect to sign reversal of M, with components

$$\chi^{(2)}(\pm M) = \chi^{(2)}_{\text{even}}(\pm M) \pm \chi^{(2)}_{\text{odd}}(\pm M). \quad (4.3.12)$$

The set of nonvanishing even tensor elements is for isotropic surfaces identical for the magnetic and nonmagnetic cases, and is given by

$$\chi^{(2)}_{\text{even}}(\pm M) = \begin{pmatrix} 0 & 0 & 0 & 0 & \chi_{xxz} & 0 \\ 0 & 0 & 0 & \chi_{yyz} & 0 & 0 \\ \chi_{zxx} & \chi_{zyy} & \chi_{zzz} & 0 & 0 & 0 \end{pmatrix}. \quad (4.3.13)$$

The odd tensor consists of different elements for the magnetization M parallel and perpendicular to the plane of incidence (xz plane). If we choose $M = M_x$ the odd tensor takes the form

$$\chi^{(2)}_{\text{odd}}(M_x) = \begin{pmatrix} 0 & 0 & 0 & 0 & 0 & \chi_{xxy} \\ \chi_{yxx} & \chi_{yyy} & \chi_{yzz} & 0 & 0 & 0 \\ 0 & 0 & 0 & \chi_{zyz} & 0 & 0 \end{pmatrix} \quad (4.3.14)$$

whereas for $M = M_y$ it changes to

$$\chi^{(2)}_{\text{odd}}(M_y) = \begin{pmatrix} \chi_{xxx} & \chi_{xyy} & \chi_{xzz} & 0 & 0 & 0 \\ 0 & 0 & 0 & 0 & 0 & \chi_{yxy} \\ 0 & 0 & 0 & 0 & \chi_{zzx} & 0 \end{pmatrix} \quad (4.3.15)$$

In the nonmagnetic case the odd tensor vanishes.

Table 4.5 Nonvanishing tensor elements of $\chi^{(2)}$ that contribute to SHG on magnetized isotropic surfaces for different magnetization directions and polarization combinations. Mix corresponds to 45° and lower-case (capital) letters denote polarizations of fundamental (second harmonic). The xz plane is chosen as plane of incidence and the tensor elements are classified as even and odd according to sign reversal of $\mathbf{M}$

	s → S	s → P	p → S	p → P	mix → S	mix → P
even isotropic	–	χ_{zyy}	–	χ_{zzz}, χ_{zxx} χ_{xzx}	χ_{yzy}	χ_{zzz}, χ_{zxx} χ_{zzz}, χ_{zyy}
odd $\mathbf{M} \parallel x$	χ_{yyy}	–	χ_{yxx}, χ_{yzz}	–	χ_{yxx}, χ_{yyy} χ_{yzz}	χ_{xxy}, χ_{zyz}
odd $\mathbf{M} \parallel y$	–	χ_{xyy}	–	χ_{xxx}, χ_{xzz} χ_{zzx}	χ_{yxy}	χ_{xxx}, χ_{xyy} χ_{xzz}, χ_{zzx}

Obviously, the SHG becomes much more complicated when the surface is magnetized. Table 4.5 lists the tensor elements that appear for various polarization combinations and an isotropic surface with and without magnetization.

This provides the tool to determine unambiguously the strength and direction of surface magnetization as well as the magnetic order–disorder transition at the Curie temperature with SHG.

The general complexity of the problem can be reduced by performing time-resolved SHG measurements on magnetized samples for opposite magnetization directions. This takes advantage of the fact that even tensor elements are almost unaffected by the magnetization, whereas the odd ones scale linearly with $\mathbf{M}$.[75] Hence we can write in first approximation

$$\chi_{ijk}^{\text{even}}(\pm M) = \chi_{0,ijk}^{\text{even}} \qquad (4.3.16)$$

and

$$\chi_{ijk}^{\text{odd}}(\pm M) = \pm \chi_{0,ijk}^{\text{odd}} |M|. \qquad (4.3.17)$$

By combining the individual tensor elements and corresponding Fresnel coefficients to an effective even and odd contribution denoted by $A\chi_0^{\text{even}}$ and $B\chi_0^{\text{odd}}$, the sum and difference of the SHG signals for opposite magnetization directions,

$$I^{\pm}(2\omega, |M|) = I(2\omega, M) \pm I(2\omega, -M), \qquad (4.3.18)$$

take the form

$$I^{+}(2\omega, M) = 2I_0^2(\omega)\left(|A\chi_0^{\text{even}}|^2 + |B\chi_0^{\text{odd}}M|^2\right), \qquad (4.3.19)$$

$$I^{-}(2\omega, M) = 4I_0^2(\omega)|A\chi_0^{\text{even}}B\chi_0^{\text{odd}}M|\cos\phi, \qquad (4.3.20)$$

where the intensity of incident fundamental is denoted by $I_0(\omega)$. The resulting phase of A, B, χ_0^{even} and χ_0^{odd} depends on $\mathbf{M}$ and is given by ϕ.

For Ni, measurements of transient linear reflectivities and theoretical calculations[76] reveal that the Fresnel factors A and B as well as the magnetization-independent factors χ_0^{even} and χ_0^{odd} in eqns (4.3.19) and (4.3.20) vary only a few per cent with electron temperature below the Curie temperature T_C. This can be understood by the large width ($\approx 0.3\,\text{eV}$) of vacant d states in the minority band[77] in comparison to kT_C. Thus we conclude that variations of $I^-(2\omega, M)$ observed in transient SHG measurements on ferromagnetic Ni are caused by a time dependence of $\mathbf{M}$.[78]

Experimental results for Ni

Experiments were performed on polished polycrystalline nickel (purity 99.99%) at room temperature in air. The samples were magnetized to saturation with the magnetization direction oriented perpendicular to the plane of incidence. P-polarized 150 fs/800 nm pump and probe pulses generated at a repetition rate of 15 kHz by a commercial Ti:sapphire regenerative amplifier system (Coherent Mira 900/RegA 9000) were utilized to excite the electrons and probe the electron temperature and magnetization by detecting the total SHG yield. Angles of incidence were 20° for the pump and 45° for the probe beam. As in the case of noble metals, SH radiation was found in three outgoing directions (cf. Fig. 4.17). Again, similar experimental measures (darkbox, spatial and spectral filters) guaranteed that only SH radiation generated by the probe beam at the sample surface was led to the photomultiplier. Owing to the high stability and high repetition rate of the laser system, the signal-to-noise ratio was sufficiently improved by averaging some 10^5 shots at each time delay. The measurements were carried out for a large variety of pump fluences leading to different initial electron temperatures. The intensity ratio of pump and probe pulses was about 3:1 for the highest pump fluences of about $7\,\text{mJ}\,\text{cm}^{-2}$.

From the time dependence of the SHG yield for opposite magnetization directions, $I^{\pm}(2\omega, M)$ was calculated (cf. eqn (4.3.18)), and the relative change

$$\Delta I^{\pm}(t) = \left[I^{\pm}(t) - I_0^{\pm}\right]/I_0^{\pm} \qquad (4.3.21)$$

normalized to the SHG yield $I_0^{\pm}$ of the probe pulse alone was formed for each time delay. The final results for seven selected pump fluences are shown in Fig. 4.26.[78]

The most striking feature of Fig. 4.26 is the smaller effect observed for $\Delta I^+(t)$ compared to $\Delta I^-(t)$ with the latter depending on the magnetization. A more subtle difference appears in Fig. 4.27, which is a magnified version of the short-time range in Fig. 4.26, showing the data for three selected pump fluences. Note that the minima of $\Delta I^-(t)$ occur about 50 fs faster than those of $\Delta I^+(t)$ at $(280 \pm 30)\,\text{fs}$. We interpret the minimum of the $\Delta I^+(t)$ curves as the time at which the electron temperature is established.

To understand the different behaviour of the curves in (a) and (b) of Figs 4.26 and 4.27, we have to analyse $\Delta I^{\pm}(t)$ in detail. First, we assume that the time

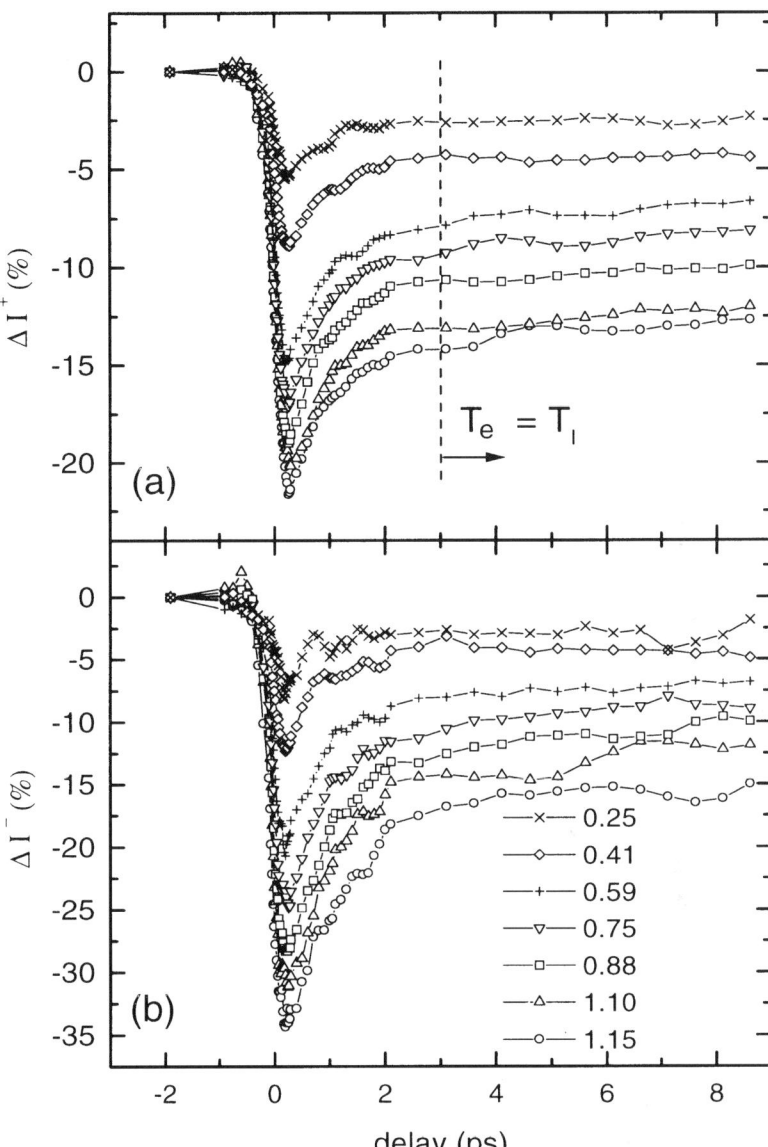

Fig. 4.26. Time dependence of (a) sums and (b) differences of SHG yields obtained for opposite magnetization directions, normalized to the SHG intensity without pump pulses (cf. eqn (4.33)). The various data sets belong to different pump fluences, given in relative units, calibrated by $1.0 \equiv 6\,\mathrm{mJ\,cm^{-2}}$. As indicated, equilibrium between electron and lattice temperatures is reached at about 3 ps.

dependence of magnetization is determined by the electron temperature relaxation, $M(t) = M(T_e(t))$. Secondly, we approximate the classical magnetization

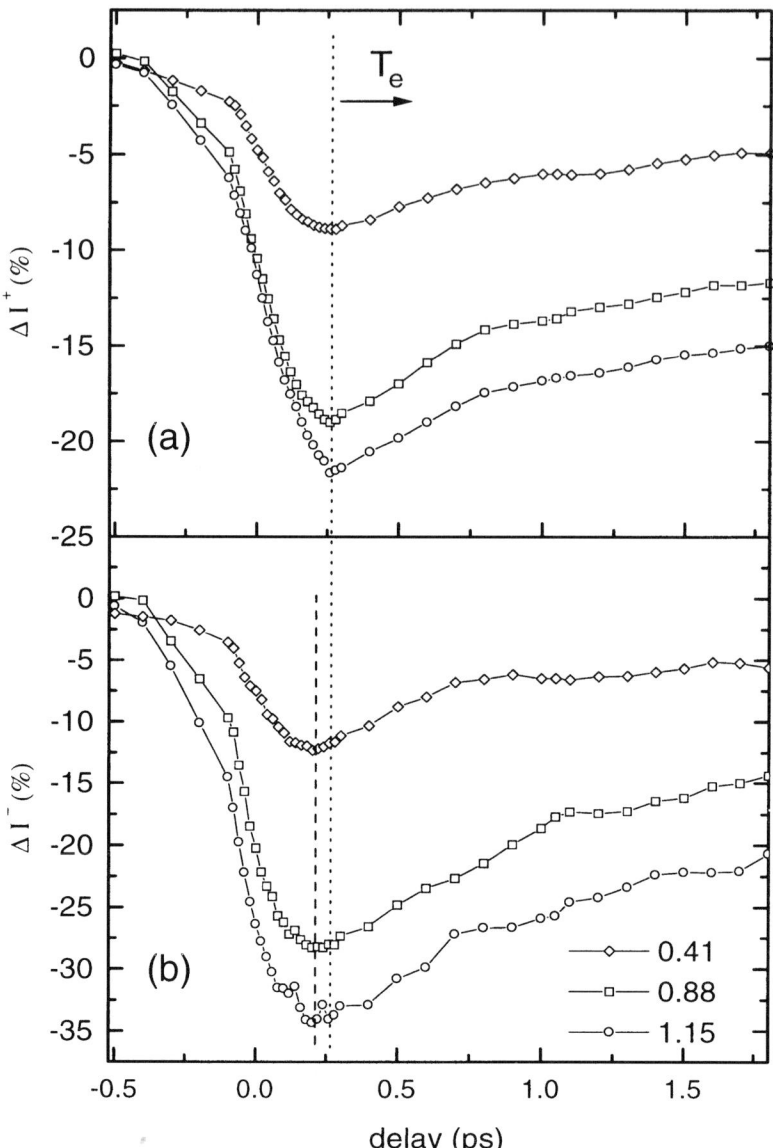

Fig. 4.27. Magnified version of Fig. 4.26 for three selected fluences (calibration: $1 \equiv 6\,\text{mJ cm}^{-2}$). The minima of $\Delta I^-(t)$, indicated by the dashed line, occur about 50 fs earlier than those of $\Delta I^+(t)$ at $(280 \pm 30)\,\text{fs}$ (dotted line).

curve[79] within the temperature range covered in our experiments ($290\,\text{K} \leq T_e \leq 550\,\text{K}$) by

$$M(T_e) = M(T_0)[1 - \text{const.}(T_e - T_0)]^{1/2}, \qquad (4.3.22)$$

where T_0 denotes the mean electron temperature monitored by the probe pulse

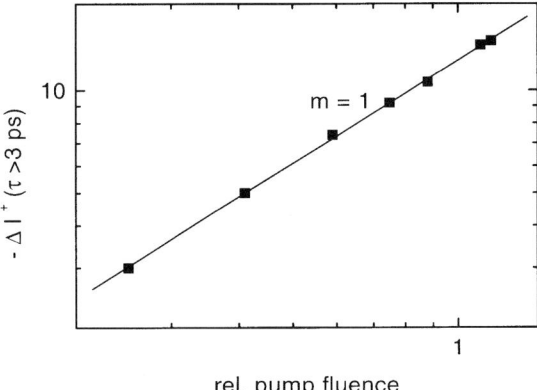

Fig. 4.28. Fluence dependence of ΔI^+, averaged over all values for $t > 3\,\text{ps}$ and normalized to values for the probe pulse alone. The linear slope and the absence of any offset warrant the proportionality between ΔI^+ and T_e (cf. eqn (4.35)).

at negative delay. Inserting this approximation into eqns (4.3.19) and (4.3.20) we obtain

$$\Delta I^+(t) = \text{const.}[T_0 - T_e(t)], \tag{4.3.23}$$

$$\Delta I^-(t) = \frac{M(T_e(t))\cos\phi}{M(T_0)} - 1. \tag{4.3.24}$$

Hence, the curves in Fig. 4.26(a) represent the time evolution of electron temperature, while those in Fig. 4.26(b) describe the transient magnetization, provided ϕ is constant. A prerequisite for the validity of these equations is a thermalized electron distribution. That $\Delta I^+(t)$ is indeed proportional to T_e is verified by the slope $m = 1$ in Fig. 4.28, where the averaged values for delay times $\geq 3\,\text{ps}$ in Fig. 4.26(a) are plotted against pump fluence. Thus, in thermal equilibrium with the lattice, T_e obeys the proportionality in eqn (4.35) and justifies the approximation made for $M(T_e)$ at longer delay times.[78]

According to eqns (4.3.23) and (4.3.24) a plot of $(\Delta I^- + 1)$ versus $-\Delta I^+$ should reproduce the classical $M(T)$ curve for nickel[79] normalized to $M(T_0)$. Such comparison is shown in Fig. 4.29(a) for delay times $\geq 0.3\,\text{ps}$ when the electron temperature is established. For this time range the data indeed fall within $\pm 5\%$ on the magnetization curve, indicated by the solid line. The good agreement demonstrates that ϕ is independent of electron temperature. Furthermore it provides an intrinsic electron temperature calibration for excitation with various pump fluences.

Since agreement with the classical magnetization curve also covers the range between 0.3 ps and 3 ps, where T_e is *not* in equilibrium with the lattice, we conclude that the electron temperature governs the magnetization. On the other hand, we can decide that the electrons are fully thermalized as soon as delay times are reached for which the points fall on the magnetization curve. From the coincidence of the minima in Fig. 4.27(a) and the agreement with the magnetiza-

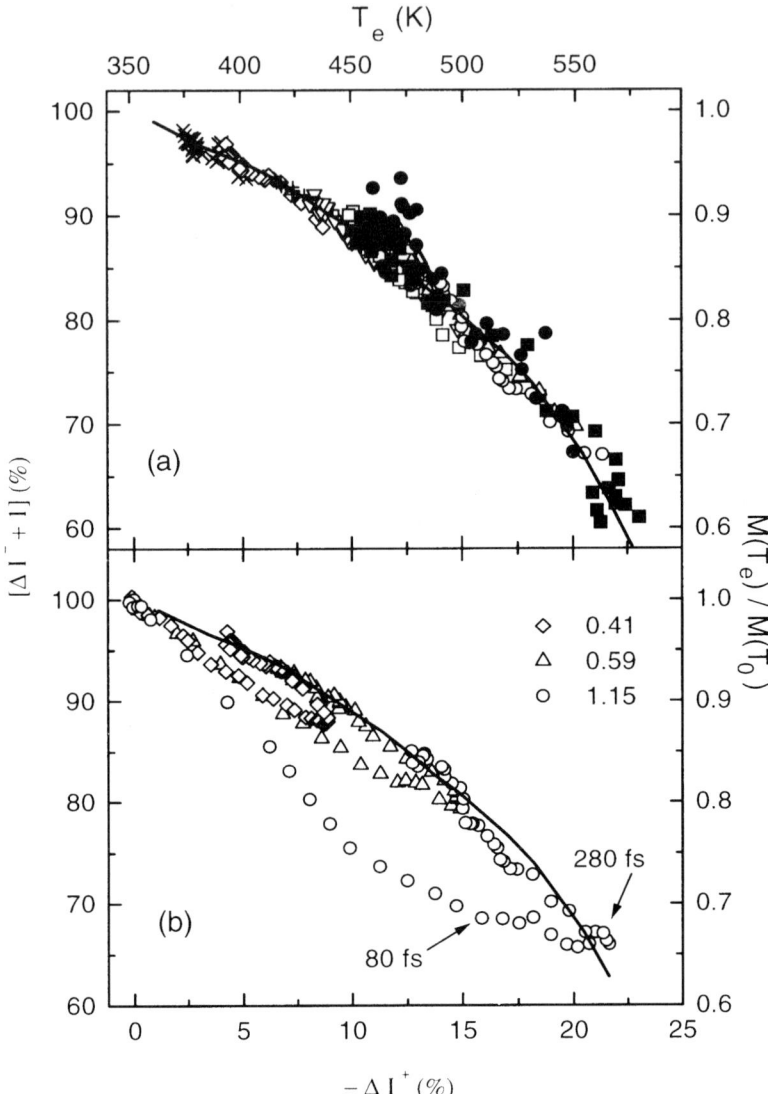

Fig. 4.29. Comparison of the classical magnetization curve (solid line) according to Weiss and Forrer[79] with measured data for all pump fluences between 1.5 and 7 mJ cm^{-2}. (a) Values for $t > 0.3$ ps with a well defined electron temperature. The good agreement demonstrates that the magnetization is governed by T_e, even in the time range where electrons and lattice are not in equilibrium. (b) Similar plot as in (a) but for data of three selected fluences covering the whole time range. Systematic deviations from the magnetization curve occur at short times when the electrons are still not thermalized.

tion curve in Fig. 4.29(a) the electron thermalization time of Ni is unambiguously determined to be (280 ± 30) fs, in excellent agreement with the 260 fs reported elsewhere.[80]

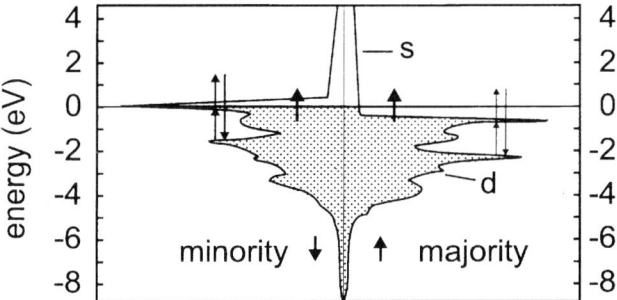

Fig. 4.30. Averaged density of states for Ni according to Callaway and Wang.[77] While single-photon transitions excite comparable rates of minority and majority spins, SHG monitors predominantly minority spins due to the resonance enhancement by the unoccupied peak in the minority band.

In Fig. 4.29(b) all data for three pump fluences, including those for short times, are plotted and compared to the magnetization curve. The strong deviation of the points for $t < 0.3$ ps can be attributed to nonequilibrium electrons and, in view of eqn (4.26), to a rapid phase change. The deviation vanishes once thermalization of the electron gas is completed. That the data fall below the magnetization curve for $t < 0.3$ ps is also in agreement with Fig. 4.27(b) where ΔI^- was observed to respond about 50 fs faster than ΔI^+.

The fact that a plot of $(\Delta I^- + 1)$ versus $-\Delta I^+$ reproduces for $t > 0.3$ ps the classical magnetization curve for nickel[79] so well needs a comment, since it can be seen from eqn (4.3.19) that ΔI^+ is actually proportional to $|M|^2$. Therefore, ΔI^+ is independent of any direction and monitors, at arbitrary delay times, only the absolute amount of momentary population difference between minority and majority spins. This difference will, of course, depend on time and diminish with higher excitation energies (cf. Fig. 4.30). Hence, it is plausible that ΔI^+ reflects the development of electron temperature out of a highly nonequilibrium state. ΔI^-, on the other hand, is sensitive to the direction of M and samples preferentially the minority spins of the vacant d states as illustrated in Fig. 4.30. The faster response of ΔI^- can be understood in terms of the restoration of spin–orbit coupling after excitation into the conduction band. Relaxing electrons recouple much faster for minority spins than for majority spins[81,82] due to the width of the vacant d states. This width amounts to about 0.3 eV (see Fig. 4.30), which is large compared to the width of the Fermi distribution for $T_e \leq T_C$.

Comparison with literature

First of all, it is worth noting that the relaxation times shown in Fig. 4.26 are not compatible with usual spin–lattice relaxation times, τ_{sl}. In our experiments, electron temperature and magnetization develop out of highly excited states with electrons and lattice in extreme nonequilibrium. In contrast, τ_{sl} denotes the time it takes for the spin system to adjust to a new temperature while the electron bath and the lattice still remain in perfect equilibrium. Hence, the values of $\tau_{sl} = (100 \pm 80)$ ps measured by Vaterlaus et al.[83,84] for ferromagnetic Gd and the limits reported for Fe[84,85] and Ni[86] must be considered in a different context.

Concerning femtosecond time-resolved magnetization dynamics, we are aware of only one investigation in the literature. It was carried out by Beaurepaire et al.[80] on Ni films and the results ought to be comparable to our data. The authors utilized the *linear* magneto-optical Kerr effect to detect hysteresis loops for different time delays between pump and probe pulses at one specific pump fluence. By comparing the time dependence of the remanence with the classical magnetization curve, they defined a spin temperature T_S and analysed its time evolution within the framework of a phenomenological three-temperature model.[83] The required knowledge of electron temperature relaxation was obtained in ref. 80 by an independent measurement of the change of transmission with time, which was assumed to scale linearly with $T_e(t)$. From analysis of their data, Beaurepaire et al. conclude that the magnetization breakdown is delayed by about 2 ps with respect to the establishment of electron temperature. This is in striking contrast to our results in Figs 4.26 and 4.27. The expected response time of itinerant magnetism is of the order of T_C^{-1},[87] amounting to about 80 fs for Ni, more in line with our data. No final statement about these contradicting observations is possible, since too many parameters differ in the experiments reported here and in ref. 80. We want to emphasize, however, that the SHG technique has the great advantage of monitoring transient electron temperature relaxation and transient magnetization *simultaneously*, without any further need for additional calibration measurements. Another important point is that in our experiments the use of various pump fluences, generating different initial electron temperatures, served as a crucial test of the predicted proportionality between $\Delta I^+(t)$ and $T_e(t)$, which was then used to prove the linear dependence of $\Delta I^-(t)$ on $M(T_e(t))$. In our analysis of the time-resolved SHG response to transient magnetization changes, all entering assumptions were experimentally justified. Hence, our experiment was more complete, which gives us confidence in our result.

4.4 Summary

Three types of experiments have been discussed: (1) Those which aimed to measure the energy relaxation of optical excited electrons in noble metals—to this purpose, we investigated time-resolved pump–probe linear and second-order reflectivities. (2) Evidence for the sensitivity of polarized SHG to the electronic structure near the Fermi level was presented. (3) It was shown that pump–probe SHG can provide information about ultrafast electron and magnetization relaxation.

Regarding (1), transient electron temperatures were derived from time-resolved linear reflectivities on bulk noble metals by using a theoretical model of the electron temperature dependence of the linear dielectric function in connection with the two-temperature model. Sensitivity to transient electron temperatures was attained by measuring with photon energies close to the interband transition threshold. Good agreement between experimental results and model calculations justified the theoretical approach. From this analysis electron–phonon coupling constants of $g(Cu) = 4.75$, $g(Au) = g(Ag) = 1.1$, in units of 10^{16} W m^{-3} K^{-1}, were derived.

The theoretical predicted saturation behaviour of linear reflectivities at high electron temperatures for photon energies smaller than the interband transition threshold was demonstrated and further supported the validity of our model for $\varepsilon(T_e)$.

Systematic measurements of transient linear reflectivities on gold films of various thicknesses led to the determination of the penetration depth of deposited energy. Conspicuous changes in the observed relaxation behaviour and comparison of the transient reflectivities with theoretical predictions by the two-temperature model yielded a range of 100 nm for ballistic electrons excited by 200 fs pulses, in agreement with ref. 36.

Pump–probe SHG on bulk Cu, Ag and Au samples was measured using a fundamental photon energy of 2 eV. For Cu and Au, this photon energy was slightly below the interband transition threshold, which caused a sensitivity enhancement of SHG to transient electron temperatures by one (Cu) or two (Au) orders of magnitude compared to the linear case. For Ag, where the 4 eV photon energy of the second harmonic was resonant with the interband transition threshold, no such enhancement was found.

SHG data were analysed by means of a phenomenological model,[30] which factorizes contributions from the linear dielectric functions at ω and 2ω and from second-order susceptibility. Knowledge of the variation of the dielectric functions with electron temperature, obtained from transient linear reflectivities, allowed us to infer the temperature dependence of $\chi^{(2)}$. It indicates similar electron temperature relaxation at the surface and in the bulk. The electron temperature dependence of $\chi^{(2)}$ obtained here will, together with its frequency dependence reported in ref. 67, serve as a crucial test for any theoretical model about the nonlinear susceptibility.

Pertaining to (2), a set of stationary measurements with polarized fundamental light and polarized SHG detection was carried out on the s/p- and d-band metals Cu, Al, Ti, Cr and Ni. While the S-polarized SHG remains the same, a striking difference of the yield pattern for P-polarized SH light was found when varying the polarization of the fundamental. This provides a fingerprint of the electronic structure near the Fermi level. A closer analysis of these data along the lines of ref. 72 would be desirable.

Finally, with regard to (3) time-resolved SHG was performed on magnetized polycrystalline Ni to investigate the magnetization dynamics following optical excitations by femtosecond laser pulses. These measurements were done for a variety of pump fluences to obtain the electron and spin relaxation for various degrees of electron temperature. The analysis of the time dependence of SHG was built on the model that sums and differences of the yield for opposite magnetization directions reflect electron temperature relaxation and magnetization dynamics, respectively. Independently of pump fluence, the electron thermalization time was found to be (280 ± 30) fs in agreement with ref. 80. Once the electron temperature is established, the data follow the classical magnetization curve.[79] This result demonstrates for the first time that the electron temperature governs the magnetization curve and that the magnetic response is not delayed with respect to the electronic one. The magnetic response might even be somewhat faster than the electron thermalization, which can be attributed to different lifetimes of excited majority and minority electrons, caused by the large density of vacant minority-spin states in Ni.

In summary, time-resolved SHG was demonstrated to be a powerful and versatile technique for monitoring ultrafast electron dynamics at metal surfaces. There are two main advantages of SHG: its surface sensitivity and the strong response to magnetization, which both originate from the sensitivity of the second-order susceptibility to electronic symmetry. When comparing the information provided by measurements of linear and second-order reflectivities, time-resolved SHG is superior to transient linear reflectivity whenever the investigated dynamic processes influence the electronic symmetry. In cases where the electronic symmetry remains unaffected, however, linear optical techniques are to be preferred since their data interpretation is far more straightforward as for SHG data.

Acknowledgements

We gratefully acknowledge stimulating discussions with K. H. Bennemann, W. Hübner, J. Güdde, R. Knorren and M. Garcia. We also thank P. West for his assistance with computer installations and D. Grosenick for his help in the early SHG experiments on noble metals. K. Böhmer permitted us to present SHG data from his thesis. The Max-Born-Institut at Berlin–Adlershof provided beam time at a CPM-Laser for pump–probe SHG measurements. The project was supported by the Deutsche Forschungsgemeinschaft, Sfb 290.

References

1. J. Reif, P. Tepper, E. Matthias, E. Westin, and A. Rosèn, *Appl. Phys.* B **46**, 131 (1988).
2. H. W. K. Tom, G. D. Aumiller, and C. H. Brito-Cruz, *Phys. Rev. Lett.* **60**, 1438 (1988).
3. H. W. K. Tom, T. F. Heinz, and Y. R. Shen, *Phys. Rev. Lett.* **51**, 1983 (1983).
4. H. W. K. Tom and G. D. Aumiller, *Phys. Rev.* B **33**, 8818 (1986).
5. R. Stolle, G. Marowsky, M. Pinnow, and O. Befort, *Appl. Phys.* B **58**, 317 (1994).
6. T. F. Heinz, C. K. Chen, D. Ricard, and Y. R. Shen, *Phys. Rev. Lett.* **48**, 478 (1982).

REFERENCES

7. R. P. Pan, H. D. Wei, and Y. R. Shen, *Phys. Rev.* B **39**, 1229 (1989).
8. W. Hübner and K. H. Bennemann, *Phys. Rev.* B **40**, 5973 (1989).
9. J. Reif, J. C. Zink, C. M. Schneider, and J. Kirschner, *Phys. Rev. Lett.* **67**, 2878 (1991).
10. M. Fiebig, D. Fröhlich, B. B. Krichevtsov, and R. V. Pisarev, *Phys. Rev. Lett.* **73**, 2127 (1994).
11. V. V. Pavlov, R. V. Pisarev, A. Kirilyuk, and T. Rasing, *Phys. Rev. Lett.* **78**, 2004 (1997).
12. H. M. van Driel, *Appl. Phys.* A **59**, 545 (1994).
13. L. E. Urbach, K. L. Percival, J. M. Hicks, E. W. Plummer, and H. L. Dai, *Phys. Rev.* B **45**, 3769 (1992).
14. K. J. Song, D. Heskett, H. L. Dai, A. Liebsch, and E. W. Plummer, *Phys. Rev. Lett.* **61**, 1380 (1988).
15. J. M. Hicks, L. E. Urbach, E. W. Plummer, and H.-L. Dai, *Phys. Rev. Lett.* **61**, 2588 (1988).
16. T. Götz, M. Buck, C. Dressler, F. Eisert, and F. Träger, *Appl. Phys.* A **60**, 607 (1995).
17. B. Lamprecht, A. Leitner, and F. R. Aussenegg, *Appl. Phys.* B **64**, 269 (1997).
18. J. R. Power, J. D. O'Mahony, S. Chandola, and J. F. McGilp, *Phys. Rev. Lett.* **75**, 1138 (1995).
19. S. Janz, D. J. Bottomley, H. M. van Driel, and R. S. Timsit, *Phys. Rev. Lett.* **66**, 1201 (1991).
20. C. K. Chen, A. R. B. de Castro, and Y. R. Shen, *Phys. Rev. Lett.* **46**, 145 (1981).
21. X. D. Zhu, Th. Rasing, and Y. R. Shen, *Phys. Rev. Lett.* **61**, 2883 (1988).
22. J. Woll, G. Meister, U. Barjenbruch, and A. Goldmann, *Appl. Phys.* A **60**, 173 (1995).
23. D. Heskett, L. E. Urbach, K. J. Song, E. W. Plummer, and H. L. Dai, *Surf. Sci.* **197**, 225 (1988).
24. B. Koopmans, A. M. Janner, H. A. Wierenga, T. Rasing, G. A. Sawatzky, and F. van der Woude, *Appl. Phys.* A **60**, 103 (1994).
25. A. Kirilyuk, T. Rasing, R. Mégy, and P. Beauvillain, *Phys. Rev. Lett.* **77**, 4608 (1996).
26. C. V. Shank, R. Yen, and C. Hirlimann, *Phys. Rev. Lett.* **51**, 900 (1983).
27. P. Saeta, J.-K. Wang, Y. Siegal, N. Bloembergen, and E. Mazur, *Phys. Rev. Lett.* **67**, 1023 (1991).
28. K. Sokolowski-Tinten, J. Bialkowski, and D. von der Linde, *Phys. Rev.* B **51**, 14186 (1995).
29. Y. M. Chang, L. Xu, and H. W. K. Tom, *Phys. Rev. Lett.* **78**, 4649 (1997).
30. J. E. Sipe, D. J. Moss, and H. M. van Driel, *Phys. Rev.* B **35**, 1129 (1987).
31. W. S. Fann, R. Storz, H. W. K. Tom, and J. Bokor, *Phys. Rev.* B **46**, 13592 (1992).
32. C.-K. Sun, F. Vallee, L. H. Acioli, E. P. Ippen, and J. G. Fujimoto, *Phys. Rev.* B **50**, 15337 (1994).
33. S. I. Anisimov, B. L. Kapeliovich, and T. L. Perel'man, *Sov. Phys. JETP* **39**, 375 (1975).
34. N. W. Ashcroft and N. D. Mermin, *Solid State Physics* (Saunders College Publishing, Philadelphia, 1976).
35. D. Pines and P. Nozières, *The Theory of Quantum Liquids* (W. A. Benjamin, New York, 1966).
36. C. Suarez, W. E. Bron, and T. Juhasz, *Phys. Rev. Lett.* **75**, 4536 (1995).
37. T. Juhasz, H. E. Elsayed-Ali, X. H. Hu, and W. E. Bron, *Phys. Rev.* B **45**, 13819 (1992).
38. R. Rosei and D. W. Lynch, *Phys. Rev.* B **5**, 3883 (1972).

39. S. S. Jah, and C. S. Warke, *Phys. Rev.* **153**, 751 (1967).
40. K. C. Rustagi, *Il Nuovo Cimento* **LIII**, 346 (1968).
41. W. Greiner, L. Neise, and H. Stöcker, *Thermodynamik und Statistische Mechanik* (Verlag Harri Deutsch, 1987).
42. P. B. Johnson and R. W. Christy, *Phys. Rev.* B **6**, 4370 (1972).
43. Ch. Kittel, *Einführung in die Festkörperphysik* (R. Oldenburg Verlag, München, Wien, 8. Auflage, 1989).
44. E. D. Palik (ed.), *Handbook of Optical Constants of Solids*, Part I (Academic Press, Orlando, FL, 1985).
45. H. Ehrenreich and H. R. Philipp, *Phys. Rev.* **128**, 1622 (1962).
46. R. W. Schoenlein, W. Z. Lin, J. G. Fujimoto, and G. L. Eesley, *Phys. Rev. Lett.* **58**, 1680 (1987).
47. Y. Siegal, E. N. Glezer, and E. Mazur, *Phys. Rev.* B **49**, 16403 (1994).
48. S. D. Brorson, A. Kazeroonian, J. S. Moodera, D. W. Face, T. C. Cheng, E. P. Ippen, M. S. Dresselhaus, and G. Dresselhaus, *Phys. Rev. Lett.* **64**, 2172 (1990).
49. H. E. Elsayed-Ali, T. B. Norris, M. A. Pessot, and G. A. Mourou, *Phys. Rev. Lett.* **58**, 1212 (1987).
50. O. B. Wright, *J. Physique* IV, **C7**, 701 (1994).
51. O. B. Wright, *Phys. Rev.* B **49**, 9985 (1994).
52. R. H. M. Groeneveld, R. Sprik, and Ad. Lagendijk, *Phys. Rev. Lett.* **64**, 784 (1990).
53. R. H. M. Groeneveld, R. Sprik, and Ad. Lagendijk, *Phys. Rev.* B **45**, 5079 (1992).
54. H. E. Elsayed-Ali, T. Juhasz, G. O. Smith, and W. E. Bron, *Phys. Rev.* B **43**, 4488 (1991).
55. T. Juhasz, H. E. Elsayed-Ali, G. O. Smith, C. Suarez, and W. E. Bron, *Phys. Rev.* B **48**, 15488 (1993).
56. C.-K. Sun, F. Valleé, L. Acioli, E. P. Ippen, and J. G. Fujimoto, *Phys. Rev.* B **48**, 12365 (1993).
57. A. A. Maznev, J. Hohlfeld, and J. Güdde, *J. Appl. Phys.* **82**, 5082 (1997).
58. S. D. Brorson, J. G. Fujimoto, and E. P. Ippen, *Phys. Rev. Lett.* **59**, 1962 (1987).
59. P. B. Corkum, F. Brunel, N. K. Sherman, and T. Srinivasan-Rao, *Phys. Rev. Lett.* **61**, 2886 (1988).
60. M. Born and E. Wolf, *Principles of Optics* (Pergamon Press, Oxford, 1959).
61. J. Hohlfeld, D. Grosenick, U. Conrad, and E. Matthias, *Appl. Phys.* A **60**, 137 (1995).
62. J. Hohlfeld, U. Conrad, and E. Matthias, *Appl. Phys.* B **63**, 541 (1996).
63. M. Gitterman and V. Halpern, *Qualitative Analysis of Physical Problems* (Academic Press, 1981).
64. O. Keller, *Phys. Rev.* B **33**, 990 (1986).
65. T. Luce, W. Hübner, and K.-H. Bennemann, *Z. Phys.* B **102**, 223 (1997).
66. K. Sokolowski-Tinten, J. Bialkowski, and D. von der Linde, *Phys. Rev.* B **51**, 14186 (1995).
67. G. Petrocelli, S. Martellucci, and R. Francini, *Appl. Phys.* A **56**, 263 (1993).
68. C. M. Li, L. E. Urbach, and H. L. Dai, *Phys. Rev.* B **49**, 2104 (1994).
69. N. E. Christensen and B. O. Seraphin, *Phys. Rev.* B **4**, 3321 (1971).
70. G. L. Eesley, *Phys. Rev. Lett.* **51**, 2140 (1983).
71. K. Pedersen and O. Keller, *J. Opt. Soc. Am.* B **6**, 2412 (1989).
72. W. Hübner, K. H. Bennemann, and K. Böhmer, *Phys. Rev.* B **50**, 17597 (1994).
73. The data were taken from the dissertation of K. Böhmer, FU Berlin, 1994.
74. D. A. Papaconstantopoulos, *Handook of the Band Structure of Elemental Solids* (Plenum Press, New York, 1986).

REFERENCES

75. U. Postogowa, W. Hübner, and K. H. Bennemann, *Surf. Sci.* **307–309,** 1129 (1994).
76. R. Knorren, private communication.
77. J. Callaway and C. S. Wang, *Phys. Rev.* B **7,** 1096 (1973).
78. J. Hohlfeld, E. Matthias, R. Knorren, and K. H. Bennemann, *Phys. Rev. Lett.* **78,** 4861 (1997); *Phys. Rev. Lett.* **79,** 960 (1997).
79. P. Weiss and R. Forrer, *Ann. Phys.* **5,** 153 (1926).
80. E. Beaurepaire, J. C. Merle, A. Daunois, and J.-Y. Bigot, *Phys. Rev. Lett.* **76,** 4250 (1996).
81. D. R. Penn, S. P. Apell, and S. M. Girvin, *Phys. Rev.* B **32,** 7753 (1985).
82. M. Aeschlimann, M. Bauer, S. Pawlik, W. Weber, R. Burgermeister, D. Oberli, and H. C. Siegmann, *Phys. Rev. Lett.* **79,** 5158 (1997).
83. A. Vaterlaus, T. Beutler, and F. Meier, *Phys. Rev. Lett.* **67,** 3314 (1991).
84. A. Vaterlaus, T. Beutler, D. Guarisco, M. Lutz, and F. Meier, *Phys. Rev.* B **46,** 5280 (1992).
85. A. Vaterlaus, D. Guarisco, M. Lutz, M. Aeschlimann, M. Stampanoni, and F. Meier, *J. Appl. Phys.* **67,** 5661 (1990).
86. M. B. Agranat, S. I. Ashitkov, A. B. Granovskii, and G. I. Rukman, *Zh. Eksp. Teor. Fiz.* **86,** 1376 (1984) [*Sov. Phys. JETP* **59,** 804 (1984)].
87. K.-H. Bennemann, private communication.

5
ELECTRONIC THEORY FOR NONLINEAR MAGNETO-OPTICS

W. Hübner

Max-Planck-Institut für Mikrostrukturphysik, Halle, Germany

5.1 Development of nonlinear magneto-optics

The quantum theory of nonlinear optics was first considered by Maria Göppert-Mayer[1] in her Ph.D. thesis entitled 'Elementarakte mit zwei Quantensprüngen' (Elementary acts with two quantum jumps) at Leipzig University. In the 1960s Bloembergen, Pershan[2] and Kelley[3] used a general form of nonlinear response theory in order to describe nonlinear optical processes, while Jha[4] started from Boltzmann's equation and Pershan[5] solved nonlinear wave equations for anharmonic oscillator models within classical electrodynamics. In all cases theoretical research focused on the prediction of new higher-order nonlinear effects. For this purpose a large number of nonlinear tensors has been discussed and group-theoretically classified.[6,7] Based on these theories the angular dependence of the nonlinear light intensity has been calculated for certain experimental situations, in particular for dielectric media. Only since 1986 has *optical second harmonic generation on metal surfaces* become a field of interest and increasing research. Nonlinear intensities have been calculated for *jellium* within the hydrodynamic model[8] and in the framework of density-functional theory.[9,10] Nonlinearities have also been incorporated in quasiparticle theories but only for the bulk of ferromagnetic insulators.[11]

The nonlinear magneto-optical Kerr effect (NOLIMOKE) describes the rotation of the polarization plane for second harmonic generation (SHG) in reflection from a ferromagnetic sample surface. This effect reflects the symmetry of the surface structure and surface magnetism and has been proposed as an ultrafast spectroscopic probe for the investigation of two-dimensional magnetism.[12-14] NOLIMOKE is very well suited for a direct investigation of surface structures and electronic and magnetic properties. Thus, two-dimensional ferromagnetic structures at surfaces, at interfaces, in thin magnetic films and in multilayers can be studied. In combination with optical 'pump and probe' techniques the nonlinear magneto-optical Kerr effect allows a time-dependent study of 2D magnetism on the pico- through femtosecond timescale. Since it seems conceivable to apply NOLIMOKE for layer-by-layer magneto-optical recording, this effect is interesting for the development of magneto-optical storage media. Note that in contrast to non-optical studies the nonlinear Kerr effect is nondestructive and not restricted to the study of remanent magnetization. As NOLIMOKE is caused by the interplay of spin–orbit and exchange interaction it becomes important for the determination of the spin–orbit interaction-induced magnetocrystalline anisotropy. Furthermore, the nonlinear magneto-optical Kerr effect can be used for measuring the spin–lattice relaxation time.[15] Note that the spin–lattice relaxation time is the ultimate time and speed limit for magneto-optical recording technologies. This relaxation time contains spin–orbit coupling via phonon–magnon interaction (in transition metals).

Motivated by these theoretical studies first experiments on Fe were performed in 1991.[16] The (110) surface of single-crystal Fe in single-domain state was studied under ultrahigh-vacuum conditions. Upon reversing the direction of the in-plane magnetization with the help of a weak external magnetic field defining the quantization axis, a drastic change of the SHG yield was observed thus confirming the theoretically predicted existence and order of magnitude of the nonlinear Kerr effect as well as its sensitivity to two-dimensional magnetism.

Subsequently the theoretical interest focused on the electronic theory for the Kerr rotation angle in second harmonic generation, where theory predicts a drastic enhancement by several orders of magnitude compared to the angle in linear magneto-optics.[17] Experiment confirmed this prediction and even found arbitrarily large rotations in a multilayer thin-film geometry of Fe. In combination with a recently performed symmetry analysis the enhanced nonlinear Kerr rotation is an ideal sensor of the magnetic easy axis at interfaces (also hidden ones).

Owing to the additional degree of freedom, nonlinear optics senses directly the electronic structure of interfaces, their thickness dependence and the substrate influence at heterogeneous interfaces, and thus probes their magnetism directly, in contrast to linear optics, which only measures film-averaged quantities. A striking example for this drastically enhanced sensitivity is the observation of spin-polarized quantum-well state oscillations in magnetic/nonmagnetic sandwich structures, whose amplitude is of the order of one rather than in the sub-percent range as for linear magneto-optics. The theoretical explanation of this phenomenon relies on nonlinear interface excitations that are absent in linear optics. Thus nonlinear magneto-optics may even detect the giant-magnetoresistance oscillations more directly than electrical measurements.

Of particular importance for future theoretical and experimental research will be the nonlinear femtosecond dynamics of hot electrons, which may cause new ultrafast magneto-optical switching mechanisms of technological relevance.

This chapter is organized as follows. In sec. 5.2 we give the classical theory of nonlinear optics and formulate the theory of second and third harmonic Mie scattering from microspheres, which is a purely classical phenomenon and may help to analyse small particles with respect to their size, shape and magnetism. Applications may involve medicine and environmental pollution.

In sec. 5.3 we review the symmetry properties of nonlinear optics and magneto-optics, in particular of the enhanced Kerr rotation, on metallic nonmagnetic and ferromagnetic surfaces, in unconventional superconductors and in antiferromagnets.

Section 5.4 is devoted to the electronic theory of nonlinear magneto-optics. There we derive the microscopic expressions for the nonlinear magneto-optical tensors and discuss results for the nonlinear Kerr spectra calculated from semiempirical as well as first-principles electronic band structures. Special emphasis is given to information on the geometric, electronic and magnetic surface and interface properties of ferromagnetic metals revealed by these spectra and not otherwise available in this pronounced way such as quantum-well oscillations.

Finally, in sec. 5.5 we briefly discuss time-resolved nonlinear optics as a rapidly growing field of future research, where dramatic new contributions of technological relevance (switching and coherent control) are to be expected from nonlinear optical and magneto-optical spectroscopies on the picosecond and femtosecond timescales.

5.2 Classical theory for nonlinear optics

The theory of *linear* response has become a very successful theoretical tool for the description of a large variety of experiments in many different areas of physics. If external 'fields' X act upon a system originally in thermodynamic equilibrium, they trigger a 'response' Y in one or more of the characteristic variables of the system, which for 'small' external perturbations is *directly proportional* to the applied field:

$$Y = L \cdot W. \tag{5.2.1}$$

Here the vector Y is the response of the system to the external field X, and L denotes the susceptibility tensor, which describes the response in terms of the microscopic material properties of the system and can be calculated in many cases from first principles. It is frequently the case that an external field triggers, besides the main response, further responses of the system, which are given by the off-diagonal elements of the tensor L. Then for all elements i, j of the susceptibility tensor L the following symmetry relation is valid (the sign being plus or minus according to the symmetry of the external field under time reversal):

$$L_{ij} = \pm L_{ji}. \tag{5.2.2}$$

These are the famous Onsager relations.

Classical electrodynamics was originally formulated as a *linear* theory.

Maxwell's equations are linear in the fields and densities of the free charges and currents

$$\nabla \cdot \mathbf{D} = \rho_{\mathrm{f}},$$
$$\nabla \cdot \mathbf{B} = 0,$$
$$\nabla \times \mathbf{E} + \frac{\partial \mathbf{B}}{\partial t} = 0, \qquad (5.2.3)$$
$$\nabla \times \mathbf{H} - \frac{\partial \mathbf{D}}{\partial t} = \mathbf{j}_{\mathrm{f}}.$$

The corresponding linear response relations for the dielectric polarization $\mathbf{P}$ and the magnetization $\mathbf{M}$ are

$$\frac{\mathbf{P}}{\varepsilon_0} = \chi^{(1)} \mathbf{E}, \qquad (5.2.4)$$
$$\mathbf{M} = \chi_M^{(1)} \mathbf{H},$$

where ε_0 is the vacuum permittivity. The diagonal elements of the tensor $\chi^{(1)}$ describe the response of the system to an external excitation by light of frequency ω while its off-diagonal elements form the basis of linear magneto-optics. In general they are smaller than the diagonal elements by a factor of $\lambda_{\mathrm{so}}/\hbar\omega$, where λ_{so} is some spin–orbit interaction parameter, and satisfy Onsager's relations.

The combination of Maxwell's third and fourth equations yields (in the absence of free currents) the optical wave equation for time-dependent fields, which is a linear and homogeneous partial differential equation of second order ($\varepsilon^{(1)} = 1 + \chi^{(1)}$):

$$\nabla \times \nabla \times \mathbf{E}^{(j)}(j\omega) + \frac{\varepsilon^{(1)}}{c^2} \frac{\partial^2}{\partial t^2} \mathbf{E} = 0. \qquad (5.2.5)$$

Note that the magnetic permeability $\mu(\omega)$ of the medium has been set equal to unity since optical frequencies are well above any magnetic resonance frequencies, which are typically in the GHz range. This approximation is usually an excellent one even for ferromagnetic systems and might only fail in the case of the so-called paramagnetic Kerr effect (the magneto-optical response in the presence of a slowly varying external magnetic field). The susceptibility tensors L (and in particular also $\chi^{(1)}$ and $\chi_M^{(1)}$) depend on the frequency of the external perturbation and on the stationary values of possible additional external fields acting on the system. It turns out that the response of a physical system to a small external perturbation is given by its microscopic average statistical equilibrium fluctuations. Therefore it is sufficient (according to the fluctuation–dissipation theorem) to know the spectrum of the equilibrium fluctuations in order to understand the microscopic and thus also the macroscopic response of the system to external excitations. So it is possible to express the susceptibilities by statistical correlation functions of the response quantities and to calculate them. In the case of nonlinear optics and magnetism these are the density–density (see first part of sec. 5.2.2), current–current, or spin–spin correlation functions. The corresponding quantum-mechanical formalism ('Kubo formalism'[18]) for the microscopic calculation of these *linear* response functions has been known for a long time and has been applied successfully to countless examples.

If the external perturbation becomes stronger the response of the system is no longer linear in the external excitation but rather contains *nonlinearities* and is given by the series

$$Y = L_1 \cdot X + L_2 : X \cdot X + L_3 : X \cdot X \cdot X + \cdots, \qquad (5.2.6)$$

where the susceptibilities L_i ($i = 1, 2, 3, \ldots$) are tensors or rank $(i + 1)$. In principle, for strong perturbations, the nonlinear terms may become comparable to the linear response. In the extreme nonlinear case any small change of the strong external fields may lead to arbitrarily large changes of the system response and thus may cause chaotic behaviour. In most cases, however, the effect of the higher nonlinearities L_i will quickly decrease with i to yield a considerable contribution to the system response only from the first L_i of the series.

Frequently, the nonlinear susceptibilities are experimentally not accessible. But for microwaves and optical excitations one is fortunately able to study besides L_1 also the influence of L_2 and L_2 (in some exceptional cases up to about L_{100}) since the development of maser and laser has allowed the experimental realization of strong electromagnetic excitation fields.

From the point of view of theoretical physics it is attractive to develop a microscopic theory of nonlinear optics and nonlinear magneto-optics in analogy to traditional linear optics and magneto-optics. This requires two steps:

1. The classical electrodynamics of the macroscopic fields has to be extended allowing for nonlinearity.
2. It is necessary to develop a nonlinear response theory from which the nonlinear susceptibilities can be calculated microscopically (quantum mechanically).

In this chapter we mostly consider the lowest-order nonlinearity L_2. To introduce this nonlinearity into classical electrodynamic theory one has to perform two extensions:

1. The nonlinear response is introduced in the material equation for the polarization (by virtue of the nonlinear susceptibility $\chi^{(2)}$):

$$\frac{P}{\varepsilon_0} = \chi^{(1)} E + \chi^{(2)} : E \cdot E. \qquad (5.2.7)$$

2. The wave equation for time-dependent fields is extended by a nonlinear source term acting as an inhomogeneity:

$$\nabla \times \nabla \times E^{(2)}(2\omega) + \frac{\varepsilon(\omega)}{c^2} \frac{\partial^2}{\partial t^2} E^{(2)}(2\omega) = -\frac{1}{c^2} \frac{\partial^2}{\partial t^2} P^{(2)}. \qquad (5.2.8)$$

In analogy to the treatment of linear response, the calculation of the nonlinear susceptibility requires the computation of a higher-order correlation function, viz. the density–density–density or current–current–current correlation function, which will later be discussed in some detail.

The nonlinear intensity I_ω is given by

$$I_{2\omega} \sim [E(\omega)]^2 \sim I_\omega^2. \qquad (5.2.9)$$

Therefore the nonlinear intensity may become as large as the linear one if the fields are large enough. This condition, however, would require huge field amplitudes ($\approx V/\text{Å}$) which usually cannot be reached in experiments or are only available for very short times and would destroy the material. Thus, in usual experiments one has

$$I_{2\omega} \ll I_\omega. \qquad (5.2.10)$$

Consequently it is often difficult to measure the nonlinear response of the system. Fortunately, however, one benefits in many cases from the fact that the linear and nonlinear susceptibilities are tensors of different rank (second or third) which possess different symmetry properties. So it becomes possible to measure the nonlinear susceptibilities even if they are absolutely small, since their signal-to-noise ratio is large due to the absence of any appreciable background. On the other hand, these different properties of the linear and nonlinear susceptibilities can be utilized to determine the symmetry of the system under consideration.

As will be shown, in nonlinear optics the second-order susceptibility vanishes in inversion-symmetric media within the electric dipole approximation. Therefore all contributions to $\chi^{(2)}$ result from the surface of the solids or the interfaces of thin films where inversion symmetry is inherently broken due to the incompleteness of the unit cell. Moreover, in the case of a rotation axis, the N-th harmonic in the M-th multipole order is capable of detecting an $(N + M)$-fold rotational symmetry. Thus, within the electric dipole approximation, second harmonic generation can resolve the rotational symmetry of (111) surfaces while linear optics cannot. Therefore the nonlinear optical susceptibility serves as an excellent probe for the investigation of the geometrical, electronic and magnetic properties of surfaces and interfaces.

Similar arguments apply to the frequency dependence of the susceptibilities. The nonlinear susceptibility describes three-photon processes and thus is clearly much more sensitive to details of the electronic material properties than the linear response function, which describes only two-photon processes. A particularly striking example of that behaviour is the nonlinear response of a heterogeneous transition metal/noble metal interface. There the non-spin-polarized d electrons of the noble metal substrate become visible in the nonlinear *magneto-*optical response via the spin polarization of the transition-metal d states acting as intermediate states at the Fermi level. The nonlinear magneto-optical response is drastically enhanced by the large density of 'nonmagnetic' noble-metal d electrons well below the Fermi energy. This effect has no analogy in linear optics.

In general one may make the following remarks concerning nonlinear susceptibilities:

1. Nonlinear susceptibilities give information about more details of a physical system (e.g. frequency dependence of NOLIMOKE spectrum, separation of film and substrate electronic influence).
2. They allow for a finer tuning of these details to the particular symmetry than do linear theories (e.g. 3-fold axes on (111) interfaces, lateral nanometre resolution for optical wavelengths due to local field enhancement).
3. They yield fundamentally new effects and insights (e.g. 90° Kerr rotation, nonmonotonic electron temperature dependence of the signal, quantum-well oscillations of amplitude unity and with two periods, magneto-optical access to buried interfaces and sandwich layers, detection of antiferromagnetic domain structures).

Of course, the previous statements not only hold for nonlinearity in optics or magneto-optics, but are valid for many other situations exhibiting nonlinear response such as elasticity, electrical resistivity, specific heat, spin susceptibility and a variety of other cases.

5.2.1 Oscillator model for harmonic generation

Classical electrodynamics (Maxwell theory) describes a spatial average of the electromagnetic fields over a range Δx which, on the one hand, is small compared to the optical wavelength but, on the other hand, is large compared to atomic structures, i.e.

$$1\,\text{nm} \leq \Delta x \leq 100\,\text{nm}. \quad (5.2.11)$$

In this subsection we derive optical second and third harmonic generation from Maxwell's equations (see also refs 5, 19). For the usual electromagnetic fields E, B, D and H and the free currents j_f and charges ρ_f (in SI units) these equations are given by eqns (5.2.3) with the following relations between the fields in matter to those in vacuum:

$$D_\alpha = \varepsilon_0 E_\alpha + \left(P_\alpha - \sum_\beta \frac{\partial Q_{\alpha\beta}}{\partial x_\beta} + \cdots \right),$$

$$H_\alpha = \frac{1}{\mu_0} B_\alpha - (M_\alpha + \cdots). \quad (5.2.12)$$

Here, P, Q and M are the macroscopic electric dipole moment, electric quadrupole moment and magnetic dipole moment per unit volume, ε_0 and μ_0 are the vacuum permittivity and permeability ($\varepsilon_0 = 8.8544 \times 10^{-12}\,\text{C}^2\,\text{m}^{-2}\,\text{N}^{-2}$ and $\mu_0 = 1.2566 \times 10^{-6}\,\text{N}\,\text{A}^{-2}$).

In *linear* optics one has the relations

$$D_i = \sum_k \varepsilon_{ik} E_k \quad \text{and} \quad B_i = \sum_k \mu_{ik} H_k, \tag{5.2.13}$$

where $\boldsymbol{\varepsilon}$ and $\boldsymbol{\mu}$ are the tensors of the dielectric function and magnetic induction. In *nonlinear* optics the displacement field $\boldsymbol{D}$ is expanded in powers of the electric field $\boldsymbol{E}$

$$D_\alpha = \sum_\beta \varepsilon^{(1)}_{\alpha\beta} E_\beta + \sum_{\beta\gamma} \varepsilon^{(2)}_{\alpha\beta\gamma} E_\beta E_\gamma + \cdots \tag{5.2.14}$$

with the linear and nonlinear (here written up to second order) tensors of rank two and three $\boldsymbol{\varepsilon}^{(1)}$ and $\boldsymbol{\varepsilon}^{(2)}$. Within the *electric dipole approximation* the polarization $\boldsymbol{P}$ and the susceptibility $\boldsymbol{\chi}$ are related by

$$D_\alpha = \varepsilon_0 E_\alpha + P_\alpha = (1 + \chi)\varepsilon_0 E_\alpha. \tag{5.2.15}$$

Neglecting field gradient terms this equation yields up to second order

$$P_\alpha = \sum_\beta \chi^{(1)}_{\alpha\beta} E_\beta + \sum_{\beta\gamma} \chi^{(2)}_{\alpha\beta\gamma} E_\beta E_\gamma + \cdots . \tag{5.2.16}$$

Under the assumption of an electric field oscillating in time and frequency ω

$$E_\alpha(t) = E_{0\alpha} + E_{1\alpha} \sin \omega t, \tag{5.2.17}$$

one obtains for the electric polarization up to second order in the electric field

$$P_\alpha = \sum_\beta \chi^{(1)}_{\alpha\beta} E_{0\beta} + \sum_\beta \chi^{(1)}_{\alpha\beta} E_{1\beta} \sin \omega t + \sum_{\beta\gamma} \chi^{(2)}_{\alpha\beta\gamma} E_{0\beta} E_{0\gamma}$$

$$+ \sum_{\beta\gamma} \chi^{(2)}_{\alpha\beta\gamma} (E_{0\beta} E_{1\gamma} + E_{1\beta} E_{0\gamma}) \sin \omega t$$

$$+ \sum_{\beta\gamma} \chi^{(2)}_{\alpha\beta\gamma} E_{1\beta} E_{1\gamma} \sin^2 \omega t. \tag{5.2.18}$$

The fourth term on the right-hand side of this equation describes the 'linear' electro-optical effect (Pockels effect) while the fifth term is responsible for optical second harmonic generation (SHG)

$$\sin^2 \omega t = \tfrac{1}{2}(1 - \cos 2\omega t). \tag{5.2.19}$$

Thus one gets the oscillation with frequency 2ω.

Assuming a nonlinear source term

$$P^{\mathrm{NLS}}(\omega) = \chi^{\mathrm{NL}} \left(E\left(\frac{\omega}{2}\right) \right)^2 = \chi^{\mathrm{NL}} \left(A\left(\frac{\omega}{2}\right) \right)^2 e^{-2ik(\frac{\omega}{2})z}, \tag{5.2.20}$$

with

$$k\left(\frac{\omega}{2}\right) = \frac{\omega}{2c} \sqrt{\varepsilon\left(\frac{\omega}{2}\right)} \tag{5.2.21}$$

(where ε is the linear dielectric function), the third and fourth of Maxwell's equations become within the electric dipole approximation

$$\nabla \times \boldsymbol{H}(\omega) = +\frac{i\omega}{c}[\varepsilon(\omega)\boldsymbol{E}(\omega) + 4\pi\boldsymbol{P}^{\text{NLS}}(\omega)],$$

$$\nabla \times \boldsymbol{E}(\omega) = -\frac{i\omega}{c}\boldsymbol{H}(\omega). \quad (5.2.22)$$

They may now be combined to give

$$\frac{\partial^2}{\partial z^2}\boldsymbol{E}(\omega) = -\left(\frac{i\omega}{c}\right)^2 \varepsilon(\omega)\boldsymbol{E}(\omega) - 4\pi\left(\frac{\omega}{c}\right)^2 \chi^{\text{NL}}\left(\boldsymbol{A}\left(\frac{\omega}{2}\right)\right)^2 e^{-2ik(\frac{\omega}{2})z}. \quad (5.2.23)$$

Employing the boundary condition $\boldsymbol{E}(\omega) = 0$ at $z = 0$ one finally obtains the following result for $\boldsymbol{E}(\omega)$:

$$\boldsymbol{E}(\omega) = -\frac{4\pi\chi^{\text{NL}}\boldsymbol{A}\left(\frac{\omega}{2}\right)}{\varepsilon(\omega) - \varepsilon\left(\frac{\omega}{2}\right)}\left(e^{-2ik(\frac{\omega}{2})z} - e^{-ik(\omega)z}\right). \quad (5.2.24)$$

From this simple formula two important conclusions can be drawn:

1. The maximum SHG yield is obtained for $\varepsilon(\omega) = \varepsilon(\omega/2)$. This is the famous 'phase matching' condition, which corresponds to the condition $k(\omega) = k(\omega/2)$ for the directions of the beams.
2. The SHG intensity is proportional to $|\boldsymbol{E}(\omega)|^2$ and thus proportional to z^2.

Classical oscillator model

The classical oscillator is used in many cases in physics as a simple model system that can be handled analytically and yet allows for important conclusions anticipating results of an improved, quantum-mechanical treatment. In electrodynamics the oscillator model assumes that the valence electrons of an atom or solid are bound to the nucleus and the core electrons by an oscillator potential. They can be displaced from their equilibrium position by an external electric field. This displacement of their position corresponds to a polarization of the atom or crystal in the electric field.

As the simplest description of *non*-linear media we consider the Hamiltonian function of a classical *an*harmonic oscillator in one dimension in the electric field E of the light:

$$\ddot{x} + \Gamma\dot{x} + \omega_0^2 x + ax^2 + bx^3 = \frac{e}{m}E(t). \quad (5.2.25)$$

Note that x is the radial displacement of the charges at a given point of a sphere, and the factors Γ, ω_0^2, a and b reflect only properties of the medium and do not depend on the angles θ and ϕ. By using the ansatz

$$x(t) = x^{(1)}(t) + x^{(2)}(t) + x^{(3)}(t), \quad (5.2.26)$$

eqn (5.2.25) is separated into

$$\ddot{x}^{(1)} + \Gamma\dot{x}^{(1)} + \omega_0^2 x^{(1)} = \frac{e}{m} E(t), \tag{5.2.27}$$

$$\ddot{x}^{(2)} + \Gamma\dot{x}^{(2)} + \omega_0^2 x^{(2)} + a(x^{(1)})^2 = 0, \tag{5.2.28}$$

$$\ddot{x}^{(3)} + \Gamma\dot{x}^{(3)} + \omega_0^2 x^{(3)} + 2ax^{(1)}x^{(2)} + b(x^{(1)})^3 = 0. \tag{5.2.29}$$

The first equation represents the well known harmonic oscillator, yielding

$$x^{(1)}(\omega) = \frac{\frac{e}{m} E}{(\omega_0^2 - \omega^2 - i\omega\Gamma)} e^{-i\omega t} \equiv \tilde{x}^{(1)} \exp^{-i\omega t}. \tag{5.2.30}$$

The second equation represents the harmonic oscillator with the driving force $-a(x^{(1)})^2$. Thus, we get

$$x^{(2)} = \frac{-a(\tilde{x}^{(1)})^2}{\left(\omega_0^2 - (2\omega)^2 - i2\omega\Gamma\right)} e^{-i2\omega t}$$

$$= \frac{-a \frac{e^2}{m^2} E^2}{\left(\omega_0^2 - (2\omega)^2 - i2\omega\Gamma\right)(\omega_0^2 - \omega^2 - i\omega\Gamma)^2} e^{-i2\omega t}. \tag{5.2.31}$$

With the approximation

$$\frac{-a}{\left(\omega_0^2 - (2\omega)^2 - i2\omega\Gamma\right)} \equiv 1 \tag{5.2.32}$$

we neglect the constant factor a and the resonance at 2ω. As a consequence, the second harmonic displacement yields $x^{(2)} = (x^{(1)})^2$. Analogously $x^{(3)}$ can be obtained from eqn (5.2.29) by using

$$2ax^{(1)}x^{(2)} + b(x^{(1)})^3 = \frac{(b - 2a^2)(\tilde{x}^{(1)})^3}{\left(\omega_0^2 - (2\omega)^2 - i2\omega\Gamma\right)} e^{-i3\omega t}.$$

This results in

$$x^{(3)} = \frac{a^2 - b}{\left(\omega_0^2 - (2\omega)^2 - i2\omega\Gamma\right)\left(\omega_0^2 - (3\omega)^2 - i3\omega\Gamma\right)} (x^{(1)})^3$$

$$= \frac{(a^2 - b)\frac{e^3}{m^3} E^3}{\left(\omega_0^2 - (2\omega)^2 - i2\omega\Gamma\right)\left(\omega_0^2 - (3\omega)^2 - i3\omega\Gamma\right)} \frac{e^{-i3\omega t}}{(\omega_0^2 - \omega^2 - i\omega\Gamma)^3}.$$

$$\tag{5.2.33}$$

Again, we get $x^{(3)} = (x^{(1)})^3$ by neglecting the frequency resonances for 2ω and 3ω, i.e.

$$\frac{a^2 - b}{\left(\omega_0^2 - (2\omega)^2 - i2\omega\Gamma\right)\left(\omega_0^2 - (3\omega)^2 - i3\omega\Gamma\right)} \equiv 1. \quad (5.2.34)$$

In conclusion, the identities in eqns (5.2.32) and (5.2.34) motivate the relation $\sigma^{(n)} = (\sigma^{(1)})^n$ for the charges as sources in the Mie theory of second harmonic generation (SHG) and third harmonic generation (THG). This relation will be used in the next subsection where we only deal with the size dependence of the radiated intensities from a spherical (metal) particle at a fixed frequency and do not determine their absolute value. Thus, neglecting the resonance terms in 2ω and 3ω should not have important consequences for the angular distribution of SHG and THG scattered from spheres in the Mie range. Note that the neglect of the factors a and b, which do not depend on the position at the boundary, changes only the absolute value of the intensities, but does not affect their angular dependence.

5.2.2 Nonlinear Mie theory for spherical particles

Motivation

The optical analysis of the size and shape of small (metal) particles is one of the classical problems. Here, we investigate second and third harmonic light generation by small particles, since one expects a more sensitive dependence of the integrated intensities and the resonances in the scattering profile on particle size and shape for both small spheres (in the nanometre range) and spheres with radius similar to or larger than the wavelength of the incident light (in the *Mie* range).

Optical SHG requires the breakdown of inversion symmetry and thus is very sensitive to surfaces and the atomic structure of the spheres, in contrast to optical transmission spectroscopy. Compared with resonance-type spectroscopies like surface enhanced Raman scattering (SERS) and coherent anti-Stokes Raman scattering (CARS), higher harmonic light generation has the advantage of local field enhancements already for particle sizes much smaller than the wavelength. Quite generally, SHG and THG studies yield more and new information on the interaction light with small particles than many other methods previously used. In view of increasing experimental activity in the research field of small particles, nanostructured materials and clusters, it is of considerable interest to develop the theory for SHG and THG. Our results for the Mie enhancement of SHG might also help to understand the behaviour of SH light at roughened surfaces.

So far the nonlinear optical properties of spheres in the Mie size range have been theoretically investigated by Hua and Gersten,[20], Hayata and Koshiba[21] and Östling et al.[22] Here we extend this classical model to calculate in particular the angular Mie scattering profile in nonlinear optics.[23] In this model, the nonlinear surface charges $\sigma^{(2)}$ and $\sigma^{(3)}$, respectively, determining the nonlinear polarization in the case of metals, since $\sigma^{(n)} = \mathbf{n} \cdot \mathbf{P}^{(n)}$ ($\mathbf{n}$ is the vector normal to the surface of the sphere), are expressed by the second and third power of the linear surface charge $\sigma^{(1)}$ induced by the electric field incident onto the sphere. This approach is suggested qualitatively and quantitatively by the anharmonic oscillator model (see previous subsection). Since it neglects some proportionality factors, it is designed to yield the angular scattering profile in nonlinear Mie scattering rather than the conversion efficiency of the microspheres.

However, the classical theory of the anharmonic oscillator is sufficient here, because the interference of higher multipoles, which is the characteristic property of object in the Mie size range, is a purely classical electromagnetic phenomenon, and does not require quantum theory. In contrast to Hua et al.,[20] we easily include the contributions of many higher-order multipoles, and our model applies as well to THG. Concerning this we concentrate our analysis on effects depending on the interplay between the multipoles. In particular we will compare the interference and resonance phenomena known from linear Mie theory[24] with the higher harmonic case to provide an insight into the size dependence of the radiated intensities in SHG and THG. We calculate important measurable quantities such as the angular-dependence of the scattered intensities, the degree of polarization, the ratio of forward to backward scattering, and the resonances of the total intensities as a function of cluster size in the case of SHG and THG.

As expected physically our results show an enhanced size sensitivity of the higher harmonics compared to linear Mie scattering, thus enhancing the mentioned characteristic features of the linear *Mie effect* in higher harmonic scattering. Concerning the intrinsic selection rules for higher harmonics, the multipoles contributing to SHG are very different from the linear ones, resulting in distinct changes of the angular dependence. On the other hand, we find striking similarities between linear Mie scattering and THG, which reflect the fact that the scattering is dominated by the same multipoles in both cases. Owing to the more pronounced forward scattering, backward scattering decreases more strongly in the higher harmonics. In agreement with earlier calculations by Östling et al.,[22] we find for spheres with sizes $a > \lambda_p/2$ in a wide frequency range a dramatic enhancement of the *total* intensities by a factor of 5000 in SHG and of 200 000 in THG with respect to a plane metal surface. This holds especially if the diameter of the sphere roughly equals λ_p, where λ_p is the wavelength corresponding to the plasma frequency ω_p. It is of particular interest to recall that our theory shows clearly that nonlinear scattering from small particles is more sensitively dependent on particle size and on the complex refractive index than linear Mie scattering. Experiments must show that the new spectral information yielded by nonlinear scattering is not smeared out. However, note that experiments indicate that SHG intensities measured on deposited clusters reflect differences in shape and size for particles as small as 1 nm, for which no linear signal can be extracted at all.[25]

Determination of the surface charge $\sigma^{(1)}(\omega)$

In order to determine the sources of the radiation generated in second and third harmonic $\sigma^{(2)}$ and $\sigma^{(3)}$, we start with linear Mie theory and calculate the linearly scattered electric field components and the radial component of the polarization $\mathbf{n} \cdot \mathbf{P} = \sigma(\theta, \phi)$ at the surface of the sphere. Then, we determine the radiated intensities in second and third harmonic by matching the electromagnetic fields and the nth power of $\sigma^{(1)}(\theta, \phi)$ using the appropriate boundary conditions. The surface charge model needs to be checked by further studies.

The surface charge density from which the sources of the higher harmonics are calculated is determined from linear Mie theory. For this we expand the fields in the form

$$\mathbf{E}_i(\mathbf{x}) = \sum_{l,m} C(l) \left[a_M^i(l,m) f_l^i(k_1 r) \mathbf{X}_{l,m}(\theta, \phi) \right.$$
$$\left. + \frac{m}{|m|} a_E^i(l,m) \frac{1}{\varepsilon(\omega)k} \nabla \times f_l^i(k_1 r) \mathbf{X}_{l,m}(\theta, \phi) \right]. \quad (5.2.35)$$

Here, $\mathbf{X}_{l,m}$ is a vector spherical harmonic as introduced by Jackson[26] with $C(l) = i^l \sqrt{4\pi(2l+1)}$, $k = \omega/c$ and $a_M^i(l,m)$ and $a_E^i(l,m)$ refer to the magnetic (transverse electric) and electric (transverse magnetic) multipoles. The index i specifies the incident ($i \equiv \text{inc}$), the scattered ($i \equiv \text{sc}$), or the internal ($i \equiv \text{in}$) fields. For incident waves of positive and negative helicity we have $a_M^{\text{inc}}(l, \pm 1) = a_E^{\text{inc}}(l, \pm 1) = 1$. Here we use a superposition of both to get linear polarization. The spherical Hankel functions of the first kind $f_l^{\text{sc}}(kr) = h_l^{(1)}(kr)$ and Bessel functions $f_l^{\text{inc,in}}(kr) = j_l(kr)$ describe the radial part of the field outside and inside the sphere. The magnetic field is given by the Maxwell equation for harmonic fields

$$\mathbf{B} = -(i\omega/c) \nabla \times \mathbf{E}. \quad (5.2.36)$$

Using the boundary conditions at the surface of the sphere

$$\mathbf{n} \times (\mathbf{E}_{\text{sc}} + \mathbf{E}_{\text{inc}}) = \mathbf{n} \times \mathbf{E}_{\text{in}}, \quad (5.2.37)$$

and

$$\mathbf{n} \times (\mathbf{B}_{\text{sc}} + \mathbf{B}_{\text{inc}}) = \mathbf{n} \times \mathbf{B}_{\text{in}}, \quad (5.2.38)$$

we obtain the expansion coefficients of the scattered wave:

$$a_E^{\text{sc}}(l, \pm 1) = \frac{j_l(kr) \frac{\partial}{\partial r}[r j_l(k_1 r)] - \varepsilon(\omega) j_l(k_1 r) \frac{\partial}{\partial r}[r j_l(kr)]}{\varepsilon(\omega) j_l(k_1 r) \frac{\partial}{\partial r}[r h_l^{(1)}(kr)] - h_l^{(1)}(kr) \frac{\partial}{\partial r}[r j_l(k_1 r)]} \bigg|_{r=a} \quad (5.2.39)$$

and

$$a_M^{\text{sc}}(l, \pm 1) = \frac{j_l(kr) \frac{\partial}{\partial r}[r j_l(k_1 r)] - j_l(k_1 r) \frac{\partial}{\partial r}[r j_l(kr)]}{j_l(k_1 r) \frac{\partial}{\partial r}[r h_l^{(1)}(kr)] - h_l^{(1)}(kr) \frac{\partial}{\partial r}[r j_l(k_1 r)]} \bigg|_{r=a} \quad (5.2.40)$$

CLASSICAL THEORY FOR NONLINEAR OPTICS 281

at the surface of the sphere with radius a. From the continuity of the electrical displacement at the surface of a perfect conductor,

$$\mathbf{n} \cdot (\mathbf{D}_{sc} + \mathbf{D}_{inc}) = \mathbf{n} \cdot \mathbf{D}_{in}, \qquad (5.2.41)$$

the surface charge results as

$$\sigma^{(1)}(\theta, \phi) = \frac{1}{4\pi} \text{Re}[(\mathbf{E}_{sc} + \mathbf{E}_{inc} - \mathbf{E}_{in}) \cdot \mathbf{n}] e^{-i\omega t}. \qquad (5.2.42)$$

Here, $\mathbf{n} = \mathbf{r}/|\mathbf{r}|$ and Re denotes the real part. Using eqn (5.2.35) and expanding in spherical harmonics one gets

$$\sigma^{(1)}(\theta, \phi) = \frac{1}{2} \sum_{l, m = \pm 1} a_{l,m}^{(1)} Y_{l,m}(\theta, \phi) e^{-i\omega t} + \text{c.c.} \qquad (5.2.43)$$

Using the orthogonality of the spherical harmonics one obtains for the expansion coefficients

$$a_{l, \pm 1}^{(1)} = \frac{1}{4\pi} \left(1 - \frac{1}{\varepsilon(\omega)}\right) \frac{C(l) i \sqrt{l(l+1)}}{ka}$$
$$\times (j_l(ka) + a_E^{sc}(l, \pm 1) h_l^{(1)}(ka)), \qquad (5.2.44)$$

where the coefficients a_E^{sc} are given by eqns (5.2.39) and (5.2.40). This then completes the determination of the surface charge $\sigma^{(1)}$.

Determination of $(\sigma^{(1)})^n$ as sources of the higher harmonics

Using the anharmonic oscillator model for the displacement of the surface charge we find the nonlinear radiated fields from the boundary conditions with $\sigma^{(n)} = (\sigma^{(1)})^n$. Hence, we proceed by determining the sources of the higher harmonics via expanding the nth power of the surface charge in eqn (5.2.43) in terms of the spherical harmonics:

$$(\sigma^{(1)}(\theta, \phi))^n = \frac{1}{2} \sum_{l, m} a_{l,m}^{(n)} Y_{l,m}(\theta, \phi) e^{-ni\omega t} + \text{c.c.} \qquad (5.2.45)$$

Neglecting time-independent terms and using eqn (5.2.43) we obtain the coefficients in the case of second harmonic generation as

$$a_{l,2}^{(2)} = \frac{1}{2} \sum_{l_1=1}^{\infty} \sum_{l_2=1}^{\infty} a_{l_1,1}^{(1)} a_{l_2,1}^{(1)} \int Y_{l,2}^* Y_{l_1,1} Y_{l_2,1} \, d\Omega,$$

$$a_{l,-2}^{(2)} = \frac{1}{2} \sum_{l_1=1}^{\infty} \sum_{l_2=1}^{\infty} a_{l_1,-1}^{(1)} a_{l_2,-1}^{(1)} \int Y_{l,-2}^* Y_{l_1,-1} Y_{l_2,-1} \, d\Omega, \qquad (5.2.46)$$

$$a_{l,0}^{(2)} = \frac{1}{2} \sum_{l_1=1}^{\infty} \sum_{l_2=1}^{\infty} a_{l_1,1}^{(1)} a_{l_2,-1}^{(1)} \int Y_{l,0}^* Y_{l_1,1} Y_{l_2,-1} \, d\Omega,$$

and for third harmonic generation (also neglecting terms with $e^{-i\omega t}$) as

$$a_{l,1}^{(3)} = \frac{1}{2} \sum_{l_1=1}^{\infty} \sum_{l_2=1}^{\infty} a_{l_1,-1}^{(1)} a_{l_2,-2}^{(2)} \int Y_{l,1}^* Y_{l_1,1} Y_{l_2,2} \, d\Omega,$$

$$a_{l,-1}^{(3)} = \frac{1}{2} \sum_{l_1=1}^{\infty} \sum_{l_2=1}^{\infty} a_{l_1,1}^{(1)} a_{l_2,-2}^{(2)} \int Y_{l,-1}^* Y_{l_1,1} Y_{l_2,-2} \, d\Omega,$$

$$a_{l,3}^{(3)} = \frac{1}{2} \sum_{l_1=1}^{\infty} \sum_{l_2=1}^{\infty} a_{l_1,-1}^{(1)} a_{l_2,-2}^{(2)} \int Y_{l,3}^* Y_{l_1,1} Y_{l_2,2} \, d\Omega,$$

$$a_{l,-3}^{(3)} = \frac{1}{2} \sum_{l_1=1}^{\infty} \sum_{l_2=1}^{\infty} a_{l_1,-1}^{(1)} a_{l_2,2}^{(2)} \int Y_{l,-3}^* Y_{l_1,-1} Y_{l_2,-2} \, d\Omega.$$

(5.2.47)

The integrals can be expressed by the 3j-symbols and yield the coupling of the multipoles. Because of angular momentum conservation, only coefficients with $m = 0, \pm 2$ in SHG and $m = \pm 1, \pm 3$ in THG are different from zero.

Radiated higher harmonic fields

Now, from $(\sigma^{(1)})^n$ we determine the higher harmonic electric fields by matching the electric fields inside and outside the sphere. Thus, using the anharmonic oscillator model we obtain the nonlinear fields from $\sigma^{(n)}$, where $\sigma^{(n)} \approx (\sigma^{(1)})^n$. Then

$$\mathbf{n} \cdot (\mathbf{D}_{\text{out}}^{(n)} - \mathbf{D}_{\text{in}}^{(n)}) = 4\pi (\sigma^{(1)})^n, \quad (5.2.48)$$

and

$$\mathbf{n} \times (\mathbf{E}_{\text{out}}^{(n)} - \mathbf{E}_{\text{in}}^{(n)}) = 0. \quad (5.2.49)$$

As a result of spherical symmetry, only transverse magnetic waves are generated by the oscillating surface charge. Thus, the fields in the nonlinear case are

$$\mathbf{E}_i^{(n)}(\theta, \phi) = \sum_{l,m} \frac{m}{|m|} A_{\text{E}}^{(n)}(l,m) \frac{1}{\varepsilon(n\omega)k} \cdot \nabla \times f_l^i(k_1 r) \mathbf{X}_{l,m}(\theta, \phi), \quad (5.2.50)$$

where $i \equiv $ out or $i \equiv$ in, respectively, and $k = n\omega/c$, $k_1 = \sqrt{\varepsilon(n\omega)}\,k$, $f_l^{\text{in}}(kr) = j_l(k_1 r)$ and $f_l^{\text{out}} = h_l^{(1)}(kr)$. The boundary conditions give the coefficients of the radiated field

$$A_{\text{E}}^{(n)}(l,m) = \frac{\dfrac{\partial}{\partial r}[rj_l(k_1 r)]}{\varepsilon(n\omega) j_l(k_1 r) \dfrac{\partial}{\partial r}[rh_l^{(1)}(kr)] - h_l^{(1)}(kr) \dfrac{\partial}{\partial r}[rj_l(k_1 r)]}$$

$$\times \frac{\pi k a}{\sqrt{l(l+1)}} a_{l,m}^{(n)}. \quad (5.2.51)$$

In this case we have $k_1 = \sqrt{\varepsilon(n\omega)}\,k$. This completes the determination of the higher harmonic electric fields.

Calculation of the scattering profile, degree of Mie asymmetry and integrated intensities

To study the angular dependence of the radiated fields we use the quantity $|E_\phi(\theta, \phi)|^2 + |E_\theta(\theta, \phi)|^2$ according to Born and Wolf,[27] where $E_\theta(\theta, \phi)$ and $E_\phi(\theta, \phi)$ are the tangential components of $\boldsymbol{E}_{\text{sc}}(\theta, \phi)$ in the linear case and of $\boldsymbol{E}_{\text{out}}(\theta, \phi)$ in the nonlinear case. This definition is equivalent to the absolute value of the radial part of the Poynting vector $|\boldsymbol{n} \cdot (\boldsymbol{E} \times \boldsymbol{H})|$. The following formulae represent $|E_\phi(\theta, \phi)|^2$ and $|E_\theta(\theta, \phi)|^2$ in the far-field approximation. We obtain after evaluating the m-summation in the linear case

$$|E_{\text{sc},\phi}(\theta,\phi)|^2 = \left| \sum_{l=1}^{\infty} C(l) \left(\frac{dP_l^1(\cos\theta)}{d\theta} a_M^{\text{sc}}(l,1) + \frac{P_l^1(\cos\theta)}{\sin\theta} a_E^{\text{sc}}(l,1) \right) \right|^2 \sin^2\phi,$$

$$|E_{\text{sc},\phi}(\theta,\phi)|^2 = \left| \sum_{l=1}^{\infty} C(l) \left(\frac{P_l^1(\cos\theta)}{\sin\theta} a_M^{\text{sc}}(l,1) + \frac{dP_l^1(\cos\theta)}{d\theta} a_E^{\text{sc}}(l,1) \right) \right|^2 \cos^2\phi.$$

(5.2.52)

For second harmonic generation we get

$$\left| E_{\text{out},\phi}^{(2)}(\theta,\phi) \right|^2 = \left| \sum_{l=1}^{\infty} \sqrt{4\pi(2l+1)} \left(\frac{dP_l^0(\cos\theta)}{d\theta} A_E^{(2)}(l,0) \right. \right.$$

$$\left. \left. + \frac{dP_l^2(\cos\theta)}{d\theta} 2K(l) A_E^{(2)}(l,2) \cos(2\phi) \right) \right|^2, \quad (5.2.53)$$

$$\left| E_{\text{out},\phi}^{(2)}(\theta,\phi) \right|^2 = \left| \sum_{l=2}^{\infty} \sqrt{4\pi(2l+1)} \frac{dP_l^0(\cos\theta)}{d\theta} A_E^{(2)}(l,2) \right|^2 \sin^2(2\phi).$$

Finally, for third harmonic generation we get

$$\left| E_{\text{out},\phi}^{(3)}(\theta,\phi) \right|^2 = \left| \sum_{l=1}^{\infty} \sqrt{4\pi(2l+1)} \left(\frac{P_l^1(\cos\theta)}{\sin\theta} A_E^{(3)}(l,1) \sin\phi \right. \right.$$

$$\left. \left. + \frac{P_l^3(\cos\theta)}{\sin\theta} K(l) A_E^{(3)}(l,3) \sin(3\phi) \right) \right|^2,$$

(5.2.54)

$$\left| E_{\text{out},\phi}^{(3)}(\theta,\phi) \right|^2 = \left| \sum_{l=1}^{\infty} \sqrt{4\pi(2l+1)} \left(\frac{dP_l^1(\cos\theta)}{d\theta} A_E^{(3)}(l,0) \cos\phi \right. \right.$$

$$\left. \left. + \frac{dP_l^3(\cos\theta)}{d\theta} K(l) A_E^{(3)}(l,3) \cos(3\phi) \right) \right|^2,$$

where $K(l)$ are l- and m-dependent factors. Note that the Legendre polynomials p_l^m with $m = 0, 2$ are identically zero for $\theta = 0, \pi$. This important result means that there is no direct scattering in forward nor backward direction in the second harmonic case. Furthermore the ϕ dependence of the linear scattering and THG is described by the interval $(0, \pi)$ and by $(0, \pi/2)$ in SHG according to the symmetries of the sine and cosine terms.

To determine the asymmetry of forward and backward scattering in the *Mie* range we introduce the quantity

$$R = \frac{I_{\text{forw}} - I_{\text{back}}}{I_{\text{forw}} + I_{\text{back}}}. \tag{5.2.55}$$

We call this the 'degree of Mie asymmetry'. In the linear case I_{forw} and I_{back} are the scattering intensities taken at $\theta = 0$ and $\theta = \pi$ respectively. As these quantities are identically zero in SHG, we use for I_{forw} and I_{back} the maxima of the scattering intensities along the direction of propagation of the incident wave for $\phi = 0, \pi/4, \pi/2$ and θ covering the interval $(0, \pi)$. In THG the angular dependence of the radiated intensities is more complicated as compared to the linear case and we take the maximum for $\phi = 0$ and θ ranging from 0 to π.

Finally, we calculate the angle-integrated scattered intensities. We obtain in the linear case

$$Q_{\text{sc}} = \frac{1}{\pi(ka)^2} \sum_{l,m} \frac{2l+1}{l(l+1)} \left(|a_{\text{E}}^{\text{sc}}(l,m)|^2 + |a_{\text{M}}^{\text{sc}}(l,m)|^2 \right), \tag{5.2.56}$$

and for the nth harmonic

$$Q_{\text{out}}^{(n)} = \frac{1}{\pi(ka)^2} \sum_{l,m} |A_{\text{E}}^{(n)}(l,m)|^2. \tag{5.2.57}$$

Here, Q_{sc} and $Q_{\text{out}}^{(n)}$ are given in units of the geometric cross-section of the sphere πa^2. In this formulation the optical theorem, which links the linear extinction efficiency to the normalized scattering amplitude in the forward direction, has the form

$$Q_{\text{ext}} = \frac{2}{\pi(ka)^2} \left(|E_{\text{sc},\theta}(\theta=0)|^2 + |E_{\text{sc},\phi}(\theta=0)|^2 \right), \tag{5.2.58}$$

with

$$Q_{\text{ext}} = \frac{1}{\pi(ka)^2} \sum_{l,m} (2l+1) \{ \text{Re}[a_{\text{E}}^{\text{sc}}(l,m)] + \text{Re}[a_{\text{M}}^{\text{sc}}(l,m)] \}. \tag{5.2.59}$$

Numerical results

In this section, we present numerical results for experimentally accessible quantities, in particular the *angular dependence* of the radiated intensities obtained using eqns (5.2.52)–(5.2.54). The *degree of Mie asymmetry R* calculated from eqn (5.2.55) reflects the strength of the asymmetry along the direction of propagation according to the *Mie effect*. The *integrated intensities* yield the resonances as a function of cluster size. To get an overview of the possible

Table 5.1 Complex refractive indices $N = \tilde{n} + i\tilde{n}'$ for iron and nickel used in this work at ω, 2ω, and 3ω. The wavelength is $\lambda = 617$ nm and $\omega = (2\pi/\lambda)c$

	Iron		Nickel	
	$\tilde{n}$	$\tilde{n}'$	$\tilde{n}$	$\tilde{n}'$
ω	2.88	3.05	1.99	4.02
2ω	1.69	2.06	2.01	2.18
3ω	1.49	1.41	1.29	1.89

phenomena appearing in nonlinear Mie scattering we compare the higher harmonic results with linear Mie scattering. The restriction to metallic particles caused by our theory is reached by using complex refractive indices.

To be specific we analyse the ka dependence of the quantities at the refractive indices of Fe and Ni at an optical wavelength 617 nm measured by Johnson and Christy.[28] In the linear case, for comparison with calculations by other authors, we also use the refractive index of water droplets, as given by Bohren and Huffman.[29] The values for Fe and Ni are listed in Table 5.1. The refractive index is constant in all figures unless specified otherwise. Thus varying the size parameter ka means varying the size of the sphere.

In general, the shape of the plots is governed by the values of the coefficients in the series expansions (5.2.52)–(5.2.54). It is well known[26] that in the linear series (5.2.52) only terms $l \leq ka$ contribute significantly. For $l > ka$ the terms decrease very rapidly, whereas for $l \ll ka$ they have comparable magnitudes. We restrict our calculation to $l < [\text{Max}(\tilde{n}ka, \tilde{n}'ka) + 15]$ with a maximum value of $l = 50$, where $\tilde{n}$ and $\tilde{n}'$ are the real and imaginary parts of the complex refractive index. This gives satisfactory convergence of the multipole expansion up to $ka < 10$ in linear optics, SHG, and THC.

To check the numerical accuracy we compare the linear results with those of ref. 29 and find excellent agreement. Furthermore, we check the optical theorem, which must be satisfied in the linear case. The quantity

$$\Delta Q \equiv \left| Q_{\text{ext}} - 2\left(|E_{\text{sc},\theta}|^2 + |E_{\text{sc},\phi}|^2\right)_{\theta=0} / [\pi(ka)^2] \right|$$

obtained from eqn (5.2.58) should be exactly zero, which is fulfilled to an accuracy better than 10^{-12} for all size parameters applied in this chapter. It reveals the quality of the calculation for the linear scattering, which is important, since it is the basis of the calculation of the higher harmonics and serves as a reference for the higher harmonic results.

Angular dependence of the intensities First, we show polar plots of $|E_\phi|^2 + |E_\theta|^2$ (see Figs 5.1–5.3). The geometry of the scattering is specified in the inset of Fig. 5.1 with the direction of propagation of the incident wave being parallel to the positive z axis and polarization along the positive x axis. In Figs. 5.1–5.3 the clusters are located at $(x = 0, y = 0, z = 0)$ above the yz plane drawn in the

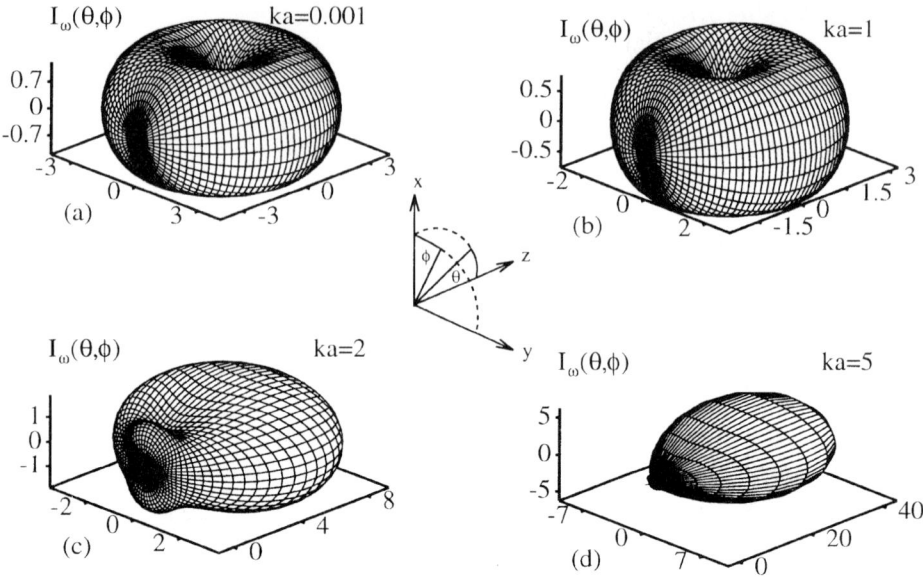

Fig. 5.1. Polar plots of linear scattered intensities for Fe: (a) scattered linear intensity $I_\omega(\theta, \phi)$ in units of 10^{-12} with $ka = 0.001$; (b) $I_\omega(\theta, \phi)$ in units of 10^{-4} with $ka = 1$; (c) $I_\omega(\theta, \phi)$ in units of 1 with $ka = 2$; and (d) $I_\omega(\theta, \phi)$ with $ka = 5$ in units of 1. The inset shows the scattering geometry. The direction of the incident light is determined by $\theta = 0$ with polarization along the axis defined by $\phi = 0$ and $\theta = \pi/2$. The clusters here and in the following two figures are located at $(x = 0, y = 0, z = 0)$, above the indicated xy plane.

figures. Figure 5.1(a) shows *Rayleigh* scattering in the linear case according to the dipole term with $l = 1$. The characteristic $\cos^2 \theta$ dependence appears along the xz plane. The other plots of linear optics (Fig. 5.1) show the well known angular dependence.[24,27,29] For a value of $ka = 1$ an asymmetry of forward and backward scattering appears according to the Mie effect. The ratio of forward to backward scattering $I_{\text{forw}}/I_{\text{back}}$ increases strongly with increasing size parameter ka beginning at a value of $ka \approx 1$. 'New' maxima due to the excitation of higher-order multipoles grow out in the backward direction and move to the forward direction with increasing ka. The ϕ dependence is not as striking and not as complicated as the θ dependence since we have a superposition of the form $A(\theta) \cos^2 \phi + B(\theta) \sin^2 \phi$. The scattering behaviour is dominated by the strong increase of the intensities in the forward direction described by the θ dependence. The fact that the ϕ dependence is described by the interval $(0, \pi)$ in the linear case and harmonics with odd order is most important. This differs from SHG and harmonics with even order where the ϕ dependence is fully described by the interval $(0, \pi/2)$.

The polar plots in the case of THG are quite similar to the linear case up to values of $ka \approx 2$. The differences between Figs 5.1(b) and 5.2(b) with $ka = 1$ reflect the stronger increase in the ratio of forward to backward scattering in

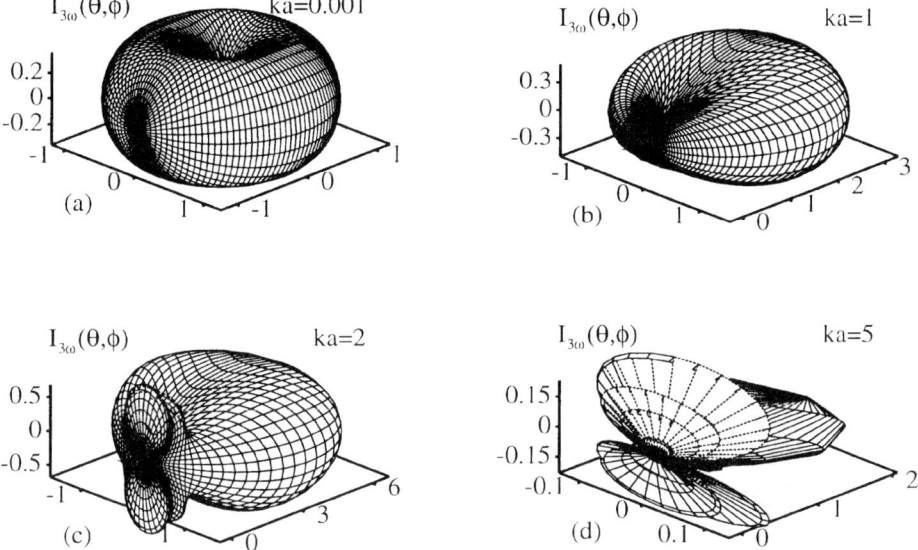

Fig. 5.2. Polar plots of the third harmonic (THG) intensities for Fe: (a) scattered 3ω intensity $I_{3\omega}(\theta,\phi)$ in units of 10^{-15} with $ka = 0.001$; (b) $I_{3\omega}(\theta,\phi)$ in units of 1 with $ka = 1$; (c) $I_{3\omega}(\theta,\phi)$ in units of 10^{-5} with $ka = 2$; and (d) $I_\omega(\theta,\phi)$ in units of 10^{-5} with $ka = 5$.

THG with increasing ka. The plots in Figs 5.1(c) and 5.2(c) with $ka = 2$ are very similar apart from one more maximum appearing in the third harmonic case for $\theta \approx \pi/2$. For $ka = 5$ the ratio of the intensities perpendicular and parallel to the direction of propagation are much larger in third harmonic than in the linear case. Note the different scales of the axes for different ka. The terms for $m = 3$ in THG are negligible compared to the terms with $m = 1$. So the differences in the magnitudes of the linear and third harmonic intensities are caused by the coefficients $a_E^{sc}(l,1)$, $a_M^{sc}(l,1)$ and $A_E^{(3)}(l,1)$ only.

Figure 5.3 indicates that the angular dependence of the intensities in the second harmonic is very different from the linear and third harmonic cases. The vanishing direct forward and backward intensities and the $\cos^2(2\phi)$ and $\sin^2(2\phi)$ behaviour produce the club-shaped structure. But the main overall features of the linear and THG plots appear also in SHG. The plots become asymmetric with respect to θ in the range of $ka \approx 1$. The ratio of forward to backward scattering increases with ka and is between the linear and third harmonic values, which can be seen from the values listed later in Table 5.2 for different size parameters ka. In general, harmonics with even order will show an angular dependence like SHG because the Legendre polynomials of $m \neq 1$ vanish at $\theta = 0, \pi$, and ϕ will appear in the cosine and sine terms in connection with $n = 0, 2, \ldots, 2p$ (p integer). Analogous harmonics of odd order will behave similarly to the linear case (see eqns (5.2.52)–(5.2.54)).

In conclusion, we find that it is sufficient to analyse the angular dependence

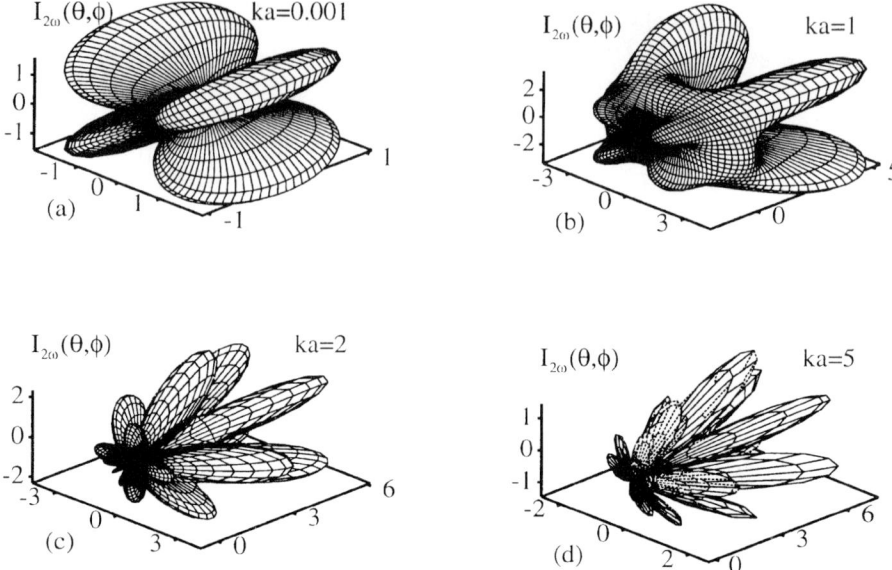

Fig. 5.3. Polar plots of the second harmonic (SHG) intensities for Fe: (a) scattered 2ω intensity $I_{2\omega}(\theta,\phi)$ in units of 10^{-18} with $ka = 0.001$; (b) $I_{2\omega}(\theta,\phi)$ in units of 1 with $ka = 1$; (c) $I_{2\omega}(\theta,\phi)$ in units of 10^2 with $ka = 2$; and (d) $I_{\omega}(\theta,\phi)$ in units of 10^6 with $ka = 5$.

for size parameters up to 5 to get a qualitative description of the resonances and interferences appearing in the Mie range. For $ka < 5$ the l values of the dominating terms are a little smaller than ka and no terms with $l \ll ka$ exist. Terms with $l \leq ka$ can be very different from each other, in contrast to terms with $l \ll ka$, which have comparable magnitudes. Thus, a pronounced transition range from pure dipole scattering to pure Mie scattering exists. Of course we cannot reach the transition from Mie scattering to the optical limit of reflection, due to the numerical limitation of the amount of multipoles by $l_{max} = 50$. In sec. 5.2.5 we will see that only for much larger sizes do the scattering patterns become simple again since only a few pronounced structures survive.

Forward vs backward scattering By computing R as defined in eqn (5.2.55), which is a measure of the difference between the forward and backward intensities, as a function of the size parameter ka and the real or imaginary part of the refractive index $N = \tilde{n} + i\tilde{n}'$, we want to study the development of the asymmetry of forward to backward intensities. Of particular interest will be enhanced backward scattering.

Figure 5.4(a) shows the increase of forward scattering with the size of the sphere in linear scattering. The oscillations correlate with intensity maxima in the backward direction resulting from maxima of the coefficients (see ref. 30). Enhanced backward scattering appears only in the small range of $0.5 < ka < 1$ in

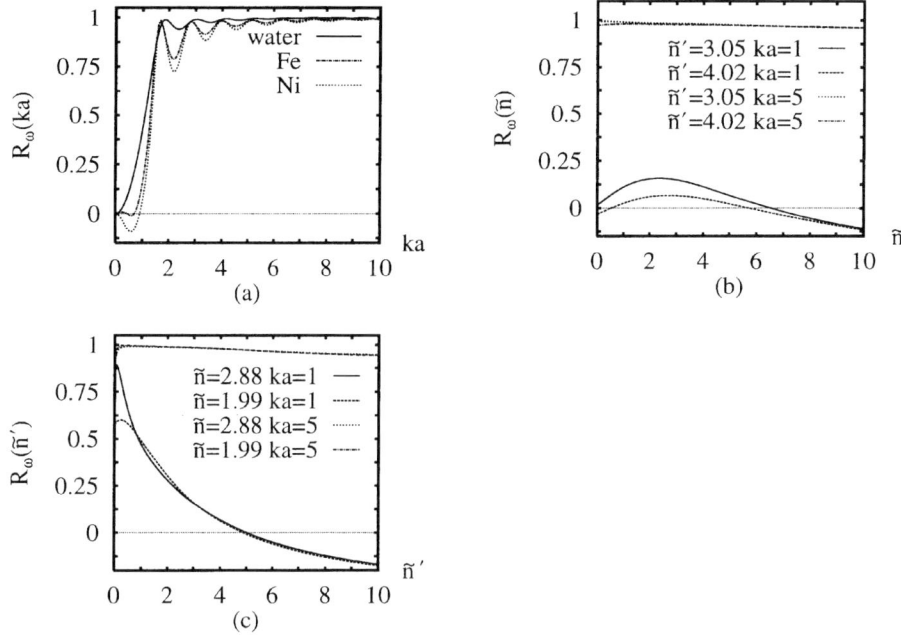

Fig. 5.4. Degree of Mie asymmetry R (defined as $R = (I_{\text{forw}} - I_{\text{back}})/(I_{\text{forw}} + I_{\text{back}})$, see eqn (5.2.55)) in the linear case (a) as a function of the size parameter ka for refractive indices referring to Fe, Ni and water droplets, (b) as a function of the real part of the refractive index $\tilde{n}$, and (c) as a function of the imaginary part of the refractive index $\tilde{n}'$ for size parameters $ka = 1$ and $ka = 5$ using values of $\tilde{n}$, $\tilde{n}'$ for Fe and Ni.

the case of iron. The curve for water droplets shows no enhanced backscattering, but the overall behaviour is the same as for metals. To get a rough impression of the way in which the quantity R changes for different refractive indices representing other materials, we scan R as a function of $\tilde{n}$ in Fig. 5.4(b) and $\tilde{n}'$ in Fig. 5.4(c). This shows that increasing the imaginary part of the refractive index diminishes the forward scattering. In the case $ka = 1$ and $\tilde{n}' > 5$ an enhancement of the backward scattering occurs. The $\tilde{n}$ dependence of R is similar.

In the higher harmonic case (Fig. 5.5) the Mie effect is strongly enhanced. The limit of dominating forward scattering ($R = 1$) is obtained earlier than in the linear case. The higher the order, the stronger is the enhancement of the forward scattering in this range of size parameters. This may change for larger size parameters. Regions of enhanced backward scattering are hard to find. In the case of second harmonic generation, they exist only for small $\tilde{n}$ or $\tilde{n}'$ around 5 and small ka, whereas we have not so far been able to find enhanced backward scattering for metal clusters ($\tilde{n}' \gg 0$) in third harmonic generation.

Integrated intensities Mie resonances of the radiated intensities integrated over

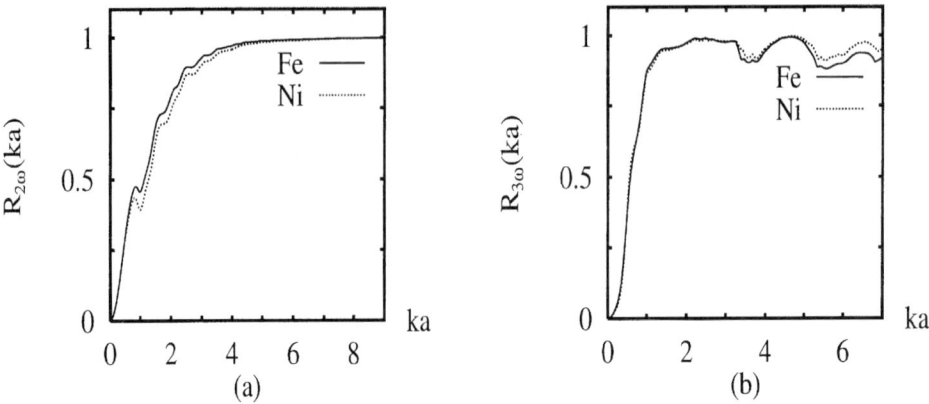

Fig. 5.5. Degree of Mie asymmetry R as a function of the size parameter ka for (a) SHG and (b) THG with the refractive indices referring to Fe and Ni, respectively.

4π can be plotted as functions of the sphere radius a or the absolute value of the wavevector k of the incident light. Here we only refer to increasing or decreasing sphere sizes.

In Figure 5.6, reproducing results from [29], we show the scattering efficiency Q_{sc} of water droplets as a function of the size parameter. The main features are the dominant interference structure built up by interferences between the incident wave and forward scattered light and the fine structure ('ripple' structure) reflecting resonant surface modes. The latter correlates with resonances in the coefficients $a_E^{sc}(l,m)$ and $a_M^{sc}(l,m)$. They are resonant if their imaginary part is zero. Furthermore, the results exhibit the 'optical paradox' $\lim_{ka \to \infty} Q_{ext}(ka) = 2$. For large size parameters the distance of the ripples can be expressed directly by the refractive index. Finite imaginary parts of the

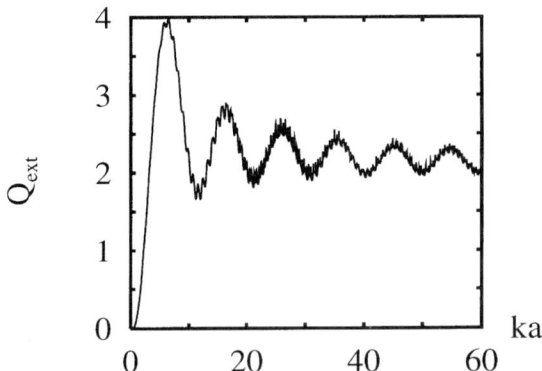

Fig. 5.6. The linear extinction $Q_{ext}(ka)$ as a function of the size parameter ka and varying the radius of the sphere for water droplets with complex refractive index $N = 1.33 + i \cdot 10^{-7}$ (see ref. 5).

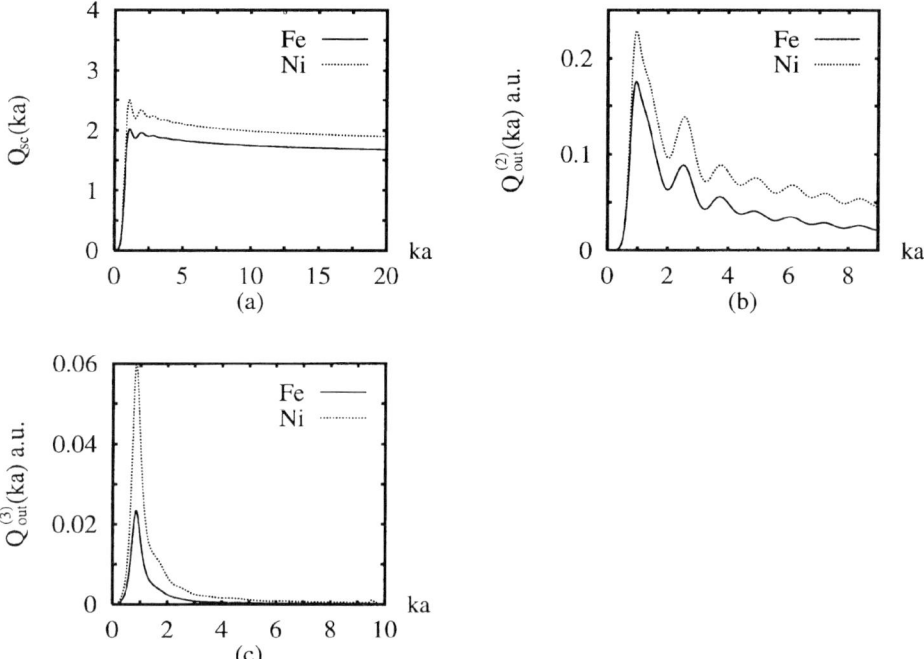

Fig. 5.7. Scattering efficiencies Q_{sc} and $Q_{out}^{(n)}$ as a function of size parameter ka for (a) linear scattering, (b) SHG, and (c) THG in the case of Fe and Ni, respectively.

refractive index $\tilde{n}'$ would damp the resonance. Then the values of the real parts of the coefficients at the resonant point would be smaller than their maximum value 1 obtained for $\tilde{n}' = 0$. Even for $\tilde{n}' \approx 0.1$ the ripple structure will be damped out. Accordingly, there is no ripple structure in the scattering efficiency of Fe and Ni as a function of size as shown in Fig. 5.7(a).

In higher harmonics no structure correlated with resonances of the coefficient $A_E^{(n)}$ can be detected even for vanishing imaginary part of the refractive index up to a numerical accuracy of 10^{-10}. Each coefficient $a_{l,m}^{(n)}$ is a combination of all linear electric multipole coefficients determined by eqns (5.2.46) and (5.2.47), respectively. Correspondingly, the size dependence of the coefficients $A_E^{(n)}(l, m)$ has many small peaks. In contrast, the size resonances caused by multipole combinations, and known as *interference structure* in linear scattering, are more pronounced in SHG in the case of metals (compare Figs 5.7(a) and (b)). In THG (Fig. 5.7(c)) the first peak is even more dominant and the following resonances can only be found for enhanced resolution. In both cases, the peak positions have a weak dependence on the refractive index, but the number of peaks in SHG is twice the peak number in THG for $0 < ka < 10$. Decreasing the imaginary part of the refractive index down to zero changes the behaviour drastically. Instead of a dominating first peak, the scattering efficiency now shows a continuous increase for $0 < ka < 20$ with oscillations stronger than in

Table 5.2 Ratio of forward to backward intensities as a function of ka in the linear case, SHG and THG

	$I_{\text{forw}}/I_{\text{back}}$		
ka	Linear	SHG	THG
1	1.4	2.7	15
2.1	8.7	10	113
4.8	64	136	206

the case of $\tilde{n}' > 1$. In both harmonics and in the linear case the absolute values grow with the absolute value of the refractive index if metals are considered.

We find that in the nonlinear case the forward Mie scattering is more strongly enhanced (see Table 5.2). The nonlinear optical response yields a stronger size sensitivity. Higher multipoles contribute already at smaller size parameters ka. For example, we find that the light intensities perpendicular to the forward direction divided by the light intensities parallel to it are much larger in the third harmonic than in the linear case. Since the values of the Legendre polynomials with azimuthal quantum numbers $m = 0, \pm 2$ vanish for $\theta = 0, \pi$ and due to the ϕ dependence of the form $A\cos^2(2\phi) + B\sin^2(2\phi)$ (see eqn (5.2.53)), the angular dependence of the second harmonic intensities is very different from the linear and third harmonic distributions. In particular, direct forward and backward scattering vanishes. In contrast, the results for linear Mie scattering (terms with $m = \pm 1$ only) and THG (terms with $m = \pm 1, \pm 3$) are similar. We find that the terms with $m = \pm 3$ in THG and $m = 0$ in SHG are negligible compared to the other contributions. The more pronounced Mie effect prevents enhanced backward scattering in higher harmonics (Fig. 5.5).

Summary: nonlinear Mie scattering from metallic spheres

As expected physically our results show an enhanced size sensitivity of the higher harmonics compared to linear Mie scattering, thus enhancing the mentioned characteristic features of the linear Mie effect in higher harmonic scattering.

Concerning the intrinsic selection rules for higher harmonics, the multipoles contributing to SHG are very different from the linear ones, resulting in distinct changes of the angular dependence. On the other hand, we find striking similarities between linear Mie scattering and THG, which reflect the fact that the scattering is dominated by the same multipoles in both cases.

Owing to the more pronounced forward scattering, backward scattering decreases more strongly in the higher harmonics.

In agreement with earlier calculations by Östling et al. we find for spheres with sizes $a > \lambda_p/2$ in a wide frequency range a dramatic enhancement of the *total* intensities by a factor of 5000 in SHG and of 200 000 in THG with respect to a plane metal surface. This holds especially if the diameter of the sphere roughly equals λ_p, where λ_p is the wavelength corresponding to the plasma frequency ω_p.

It is of particular interest to recall that our theory shows clearly that nonlinear scattering from small particles is more sensitively dependent on particle size and on the complex refractive index than linear Mie scattering.

Experiments must show that the new spectral information yielded by nonlinear scattering is not smeared out. However, note that experiments indicate that SHG intensities measured on deposited clusters reflect differences in shape and size for particles as small as 1 nm, for which no linear signal can be extracted at all.[25] Of course, an alternative analysis not using $\sigma^{(n)} = (\sigma^{(1)})^n$ is necessary.

5.2.3 Nonlinear Mie theory for particles of arbitrary shape

Here we extend our theory for the nonlinear response of a spherical particle[22,23] to a particle of arbitrary shape.[31] The discontinuity of the electrical displacement at the surface can be expressed in terms of the boundary conditions for the radial component of the electrical fields. Using also the continuity of the tangential component of the electric field, the radiated fields can be fully determined. This is a simple task for spherical particles if spherical coordinates are used. In the case of nonspherical particles, however, matrix equations come into play, where the size of the matrices depends on the ratio of the particle dimension to the wavelength of the light and also on the particle shape.

To obtain the results for linear scattering, which are necessary to determine the induced surface charge $\sigma^{(1)}$, a formalism by Barber and Yeh[32] is used. This is based on an integral equation method by Waterman[33,34] and Schelkunoff's equivalence principle,[35] yielding matrix equations. This choice facilitates the task because Barber and Yeh use nearly the same representation of the fields with spherical coordinates and their theory is valid for arbitrary cluster shapes and arbitrary complex indices of refraction.

In some approximation for the special case of ellipsoidal particles the largest multipole coming into play is given by the major axis. This fact explains (i) the interference patterns in linear and nonlinear Mie scattering which are approximately determined by the largest particle dimension and (ii) the local field enhancement at those points, since at the same time they are the points of maximum curvature.

Determination of the source $\sigma^{(n)}$

To determine the source $\sigma^{(n)}$ we again use the approximation $\sigma^{(n)} = (\sigma^{(1)})^n$. To obtain the nth power of the linear charge we proceed as follows. The linear surface charge is defined as

$$\sigma^{(1)}(\theta, \varphi) = \mathbf{n} \cdot \mathbf{P}^{(1)}(\theta, \varphi) \qquad (5.2.60)$$

can can be expressed as

$$\sigma^{(1)}(\theta, \varphi) = \frac{1}{4\pi} \operatorname{Re}\left[(1 - \varepsilon^{-1})(\mathbf{E}_{\text{sc}} + \mathbf{E}_{\text{inc}}) \cdot \mathbf{n}\right] e^{-i\omega t}. \qquad (5.2.61)$$

Here we denote the scattered field by the index 'sc' and the incident field by 'inc'. Therein we refer to the boundary conditions of the normal component of the dielectric displacement in the linear case

$$\mathbf{n} \cdot (\mathbf{D}_{sc} + \mathbf{D}_{inc}) = \mathbf{n} \cdot \mathbf{D}_{in}$$

(the index 'in' states the field inside the cluster) and to the relation

$$\mathbf{D} = \varepsilon \mathbf{E} = \mathbf{E} + 4\pi \mathbf{P}.$$

Furthermore, $\sigma^{(1)}$ is expanded in terms of spherical harmonics

$$\sigma^{(1)}(\theta, \varphi) = \frac{1}{2} \sum_{l, m = \pm 1} a_{l,m}^{(1)} Y_{l,m}(\theta, \varphi) e^{-i\omega t} + \text{c.c.} \quad (5.2.62)$$

To calculate the coefficients the fields $\mathbf{E}_{out}$ and $\mathbf{E}_{sc}$, which enter via eqn (5.2.61), are expanded in terms of vector spherical harmonics as introduced by Jackson[26] in the form

$$\mathbf{E}_i^{(n)}(\mathbf{x}) = \sum_{l,m} C(l) \left[K_M(l,m) f_l^i(k_1 r) \mathbf{X}_{l,m}(\theta, \varphi) \right.$$

$$\left. + \frac{m}{|m|} K_E(l,m) \frac{1}{\varepsilon(n\omega)k} \nabla \times f_l^i(k_1 r) \mathbf{X}_{l,m}(\theta, \varphi) \right]. \quad (5.2.63)$$

Therein, $\mathbf{X}_{l,m}(\theta, \varphi) = L Y_{l,m}(\theta, \varphi) / \sqrt{l(l+1)}$ is a vector spherical harmonic ($\mathbf{L} = 1/i(\mathbf{r} \times \nabla)$ is the angular momentum operator, $C(l) = i^l \sqrt{4\pi(2l+1)}$, $k = n\omega/c$ and $k_1 = \sqrt{\varepsilon(n\omega)}\,k$. The multipole coefficients $K_M^{(n)}(l,m)$ and $K_E^{(n)}(l,m)$ refer to the *magnetic* (transverse electric TE) and *electric* (transverse magnetic TH) multipoles. The index i specifies the fields external ($i \equiv$ out) or internal ($i \equiv$ in) to the cluster. The spherical Hankel functions $f_l^{out}(kr) = h_l^{(1)}(kr)$ and Bessel functions $f_l^{in}(kr) = j_l(k_1 r)$ describe the normal projection of the field inside and outside the particle. In the linear case the external field splits into the incident and the scattered field $\mathbf{E}_{inc}$ and $\mathbf{E}_{sc}$, respectively. The coefficients of $\mathbf{E}_{inc}$ are known from the input field, while $\mathbf{E}_{sc}$ can be calculated by using the results from the theory for the linear problem by Barber and Yeh.[32]

Determination of the radiated field $\mathbf{E}_{out}^{(n)}$

To calculate the radiated field $\mathbf{E}_{out}^{(n)}$ it is more convenient to introduce the abbreviations

$$K_M'(l,m) \equiv C(l) K_M(l,m), \qquad K_E'(l,m) \equiv \frac{m}{|m|} \frac{1}{\varepsilon(\omega)k} C(l) K_E(l,m)$$

$$(5.2.64)$$

in the sum representation of eqn (5.2.63). This leads to

$$E_i^{(n)}(x) = \sum_{l,m} \left[K'_M(l,m) f^i(k_1 r) X_{l,m}(\theta, \varphi) + K'_E(l,m) \nabla \times f_l(k_1 r) X_{l,m}(\theta, \varphi) \right]. \tag{5.2.65}$$

In the nth harmonic case the source $\sigma^{(n)}$ acts as the discontinuity of the normal part of the dielectric displacement

$$\mathbf{n} \cdot (\mathbf{D}_{\text{out}}^{(n)} - \mathbf{D}_{\text{in}}^{(n)}) = 4\pi \sigma^{(n)}. \tag{5.2.66}$$

Furthermore we will use the continuity condition of the tangential component of the electric fields

$$\mathbf{n} \times (\mathbf{E}_{\text{out}}^{(n)} - \mathbf{E}_{\text{in}}^{(n)}) = 0. \tag{5.2.67}$$

For particles of arbitrary shape the radius vector $\mathbf{e}_r$ is no longer parallel to $\mathbf{n}$, and the radius r rather becomes a function of θ and φ at the boundary since every point of the surface is determined by the angles. Furthermore the normal vector $\mathbf{n}$ now has the form

$$\mathbf{n}(\theta, \phi) = n_r(\theta, \varphi) \mathbf{e}_r + n_\theta(\theta, \varphi) \mathbf{e}_\theta + n_\varphi(\theta, \varphi) \mathbf{e}_\varphi.$$

Here, $(\mathbf{e}_r, \mathbf{e}_\theta, \mathbf{e}_\varphi)$ are the basic vectors corresponding to spherical coordinates. As a result eqn (5.2.67) splits into three equations, one for every component:

$$\begin{aligned}
\mathbf{n} \cdot (\mathbf{D}_{\text{out}}^{(n)} - \mathbf{D}_{\text{in}}^{(n)}) &= 4\pi \sigma^{(n)}, \\
\mathbf{e}_r \cdot \left[\mathbf{n} \times (\mathbf{E}_{\text{out}}^{(n)} - \mathbf{E}_{\text{in}}^{(n)}) \right] &= 0, \\
\mathbf{e}_\theta \cdot \left[\mathbf{n} \times (\mathbf{E}_{\text{out}}^{(n)} - \mathbf{E}_{\text{in}}^{(n)}) \right] &= 0, \\
\mathbf{e}_\varphi \cdot \left[\mathbf{n} \times (\mathbf{E}_{\text{out}}^{(n)} - \mathbf{E}_{\text{in}}^{(n)}) \right] &= 0.
\end{aligned} \tag{5.2.68}$$

By using the series representations

$$\begin{aligned}
\mathbf{E}_{\text{out}}^{(n)} &\equiv \sum_{l,m} \left[A_E^{(n)} h_l^{(1)}(kr) X_{l,m} + A_M^{(n)} \nabla \times h_l^{(1)}(kr) X_{l,m} \right], \\
\mathbf{E}_{\text{in}}^{(n)} &\equiv \sum_{l,m} \left[B_E^{(n)} j_l(k_1 r) X_{l,m} + B_M^{(n)} \nabla \times j_l(k_1 r) X_{l,m} \right],
\end{aligned} \tag{5.2.69}$$

multiplying every equation by $Y_{l',m'}$ where (l',m') runs over all (l,m) combinations used in the series representations of the fields and the source, and integrating over the solid angle, we get the following matrix equations:

$$\begin{aligned}
S^1 A_M^{(n)} + S^2 A_E^{(n)} - S^3 B_M^{(n)} - S^4 B_E^{(n)} &= 2\pi \mathbf{a}^{(n)}, \\
M_r^1 A_M^{(n)} + M_r^2 A_E^{(n)} - M_r^3 B_M^{(n)} - M_r^4 B_E^{(n)} &= 0, \\
M_\theta^1 A_M^{(n)} + M_\theta^2 A_E^{(n)} - M_\theta^3 B_M^{(n)} - M_\theta^4 B_E^{(n)} &= 0, \\
M_\varphi^1 A_M^{(n)} + M_\varphi^2 A_E^{(n)} - M_\varphi^3 B_M^{(n)} - M_\varphi^4 B_E^{(n)} &= 0.
\end{aligned} \tag{5.2.70}$$

The coefficients are denoted by the vectors

$$A_M^{(n)} = \begin{pmatrix} A_M^{(n)}(1,-1) \\ A_M^{(n)}(1,0) \\ \vdots \\ A_M^{(n)}(l,m) \\ A_M^{(n)}(l,m+1) \\ \vdots \\ A_M^{(n)}(l+1,m) \\ \vdots \\ (l_{max}, l_{max}) \end{pmatrix}, \quad a^{(n)} = \begin{pmatrix} a_{l,-1}^{(n)} \\ a_{1,0}^{(n)} \\ \vdots \\ a_{l,m}^{(n)} \\ a_{l,m+1}^{(n)} \\ \vdots \\ a_{l+1,m}^{(n)} \\ \vdots \\ a_{l_{max},l_{max}}^{(n)} \end{pmatrix},$$

where the coefficients in the vector $A_E^{(n)}$, $B_M^{(n)}$ and $B_E^{(n)}$ are ordered in analogy to $A_M^{(n)}$. Thus eqn (5.2.70) becomes a matrix equation of the form

$$\begin{pmatrix} S^1 & S^2 & S^3 & S^4 \\ M_r^1 & M_r^2 & M_r^3 & M_r^4 \\ M_\theta^1 & M_\theta^2 & M_\theta^3 & M_\theta^4 \\ M_\varphi^1 & M_\varphi^2 & M_\varphi^3 & M_\varphi^4 \end{pmatrix} \begin{pmatrix} A_M^{(n)} \\ A_E^{(n)} \\ B_M^{(n)} \\ B_E^{(n)} \end{pmatrix} = \begin{pmatrix} 2\pi a^{(n)} \\ 0 \\ 0 \\ 0 \end{pmatrix}, \quad (5.2.71)$$

which can be solved by standard numerical techniques. Moreover it is only necessary to calculate the coefficients $A_E^{(n)}$ and $A_M^{(n)}$, since they fully determine the nth harmonic radiated field. This completes the solution of the nonlinear Mie scattering problem for arbitrary cluster shapes and arbitrary complex index of refraction.

Note that the matrix equations and the many equations presented in this chapter are necessary to describe the interesting physics. They are a consequence of the nonsphericity of the particles, implying boundary conditions that vary on the surface of the particle. This leads to the coupling of different electromagnetic modes. This is in contrast to the radiation by a spherical cluster. This feature is of general validity independent of the approximations made for the sources. In the case of spherical particles all equations will collapse to those of the spherical case.

In principle all modes must be summed in the series in eqn (5.2.63) to represent nonspherical particles by spherical coordinates. So one limitation of the theory is the finite number of (l, m) combinations that can be taken into account. As a consequence the matrices M and S in eqn (5.2.70) will become very large for strong deviations from the spherical shape. This purely numerical limitation is natural for particle shapes which cannot be described by coordinates that decouple the Helmholtz equation.

On the other hand the theory has the advantage of very compact form since the source for the nonlinear response is approximated just by $\sigma^{(n)} = (\sigma^{(1)})^n$ according to Östling et al.[22] and it only needs the boundary conditions for the electric fields from eqns (5.2.66) and (5.2.67) for the derivation of the radiated fields from the source. The theory is especially valid for particles of sizes in the Mie range since it takes into account completely the combinations of the multipoles in the higher harmonic case, which are characteristic for the radiation of a particle in the Mie range. The approximation $\sigma^{(n)} = (\sigma^{(1)})^n$ for the source term should be critically assessed, in particular for metals, if one is interested in details of the scattering profiles and effects due to the geometry of the cluster.

5.2.4 Magnetic nonlinear Mie theory

The incorporation of ferromagnetism in nonlinear Mie theory is of importance for the description of magnetic microspheres. Higher harmonic Mie scattering from these particles grown at or deposited on suitable (mostly insulating) substrate surfaces yields a competitive experimental method for the characterization of the geometrical, electronic, and magnetic properties of magnetic clusters in the Rayleigh and Mie size ranges. Nonlinear magnetic Mie scattering might be an optical tool, which is superior to the conventional Stern–Gerlach apparatus. The latter are difficult to handle, since they require time-of-flight measurements[36] of monodisperse clusters in the gas phase. By contrast, nonlinear optics is very sensitive to shape and curvature of nanoparticles and in particular to magnetism. Moreover, it yields ultrafast spectroscopic information about the electronic states.

In order to conceive a theory for nonlinear *magnetic* Mie scattering, which does not exist so far, one may start from the surface polarization and consequently the surface charge of eqns (5.2.XX) and (5.2.XX) yielding the sources for linear as well as nonlinear Mie scattering. Since in ferromagnetic particles the vector of the polarization become tilted due to the inequality of the number of majority and minority charge carriers, one may superimpose two different (for simplicity spherical) source terms for both types of electrons. Consequently, a tilted polarization will result and finally, after carrying out the full Mie theory for the microparticles, give an *asymmetry* of the angular scattering distributions in both the Rayleigh and Mie cases. This becomes clear from Fig. 5.8 for a size parameter $ka = 3$.

5.2.5 Magic angle in third harmonic generation

Mie scattering is an important and powerful tool for the analysis of the linear optical response by particles with sizes larger than or equal to the wavelength of the incident light. In this range the absolute value of the intensities as well as their angular dependence show a very complex behaviour caused by the high sensitivity of the excited electromagnetic surface modes on variations of the macroscopic parameters size, shape and refractive index. With the recent development of lasers generating ultrashort pulses with high intensities, higher

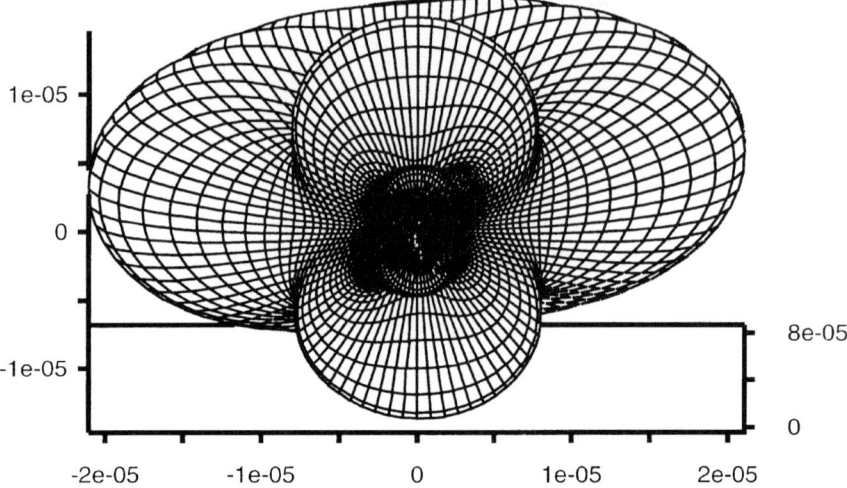

Fig. 5.8. Magnetic third harmonic Mie scattering for spheres of size $ka = 3$.

harmonic radiation becomes a novel spectroscopic probe for the characterization of microparticles.

Our results for SHG and THG from particles with size parameter $ka < 10$ showed that in this size range the linear Mie effects described by several authors[27,29] are even more pronounced in the higher harmonics. Upon increasing ka, the intensity scattered in the forward direction, which dominates the angular intensity profile in the linear case, is amplified for $ka < 10$. In all harmonics an increasing number of maxima appear next to the major directions in combination with an increased number of multipoles. In nonlinear Mie scattering the intensity profile of the higher harmonics is much more complicated due to the interference of multipoles of higher order than in the linear case.[23]

Here we present theoretical data of the angular resolved intensity profile of water droplets with size parameters $ka < 80$. Intensity profiles with stable angular dependence (called 'magic angles' in analogy to rainbow scattering) are detected over a wide range of size parameters in contrast to the well known behaviour in linear Mie scattering. In good agreement, both theory and the experiment by Kasparian et al.[37] yield that the absolute values of the intensities in forward and backward scattering are of the same order of magnitude, in contrast to linear Mie scattering. Moreover, with increasing size parameter the scattering patterns are stabilized rather than getting more and more complicated.

To take into account the fluctuation of particle sizes in our experiment we calculated the radiated intensities as an average over a wide ka range in steps of $\Delta(ka) = 0.025$. Though this value is much larger than the value necessary to obtain the complete fine structure in the linear case, $\Delta(ka) = 10^{-7}$,[38] we chose

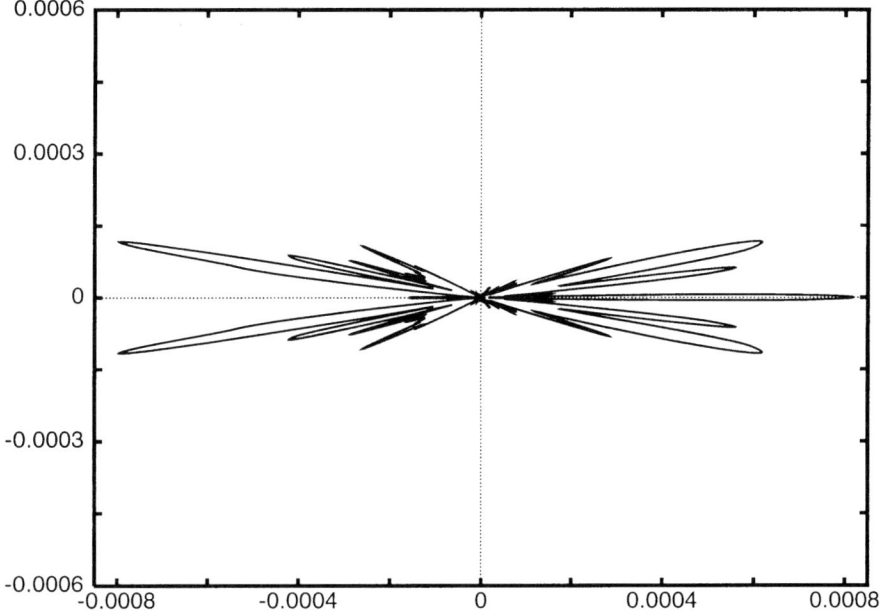

Fig. 5.9. Angular dependence of the THG intensities for a droplet size $ka \approx 76$.

this value in order to combine both high accuracy and a sufficient account of the indefiniteness of the parameters particle size and shape and laser wavelength.

Figure 5.9 shows the calculated angular dependence of the third harmonic generation for droplet sizes around 10.2 μm (size parameter $ka = 76$ with a width of $\Delta ka = 4$ for parallel polarization). Both theory and experiment show that the intensities in forward and backward directions are of the same order. This is an important difference to linear Mie scattering where the forward intensity increases remarkably with increasing size parameter. In addition the intensity data in backward scattering show, in good agreement with experiment, two peaks the size of which decreases with deviation from the backward direction.

The stability of the intensity patterns for ka around 75 is in sharp contrast to the increasing complexity of linear and higher harmonic Mie scattering for $ka < 10$. Two reasons are responsible for this behaviour. (i) First, the increased number of required multipoles (three times as many as in the linear case for the same ka) does not lead to a much more complex structure with a great number of maxima at more or less randomly distributed angles and values but rather stabilizes the angles of the intensity maxima which are sharply peaked near the forward and backward directions. (ii) Secondly, averaging over a wide range of size parameters complying to a broad variation of the size parameters in one experimental data set does not destroy the pronounced structure of the angular dependent intensity profile. Another important fact is that the absolute values of the intensities in forward and backward directions are of the same order, in contrast to linear Mie scattering where the forward peak shows a strong increase with size parameter (for $ka = 80$ the ratio of the linear intensities in forward and backward directions is around 6700). This behaviour opens up new possibilities to apply THG for the characterization of particles in a size range between Rayleigh scattering and geometrical optics and gives new perspectives for the remote sensing of environmental pollution of stratosphere and troposphere.

5.3 Symmetry and nonlinear Kerr rotation

While in classical theory such as Mie theory the description of material properties (mostly by optical constants) is secondary to the calculation of material independent nonlinear optical and magneto-optical properties (such as the Mie enhancement for certain particle sizes), the microscopic theory of nonlinear optics (here second harmonic generation) and of the nonlinear magneto-optical Kerr effect at metallic surfaces essentially requires two ingredients that are tightly interconnected, viz.,

1. the symmetry of a particular crystal surface and the specification of the different polarization configurations of incident fundamental and reflected second harmonic light, and
2. the nonlinear susceptibility tensors involving band structure and optical transition matrix elements.

The ingredient (1) is obtained from group theory while quantum-mechanical response theory and electronic structure calculations provide the ingredient (2). In this section, we perform the symmetry analysis of second harmonic generation and nonlinear magneto-optics (in particular the nonlinear Kerr rotation). The required input from microscopic theory will be discussed in detail in sec. 5.4.

5.3.1 Polarization dependence of second harmonic generation

Motivation

Recently, the nonlinear optical response of metallic surfaces has been studied intensively. The theoretical analysis uses typically the free-electron approximation or the jellium model.[39,40] Clearly, this symmetry analysis must be extended for studying the nonlinear optical response of transition metals and for analysing d- vs. s-electron contributions to SHG. As a result of the different localization lengths of s and d wavefunctions and the associated different sensitivities to symmetry breaking at surfaces and interfaces, we find that the wavelength and polarization dependences of the SH light on the incoming light are rather different for noble metals such as Cu, Ag, or Au and for transition metals such as Ni, Co, or Fe, in particular at lower frequencies.

Theory

The calculation of the SHG fields reflected from metal surfaces involves (i) the determination of the nonlinear polarization $P^{(2)}(2\omega)$, which acts as a source term in the classical wave equation (5.2.8), in terms of the incident photon field (see eqn (5.2.7)) and (ii) the solution of the latter for the reflected field radiated by the source. While step (i) requires the computation of the nonlinear susceptibilities, step (ii) is performed within classical electrodynamics taking into account the proper boundary conditions for surfaces (in analogy to those for spheres, which have been discussed in sec. 5.2). Usually one follows the procedure by Sipe *et al*[41] and starts with a radiating sheet of polarization $P^{(2)}(2\omega)$ positioned infinitesimally above the surface. The nonlinear field $E^{(2\omega)}$ results then from the superposition of the field directly radiated into vacuum with the fields radiated first downwards to the surface and getting reflected there before radiating into the vacuum. This procedure yields the reflected SHG fields as the nonlinear susceptibility, sandwiched between the classical linear electrodynamic factors (consisting of Fresnel and transmission coefficients) at frequencies ω and 2ω (of course, times the incident field squared). The coefficients at frequency ω enter squared (see eqn (5.3.17)). To be specific, we start here from the result for the reflected light at frequency 2ω given in a compact form by Böhmer *et al*.[42-44] for C_{4v} symmetry without magnetization as

$$E^{(2\omega)}(\Phi, \varphi) = 2\mathrm{i}\left(\frac{\omega}{c}\right)|E_0^{(\omega)}|^2 \begin{pmatrix} A_p F_c \cos\Phi \\ A_s \sin\Phi \\ A_p N^2 F_s \cos\Phi \end{pmatrix}$$

$$\times \begin{pmatrix} 0 & 0 & 0 & | & 0 & \chi^{(2)}_{xzx} & 0 \\ 0 & 0 & 0 & | & \chi^{(2)}_{xzx} & 0 & 0 \\ \chi^{(2)}_{zxx} & \chi^{(2)}_{zxx} & \chi^{(2)}_{zzz} & | & 0 & 0 & 0 \end{pmatrix} \begin{pmatrix} f_c^2 t_p^2 \cos^2\varphi \\ t_s^2 \sin^2\varphi \\ f_s^2 t_p^2 \cos^2\varphi \\ 2f_s t_p t_s \cos\varphi \sin\varphi \\ 2f_c f_s t_p^2 \cos^2\varphi \\ 2f_c t_p t_s \cos\varphi \sin\varphi \end{pmatrix}. \quad (5.3.1)$$

Here, Φ and φ denote the angles of polarization of the reflected frequency-doubled and of the incident light (see Fig. 5.10). $f_{c,s}$ are Fresnel coefficients and $t_{s,p}$ are the linear transmission coefficients. The complex indices of refraction at frequencies ω and 2ω are $n = n_1 + \mathrm{i}k_1$ and $N = n_2 + \mathrm{i}k_2$. The projections of the incident wavevector $\mathbf{k}$ on the spatial coordinates inside the medium are $f_s = \sin\theta/n$ and $f_c = \sqrt{1-f_s^2}$. The corresponding quantities for the reflected SHG light are $F_s = \sin\Theta/N$ and $F_c = \sqrt{1-F_s^2}$, where θ and Θ denote the angle of incidence of the incoming light and the angle of reflection of the

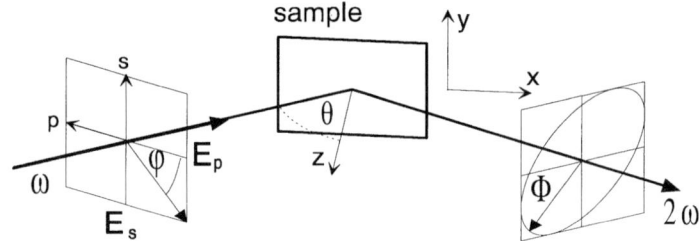

Fig. 5.10. The polarization and geometry of the incoming ω and reflected 2ω light, respectively. φ and Φ are the polarizations of the incident light and the reflected frequency-doubled photons. $\varphi = 0°$ corresponds to p polarization and $\varphi = 90°$ to s polarization. θ denotes the angle of incidence. The crystal axes x and y are in the crystal surface plane whereas z is parallel to the surface normal.

reflected SHG light, respectively. The linear transmission coefficients are given by[41–43]

$$t_p = \frac{2\cos\theta}{n\cos\theta + f_c}, \qquad t_s = \frac{2\cos\theta}{\cos\theta + nf_c},$$
$$T_p = \frac{2\cos\Theta}{N\cos\Theta + F_c}, \qquad T_s = \frac{2\cos\Theta}{\cos\Theta + NF_c}. \qquad (5.3.2)$$

The corresponding amplitudes A_p and A_s in eqn (5.3.1) are

$$A_p = \frac{2\pi T_p}{\cos\Theta}, \quad \text{and} \quad A_s = \frac{2\pi T_s}{\cos\Theta}, \qquad (5.3.3)$$

respectively. From eqn (5.3.1) one gets directly the intensity $I(\Phi, \varphi)$ of the reflected SHG light: $I(\Phi, \varphi) = |E^{(2\omega)}(\Phi, \varphi)|^2$. Thus, one obtains for p-polarized SHG with vacuum permittivity ε and vacuum permeability μ_0 for α-polarized incident light

$$I_\alpha(\text{p-SH}) = 2\sqrt{\frac{\varepsilon_0}{\mu_0}} \left|\left(2i\frac{\omega}{c}\right)\right|^2 |E_0(\omega)|^4$$
$$\times \left| A_p \{ [F_c \chi^{(2)}_{xzx} 2f_c f_s + N^2 F_s (\chi^{(2)}_{zxx} f_c^2 + \chi^{(2)}_{zzz} f_s^2)] t_p^2 \cos^2\varphi \right.$$
$$\left. + N^2 F_s \chi^{(2)}_{zxx} t_s^2 \sin^2\varphi \} \right|^2, \qquad (5.3.4)$$

and for s-polarized SHG

$$I_\alpha(\text{s-SH}) = 2\sqrt{\frac{\varepsilon_0}{\mu_0}} \left|\left(2i\frac{\omega}{c}\right)\right|^2 |E_0(\omega)|^4 \left| A_s \chi^{(2)}_{xzx} 2f_s t_p t_s \cos\varphi \sin\varphi \right|^2. \qquad (5.3.5)$$

Here, $I_\alpha(\beta\text{-SH})$ denotes the intensity of β-polarized reflected light at frequency

2ω resulting from α-polarized incoming light ($\alpha = p, s$, $\varphi = \pi/4$). In particular, one gets for p-polarized incident and p-polarized reflected SHG light

$$I_p(\text{p-SH}) = 2\sqrt{\frac{\varepsilon_0}{\mu_0}} \left|\left(2\frac{\omega}{c}\right)\right|^2 |E_0(\omega)|^4$$

$$\times \left| A_p t_p^2 [F_c \chi_{xzx}^{(2)} 2f_c f_s + N^2 F_s (\chi_{zxx}^{(2)} f_c^2 + \chi_{zzz}^{(2)} f_s^2)] \right|^2, \quad (5.3.6)$$

and for s-polarized incident and p-polarized reflected SHG light

$$I_s(\text{p-SH}) = 2\sqrt{\frac{\varepsilon_0}{\mu_0}} \left|\left(2\frac{\omega}{c}\right)\right|^2 |E_0(\omega)|^4 \left| A_p N^2 F_s \chi_{zxx}^{(2)} t_s^2 \right|^2, \quad (5.3.7)$$

and furthermore for 'mixed'-polarized (45° polarization) incident and s-polarized SHG light

$$I_{\varphi=\pi/4}(\text{s-SH}) = 2\sqrt{\frac{\varepsilon_0}{\mu_0}} \left|\left(\frac{\omega}{c}\right)\right|^2 |E_0(\omega)|^4 \left| A_s \chi_{xzx}^{(2)} 2f_s t_p t_s \right|^2. \quad (5.3.8)$$

The response in the latter two configurations is determined by one single tensor element $\chi_{zxx}^{(2)}$ and $\chi_{xzx}^{(2)}$, respectively. Then we get for an incidence angle of 45° for the ratio $I_p(\text{p-SH})/I_s(\text{p-SH})$ the result

$$\frac{I_p(\text{p-SH})}{I_s(\text{p-SH})} = \left| \frac{(F_c \chi_{xzx}^{(2)} 2f_c f_s + N^2 F_s \chi_{zzz}^{(2)} f_s^2)^2 t_p^2}{N^2 F_s \chi_{zxx}^{(2)} t_s^2} \right|^2, \quad (5.3.9)$$

where the contribution form $\chi_{zxx}^{(2)}$ has been neglected in the numerator. Note that this tensor element has been experimentally determined from $I_{\varphi=\pi/4}(\text{s-SH})$ and has been found to be at least three times smaller than the other two tensor elements (see below). Equation (5.3.8) holds also for noble metals.

It follows now from eqn (5.3.9) that s and d electrons contribute rather differently to the intensity of SHG light.

First, for noble metals like Cu, Ag and Au, one gets using eqn (5.3.7) due to $\chi_{ijl}^{(2)} \sim \langle s|i|s\rangle \langle s|j|s\rangle \langle s|l|s\rangle$ and $\langle s|x|s\rangle$ that

$$I_s(\text{p-SH}) = 0,$$

if the light frequency ω is such that $\hbar\omega < (\varepsilon_F - \varepsilon_d^{\max})$, implying that no transitions involving d states occur. Note, $\chi_{zxx}^{(2)}$ in eqn (5.3.7) involves the matrix elements $\langle s|x|s\rangle$, and these are nearly zero. Hence, for noble metals

$$\frac{I_s(\text{p-SH})}{I_p(\text{p-SH})} \to 0, \quad (5.3.11)$$

if the optical response does not involve d-state transitions. Thus, eqns (5.3.XX) and (5.3.XX) suggest a very sensitive test of the quantity $(\varepsilon_F - \varepsilon_d^{\max})$ and of many-body contributions to this. Comparison with results from band-structure calculations may help to identify correlation effects.

If the frequency ω increases, then for $\hbar\omega \geq (\varepsilon_F - \varepsilon_d^{max})$ also s → d and d → d transitions contribute and

$$I_s(\text{p-SH}) \sim \left|\sum_{ijl} t_s^2 F_s \chi_{zxx}^{(2)d}\right|^2 + \left|\sum_{ijl} t_s^2 F_s \chi_{zxx}^{(2)s}\right|^2 + \cdots. \quad (5.3.12)$$

Hence, for increasing frequency ω, $I_s(\text{p-SH})$ is no longer (nearly) zero, increases, and $I_s(\text{p-SH}) \to I_p(\text{p-SH})$ as for transition metals.

Secondly, for transition metals like Fe, Co, Ni and Cr, we estimate, in contrast to noble metals, for all frequencies ω that

$$I_p(\text{p-SH}) \geq I_s(\text{p-SH}). \quad (5.3.13)$$

For example, for the Ni(001) surface, using $\hbar\omega = 2.056\,\text{eV}$ ($\lambda = 603\,\text{nm}$) and for the optical constants[28] $n_1 = 1.98$, $k_1 = 3.92$, $n_2 = 2.02$ and $k_2 = 2.18$, we find using eqn (5.3.9) for $\theta = 45°$

$$0.75 \leq \frac{I_p(\text{p-SH})}{I_s(\text{p-SH})} \leq 6.7, \quad (5.3.14)$$

if we allow the relative phase of the tensor elements $\chi_{zzz}^{(2)}$ and $\chi_{xzx}^{(2)}$ to vary between 0 and 2π. This result gives already a reasonable account of the experimentally observed ratio of 1.25.

Note that the estimate of the ratio $I_p(\text{p-SH})/I_s(\text{p-SH})$ was obtained only on the basis of symmetry considerations for the matrix elements in $\chi_{ijl}^{(2)}$ and by approximating these by atomic-orbital matrix elements. Of course, for discussing systematically the nonlinear optical response of different noble and transition metals, we have to include density-of-states effects. How much this matters can be seen from results for $\chi_{zxx}^{(2)}/\chi_{xzx}^{(2)}$. This ratio should be equal to unity if the density of states plays no role. This corresponds to the case of perfect Kleinman symmetry.[45] However, note that in $\chi_{ijl}^{(2)}$ the indices j,l refer to ω photon transitions, while i belongs to the 2ω photon transition. This fact can be rather important, for example, for analysing differences between noble metals and transition metals. For transition metals $\chi_{xjl}^{(2)}$ is rather unfavourable for the SHG yield, since s → s transitions at frequency 2ω may occur and reduce the intensity $I_{\varphi=\pi/4}(\text{s-SH})$ due to $\langle s|x|s\rangle \ll \langle d|x|d\rangle$—see for illustration Figs 5.11(c) and (d). In contrast, for noble metals, $\chi_{xjl}^{(2)}$ is favourable for the SHG yield, if the transitions at frequency ω involve s → s transitions only and if d → s transitions at frequency 2ω come into play. The situation is illustrated in Figs 5.11(a) and (b). Thus, the deviation from Kleinman's symmetry of the ratio $\chi_{zxx}^{(2)}/\chi_{xzx}^{(2)}$ is interesting, a fingerprint of density-of-states effects, and reflects to some extent the contribution of d electrons to the optical transitions and the degree of localization of the Fermi level electrons. Experimentally this ratio ranges from less than 0.013 for Al (with no d electrons) over 0.085 for Cu (where the upper d-band edge is 1.9 eV below the Fermi level) and 0.33 for Au (where the upper d-band edge is 1.8 eV below the Fermi level) to 0.35 for Ni[42] (with a large

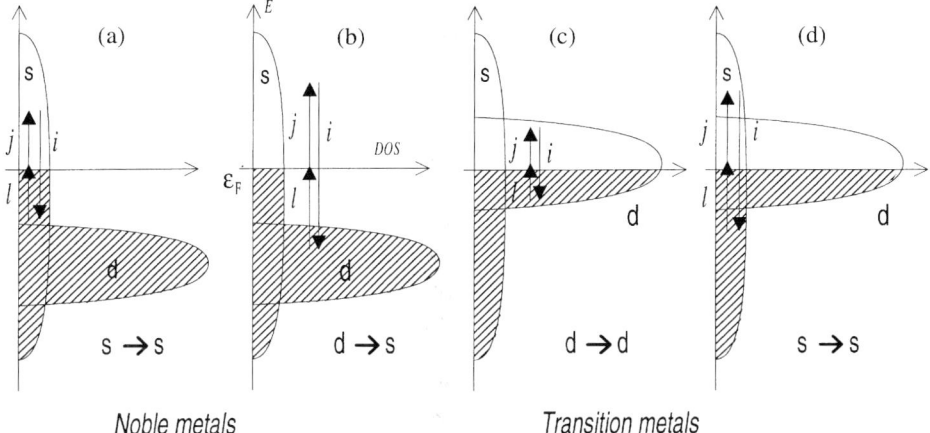

Fig. 5.11. Illustration of second harmonic generation from noble metals ((a) and (b)) and transition metals ((c) and (d)). For noble metals, in case (a) no d electrons can be optically excited, which in contrast is possible in case (b). In case (c) for transition metals, predominantly d electrons contribute to the SHG yield, whereas in case (d) the excitation starts from the s band.

density of d electrons at the Fermi level). For the determination of the ratio $\chi^{(2)}_{zxx}/\chi^{(2)}_{xzx}$, see remarks below.

As can be seen from the large range of values for the ratio $I_p(\text{p-SH})/I_s(\text{p-SH})$ in eqn (5.3.14), it is very important for a quantitative analysis to include not only the symmetry properties of the dipole matrix elements, but also the phases, since the tensor elements $\chi^{(2)}_{ijl}$ are complex quantities. As a result, important interference contributions may occur and change the results obtained from a coherent superposition of the partial intensities given by $|\chi^{(2)}_{ijl}|$ alone. For a full analysis one has to perform an electronic calculation of $\chi^{(2)}_{ijl}$ as done previously by Pustogowa et al.[46,47] To compare theory and experiment, the latter has to be performed for several optical geometries to allow for a complete determination of the absolute values and phases of $\chi^{(2)}_{ijl}$.

To see how the two relative phases between the three tensor elements $\chi^{(2)}_{zzz}$, $\chi^{(2)}_{xzx}$ and $\chi^{(2)}_{zxx}$ entering $I_\alpha(\text{p-SH})$ come into play, we rewrite eqn (5.3.4) as

$$I_\alpha(\text{p-SH}) = \left|\left(2i\frac{\omega}{c}\right)\right|^2 |E_0(\omega)|^4$$
$$\times \left| A_p \chi^{(2)}_{zzz} \left\{ \left[F_c \frac{\chi^{(2)}_{xzx}}{\chi^{(2)}_{zzz}} 2f_c f_s + N^2 F_s \left(\frac{\chi^{(2)}_{xzx}}{\chi^{(2)}_{zzz}} f_c^2 + f_s^2 \right) \right] t_p^2 \cos^2 \varphi \right.$$
$$\left. + N^2 F_s \frac{\chi^{(2)}_{zxx}}{\chi^{(2)}_{zzz}} t_s^2 \sin^2 \varphi \right\} \right|^2,$$

with

$$\frac{\chi^{(2)}_{xzx}}{\chi^{(2)}_{zzz}} = \left|\frac{\chi^{(2)}_{xzx}}{\chi^{(2)}_{zzz}}\right| e^{i\varphi_1} \quad \text{and} \quad \frac{\chi^{(2)}_{zxx}}{\chi^{(2)}_{zzz}} = \left|\frac{\chi^{(2)}_{zxx}}{\chi^{(2)}_{zzz}}\right| e^{i\varphi_2}. \quad (5.3.16)$$

Now, the two absolute ratios $|\chi^{(2)}_{xzx}/\chi^{(2)}_{zzz}|$ and $|\chi^{(2)}_{zxx}/\chi^{(2)}_{zzz}|$ and the two relative phases φ_1 and φ_2 remain to be determined. As can be seen from eqns (5.3.7) and (5.3.8), the intensities $I_{\varphi=\pi/4}(\text{s-SH})$ and $I_s(\text{p-SH})$ depend exclusively on one single nonlinear tensor element each, $\chi^{(2)}_{xzx}$ and $\chi^{(2)}_{zxx}$, respectively. Thus, the intensity ratio

$$\frac{I_{\varphi=\pi/4}(\text{s-SH})}{I_s(\text{p-SH})},$$

which may be taken from experiment, gives directly the absolute ratio $|\chi^{(2)}_{xzx}/\chi^{(2)}_{zxx}|$. Using for example for $I_{\varphi=\pi/4}(\text{s-SH})/I_s(\text{p-SH})$ the experimental value and the appropriate linear optical constants of Ni, one gets

$$|\chi^{(2)}_{xzx}/\chi^{(2)}_{zxx}| = 1/0.35 \approx 2.8. \qquad (5.3.17)$$

Similarly, we calculate (1/0.013) for Al, (1/0.085) for Cu, and (1/0.33) for Au.

To obtain the remaining relative phases φ_1 and φ_2, and the absolute ratio $|\chi^{(2)}_{xzx}/\chi^{(2)}_{zzz}|$, one may use the results for the p-polarized SHG yield for incident p-polarized light and finally obtain for Ni

$$\chi^{(2)}_{xzx} = \chi^{(2)}_{zzz} e^{i1.95\pi} \quad \text{and} \quad \chi^{(2)}_{zxx} = \chi^{(2)}_{zzz} e^{i0.505\pi}; \qquad (5.3.18)$$

see results presented in Fig. 5.12. Thus, we have completely determined the absolute ratios and the relative phases of all nonvanishing elements of the nonlinear tensor $\chi^{(2)}_{ijl}$ for surfaces of fcc crystals.

However, it is important to note that only the combination of the symmetry properties of the (real) transition matrix elements *and* the density-of-states effects will account for a quantitative understanding of the SHG yield. Furthermore, since the tensor elements $\chi^{(2)}_{ijl}$ are complex quantities, their phases cause important interference contributions to the coherent superposition of the partial SHG yields resulting from the single tensor elements alone. An *electronic* theory is required for the calculation of the full complex SHG tensor $\chi^{(2)}_{ijl}$ to explain the experiments quantitatively and in detail. It is clear from eqns (5.3.1)–(5.3.8) that experiments have to be carried out for several optical geometries to sort out the tensor elements for comparison with theory and to allow for a complete determination of their *absolute values* and *phases*.

The physics described by our theory is evident also from the results shown in Table 5.3. Here, we have estimated the dipole matrix elements which determine largely $\chi^{(2)}_{ijl}$ by using atomic s(p)- and d-electron wavefunctions. Decomposing $\chi^{(2)}_{ijl}$ according to the character of the wavefunctions, it follows that

$$\chi^{(2,s)}_{zzz} \gg \chi^{(2,s)}_{zxx}, \qquad (5.3.19)$$

since $\langle s|x|s \rangle \approx 0$, and

$$\chi^{(2,d)}_{zzz} \gtrsim \chi^{(2,d)}_{zxx}, \qquad (5.3.20)$$

since $\langle d|z|d \rangle \gtrsim \langle d|x|d \rangle$. Generally, for transition metals, $\chi^{(2,d)}_{ijl}$ is largest by far. For noble metals, $\chi^{(2,d)}_{ijl}$ begins to dominate over $\chi^{(2,s)}_{ijl}$ as $\hbar\omega$ increases and becomes larger than the energy distance of the d states from the Fermi energy

Table 5.3 Nondipole transition matrix elements at the surface involved in the calculation of $\chi^{(2)}_{ijl}$ and $I(\text{SH})$, respectively

Matrix	Value
$\langle 4s\mid x\pm iy\mid 4s\rangle$	0
$\langle 4s\mid z\mid 4s\rangle$	$-\dfrac{a_B}{Z}$
$\langle 4p_{m=\pm 1}\mid z\mid 4p_{m=\pm 1}\rangle$	$-\dfrac{69}{16}\dfrac{a_B}{Z}$
$\langle 4p_{m=0}\mid z\mid 4p_{m=0}\rangle$	$-\dfrac{69}{8}\dfrac{a_B}{Z}$
$\langle 3d_{m=\pm 2}\mid z\mid 3d_{m=\pm 2}\rangle$	$-\dfrac{105}{64}\dfrac{a_B}{Z}$
$\langle 3d_{m=\pm 1}\mid z\mid 3d_{m=\pm 1}\rangle$	$-\dfrac{105}{32}\dfrac{a_B}{Z}$
$\langle 3d_{m=0}\mid z\mid 3d_{m=0}\rangle$	$-\dfrac{105}{32}\dfrac{a_B}{Z}$
$\langle 3d_{m=0}\mid z\mid 4s\rangle$	$0.7005\dfrac{\sqrt{5}}{16}\dfrac{a_B}{Z}$
$\langle 4p_{m=\pm 1}\mid x\pm iy\mid 4p_{m=0}\rangle$	$-\dfrac{69}{8\sqrt{2}}\dfrac{a_B}{Z}$
$\langle 4p_{m=0}\mid x\pm iy\mid 4p_{m=\mp 1}\rangle$	$-\dfrac{69}{8\sqrt{2}}\dfrac{a_B}{Z}$
$\langle 3d_{m=\pm 2}\mid x\pm iy\mid 3d_{m=\pm 1}\rangle$	$-\dfrac{105}{32}\dfrac{a_B}{Z}$
$\langle 3d_{m=\mp 1}\mid x\pm iy\mid 3d_{m=\mp 2}\rangle$	$-\dfrac{105}{32}\dfrac{a_B}{Z}$
$\langle 3d_{m=\pm 2}\mid x\pm iy\mid 4s\rangle$	$\dfrac{0.7005}{8}\sqrt{\dfrac{15}{2}}\dfrac{a_B}{Z}$
$\langle 3d_{m=\pm 1}\mid x\pm iy\mid 3d_{m=0}\rangle$	0
$\langle 3d_{m=0}\mid x\pm iy\mid 3d_{m=\mp 1}\rangle$	0

ε_F, $\hbar\omega > (\varepsilon_F - \varepsilon_d^{\max})$. Furthermore, approximately $I_p(\text{p-SH})$ is essentially given by $\chi^{(2)}_{zzz}$ and $\chi^{(2)}_{xzx}$, whereas $I_s(\text{p-SH})$ is determined by $\chi^{(2)}_{xzx}$ alone. It is the case of s-polarized SH light $I_s(\text{s-SH}) = I_s(\text{p-SH}) = 0$. And $I_{\varphi=\pi/4}(\text{s-SH})$ is given by $\chi^{(2)}_{xzx}$ alone. Hence, using Table 5.4, we find quite generally that:

(1) for *noble metals* such as Cu, Au, Ag

$I_p(\text{p-SH}) \gg I_s(\text{p-SH})$, if $\hbar\omega < (\varepsilon_F - \varepsilon_d^{\max})$,

$I_p(\text{p-SH}) \approx |\chi^{(2,s)}_{ijl}|^2 + |\chi^{(2,\text{int})}_{ijl}|^2$, if $\hbar\omega < (\varepsilon_F - \varepsilon_d^{\max})$,

$I_p(\text{p-SH}) \to I_p^{\text{trans.met.}}(\text{p-SH})$, $\hbar\omega$ large;

Table 5.4 Values for the p-polarized partial SHG yields $I_{s,p}^{(s,d,int)}$(p-SH) of d and s electrons and interference contributions

Partial SHG yield	Value
$I_{p\text{-in}}^{(2)(d)}$(p-SH)	$25\left(\dfrac{105}{32}\right)^6\left(\dfrac{a_B}{Z}\right)^6$
$I_{s\text{-in}}^{(2)(d)}$(p-SH)	$9\left(\dfrac{105}{32}\right)^6\left(\dfrac{a_B}{Z}\right)^6$
$I_{p\text{-in}}^{(2)(s)}$(p-SH)	$\left(\dfrac{a_B}{Z}\right)^6$
$I_{s\text{-in}}^{(2)(s)}$(p-SH)	0
$I_{p\text{-in}}^{(2)(int)}$(p-SH)	$0.015152\left(\dfrac{a_B}{Z}\right)^6$
$I_{s\text{-in}}^{(2)(int)}$(p-SH)	$0.052908\left(\dfrac{a_B}{Z}\right)^6$

(2) for *transition metals* such as Ni, Co, Fe, Cr

$$I(\text{p-SH}) \approx |\chi_{ijl}^{(2,d)}|^2,$$
$$I_p(\text{p-SH}) \geq 2I_s(\text{p-SH}).$$

This completes the theoretical analysis of the polarization dependence of the SHG yield and shows how it reveals the electronic structure at surfaces and interfaces.

To demonstrate the different polarization dependences of the SHG yield for noble metals and transition metals, we show in Fig. 5.12 numerical results for the polarization dependence of the p-polarized SHG yield I_φ(p-SH) for Cu and Ni at an angle of incidence $\theta = 45°$ corresponding to the experimental situation.[42,43] The absolute ratios and relative phases of the tensor elements are listed in Table 5.5 as well as the used linear optical constants at frequencies ω and 2ω. First, we refer to results for Ni given by the topmost curve. These are typical for *transition metals*. For Ni we used an incident light wavelength of 603 nm. The results are in perfect agreement with experiment.[42,43] Despite the fact that we have fitted the phases φ_1 and φ_2 to experiment, this is remarkable, since the ratio $|\chi_{xzx}^{(2)}/\chi_{zzz}^{(2)}|$ has been obtained using only the matrix elements calculated with atomic wavefunctions. As mentioned the line shape is characteristic for transition metals. Similar results are expected for Cr and for Fe and Co. The SHG yield for s- and p-polarized incident light is nearly the same in transition metals due to the *isotropy* of the transition matrix elements for d electrons

$$\langle d|z|d\rangle \approx \langle d|x|d\rangle.$$

Slight intensity minima occur at polarization angles of 45° and 135°. These arise

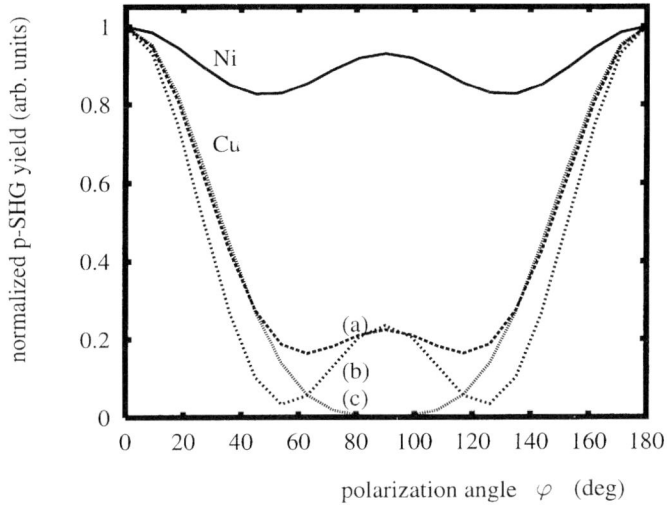

Fig. 5.12. Numerical results for the SHG yield $I_\varphi(\text{p-SH})$ for Ni and for Cu, curves (a), (b) and (c) using eqn (5.3.16). φ refers to the polarization of the incident light. For input parameters see text and Table 5.5.

from the interference of the tensor elements $\chi^{(2)}_{ijl}$ and the Fresnel factors; see eqn (5.3.4). The intensity profile is always symmetric around $\varphi = 90°$ and for transition metals approximately symmetric around $\varphi = 45°$ and $\varphi = 135°$.

Secondly, to demonstrate the typical frequency dependence of the SHG yield as a function of incident polarization for *noble metals* (NM), results for Cu are given in Fig. 5.12 by the curves (a), (b), (c). Note that the incident light polarization dependence of the SHG yield is found to be very different for noble metals such as Cu than for transition metals (TM). The $\varphi = 90°$ symmetry survives. The p-polarized SHG yield falls off to one-fourth or less at incident s

Table 5.5 Absolute values and relative phases of the nonlinear tensor elements $\chi^{(2)}_{ijl}$ and optical constants n_i, k_i of Ni and Cu corresponding to the four curves of Fig. 5.12 (λ denotes the wavelength of the incident light). The optical constants for Ni are taken from ref. 10 and for Cu from ref. 14

	Ni ($\lambda = 603$ nm)	Cu ($\lambda = 516.6$ nm) curve (a)	Cu ($\lambda = 652.5$ nm) curve (b)	Cu ($\lambda = 652.5$ nm) curve (c)
$|\chi^{(2)}_{zzx}/\chi^{(2)}_{zzz}|$	0.60	0.363	0.363	0.0363
$|\chi^{(2)}_{zxx}/\chi^{(2)}_{xzx}|$	0.35	0.085	0.085	0.085
phase φ_1	1.945π	1.02π	1.02π	1.02π
phase φ_2	0.505π	1.34π	1.34π	1.34π
$n_1(\omega)$	1.98	1.12	0.214	0.214
$k_1(\omega)$	3.92	2.60	3.67	3.67
$n_2(2\omega)$	2.02	1.53	1.34	1.34
$k_2(2\omega)$	2.18	1.71	1.81	1.81

polarization compared to its (normalized) value at incident p polarization, due to the strong *anisotropy* of the transition matrix elements for s electrons

$$\langle s|z|s\rangle \approx \langle s|x|s\rangle \approx 0.$$

The Cu curves (a) and (c) refer to wavelengths exciting and not exciting the Cu d electrons, respectively. The results are in very good agreement with experimental ones for the frequency-dependent SHG yield obtained for Cu by Petrocelli et al.[48] While curve (a), which is for an incident light wavelength of 516.6 nm, still shows shallow minima at 63° and 117°, curve (c), which is for an incident light wavelength of 652.5 nm, falls off monotonically to zero at incident s polarization (90°). The minima are due to the phases of the complex nonlinear tensor elements giving rise to interference contributions as well as due to the complex optical constants. These interferences will be absent if the d band of Cu can no longer be excited for larger wavelengths. Note the shape of curve (a) would also be characteristic of Au at 603 nm as is observed in the experiment by Böhmer et al.[42,43] These results indicate that

$$(I_s(\text{p-SH}))_{\text{NM}} \xrightarrow{\omega} (I_s(\text{p-SH}))_{\text{TM}}$$

for increasing frequency ω as claimed previously on theoretical grounds.

Note that experimentally $I_s(\text{p-SH})$ is larger for Au than for Cu. The experiments[42,43] are performed for $\hbar\omega = 2.056$ eV. Then, $\hbar\omega > (\varepsilon_F - \varepsilon_d^{\max})$, according to band-structure calculations by Papaconstantopoulos,[49] yielding $(\varepsilon_F - \varepsilon_d^{\max}) \approx 1.8$ eV for Au and 1.9 eV for Cu. Hence, one would expect for both metals Cu and Au already larger values for $I_s(\text{p-SH})$. Owing to many-body band narrowing $(\varepsilon_F - \varepsilon_d^{\max})$ is for both metals somewhat larger than what results from band-structure calculations. However, since $(\varepsilon_F - \varepsilon_d^{\max})_{\text{Au}} < (\varepsilon_F - \varepsilon_d^{\max})_{\text{Cu}}$ one also expects in agreement with experiment

$$(I_s(\text{p-SH}))_{\text{Au}} > (I_s(\text{p-SH}))_{\text{Cu}}$$

at energy $\hbar\omega = 2.056$ eV.

Curve (b) corresponds also to Cu at a wavelength of 652.5 nm as curve (c). For this curve (b), however, only the optical constants have been adapted to this wavelength, whereas the nonlinear tensor elements of wavelength 516.6 nm were kept as in curve (a). At $\varphi = 90°$, this curve reaches the same height as the Cu curve (a), but it shows deep minima at $\varphi = 54°$ and $\varphi = 126°$. Thus, this curve (b) clearly demonstrates that for the correct theoretical description of the frequency and polarization dependence of the SHG yield the frequency dependence of the nonlinear tensor elements is as essential as the frequency dependence of the optical constants at frequencies ω and 2ω.

Thus, the polarization dependence of the SHG yield shows characteristic curve shapes which allow for a clear distinction between the electronic structure of noble-metal and transition-metal surfaces. As claimed, SHG is a very sensitive probe of the electronic structure.

One very interesting application of our theory for the polarization dependence of the SHG yield is to transition-metal oxides;[50] see sec. 5.3.5. Note that late 3d transition-metal oxides, in particular NiO, are charge-transfer insulators.

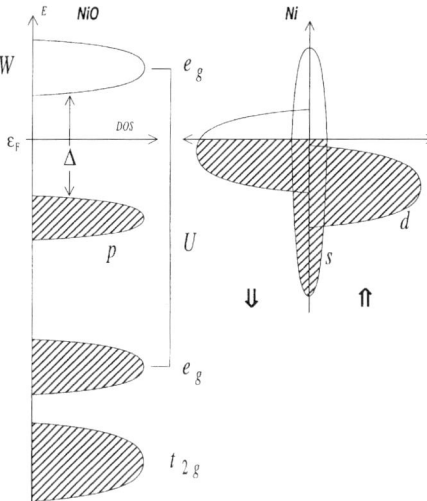

Fig. 5.13. Scheme of the electronic structure of NiO and Ni. W is the width of the upper Hubbard band, Δ is the charge-transfer gap, U denotes the on-site Coulomb repulsion, e_g and t_g are the two- and three-dimensional representations of d-electron states in cubic environment.

NiO has a gap of 4 eV separating the filled oxygen 2p band from the empty upper Hubbard portion of the Ni 3d band; see Fig. 5.13 for illustration. These transition-metal oxides are of particular interest due to the strong electron correlations. Thus, the frequency-dependent SHG could give some important information about the energy gap of the transition-metal oxides and the width of the upper Hubbard band and thus about correlations.

In our theory, we have neglected contributions to the SHG yield of higher than electric dipole order. However, we estimate them to be smaller by a factor of $\bar{\alpha}^2 \approx (1/137)^2 = 5 \times 10^{-5}$ for the magnetic dipole and of $(2\pi a_B/\lambda)^2 \approx 2.5 \times 10^{-7}$ for the electric quadrupole contribution. Here, $\bar{\alpha}$ denotes the fine-structure constant, and a_B is Bohr's radius. The resulting smallness cannot be compensated by the skin depth factor. The implied interface sensitivity of SHG is supported by the agreement of our theory with experiment.

Summarizing this symmetry analysis we state that the polarization dependence of the SHG signal already gives important information on the geometric and electronic structure of metallic surfaces. The characteristically different polarization dependence allows for a distinction between transition-metal and noble-metal surfaces although a more detailed analysis requires the additional information of the microscopic tensors. The latter yield the frequency dependence and the phases.

5.3.2 Polarization dependence of the nonlinear Kerr effect

Motivation
Here, we extend our previous theory for the polarization dependence of SHG at

noble-metal and in particular transition-metal surfaces by including the symmetry properties of NOLIMOKE imposed by *magnetism* and thus (i) calculate the polarization dependence of the nonlinear Kerr rotation, and (ii) determine the direction of the magnetization vector at interfaces of films and multilayers. In view of our previous results for the polarization dependence of SHG we expect an interesting dependence of the enhancement of the nonlinear Kerr rotation on the polarization and the easy axis. Note that the polarization is controlled by the matrix elements and thus it depends sensitively on the symmetry of the wavefunction.[51,52]

Theory

First we consider the *longitudinal configuration* with magnetization parallel to the interface, $\boldsymbol{M} \| \hat{\boldsymbol{x}}$, and where the optical plane is the xz plane. In this configuration, the nonlinear susceptibility tensor contains 10 different nonvanishing tensor elements with five of them being even under magnetization reversal and the other five being odd.[14] Thus, the symmetry breaking by the magnetization induces five more tensor elements and causes the nonmagnetic ones to become all different. The nonlinear susceptibility tensor is given explicitly by

$$\begin{pmatrix} 0 & 0 & 0 & | & 0 & \chi^{(2)}_{xzx} & \chi^{(2)}_{xxy} \\ \chi^{(2)}_{yxx} & \chi^{(2)}_{yyy} & \chi^{(2)}_{yzz} & | & \chi^{(2)}_{yyz} & 0 & 0 \\ \chi^{(2)}_{zxx} & \chi^{(2)}_{zyy} & \chi^{(2)}_{zzz} & | & \chi^{(2)}_{zyz} & 0 & 0 \end{pmatrix}. \qquad (5.3.21)$$

For the calculation of the nonlinear Kerr rotation it is convenient to assume p or s polarization of the incident light and to detect the rotated polarization plane in the reflected SH signal upon magnetization reversal. Hence, one needs to know the fields for the polarization combinations referring to the incoming and outgoing light: p → p, p → s, s → p and s → s. One obtains then for the longitudinal configuration for p-polarized SH light generated from p-polarized incident light

$$E^{(2\omega)}_p(\text{p-SH}) = 2i |E^{(\omega)}_0|^2 \, A_p t_p^2 [F_c \chi^{(2)}_{xzx} 2 f_c f_s + N^2 F_s (\chi^{(2)}_{zxx} f_c^2 + \chi^{(2)}_{zzz} f_s^2)], \qquad (5.3.22)$$

for p-polarized SH light generated from s-polarized incident light

$$E^{(2\omega)}_s(\text{p-SH}) = 2i |E^{(\omega)}_0|^2 \, A_p t_s^2 N^2 F_s \, \chi^{(2)}_{zyy}, \qquad (5.3.23)$$

for s-polarized SH light generated from p-polarized incident light

$$E^{(2\omega)}_p(\text{s-SH}) = 2i |E^{(\omega)}_0|^2 \, A_s t_p^2 (\chi^{(2)}_{yxx} f_c^2 + \chi^{(2)}_{yzz} f_s^2), \qquad (5.3.24)$$

and for s-polarized SH light generated from s-polarized incident light

$$E^{(2\omega)}_s(\text{s-SH}) = 2i |E^{(\omega)}_0|^2 \, A_s t_s^2 \chi^{(2)}_{yyy}. \qquad (5.3.25)$$

Note that *magnetism* occurs in all 10 nonvanishing elements of the tensor $\chi_{ijk}^{(2)}$ and in the complex indices of refraction at the fundamental and the SH frequency, $n(\omega)$ and $N(2\omega)$, respectively. The dominant nonlinear magneto-optical Kerr effect, however, results from $\chi_{ijk}^{(2)}$, in particular from the five tensor elements $\chi_{xxy}^{(2)}$, $\chi_{yxx}^{(2)}$, $\chi_{yyy}^{(2)}$, $\chi_{yzz}^{(2)}$ and $\chi_{zyz}^{(2)}$, which are odd upon magnetization reversal. Using these expressions for $E_\varphi^{(2\omega)}$ we find in the case of the longitudinal ($=$ meridional) Kerr configuration for the nonlinear Kerr rotation of p-polarized incoming light

$$\phi_{K,p}^{(2)} = \operatorname{Re} \frac{E_p^{(2\omega)}(\text{s-SH})}{E_p^{(2\omega)}(\text{p-SH})}$$

$$= \operatorname{Re} \frac{A_s}{A_p} \frac{\chi_{yxx}^{(2)} f_c^2 + \chi_{yzz}^{(2)} f_s^2}{F_c \chi_{xzx}^{(2)} 2 f_c f_s + N^2 F_s (\chi_{zxx}^{(2)} f_c^2 + \chi_{zzz}^{(2)} f_s^2)}, \quad (5.3.26)$$

and for s-polarized incoming light

$$\phi_{K,s}^{(2)} = \operatorname{Re} \frac{E_s^{(2\omega)}(\text{s-SH})}{E_s^{(2\omega)}(\text{p-SH})} = \operatorname{Re} \frac{A_s}{A_p} \frac{\chi_{yyy}^{(2)}}{N^2 F_s \chi_{zyy}^{(2)}}. \quad (5.3.27)$$

Note that the NOLIMOKE rotation measures the *electric field vectors* rather than intensities. Note, in the case of the transverse ($=$ equatorial) Kerr configuration ($\boldsymbol{M} \parallel \hat{\boldsymbol{y}}$, optical plane is the xz plane), that no Kerr rotation can be observed. Instead one measures an intensity change upon magnetization reversal,[16,53] whereas the total reflected SH intensity does not change upon magnetization reversal in the longitudinal configuration.

Secondly, we determined the Kerr rotation for the *polar configuration*, in which the magnetization is perpendicular to the surface. Now, the optical plane is again the xz plane. However, in contrast to the usual notion in linear optics, perpendicular incidence is not yet assumed, since SHG behaves differently for the nonlinear excitation in the interface plane and perpendicular to it. Linear optics, however, makes no such differentiation.[54] In the case of the polar configuration the nonlinear susceptibility has seven nonvanishing tensor elements, five of which are different. Three of these elements are even and two ($\chi_{xyz}^{(2)}$ and $\chi_{yzx}^{(2)}$) are odd in M. Note that the polar configuration is much more symmetric than the longitudinal Kerr configuration, thus causing more tensor elements to vanish and to be equal. In detail, the nonlinear susceptibility is given by

$$\begin{pmatrix} 0 & 0 & 0 & | & \chi_{xyz}^{(2)} & \chi_{xzx}^{(2)} & 0 \\ 0 & 0 & 0 & | & \chi_{xzx}^{(2)} & -\chi_{xyz}^{(2)} & 0 \\ \chi_{zxx}^{(2)} & \chi_{zxx}^{(2)} & \chi_{zzz}^{(2)} & | & 0 & 0 & 0 \end{pmatrix}. \quad (5.3.28)$$

The calculation of the fields for the polarization combinations p → p, p → s, s → p and s → s gives for p-polarized SH light generated from p-polarized incident light in the polar configuration

$$E_p^{(2\omega)}(\text{p-SH}) = 2i|E_0^{(\omega)}|^2 A_p t_p^2 [F_c \chi_{xxz}^{(2)} 2 f_c f_s + N^2 F_s (\chi_{zxx}^{(2)} f_c^2 + \chi_{zzz}^{(2)} f_s^2)], \quad (5.3.29)$$

for p-polarized SH light generated from s-polarized incident light

$$E_s^{(2\omega)}(\text{p-SH}) = 2i|E_0^{(\omega)}|^2 A_p N^2 F_s t_s^2 \chi_{zxx}^{(2)}, \quad (5.3.30)$$

for s-polarized SH light generated from p-polarized incident light

$$E_p^{(2\omega)}(\text{s-SH}) = -2i|E_0^{(\omega)}|^2 A_s \chi_{xyz}^{(2)} 2 f_c f_s t_p^2, \quad (5.3.31)$$

and finally for s-polarized SH light generated from s-polarized incident light

$$E_s^{(2\omega)}(\text{s-SH}) = 0. \quad (5.3.32)$$

Note that it makes no sense to consider the field $E_{\varphi=\pi/4}^{(2\omega)}(\text{s-SH})$ for both the polar and longitudinal Kerr configurations, since due to the magnetization one or even both of the two quantities $E_p^{(2\omega)}(\text{s-SH})$ and $E_s^{(2\omega)}(\text{s-SH})$ are nonzero in this case in contrast to the nonmagnetic case where both $E_p^{(2\omega)}(\text{s-SH})$ and $E_s^{(2\omega)}(\text{s-SH})$ vanish. Using the fields $E_\varphi^{(2\omega)}$ in the *polar* Kerr configuration, we obtain for the nonlinear magneto-optical rotation in the case of p-polarized incident light

$$\phi_{K,p}^{(2)} = -\text{Re} \frac{E_p^{(2\omega)}(\text{s-SH})}{E_p^{(2\omega)}(\text{p-SH})}$$

$$= \text{Re} \frac{A_s}{A_p} \frac{\chi_{xyz}^{(2)} 2 f_c f_s}{F_c \chi_{xxz}^{(2)} 2 f_c f_s + N^2 F_s (\chi_{zxx}^{(2)} f_c^2 + \chi_{zzz}^{(2)} f_s^2)} \quad (5.3.33)$$

and in the case of s-polarized incident light

$$\phi_{K,s}^{(2)} = \text{Re} \frac{E_s^{(2\omega)}(\text{s-SH})}{E_s^{(2\omega)}(\text{p-SH})} = \text{Re} \frac{A_s}{A_p} \frac{0}{N^2 F_s t_s^2 \chi_{zxx}^{(2)}}. \quad (5.3.34)$$

Note that in both the longitudinal and the polar Kerr configurations and for both p and s input polarizations the Kerr rotation contains only odd tensor elements in the numerator and only even tensor elements in the denominator as generally expected and as derived already by Pustogowa et al.[17] for not too large rotation angles. The dependence on the incident angle results here and in ref. 17 exclusively from the linear optical coefficients. Assuming $\chi_{yxx}^{(2)} < \chi_{yzz}^{(2)}$ and $\chi_{zxx}^{(2)} < \chi_{xzx}^{(2)} < \chi_{zzz}^{(2)}$, eqn (5.3.26) for the nonlinear Kerr rotation in the longitudinal geometry yields in agreement with ref. 17

$$\phi_{K,p}^{(2)} = \text{Re} \frac{A_s}{A_p N^2 F_s} \frac{\chi_{yzz}^{(2)}}{\chi_{zzz}^{(2)}} \approx \frac{1}{N \sin \theta} \frac{\chi_{yzz}^{(2)}}{\chi_{zzz}^{(2)}}, \quad (5.3.35)$$

if the same approximations are made. This fact is easily seen from the $\sin^2\Theta$ terms in the *even linear* tensor elements contributing to the denominator of $\phi_{K,p\text{-in}}^{(2)}$ and cancelling the $\sin\Theta$ originating from the *odd linear* susceptibility tensor elements. In ref. 17, however, also the magnetism in the linear optical factors and the detailed form of the nonlinear Frenel coefficients belonging to the particular choice of the configuration have been included. Thus, the effects of the $1/(N\sin\theta)$ term are suppressed, which results in a much weaker angular dependence of $\phi_{K,p}^{(2)}$.

This completes the determination of the polarization dependence of the Kerr rotation in nonlinear optics. The Kerr angle is expressed in terms of $\chi_{ijk}^{(2)}$, which may be calculated by an electronic theory. According to eqns (5.3.26), (5.3.27) and (5.3.33), (5.3.34) one may determine the easy magnetic axis from the Kerr rotation.

We now discuss the special cases of perpendicular and grazing incidence. For *perpendicular* incidence ($\theta = \Theta = 0°$) the Fresnel factors become

$$f_s = 0, \quad f_c = 1, \quad F_s = 0, \quad F_c = 1, \qquad (5.3.36)$$

and the linear transmission coefficient simplify to

$$t_p = t_s = \frac{2}{1+n}, \quad T_p = T_s = \frac{2}{1+N}. \qquad (5.3.37)$$

Thus, the corresponding amplitudes A_p and A_s in eqn (5.3.26) are

$$A_p = A_s = \frac{4\pi}{1+N}. \qquad (5.3.38)$$

Thus, the nonlinear Kerr rotation in the *longitudinal* configuration for p input polarization becomes

$$\phi_{K,p\text{-in, long}}^{(2)} = 1\,\text{Re}\,\frac{\chi_{yxx}^{(2)} \cdot 1}{0 \cdot \chi_{xzx}^{(2)} \cdot 0 + 0} \to \infty \qquad (5.3.39)$$

and for s input polarization

$$\phi_{K,s\text{-in, long}}^{(2)} = 1\,\text{Re}\,\frac{\chi_{yyy}^{(2)} \cdot 1}{0 \cdot \chi_{zyy}^{(2)}} \to \infty. \qquad (5.3.40)$$

Thus, the nonlinear Kerr rotation angle becomes arbitrarily large for perpendicular incidence. This is equally true for p and for s input polarization. Note that this divergence of the angle means according to eqn (5.3.26) a rotation by up to 90°. We use now eqns (5.3.26) and (5.3.27) to calculate the nonlinear Kerr rotation angle for Fe in the longitudinal configuration for p- and s-polarized incident light. Results for $\Phi_K^{(2)}$ are shown in Fig. 5.14. These results were obtained from our microscopic theory[17,47] for the nonlinear Kerr susceptibilities $\chi_{yzz}^{(2)}$ and $\chi_{zzz}^{(2)}$, the spin–orbit coupling constant has been kept fixed at 50 meV, and the complex indices of refraction at 1.6 eV and 3.2 eV were taken from

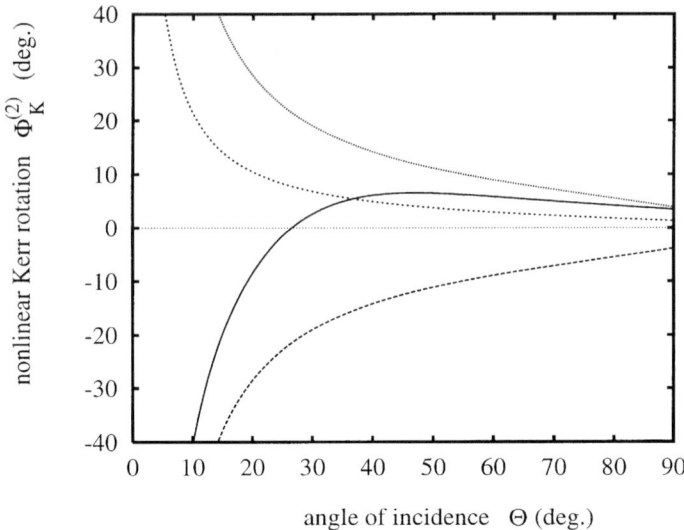

Fig. 5.14. Nonlinear Kerr rotation angles for p-polarized incident light $\phi^{(2)}_{K,p}$ (full and short-dashed curves) and for s-polarized incident light $\phi^{(2)}_{K,p}$ (long-dashed and dotted curves) for Fe at 770 nm as a function of the angle of incidence θ in the longitudinal Kerr configuration. The relative phase between $\chi^{(2)}_{zxx} = \chi^{(2)}_{zyy}$ and $\chi^{(2)}_{zzz}$ is 0.505π in the full and long-dashed curves and 1.505π in the short-dashed and dotted curves.

Johnson and Christy[28] ($n = 2.87 + i3.28$, $N = 2.12 + i2.50$). For the absolute ratios and the relative phases of the complex tensor elements we use the values $\chi^{(2)}_{xzx} = 0.60\chi^{(2)}_{zzz} e^{i1.945\pi}$, $\chi^{(2)}_{yyy} = \chi^{(2)}_{yxx} = 1.60\chi^{(2)}_{zzz} e^{i0.5\pi}$, $\lambda_{so}/\hbar\omega$ and $\chi^{(2)}_{yzz} = \chi^{(2)}_{zzz} e^{i0.5\pi}$ $\lambda_{so}/\hbar\omega$. For the remaining quantities we use $\chi^{(2)}_{zxx} = \chi^{(2)}_{zyy} = 0.0681\chi^{(2)}_{zzz} e^{i0.505\pi}$ for the full ($\Phi^{(2)}_{K,p}$) and long-dashed ($\Phi^{(2)}_{K,s}$) curves and $\chi^{(2)}_{zxx} = \chi^{(2)}_{zyy} = 0.0681\chi^{(2)}_{zzz} e^{i1.505\pi}$ for the short-dashed ($\Phi^{(2)}_{K,p}$) and dotted ($\Phi^{(2)}_{K,s}$) curves.

The results of Fig. 5.14 clearly show the divergence of the nonlinear Kerr rotation $\Phi^{(2)}_K$ for p- and s-polarized incident light in the case of perpendicular incidence and the increased enhancement for s polarization. This result is supported by experiments by Koopmans et al.[55] on Fe/Cr multilayers, which find a drastic enhancement of the nonlinear Kerr rotation in the longitudinal configuration. Upon decreasing the angle of incidence from 75° to 15°, values of $\phi^{(2)}_{K,p\text{-in}}$ from 1° to 5° and of $\phi^{(2)}_{K,s\text{-in}}$ from 5° to 15° are observed. Thus, in contrast to linear optics, there is no need to resort to uranium-based compounds, large external magnetic fields or low temperatures in order to obtain arbitrarily large Kerr rotations in nonlinear optics.

The fact that the experimental values do not increase monotonically in this range of angles of incidence is readily understood from interference effects in multilayers and is discussed in detail by Koopmans et al. As pointed out by these authors, at surfaces and in films $\phi^{(2)}_{K,p\text{-in}}$ and $\phi^{(2)}_{K,s\text{-in}}$ should be smooth functions of angle of incidence in agreement with our result.

Our theory yields that in general for finite angles of incidence the enhancement of the nonlinear Kerr rotation should be much more pronounced for s-polarized incident light, since in p polarization the denominator of the formula for $\phi_{K,p}^{(2)}$ contains several independent contributions which tend to decrease $\phi_{K,p}^{(2)}$. Furthermore, the excitation in the z direction (in p polarization) entering this denominator causes only a moderate enhancement of $\phi_{K,p}^{(2)}$. This is in agreement with the theoretical results by Pustogowa et al.[17] and with the experiments presently available.[43,55,56] Note that the microscopic theory by Pustogowa et al. treats in particular the enhancement in the longitudinal Kerr configuration yielding a $\phi_{K,p\text{-in}}^{(2)}$ of 2° to 4° for an Fe surface in the optical frequency range. This prediction is experimentally confirmed by Koopmans and Rasing. It is important to make two remarks concerning the choice of the longitudinal configuration and p input polarization by Pustogowa et al.:

1. Only in the longitudinal configuration can linear and nonlinear Kerr rotations be meaningfully compared since for the polar configuration the nonlinearity parallel to the surface depends very much on the localization of the excited electrons and their degree of jellium-like behaviour and in the transverse configuration no rotation is observed. Only an intensity change upon magnetization reversal will happen.
2. Only in the longitudinal configuration with incident p polarization can the Kerr rotation angle be defined as in linear optics with respect to the polarization of the incident photons as frequency ω. For incident s polarization as well as for the polar Kerr configuration one has to resort to the definition of $\phi_{K,p}^{(2)}$ as being one half of the angle by which the major half-axis of the second harmonic polarization ellipse is rotated upon magnetization reversal, thus not referring at all to the incident beam polarization.

In the *polar* Kerr configuration we obtain for perpendicular incidence ill-defined formulas for both p and s input polarization

$$\phi_{K,\text{p-in,pol.}} = \phi_{K,\text{s-in,pol.}} = \text{Re}\,\frac{0}{0}. \tag{5.3.41}$$

Expansion of the formulae for nearly perpendicular incidence, however, shows that there is no s-polarized second harmonic signal expected within the electric dipole approximation, but there is p-SH generated from s input polarization by exciting the tensor element $\chi_{yxx}^{(2)}$, thus leading to a vanishing $\phi_{K,s}^{(2)}$. On the other hand, for p input polarization both s-SH (numerator of $\phi_{K,p}^{(2)}$) and p-SH (denominator) yields become finite (apart from accidental zeros due to interference of the several complex tensor elements or due to a special choice of the frequency ω), thus yielding in general a finite $\phi_{K,p}^{(2)}$.

This is why NOLIMOKE is a unique tool for the determination of the easy axis in films and at interfaces and which cannot be determined by other tools. According to our analysis it is necessary for that purpose to shine light in at slightly off-perpendicular incidence. Then *perpendicular* interface magnetization (see our theory for *polar* Kerr configuration) would show no nonlinear Kerr rotation for s input but an appreciable Kerr angle for p input. On the other hand, the characteristic signature of *in-plane* magnetization (see our theory for *longitudinal* Kerr configuration) is expected to exhibit a moderately enhanced nonlinear Kerr angle for p input but a large nonlinear Kerr rotation for s input polarization. Thus, in switching from s to p input, the nonlinear Kerr rotation should increase for perpendicular and decrease for in-plane easy axis.

Finally we discuss the NOLIMOKE rotation for *grazing* incidence $\theta = \Theta = 90°$. In this case we have $\cos\theta = 0$ and $\sin\theta = 1$ and get the following Fresnel factors:

$$f_s = \frac{1}{n}, \quad f_c = \sqrt{1-\left(\frac{1}{n}\right)^2}, \quad F_s = \frac{1}{N}, \quad F_c = \sqrt{1-\left(\frac{1}{N}\right)^2}. \quad (5.3.42)$$

For grazing incidence it is meaningless to consider the transmission coefficients alone, which vanish. Instead, what matters for the Kerr rotation are the amplitudes A_p and A_s, which become

$$A_p = \frac{4\pi}{F_c}, \quad A_s = \frac{4\pi}{NF_c}. \quad (5.3.43)$$

This yields the ratio

$$\frac{A_s}{A_p} = \frac{1}{N}. \quad (5.3.44)$$

These equations together with eqns (5.3.26), (5.3.27) and (5.3.33), (5.3.34) show that the use of grazing incidence does not lead to simplified formulae for the NOLIMOKE rotation in the case of p or s input polarization.

Furthermore, it is an interesting observation that in NOLIMOKE there is a relative phase of 90° between the odd and even elements of the nonlinear susceptibility tensor $\chi^{(2)}_{ijk}$ in contrast to linear optics. This phase has already been found in the early theories by Hübner et al[13] and Pan et al.[14] and later observed in the experiment by Wierenga et al.[53] They observed $\Phi = 88°$. We discuss the microscopic origin of this relative phase in the following subsection.

In summary, we have shown that in SH the Kerr rotation depends sensitively on the light polarization, on the magnitude and direction of the magnetization and therefore on the easy axis.

Magnetic phase shift in nonlinear magneto-optics

In this subsection, we discuss the microscopic origin of the relative phase shift of 90° between the odd and even elements of the nonlinear susceptibility tensor $\chi^{(2)}_{ijk}$. First, we have to remark that this phase does not result from the fact that the nonlinear susceptibilities contain three matrix elements each yielding a factor of i rather than two in the linear case, since this difference occurs in the even as well as in the odd tensor elements. Instead, the microscopic origin is due to spin–orbit coupling, which acts as a perturbation on one of the wavefunctions in the matrix elements of the odd tensor elements alone. For a plane-wave basis, for example, the spin–orbit perturbation yields the following identity,[57,58] which can be proven by commutator algebra,

$$\langle k' | \lambda_{so}(k \times s)\nabla V | k \rangle = i\lambda_{so} V_{k'-k}(k \times k') \cdot s, \qquad (5.3.45)$$

thus giving a phase factor of i in the odd susceptibility tensor elements. This argument holds in the linear as well as in the nonlinear case, but the resulting phase of i is compensated only in the linear case by the decomposition of $\chi_{ijk}^{(1)}$ yielding another factor of i

$$\chi_{ij}^{(1)\pm} = \chi_{ij,0}^{(1)} \pm i\chi_{ij,1}^{(1)} \sin\theta. \qquad (5.3.46)$$

This factor comes from the wave equation, which is homogeneous in linear optics. The susceptibility results directly from the dielectric function, the square root of which is the eigenvalue of the wave equation, the complex index of refraction. The eigenmodes are left- or right-handed circularly polarized photons. In the nonlinear case, however, the decomposition has no factor of i

$$\chi_{ijk}^{(2)\pm} = \chi_{ijk,0}^{(2)} \pm \chi_{ijk,1}^{(2)}, \qquad (5.3.47)$$

since $\chi_{ijk}^{(2)}$ is not related to the eigenvalues of the wave equation, which in nonlinear optics is an inhomogeneous differential equation.

5.3.3 Enhancement of the nonlinear Kerr rotation

Motivation

It is the goal of this subsection to show in detail how magnetism affects the incoming light polarization and to determine the frequency-dependent Kerr rotation.[17] For this we have to extend the Fresnel formulae to the nonlinear magnetic case, which seems of general interest by itself. Since in the linear case the Kerr rotation is suppressed by bulk interband and intraband transitions (plasmons), one expects in the nonlinear case with only surface optical response an enhancement of the Kerr rotation. Indeed we find a large enhancement of the Kerr rotation of the light polarization as a general phenomenon in nonlinear magneto-optics. This might explain then also various recent experimental results on garnets and Heusler alloys.[16,59,60] Since, also with regards to the high potential for technological application,[61,62] the search for large Kerr rotations in linear optics has been a long-standing subject of intense theoretical and experimental investigations, such an enhancement of the Kerr rotation in the nonlinear magneto-optical Kerr effect (NOLIMOKE) possibly opens a new route for applications involving readily available materials, such as iron, without resorting to low temperatures, large magnetic fields, or binary and ternary magnetic alloys.[63] Thus, our theory may become the basis for a unified use of SHG as a tool for a comprehensive study of surface and interface magnetism including magnetic anisotropy, spin–orbit coupling and quantum-well states.

Theory

Using electrodynamical theory the linear and nonlinear polarizations are expressed by the susceptibilities. The influence of magnetism on SHG is shown best by determining the change of the polarization of the incoming light. For this we use the wave equation

$$\nabla \times \nabla \times \boldsymbol{E}^{(j)}(j\omega) + \frac{\varepsilon^{(1)}(j\omega)}{c^2} \frac{\partial^2}{\partial t^2} \boldsymbol{E}^{(j)}(j\omega) = -\frac{1}{\varepsilon_0 c^2} \frac{\partial^2}{\partial t^2} \boldsymbol{P}^{(2)} \delta_{2j}, \quad (5.3.48)$$

where the frequency-dependent linear dielectric tensor is given by

$$\varepsilon^{(1)}(j\omega) = \begin{pmatrix} \varepsilon_0^{(1)}(j\omega) & 0 & -\varepsilon_1^{(1)}(j\omega) \\ 0 & \varepsilon_0^{(1)}(j\omega) & 0 \\ \varepsilon_1^{(1)}(j\omega) & 0 & \varepsilon_0^{(1)}(j\omega) \end{pmatrix} \quad (5.3.49)$$

in the longitudinal geometry (see inset (a) of Fig. 5.16). ε_0 is the vacuum permittivity and $j = 1, 2$ refer to the linear and nonlinear cases. The diagonal elements $\varepsilon_0^{(1)}(j\omega)$ have purely nonmagnetic character (are symmetric under magnetization reversal), whereas the diagonal elements $\varepsilon_1^{(1)}(j\omega)$ are magnetic contributions (antisymmetric under magnetization reversal). The nonlinear surface polarization

$$\boldsymbol{P}^{(2)}(2\omega) = \varepsilon_0 \chi^{(2)}(2\omega) : \boldsymbol{E}^{(1)}(\omega) \cdot \boldsymbol{E}^{(1)}(\omega) \quad (5.3.50)$$

acts as a source term in the wave equation. To link SHG to the spin-polarized electronic structure and to understand the microscopic origin of the Kerr rotation, we use an electronic theory[47] to determine $\chi^{(2)}(\omega)$. In the linear case without the source term, eqn (5.3.48) is an homogeneous wave equation leading to left and right-handed circularly polarized waves as eigenmodes and complex refractive indices as eigenvalues.

$$N_{1t}^2 = \varepsilon_0^{(1)}(\omega) \pm i \varepsilon_1^{(1)}(\omega) \sin \theta_{1t}. \quad (5.3.51)$$

These control the magneto-optical response, in particular the Kerr rotation. Here θ_{1t} denotes the angle of refraction of the fundamental beam. Magnetism causes a different index of refraction for left- and right-handed circularly polarized light. In the nonlinear case, the solution of the wave equation in the medium includes the solution of the homogeneous equation plus one particular solution of the inhomogeneous equation. Thus, for the transmitted field $\boldsymbol{E}_{\text{trans}}^{(2)}$, one has

$$\boldsymbol{E}_{\text{trans}}^{(2)} = \boldsymbol{E}_t^{(2)} e^{i(k_{2t}r - 2\omega t)} + \boldsymbol{E}_s^{(2)} e^{i(k_s r - 2\omega t)}, \quad (5.3.52)$$

where k_{2t} is the wavevector of the refracted second harmonic field. The wavevector of the source term is twice the wavevector of the refracted fundamental beam $k_s = 2k_{1t}$ ($\theta_s = \theta_{1t}$). Substituting the field $E^{(2)}_{trans}$ into eqn (5.3.48) one gets the eigenvalues

$$N_s^2 = \varepsilon_0^{(1)}(\omega) \pm i\varepsilon_1^{(1)}(\omega)\sin\theta_s. \qquad (5.3.53)$$

Similarly to the linear case, the homogeneous part of the nonlinear wave equation yields complex refractive indices for the transmitted second harmonic part and

$$N_{2t}^2 = \varepsilon_0^{(1)}(2\omega) \pm i\varepsilon_1^{(1)}(2\omega)\sin\theta_{2t} \qquad (5.3.54)$$

for the transmitted second harmonic part. Note that the solution of the inhomogeneous wave equation no longer relates to indices of refraction, but directly to the surface response function $\chi^{(2)}$, which does not correspond to indices of refraction. These complex refractive indices will be used for the determination of the Kerr rotation.

To calculate now the Kerr rotation, we use the law of reflection and decompose $E^{(j)}(j\omega)$ into left- and right-handed circularly polarized light. The resulting fields

$$E^{(j)\pm}(j\omega) = E^{(j)\pm} e^{[ij\omega(t - N_\pm z/c)]}$$

are the eigenvectors of the homogeneous wave equation with indices of refraction N_+ and N_-, respectively, due to the magnetic birefringence. One obtains for the complex Kerr angle with real part $\phi_K^{(j)}$ and ellipticity $\varepsilon_K^{(j)}$ the result

$$\tan\psi_K^{(j)} = \phi_K^{(j)} + i\varepsilon_K^{(j)} = i\frac{E_r^{(j)+}(j\omega) - E_r^{(j)-}(j\omega)}{E_r^{(j)+}(j\omega) + E_r^{(j)-}(j\omega)}. \qquad (5.3.55)$$

Here the right-hand side of eqn (5.3.55) is the ratio of the x and z' components of the reflected field (see inset (b) of Fig. 5.16) expressed in terms of left- and right-handed circular polarization. Here we used $E_x^{(j)} \approx i(E_r^{(j)+} - E_r^{(j)-})$. The minus sign in $E_x^{(j)}$ is obvious since the Kerr rotation occurs only due to magnetic birefringence. Noting that the Kerr rotation occurs in the plane perpendicular to the direction of the reflected beam, $\psi_K^{(j)}$ obviously gives the Kerr angle. Equation (5.3.55) is valid for arbitrary angles of incidence and arbitrary magnetization direction.[64]

Further evaluation of eqn (5.3.55) requires the specification of the Kerr configuration. We choose the *longitudinal* Kerr configuration as shown in Fig. 5.16(a) for arbitrary angles of incidence. In this configuration the magnetization vector M lies parallel to the optical plane and parallel to the sample surface. We assume p-polarized incident light. Fig. 5.16(b) shows the definition of the linear and nonlinear Kerr angles with respect to the reflected beam of frequencies ω and 2ω, respectively.

To express the field amplitude $E_r^{(j)\pm}(j\omega)$ of the reflected beam by the incident field amplitude $E_i^{(j)\pm}(j\omega)$, we use for the linear case the usual Fresnel formulae

$$\frac{E_r^{(1)\pm}}{E_i^{(1)\pm}} = \frac{N_{1t}^2 \cos\theta_i - \sqrt{N_{1t}^2 - \sin^2\theta_i}}{N_{1t}^2 \cos\theta_i + \sqrt{N_{1t}^2 - \sin^2\theta_i}}. \quad (5.3.56)$$

Corresponding expressions for the nonlinear case are derived extending results by Bloembergen and Pershan.[2] Magnetism will be included in the indices of refraction. Without magnetism, $\varepsilon_1^{(1)} = 0$ and eqns (5.3.51), (5.3.53) and (5.3.54) reduce to $N^2 = \varepsilon_0^{(1)}(j\omega)$. For $M = 0$, $P^{(2)}$ perpendicular to the surface gives the maximum SHG yield resulting from breakdown of inversion symmetry. Thus, we choose for the orientation of the source term in eqn (5.3.48) $P^{(2)} = (0, 0, P^{(2)})$, since the nonlinear source consists of dipoles perpendicular to the surface.[65] The geometries chosen are useful for detecting in-plane and out-of-plane magnetization and easy axis. Then using the continuity of the components of the electric and magnetic field parallel to the surface we get

$$-\cos\theta_{2r} E_r^{(2)} = \cos\theta_{2t} E_t^{(2)} + \frac{\cos\theta_s N_s^2 \sin\theta_s}{\varepsilon_0^{(1)}(2\omega)\left[\varepsilon_0^{(1)}(2\omega) - N_s^2\right]} P^{(2)}, \quad (5.3.57)$$

and

$$k_{2r} E_r^{(2)} = k_{2t} E_t^{(2)} + \frac{k_s \sin\theta_s}{\varepsilon_0^{(1)}(2\omega) - N_s^2} P^{(2)}. \quad (5.3.58)$$

Eliminating $E_t^{(2)}$ and solving for $E_r^{(2)}$ we get[66]

$$E_r^{(2)} = \frac{1}{\varepsilon_0 c^2} \frac{P^{(2)} \sin\theta_s}{k_{2t}\cos\theta_{2r} + k_{2r}\cos\theta_{2t}} \frac{\varepsilon_0^{(1)}(2\omega)k_s \cos\theta_{2t} - N_s^2 k_{2t}\cos\theta_s}{\varepsilon_0^{(1)}(2\omega)\left[\varepsilon_0^{(1)}(2\omega) - N_s^2\right]}. \quad (5.3.59)$$

For expressing the Kerr angle in terms of the microscopic calculated susceptibilities we rewrite $E_r^{(2)}$ as a function of the complex indices of refraction N_s and N_{2t}, the angle of incidence θ_i, and the nonlinear source $P^{(2)}$. The following conditions are used: (i) $k_{2t} = N_{2t} 2\omega/c$, $k_s = N_s 2\omega/c$ and $k_{2r} = N_{2r} 2\omega/c$; (ii) the incident and reflected beams propagate in vacuum and hence $N_{2r} = N_{1r} = 1$ and $\theta_{2r} = \theta_{1r} = \theta_i$; and (iii) $\sin\theta_s = (1/N_s)\sin\theta_i$ and $\sin\theta_{2t} = (1/N_t)\sin\theta_i$. Thus, substituting these expressions into eqn (5.3.59) one gets

$$E_r^{(2)} = \frac{1}{\varepsilon_0 c^2} \frac{P^{(2)} \sin\theta_i}{\varepsilon_0^{(1)}(2\omega)\left[\varepsilon_0^{(1)}(2\omega) - N_s^2\right]}$$

$$\times \frac{\varepsilon_0^{(1)}(2\omega)\sqrt{N_{2t}^2 - \sin^2\theta_i} - N_{2t}^2 \sqrt{N_s^2 - \sin^2\theta_i}}{N_{2t}^2 \cos\theta_i + \sqrt{N_{2t}^2 - \sin^2\theta_i}}. \quad (5.3.60)$$

Expressing furthermore the refractive indices by the susceptibilities we find for the *nonlinear magnetic Fresnel formula* in p polarization the result (writing for $E_r^{(2)}$ explicitly $E_r^{(2)\pm}$)

$$E_r^{(2)\pm} = -\frac{\chi^{(2)\pm}\left[E_i^{(1)}\sin^2\theta_i\right]^2\sin\theta_i}{\varepsilon_0 c^2}\frac{1}{\left[1+\chi_0^{(1)}(2\omega)\right]\left[\chi_0^{(1)}(2\omega)-\chi^{(1)\pm}(\omega)\right]}$$

$$\times \frac{\left[1+\chi_0^{(1)}(2\omega)\right]S_2^\pm(\theta_i) - \left[1+\chi^{(1)\pm}(2\omega)\right]S_1^\pm}{\left[1+\chi^{(1)\pm}(2\omega)\right]\cos\theta_i + S_2^\pm(\theta_i)}$$

$$\equiv -\frac{P^{(2)\pm}\sin\theta_i}{\varepsilon_0 c^2}\frac{1}{F_2^\pm}\frac{F_1^\pm}{F_3^\pm}. \tag{5.3.61}$$

Here, we put $S_j^\pm(\theta_i) \equiv \sqrt{1+\chi^{(1)\pm}(j\omega)-\sin^2\theta_i}$ for $j=1,2$ and $\varepsilon_0^{(1)} = 1+\chi_0^{(1)}$ and $\varepsilon_1^{(1)} = \chi_1^{(1)}$. The explicit expressions for the $F_n^\pm$ follow by comparison ($n=1,2,3$). Here the tensors $\chi^{(1)\pm} = \chi_0^{(1)} \pm i\chi_1^{(1)}\sin\theta_i$ and $\chi^{(2)\pm} = \chi_0^{(2)} \pm \chi_1^{(2)}$ are decomposed into diagonal $\chi_0^{(1)}$ ('nonmagnetic') and off-diagonal $\chi_1^{(2)}$ ('magnetic') contributions. Owing to the source term in eqn (5.3.48) the i is missing in $\chi^{(2)}$. Substituting the fields in eqn (5.3.55) by using the Fresnel formulae, one gets for the longitudinal geometry the linear Kerr angle

$$\tan\psi_K^{(1)}(\omega) = -\frac{\chi_1^{(1)}(\omega)}{\chi_0^{(1)}(\omega)}\frac{\sin\theta_i\cos\theta_i}{\sqrt{\cos^2\theta_i + \chi_0^{(1)}(\omega)}}\frac{\cos(2\theta_i) + \chi_0^{(1)}(\omega)}{\cos(2\theta_i) + \chi_0^{(1)}(\omega)\cos^2\theta_i}, \tag{5.3.62}$$

and for the nonlinear Kerr angle

$$\tan\psi_K^{(2)}(\omega) = i\frac{\chi^{(2)+}F_1^+F_2^-F_3^- - \chi^{(2)-}F_1^-F_2^+F_3^+}{\chi^{(2)+}F_1^+F_2^-F_3^- + \chi^{(2)-}F_1^-F_2^+F_3^+}. \tag{5.3.63}$$

Equations (5.3.62) and (5.3.63) are the basis for determining the ω and θ_i dependencies of the linear and nonlinear Kerr rotation. Since $\xi_1^{(j)} \ll \chi_0^{(j)}$, eqn (5.3.63) may be expanded in power of ($\chi_1^{(j)}/\chi_0^{(j)}$) and linearized. One gets using $F_n^\pm = F_{n0} \pm F_{n1}$ ($n=1,2,3$)

$$\tan\psi_K^{(2)} = i\left(\frac{\chi_1^{(2)}}{\chi_0^{(2)}} + \frac{F_{11}}{F_{10}} - \frac{F_{21}}{F_{20}} - \frac{F_{31}}{F_{30}}\right), \tag{5.3.64}$$

with

$$F_{n0} = F_n^\pm(\chi_1^{(1)}(j\omega) = 0), \quad (n=1,2,3),$$

$$F_{11} = i\sin\theta_i\left\{\tfrac{1}{2}[1+\chi_0^{(1)}(2\omega)]\left(\frac{\chi_1^{(1)}(2\omega)}{S_{20}} - \frac{\chi_1^{(1)}(\omega)}{S_{10}}\right) - \chi_1^{(1)}(2\omega)S_{10}\right\}$$

$$F_{21} = -i\sin\theta_i\{\chi_1^{(1)}(\omega)[1+\chi_0^{(1)}(2\omega)]\} \tag{5.3.65}$$

$$F_{31} = i\sin\theta_i\left\{\chi_1^{(1)}(3\omega)\left[\cos\theta_i + \frac{1}{2S_{20}(\theta_i)}\right]\right\},$$

where

$$S_{j0}(\theta_i) \equiv \sqrt{1 + \chi_0^{(1)}(j\omega) - \sin^2(\theta_i)}.$$

Note that the first term in eqn (5.3.64) results from $\chi^{(2)+} \neq \chi^{(2)-}$, and has been previously neglected.[67] This term, however, gives the main contribution to the nonlinear Kerr rotation. It vanishes in the case of inversion symmetry in the bulk material, but not at the surface. The F_{n1} and F_{n0}, however, depend only on *linear* susceptibilities. Hence, nonlinearity exhibits a much stronger surface sensitivity, which is responsible for making NOLIMOKE a very useful tool for studying surface magnetism. It is interesting, that for all configurations the factor $1/\sqrt{\cos^2\theta_i + \chi_0^{(1)}(\omega)}$, which causes the small $\psi_K^{(1)}$ in eqn (5.3.62), is not present in the nonlinear case (eqn (5.3.64)). Despite the complex dependence of the parameters it is clear that the nonlinear Kerr rotation is for all θ_i always enhanced by a factor $\sqrt{\cos^2\theta_i + \chi_0^{(1)}(\omega)}$. Note that this enhancement can be traced back to the source term of the wave equation (5.3.48), where $\boldsymbol{P}^{(2)} \sim \chi^{(2)}$ and $\chi^{(2)}$ depends sensitively on the magnetic properties at the surface. This is the mathematical manifestation of the different character of the solution of the homogeneous and inhomogeneous wave equations (5.3.48) as pointed out already and of the different physics involved in linear and nonlinear optics. In the nonlinear case only the surface contributes to the optical response. Thus, the destructive contributions to the Kerr rotation by bulk inter- and intraband transitions are avoided. Furthermore, it follows clearly from eqn (5.3.64) that the interaction between the light and magnetism is mediated by the spin–orbit coupling. For vanishing spin–orbit coupling constant λ_{so} one gets $\chi_1^{(2)} = F_{n1} = 0$ ($n = 1, 2, 3$) and thus $\psi_K^{(2)} = 0$. All F_{n0} and $\chi_0^{(j)}$ are independent of spin–orbit coupling.

Results

To demonstrate numerically the effect of magnetism on SHG, in particular the frequency-dependence enhancement of the Kerr rotation, we now use eqns (5.3.62) and (5.3.63) assuming $\theta_i = 45°$. The functions $F_n^\pm$ ($n = 1, 2, 3$) (see eqn (5.3.61)) are determined by the susceptibilities $\chi_i^{(1)}$ ($i = 0, 1$). These and the nonlinear susceptibilities $\chi_i^{(2)}$ have been calculated previously by us for Fe.[47] The results obtained should be representative for transition metals. Only for the nonmagnetic linear susceptibility $\chi_0^{(1)}$ do we include additively interband and intraband contributions. For the intraband contribution, we assume as usual a non-spin-split conventional Drude form $\chi_{0,\text{intra}}^{(1)} = \omega_0^2/[\omega(\omega + i/\tau)]$, using $\tau = 9.12 \times 10^{-15}$ s and $\hbar\omega_0 = 0.74$ eV. We neglect Drude contributions to the magnetic susceptibilities, since the d electrons dominate the electronic spin polarization. For the nonlinear nonmagnetic susceptibility $\chi_0^{(2)}$ a Drude contribution, if appreciable at all, should occur only at much lower frequencies than in the linear case. Note that we use $\lambda_{so} = 50$ meV, which is the bulk value for Fe.

To check the accuracy of our electronic calculations we calculate first for Fe the frequency dependence of the linear polar Kerr angle $\phi_K^{(1)}(\omega) = \text{Re}\,\psi_K^{(1)}$ and

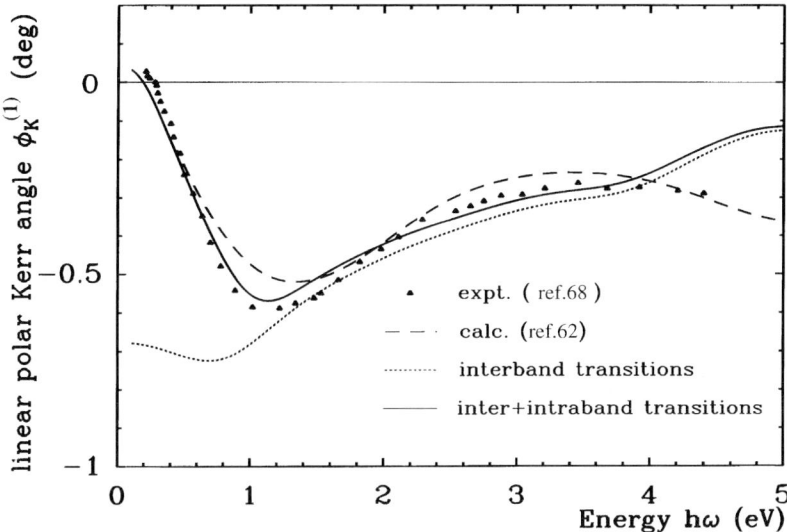

Fig. 5.15. Frequency dependence of the linear Kerr angle for the polar configuration. Theoretical results using parameters corresponding to Fe (see ref. 17) are compared with experimental data. The interband contribution is shown separately.

compare it with the experiment by Krinchik[68] and calculations by Oppeneer et al.[62] Results are shown in Fig. 5.15. Note we find excellent agreement up to $\hbar\omega = 4\,\text{eV}$. The linear Kerr angle is of the order of 0.5°. The inclusion of the intraband contribution, which changes $\phi_K^{(1)}(\omega)$ mainly for small ω, leads to a further reduction of the linear Kerr angle and is responsible for changing $\phi_K^{(1)}$ to zero. It is of interest that our analysis reproduces also the enhancement of the linear Kerr angle at the plasma frequency as already discussed by Feil and Haas.[69]

In Fig. 5.16 we show results for Fe in the case of the longitudinal Kerr geometry for the frequency dependence of the nonlinear Kerr angle $\phi_K^{(2)}(\omega) = \text{Re}\,\psi_K^{(2)}(\omega)$ for an angle of incidence of 45° and compare them with the corresponding results for $\phi_K^{(1)}(\omega)$. We obtain for the optical range (up to 2.5 eV) a considerable enhancement of $\phi_K^{(2)}(\omega)$ over $\phi_K^{(1)}(\omega)$. For $\hbar\omega > 2.5\,\text{eV}$ one has $|\phi_K^{(1)}(\omega)| \approx |\phi_K^{(2)}(\omega)| \approx 0.5°$ and for $\hbar\omega$ in the optical range one gets $\phi_K^{(1)}(\omega) \approx 0.5°$ while $\phi_K^{(2)}(\omega)$ may become 4° or larger. It is important to emphasize that the enhancement results largely from the term $\chi_1^{(2)}/\chi_0^{(2)}$, which has been neglected previously.[67] Note that the behaviour of $\phi_K^{(1)}(\omega)$ and $\phi_K^{(2)}(\omega)$ for $\omega \to 0$ is controlled by the factor $\lambda_{so}/\hbar\omega$.[47,70]

It would be interesting to analyse the nonlinear Kerr angle and its symmetry properties for different magnetic metals. For materials with large spin–orbit coupling or large magnetic moments, as in Heusler alloys, particularly large Kerr rotations could be expected.[60] Note that in materials where inversion symmetry is also absent in bulk, like in alloys, NOLIMOKE probes both surface and bulk.

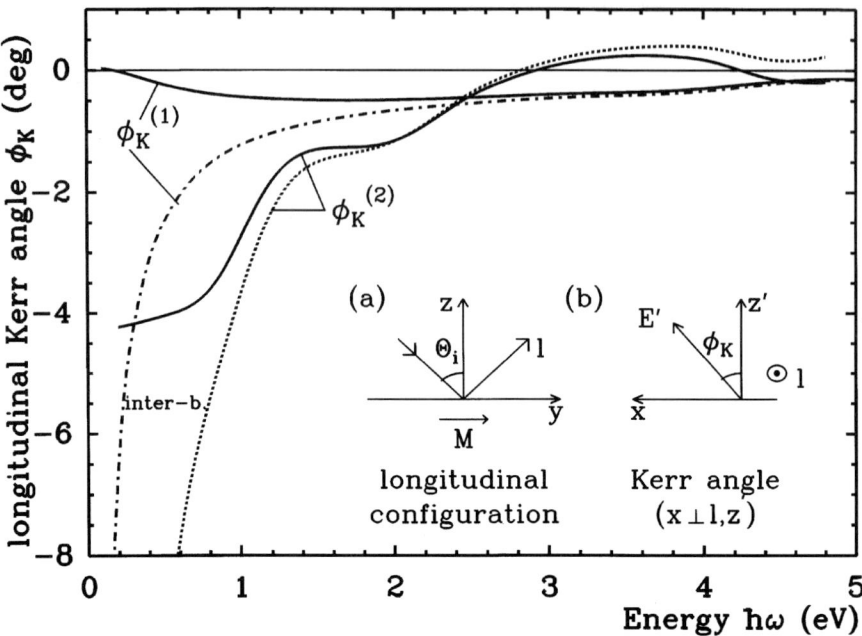

Fig. 5.16. Theoretical results for the frequency dependence of the nonlinear longitudinal Kerr angle $\phi_K^{(2)}(\omega)$ for Fe at an angle of incidence $\theta_i = 45°$ using the same parameters. For comparison also our results for the linear longitudinal Kerr angle $\phi_K^{(1)}(\omega)$ are given. The insets (a) and (b) illustrate the definitions of the longitudinal configuration and the Kerr angle. E' refers to the reflected light propagating in direction l, z' and x lie in the plane perpendicular to l, and x in the surface plane and is perpendicular to z'.

For thin magnetic films our theory would permit the determination of surface magnetization, magnetic moments and the magnetic anisotropy without involving hysteresis. Since inversion symmetry is broken at interfaces, these will contribute within the skin depth to NOLIMOKE. Our results imply that the nonlinear Kerr rotation is enhanced in all regions where inversion symmetry is broken.

Recent experiments by Böhmer and Matthias[71] on polycrystalline nickel indeed find a Kerr rotation $\phi_K^{(2)} \approx 4°$. Experiments by Rasing et al. on an Fe surface yield $\phi_K^{(2)} = 1.7°$ at 1.2 eV, in precise agreement with our theory, and thus impressively confirm our theoretical prediction of a drastic enhancement of the nonlinear Kerr rotation $\phi_K^{(2)}$ compared to the linear one ($\phi_K^{(1)} \approx 0.1°$–$0.3°$ for nickel).

Presently, NOLIMOKE is for example the only probe of magnetism at the interface of two magnetic films.[72] Experiments by Koopmans et al.[55] showed experimentally that the Kerr rotation in nonlinear optics at thin-film interfaces can become arbitrarily large due to additional multiple scattering interferences. Thus, NOLIMOKE is a unique tool to probe multilayer interfaces.

5.3.4 Nonlinear magneto-optics in s- and d-wave superconductors

Motivation

The investigation of the symmetry of the superconducting order parameters is currently one of the most exciting problems in the field of high-T_c research.[73–81] Many experiments suggest that the order parameter is anisotropic.[73–77] Furthermore, the observation of π phase shifts in corner junctions[78,80,81] is consistent with a $d_{x^2-y^2}$ symmetry of the gap function, although another experiment[79] favours a more conventional s-wave pairing. On the other hand, the occurrence of a finite tunnelling current perpendicular to the planes[82] seems to be incompatible with d-wave symmetry in a simple picture. In view of this debate, it is important to develop alternative experimental techniques that are able to discriminate between the various symmetries of the gap function. In particular, it is interesting to measure the symmetry of the gap function without the necessity of a tunnelling contact, which always includes the problems of a residual magnetic field, trapped flux in the tunnelling circuit, or singularities in the supercurrent flow at the corner.[83]

It is well known that the symmetry of the nonlinear optical susceptibility is strongly affected by a magnetization or an external applied magnetic field.[14] Thus it is of considerable interest to extend the symmetry analysis of the nonlinear magneto-optical response to the superconducting state for different symmetries of the superconducting order parameter. Although, because of gauge invariance, it is impossible to measure its phase without tunnelling contact, it is still possible to measure its symmetry, not solely its magnitude. This results from the interference of different pairing amplitudes in the dipole matrix elements of the *three* transitions in nonlinear optics. Owing to the surface sensitivity of nonlinear optics for systems with bulk inversion symmetry, one can take advantage of the broken inversion symmetry at the surface.[84,85] This is of interest, because Cooper pairing, together with the always present spin–orbit coupling, is then no longer purely singlet- or triplet-like. The interference of the singlet and triplet pairing states, which is linear in spin–orbit interaction, leads to the symmetry-sensitive contribution of the nonlinear optical response for systems in an external magnetic field. Thus, we proceed as follows:[86] (i) First we set up the superconducting BCS-type Hamiltonian and perform its group-theoretical analysis. In order to obtain the desired sensitivity of nonlinear optics to the symmetry of the gap function, this requires, as it will turn out, the simultaneous absence of inversion symmetry, presence of an external magnetic field breaking time reversal and spin–orbit interaction coupling singlet and triplet pairs. (ii) Making use of this symmetry classification and the mentioned constraints we then calculate the nonlinear magneto-optical response in s- and d-wave superconductors from the appropriate current–current–current correlation function and propose a suitable experimental geometry for the observation of this new nonlinear magneto-optical effect in superconductors.

Theory

In our theory for the nonlinear magneto-optical response of unconventional superconductors, the superconducting state is described within the BCS theory.[87] The corresponding Hamiltonian with arbitrary pairing symmetry and in a magnetic field h is given by

$$H = \sum_{k\mu} \Psi^\dagger_{k\mu} \mathcal{H}_{k\mu} \Psi_{k\mu}, \qquad (5.3.66)$$

with the four-component Nambu spinor $\Psi_{k\mu} = (c_{k\mu\uparrow}, c_{k\mu\downarrow}, c^\dagger_{-k\mu\uparrow}, c^\dagger_{-k\mu\downarrow})$. Here, $c^\dagger_{k\mu\sigma}$ is the creation operator of an electron with momentum k, band index μ and spin σ. The (4×4) matrix $\mathcal{H}_{k\mu}$ can be expressed in terms of (2×2) block matrices:

$$\mathcal{H}_{k\mu} = \begin{pmatrix} \varepsilon_{k\mu}\hat{\sigma}^0 - h\cdot\hat{\sigma} & \hat{\Delta}_{k\mu} \\ -\hat{\Delta}^*_{-k\mu} & -\varepsilon_{k\mu}\hat{\sigma}^0 - h\cdot\hat{\sigma}^* \end{pmatrix}. \qquad (5.3.67)$$

The block matrices are expanded in terms of the unit matrix $\hat{\sigma}^0$ and the vector of the Pauli matrices $\hat{\sigma}$. This notation is close to that of Sigrist and Ueda.[88] The symmetry of the superconducting order parameter is characterized by the gap function $\Delta_{\sigma\sigma';k\mu} = \langle c_{k\sigma\mu} c_{-k\sigma'\mu} \rangle$. We neglect any diamagnetic, i.e. Meissner, effect of the magnetic field, but assume a large penetration depth at the surface and no influence of the vortex structure to the optical spectrum. This seems to be reasonable at least for the excitations in the interband regime discussed in this chapter. $\Delta_{\sigma\sigma';k\mu}$ is decomposed in the usual way in singlet states ($\Delta^0_{k\mu} = \Delta^0_{-k\mu}$) and triplet states ($d_{k\mu} = -d_{-k\mu}$):

$$\hat{\Delta}_{k\mu} = \left(\Delta^0_{k\mu}\hat{\sigma} + d_{k\mu}\cdot\sigma\right)i\hat{\sigma}^y. \qquad (5.3.68)$$

The symmetry of $\Delta^0_{k\mu}$ and $d_{k\mu}$ with respect to the transition from k to $-k$ is a direct consequence of the Pauli principle. Since we consider the states at a surface, k refers to the two-dimensional in-plane momentum.

Below the transition temperature T_c, the symmetry of a system is reduced compared to the high-temperature phase. The symmetry group G of the high-temperature phase is determined by the symmetry operations which keep the Hamiltonian for $\hat{\Delta}_k = 0$ invariant. We consider a system which is for $h = 0$ invariant with respect to the group

$$G = g \times K \times U(1), \qquad (5.3.69)$$

where g, K and U(1) are the point group, time reversal operation and the gauge group of multiplication of electron creation operator by an arbitrary phase, respectively.[88-93] In the ordinary case the normal-state gauge symmetry is broken at the superconducting phase transition, i.e. the residual symmetry group is g × K. This is called a conventional superconductor. In unconventional superconductors, however, the symmetry is lower than g × K. At the transition temperature, the BCS-gap equation is an eigenvalue equation and, consequently, an eigenvector $\Delta_{\sigma\sigma';k\mu}$ belongs to one of the irreducible representations $\mathcal{D}$ of the group G. If $\mathcal{D}$ is the unit representation $\mathcal{A}_1$ (or $\mathcal{A}_{1g}$ for systems with inversion symmetry) conventional superconductivity occurs. In all other cases the superconductivity is unconventional. In order to discuss the various symmetry states of the order parameter, one has to generate all irreducible representations of the gap function, where, due to spin–orbit coupling, the spin degrees of freedom cannot be transformed independently from the spatial (orbital) coordinates. For various point groups this symmetry classification has been performed.[88–93] In all these cases the inversion operation C_i is an element of the group G. Since we are interested in the investigation of superconducting properties with surface-sensitive nonlinear optical experiments, we have to take the effect of broken inversion symmetry into account. In order to be specific, we consider the surface of a tetragonal system (bulk point group D_{4h}) with residual point group C_{4v}. This group has give irreducible representations: four of dimension 1 ($\mathscr{A}_1, \mathscr{A}_2, \mathscr{B}_1$ and $\mathscr{B}_2$) and one of dimension 2 ($\mathscr{E}$). The isotropic s_0 and anisotropic $s_{x^2+y^2}$ waves transform as $\mathscr{A}_1$, the $d_{x^2-y^2}$ wave as $\mathscr{B}_1$, the d_{xy} wave as $\mathscr{B}_2$, a $d_{x^2-y^2}d_{xy}$ wave as $\mathscr{A}_2$, and the p_x, p_y waves as $\mathscr{E}$.

In order to classify the irreducible representations of the gap function, we have to analyse the transformation properties of $\Delta_{\sigma\sigma';k\mu}$. Applying an element R of the point group to $\hat{\Delta}_{k\mu}$, following transformation of the singlet and triplet part results:

$$R\hat{\Delta}_{k\mu} = \left(\Delta^0_{\mathscr{D}_R^{(1)}k\mu} + \tilde{\mathscr{D}}_R^{(1)}d_{\mathscr{D}_R^{(1)}k\mu}\cdot\boldsymbol{\sigma}\right)i\hat{\sigma}^y. \qquad (5.3.70)$$

Here, $\mathscr{D}_R^{(1)}$ is the representation of $R \in G$ which transforms the coordinates. If one considers the transformation of a Pauli spinor with respect to a combination $R = R_0 C_i$ of the inversion operation C_i and a rotation R_0, only the rotational part has to be applied to the spinor, i.e. the representation of R in spin space is $\mathscr{D}_{R_0}^{(1/2)}$. Consequently, for the vector in spin space d_k, the representation $\tilde{\mathscr{D}}_R^{(1)} \equiv \mathscr{D}_{R_0}^{(1)}$, where the inversion operation is replaced by the identical transformation, has to be applied. Therefore, one finds in the *bulk* of a system with inversion symmetry:

$$C_i\hat{\Delta}_{k\mu} = \left(\Delta^0_{k\mu} - d_{k\mu}\cdot\hat{\boldsymbol{\sigma}}\right)i\hat{\sigma}^y,$$

since the vector d is not affected directly by the inversion operation. The minus sign results from the inversion of k to $-k$. Consequently, in the bulk, the singlet and triplet parts belong to different irreducible representations and either singlet or triplet superconductivity occurs. In contrast to this, the coexistence of singlet and triplet pairing states is possible for systems without inversion symmetry, i.e. at the *surface*. Now, the in-plane inversion operation can be realized by a rotation (rotation by π with z axis as rotation axis). This rotation transforms the vector d to $-d$ and the minus sign of the transformation $k \to -k$ is cancelled. Consequently, the irreducible representations of the gap function at the surface contain both singlet and triplet parts.

From these considerations one obtains the irreducible representations of the gap function from the simultaneous Clebsch–Gordan coupling of orbital and spin degrees of freedom. The results for the simultaneously occurring singlet

Table 5.6 Singlet and triplet parts of the gap function for the irreducible representations of the point group C_4. k_x and k_y are the components of the in-plane momentum. k_z represents an additional quantum number, which changes sign if one interchanges the two paired electrons and which is related to the layer index at the surface

Irred. representation	Singlet part	Triplet part
$\mathscr{A}_1$	const., $\cos(k_x) + \cos(k_y)$	$e_x \sin(k_y) - e_y \sin(k_x)$
$\mathscr{A}_2$	$[\cos(k_x) - \cos(k_y)]\sin(k_x)\sin(k_y)$	$e_x \sin(k_x) + e_y \sin(k_y)$
$\mathscr{B}_1$	$\cos(k_x) - \cos(k_y)$	$e_x \sin(k_y) + e_y \sin(k_x)$
$\mathscr{B}_2$	$\sin(k_x)\sin(k_y)$	$e_x \sin(k_x) - e_y \sin(k_y)$
$\mathscr{E}$	$\sin(k_x)\sin(k_z), \sin(k_y)\sin(k_z)$	$e_z \sin(k_x), e_z \sin(k_y)$

and triplet parts of the pairing amplitude are given in Table 5.6. Based on these group-theoretical classifications, we calculate now the nonlinear magneto-optical susceptibility tensor of a superconductor and focus on the interference of the simultaneously occurring singlet and triplet parts of the gap function. The optical response in second harmonic generation can be obtained from the nonlinear current–current–current correlation function:

$$\chi_{\alpha\beta\gamma}(q,\omega) = \int_{-\infty}^{\infty} \frac{d\varepsilon}{\pi} \int_{-\infty}^{\infty} \frac{d\varepsilon'}{\pi} \int_{-\infty}^{\infty} \frac{d\varepsilon''}{\pi} I_{\alpha\beta\gamma}(q,\varepsilon,\varepsilon',\varepsilon'')$$

$$\times \left\{ \frac{[f(\varepsilon'') - f(\varepsilon')]}{(\omega + i\delta - \varepsilon'' + \varepsilon')} - \frac{[f(\varepsilon') - f(\varepsilon)]}{(\omega + i\delta + \varepsilon' + \varepsilon)} \right\} \Big/ 2(\omega + i\delta) - \varepsilon' + \varepsilon$$

(5.3.71)

where $f(\varepsilon)$ is the Fermi function and the spectral function $I_{\alpha\beta\gamma}(q,\varepsilon,\varepsilon',\varepsilon'')$ is given by

$$I_{\alpha\beta\gamma}(q,\varepsilon,\varepsilon',\varepsilon'') = \text{Tr}\big(J_{-2q\alpha}\varrho(\varepsilon)J_{q\beta}\varrho(\varepsilon')J_{q\gamma}\varrho(\varepsilon'')\big). \quad (5.3.72)$$

$J_{q\alpha}$ is the αth component of the current operator

$$J_q = \sum_{k\sigma\mu\nu} j_{k\mu\nu} c_{k+(q/2)\sigma\mu} c_{k-(q/2)\sigma\nu}, \quad (5.3.73)$$

and $\varrho(\varepsilon) = -(1/\pi)\text{Im}(\varepsilon + i\delta + H)^{-1}$ is the density-of-states matrix with Hamiltonian H of eqn (5.3.67). The trace has to be performed with respect to all single-particle states, i.e. the momentum (k), band (μ,ν), spin (σ) and Nambu degrees of freedom. In the following, we consider only interband transitions $\mu \neq \nu$, and the limit of the dipole approximation $q \to 0$ can be performed without special care for plasmonic excitations.

Here, we restrict ourselves to a special interband excitation process. We consider the transition from the initial state i with energy $E_i \approx -3\,\text{eV}$, below the Fermi energy, to the intermediate state s at the Fermi level (which is the only superconducting state) and to the final state f with energy $E_f \approx 3\,\text{eV}$, above E_F, i.e. we consider $\mu = \text{f}$, $\kappa = \text{s}$ and $\nu = \text{i}$. A possible, but not necessary, origin of the states i and f might be due to the Mott–Hubbard splitting of the hybridized Cu $3d_{x^2-y^2}$ and O $2p_{x(y)}$ orbitals. Since the intermediate state is the only state with superconducting coherence, we skip the band index of the matrices $\hat{u}_k$ and $\hat{v}_k$. Performing finally the traces in eqn (5.3.72), we obtain

$$\chi_{\alpha\beta\gamma}(\omega) = \chi^{(0)}_{\alpha\beta\gamma}(\omega) + \chi^{(h)}_{\alpha\beta\gamma}(\omega) + \mathcal{O}(h^2), \tag{5.3.74}$$

where

$$\chi^{(0)}_{\alpha\beta\gamma}(\omega) = \sum_k j^\alpha_{k\mathrm{fi}} j^\beta_{k\mathrm{is}} j^\gamma_{k\mathrm{sf}} \left[|u^0_k|^2 G_1(\mathbf{k},\omega) + |v^0_k|^2 G_2(\mathbf{k},\omega) \right] \tag{5.3.75}$$

is the zero-field susceptibility tensor in second harmonic generation within the superconducting state, which gives already a contribution without spin–orbit interaction. The nonvanishing tensor elements for $\chi^{(0)}_{\alpha\beta\gamma}(\omega)$ are the same as in the normal state ($\alpha\beta\gamma \in \{zzz, zxx, zyy, xzx, xxz, yzy, yyz\}$). This results here from the transformation properties of the three matrix elements $j^\alpha_{k\mathrm{fi}} j^\beta_{k\mathrm{is}} j^\gamma_{k\mathrm{sf}}$, which transform as the corresponding combination of the coordinates $x_\alpha x_\beta x_\gamma$, even if a single matrix element (e.g. $j^\alpha_{k\mathrm{fi}}$) does not transform like x_α. The functions $G_{1(2)}(\mathbf{k},\omega)$ result from the numerous combinations of Fermi functions and energy denominators which occur by performing the traces in spin and Nambu space. $\chi^{(h)}_{\alpha\beta\gamma}(\omega)$ is the new contribution of the magneto-optical Kerr effect in the superconducting state. This can be seen from the symmetry relations of the magneto-optical susceptibility

$$\chi^{(h)}_{\alpha\beta\gamma}(\omega) = \sum_k j^\alpha_{k\mathrm{fi}} j^\beta_{k\mathrm{is}} j^\gamma_{k\mathrm{sf}} (v^0_k)^* \mathbf{v}_k \cdot \mathbf{e}_\mathrm{h} F(\mathbf{k},\omega), \tag{5.3.76}$$

which depends on the superconducting gap function not only through its magnitude but also through Δ_k itself. Owing to the additional triplet part, however, the result is still gauge-invariant. The function $F(\mathbf{k},\omega)$ corresponds to $G_{1(2)}(\mathbf{k},\omega)$ for $\chi^{(h)}_{\alpha\beta\gamma}(\omega)$. Considering a magnetic field parallel to the x axis (in-plane), the nonvanishing elements of $\chi^{(h)}_{\alpha\beta\gamma}(\omega)$ are

$$\alpha\beta\gamma \in \{yyy, xxy, xyx, yxx, zzy, zyz, yzz\}.$$

This results from the combination of the transformation properties of the normal-state matrix elements and of the symmetry-sensitive term $(v^0_k)^* \mathbf{v}_k \cdot \mathbf{e}_\mathrm{h}$. Note that the $\mathbf{k}$ dependence of the latter results from the irreducible representations given in Table 5.6.

The above matrix elements lead to a rotation of the polarization of the incident light due to the interference of the singlet and triplet states at the surface of a superconductor. Thus nonlinear magneto-optics, unlike linear optical probes, indeed provides an optical method to discriminate different superconducting paring symmetries by exclusively employing the effect of optical photons to low-energy excitations.

For the numerical calculation we discuss the tensor elements $\chi^{(0)}_{zzz}$ of second harmonic generation without magnetic field and $\chi^{(h)}_{yzz}$ which gives rise to the rotation of the polarization plane for an applied magnetic field parallel to the x

axis. For simplicity we neglect the dispersion of the initial and the final states and consider solely the k dependence that results from the superconducting gap function.

The calculations are performed for a magnetic field of 9 T (corresponding to a field-induced band splitting of 0.5 meV) and a temperature of 1.5 K. The magnitude of the singlet part of the gap function is assumed to be 5 meV. Furthermore, the magnitude of the dipole matrix elements is estimated to be 10^{-11} m.

In Figs 5.17(a) and (b), we show ω^2 Im $\chi_{zzz}^{(0)}(\omega)$ and ω^2 Im $\chi_{yzz}^{(0)}(\omega)$ for an isotropic s wave. For the conventional SHG, we find a line shape similar to the real part of a Lorentzian, which is typical for a three-level system discussed in this chapter. More interestingly, the fine structure of the peak, shown in the inset of Fig. 5.17(a), clearly shows the energy scale of the superconducting gap. Comparing this behaviour with the Kerr signal ω^2 Im $\chi_{yzz}^{(0)}(\omega)$ of Fig. 5.17(b), one finds that the interference of the singlet and triplet pairing states leads to a line shape with several pronounced zeros and with a fine structure that yields, besides the energy scale of the superconducting gap, also excitations that result from the magnetic field splitting. In all our calculations, this line shape was exclusively observed for the isotropic s wave and can be considered as a fingerprint of this symmetry.

In Figs 5.18 and 5.19, the corresponding results for the anisotropic s wave and the $d_{x^2-y^2}$ wave are shown. Although the result for the conventional SHG is similar to that of the isotropic s wave, a totally different line shape of the Kerr signal results. This is due to the symmetry-dependent prefactor in $\chi_{yzz}^{(h)}(\omega)$ and can be used to discriminate these two symmetries from the isotropic s wave. Furthermore the fine structures of these two symmetries are very different from the isotropic s wave. Owing to the occurrence of nodes in the gap, not only a peak, but a whole broad band between 3 eV and 3.01 eV is observable. This range is surprisingly given by twice the superconducting gap magnitude. Unfortunately, there are only slight differences between the two symmetries shown in Figs 5.18 and 5.19. This is due to the similar k dependence of the triplet part given in Table 5.6. Only the fine structure of the peaks displayed in the insets of Fig. 5.18(b) and 5.19(b) exhibits a clear difference, where the anisotropic s wave has a clear zero at 3.01 eV, which is more or less smeared out for the $d_{x^2-y^2}$ wave.

A simple estimate shows that the observability of the new contribution to the nonlinear Kerr effect is clearly guaranteed for the anisotropic s wave. Although, the intensities of the other symmetries are smaller than the estimated value of the usual nonlinear Kerr effect in both the normal and superconducting states, we believe that this effect is still observable for the following reason. Owing to the neglect of the dispersion of the states in our model band structure, the disappearance of the signal results from cancellations of contributions of the order of magnitude of the anisotropic s wave in the k-summation. A more realistic band structure immediately leads to larger intensities of $\chi_{yzz}^{(h)}$, while keeping the characteristics of the line shapes of the spectra.

In contrast to the anisotropic s wave and the $d_{x^2-y^2}$ wave, we find for the d_{xy}

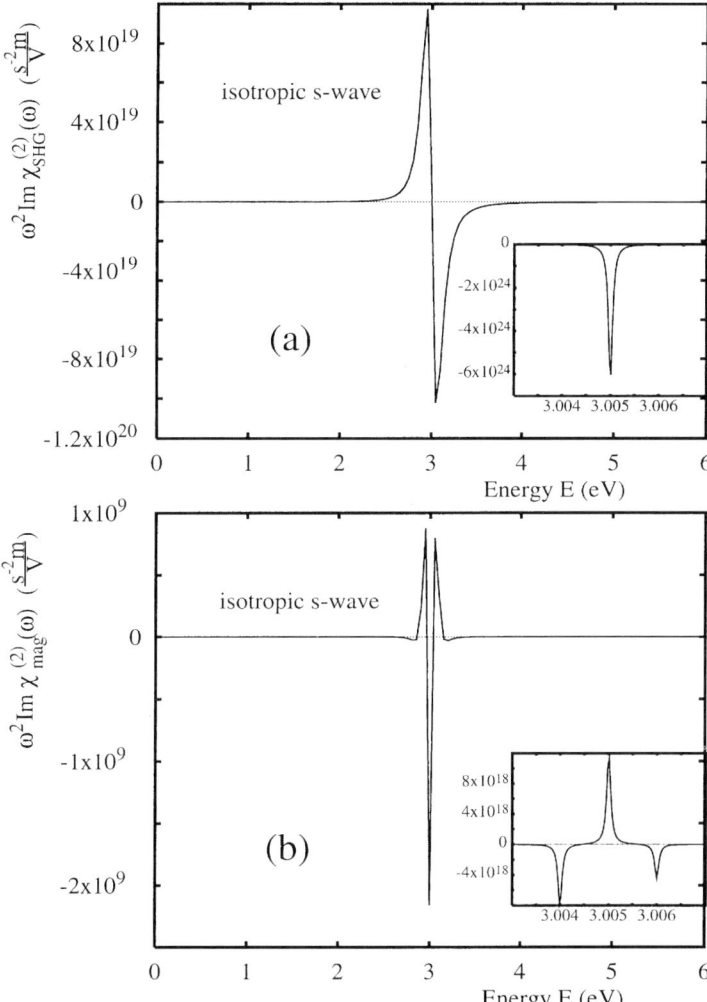

Fig. 5.17. (a) $\omega^2 \, \text{Im} \, \chi_{zzz}^{(0)}(\omega)$ and (b) $\omega^2 \, \text{Im} \, \chi_{yzz}^{(h)}(\omega)$ for an isotropic s-wave symmetry of the gap function. Note the very different line shapes, typical for the isotropic s wave. The insets show the fine structure of the results near the excitation energy of 3 eV. For better visibility the spectra are artificially broadened by a Lorentzian width δ of 50 meV in the main figures and of 0.5 meV in the insets throughout. Consequently, the inset peak heights differs from those of the main figures.

symmetry that there occurs a sign change between the magnetic and nonmagnetic optical spectrum. This is observable since the sign of the Kerr spectrum determines the direction of rotation of the polarization axis, i.e. the Kerr angle. Furthermore, the satellites cover over the range from 3 eV to 3.005 eV, i.e. only one times the gap magnitude.

For the existence of a finite Kerr signal, it is necessary to break time reversal

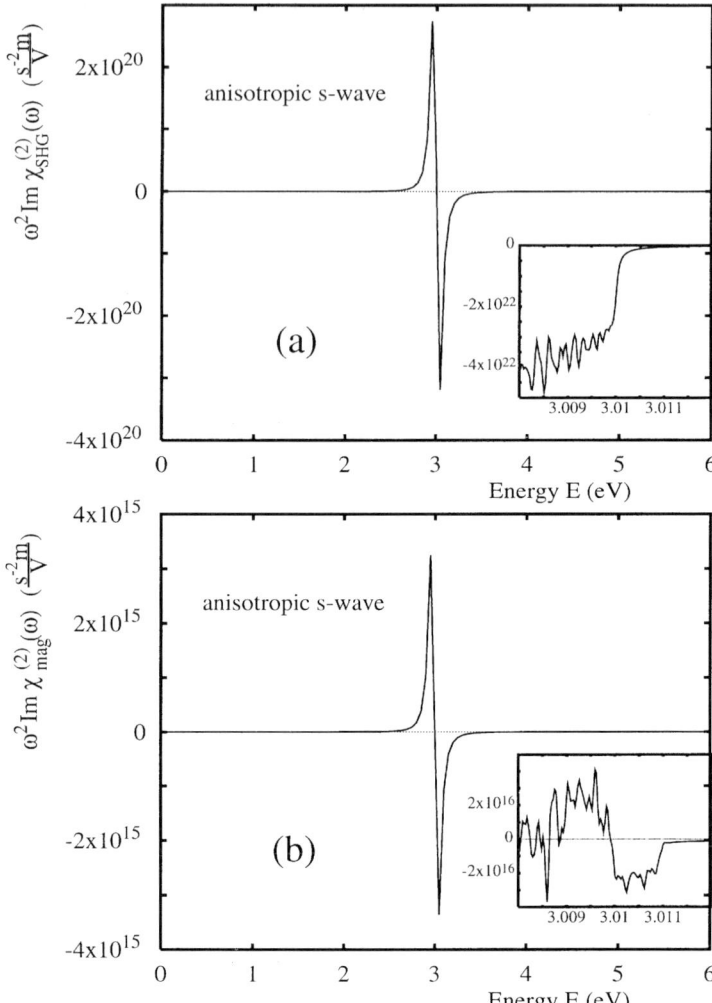

Fig. 5.18. As in Fig. 5.17, but for an anisotropic $s_{x^2+y^2}$ wave. Note the difference of the magnetic line shape compared to the isotropic s wave and the broad fine structure up to two times the magnitude of the gap, shown in the inset.

symmetry and to apply an external magnetic field. This enables one to keep any direction of the field fixed and to study the anisotropy of the effects discussed in this chapter. However, owing to the strong but short-ranged antiferromagnetic correlations, it might also be possible to take advantage of the locally broken time reversal symmetry of the high-T_c materials. Since a finite Kerr signal is expected for certain long-range ordered antiferromagnets,[94] a pump-and-probe experiment (on a timescale faster than the average lifetime of the local spin configurations $\tau_{\rm spin} \approx 10^{-11} - 10^{-13}$ s) could be able to resolve the influence of the neighbouring spins on the site that is excited by the optical excitation.

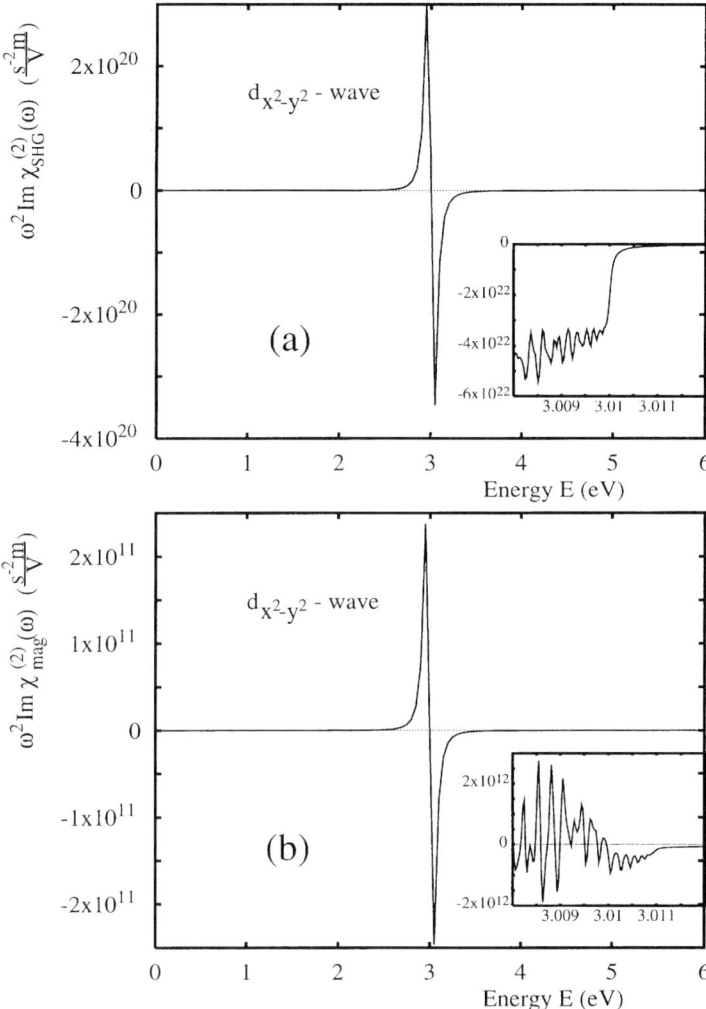

Fig. 5.19. As in Fig. 5.17, but for $d_{x^2-y^2}$ wave. The spectra are very similar to those for the $s_{x^2+y^2}$ wave, which results from the similar k dependence of the triplet pairing amplitude, given in Table 5.6. Slight differences occur for the fine structure near 3 eV.

5.3.5 Nonlinear magneto-optics in antiferromagnets

Motivation

In the last decade a great number of experimental and theoretical works have demonstrated that optical second harmonic generation (SHG) is a suitable tool not only for detecting ferromagnetism of metallic surfaces but also for determining the direction of magnetization.[14,16,43,51,60,95] Since many oxides (like oxidized transition metals) have antiferromagnetic structure, the possibility of using optical SHG also for detecting antiferromagnetic spin configurations would be especially important for the examination of oxides in catalysis. Up to now optical effects probing antiferromagnetism have only been discussed for bulk crystals.[96-012] A breakthrough in the investigation of antiferromagnetic crystal structure by means of SHG was seen with the experiments of Fiebig et al.[94] They succeeded in observing for the first time antiferromagnetic domains of bulk Cr_2O_3 by the direct coupling of light to the antiferromagnetic order parameter. In their experiments the 180° domains exhibit a pronounced intensity contrast of the transmitted SHG light. Fiebig et al. explained this phenomenon by the interference of the nonlinear electric and magnetic susceptibility tensors, which depends on the helicity of the incident circularly polarized light and on the orientation of the antiferromagnetic vector of the domain. Gros et al[103, 104] confirmed this interpretation, explaining the existence of nonvanishing elements of the electric susceptibility tensor in the antiferromagnetic state by symmetry considerations and microscopic calculations. The electronic theory reveals that this new effect results from the simultaneous action of spin–orbit coupling and tetragonal crystal distortion. For crystals with an inversion centre the nonlinear electric susceptibility tensor vanishes. In Cr_2O_3 antiferromagnetic ordering breaks the spatial inversion symmetry so that elements of $\chi_{el}^{(2\omega)}$ may be nonzero and SHG becomes allowed. Many antiferromagnetic crystals, however, possess inversion as an element of symmetry even in the antiferromagnetic phase. But at the surface or at interfaces in multilayer sandwiches (such as semiconductor or metal heterostructures) the symmetry in principle reduces. Especially space inversion symmetry disappears and nonzero elements of $\chi_{el}^{(2\omega)}$ may be found. That is why surfaces or thin films of highly symmetric antiferromagnetic crystals are particularly appropriate for examination by SHG. For linear optics Dzyaloshinskii et al.[105] recently presented a theory based on classical electrodynamics which describes a nonreciprocal optical rotation in antiferromagnetic materials consisting of ferromagnetic monolayers. This rotation is highly dependent on the parity of the number of ferromagnetic layers and, provided this number is odd, in particular on the spin direction of the first layer. Hence, it is very sensitive to surfaces of antiferromagnetic crystals, but only to those with ferromagnetic spin configuration. Moreover, the predicted rotation is small (of the order of a/λ, atomic distance over wavelength) and difficult to distinguish from additional macroscopic nonreciprocal effects in the bulk. Thus, it is a challenge to develop a theory for the direct optical probing of the antiferromagnetic order parameter of the surfaces.

At the surface the nonlinear electrical susceptibility $\chi_{el}^{(2\omega)}$ must disappear neither in the para- nor in the antiferromagnetic phase. Therefore a discrimination between these states can only be based on the existence of different nonvanishing tensor elements of $\chi_{el}^{(2\omega)}$ in the two states and not on the principal disappearance of the susceptibility tensor in one of the phases (as was the case for bulk Cr_2O_3). Thus, we determine the nonvanishing elements of $\chi_{el}^{(2\omega)}$ for different antiferromagnetic spin configurations of fcc (001), (110) and (111) surfaces.[106] Since the fcc bulk structure always remains centrosymmetric in the antiferromagnetic phase, all resulting tensor elements solely describe a surface effect. Consequently, within the electric dipole approximation any SHG signal is clearly due to the surface. We will point out that in this way for certain spin configurations surface antiferromagnetism can be detected.

Theory

For most of the possible antiferromagnetic spin configurations, unit-cell doubling occurs. Initial and doubled unit cells differ only in their symmetry of time. (They possess the same spatial symmetry, but a combination with time inversion becomes necessary in different cases.) Since we show that the inspected lattices are invariant under time reversal even in the antiferromagnetic phase, unit-cell doubling does not impair our symmetry analysis.

Maxwell's equations lead to a nonlinear electrical polarization $P_{el}^{(2\omega)}$, which is related to the incident photo field by the nonlinear electrical susceptibility $\chi_{el}^{(2\omega)}$:

$$P_{el}^{(2\omega)} = \chi_{el}^{(2\omega)} : E^{(\omega)} E^{(\omega)}. \tag{5.3.77}$$

Thereby $\chi_{el}^{(2\omega)}$ is a polar tensor of rank 3. According to von Neumann's principle this property tensor has to stay invariant under any symmetry transformation l of the lattice:[107,108]

$$\chi_{i'j'k'}^{(2\omega)} = l_{i'i} l_{j'j} l_{k'k} \chi_{ijk}^{(2\omega)}, \qquad i, j, k = x, y, z. \tag{5.3.78}$$

If time inversion $R : t \to -t$ alone or in combination with any space operation l belongs to the classifying symmetry elements, eqn (5.3.78) must be replaced by

$$\chi_{i'j'k'}^{(2\omega)} = \pm l_{i'i} l_{j'j} l_{k'k} \chi_{ijk}^{(2\omega)}, \qquad i, j, k = x, y, z, \tag{5.3.79}$$

where $-$ refers to the case when $\chi^{(2\omega)}$ changes sign under time inversion R (c tensor), and $+$ refers to the case when $\chi^{(2\omega)}$ remains invariant (i tensor).[107] In our case a symmetry analysis of the susceptibility tensor requires special care of two points: reduction of symmetry at the surface, and invariance of the considered antiferromagnetic spin configurations under time inversion. At the surface, space inversion, rotations about an axis in the surface plane, and reflections on the surface can no longer be symmetry operators. The remaining point groups, which describe the symmetry of the respective antiferromagnetic surface, usually are only subgroups of the classical of magnetic point groups characterizing the symmetry of a three-dimensional lattice. Whereas for these 'three-dimensional' point groups all allowed tensor elements have been derived already,[107,108] for the subgroups we had to calculate the nonvanishing tensor elements ourselves using eqns (5.3.78) and (5.3.79). In determining all symmetry operations of an antiferromagnetic structure it is important to bear in mind that the respective spin arrangement must remain unchanged, too. So generally some space operations l may be allowed symmetry elements only in combination with time inversion R, which corresponds to the flip of spins. All antiferromagnetic configurations we examined, however, show such a high symmetry that always a translation T about a lattice vector exists that can replace the operation R. That

Table 5.7 Nonvanishing elements of $\chi_{el}^{(2)}$ for certain spin configurations of the (001), (110) and (111) surfaces of a fcc lattice

Surface	Config.	Point group	Symmetries	Nonzero elements
(001)	para	4mm	$1, 2_z, \pm 4_z,$ $\bar{2}_x, \bar{2}_y, \bar{2}_{xy}, \bar{2}_{-xy}$	$zxx = zyy, zzz, xxz = xzx = yyz = yzy$
(001)	ferro ($M \| x$)	—	$1, \underline{2}_z, \bar{2}_x, \bar{2}_y$	$xxz = xzx, zxx, yyz = yzy, zyy, zzz$ $zzy = zyz, yzz, xxy = xyx, yxx, yyy$
(001)	AFM (a)	—	$1, \bar{2}_y$	$xxx, yyx = yxy, xyy, xxz = xzx, zxx,$ $zzx = zxz, xzz, zyy, zzz$
	(b)	—	$1, \bar{2}_x$	$xxx, xxy = xyx, yxx, yyy, xxz = xzx, zxx$
	(c)	2	$1, 2_z$	$zxx, xxz = xzx, zyy, yyz = yzy, zzz,$ $xyz = xzy, yzx = yxz, zxy = zyx$
(110)	para	mm2	$1, 2_z, \bar{2}_x, \bar{2}_y$	$zxx, xxz = xzx, zyy, yyz = yzy, zzz$
(110)	AFM	2	$1, 2_z$	$zxx, xxz = xzx, zyy, yyz = yzy, zzz,$ $xyz = xzy, yzx = yxz, zxy = zyx$
(111) >1 ML	para	3m	$1, \pm 3_z, 3(\bar{2} \perp)$	$xxx = -xyy = -yyx = -yxy, zxx = zyy,$ $xxz = xzx = yyz = yzy, zzz$
(111) >1 ML	AFM	—	$1, \bar{2}_y$	$xxx, xyy, yyx = yxy, zxx, xxz = xzx,$ $zyy, yyz = yzy, xzz, zzx = zxz, zzz$
(111) 1 ML	para	6mm	$1, 2_z, \pm 6_z, 6(2_\perp)$	$zxx = zyy, xxz = xzx = yyz = yzy, zzz$
(111) 1 ML	AFM	2	$1, 2_z$	$zxx, xxz = xzx, zyy, yyz = yzy, zzz,$ $xyz = xzy, yzx = yxz, zxy = zyx$

is, T transforms the original lattice into a spin-flipped one. Thus, RT is a symmetry element of the lattice space group. But since the nonzero elements of the physical property tensors principally follow from the point group of a crystal, and this point group is obtained from the space group by setting all translations to identity I,[107] we see that for the point groups of our surfaces $RI = R$ is a symmetry element: even in the antiferromagnetic state the three surfaces stay invariant under time reversal. Nevertheless we must not use R for deriving the tensor elements of $\chi^{(2\omega)}$. For SHG is a dynamic process with a preferred direction of time.[103,104] In this case von Neumann's principle is restricted to pure space operations of symmetry.[107,109] In this chapter there is not enough space to list the nonvanishing tensor elements for all the antiferromagnetic spin configurations of the three surfaces (001), (110) and (111) we classified. So we restrict ourselves to stating the one appropriate spin structure for every surface (see table 5.7). In all the given antiferromagnetic states, additional tensor elements appear in comparison with the paramagnetic phase.

As one sees from Table 5.7 spin configuration (c) of the (001) surface is especially suited to prove antiferromagnetism: the tensor elements $xyz = xzy$, $yzx = yxz$ and $zxy = zyx$ appear in the antiferromagnetic phase only. Both in the para- and in the ferromagnetic state ($M \| x$) they are zero. Therefore, this antiferromagnetic configuration can be detected experimentally by measuring the SHG response that is characteristic for the existence of these particular tensor elements. For that purpose you can take advantage of the polarization dependence of the SHG signal.[43,51] The reflected light at frequency 2ω is calculated by the formula

$$E^{(2\omega)}(\Phi,\phi) = 2\mathrm{i}\left(\frac{\omega}{c}\right)|E_0^{(\omega)}|^2 \begin{pmatrix} A_\mathrm{p} F_\mathrm{c} \cos\Phi \\ A_\mathrm{s} \sin\Phi \\ A_\mathrm{p} N^2 F_\mathrm{s} \cos\Phi \end{pmatrix} \chi_\mathrm{el}^{(2\omega)}$$

$$\times \begin{pmatrix} f_\mathrm{c}^2 t_\mathrm{p}^2 \cos^2\varphi \\ t_\mathrm{s}^2 \sin^2\varphi \\ f_\mathrm{s}^2 t_\mathrm{p}^2 \cos^2\varphi \\ 2 f_\mathrm{s} t_\mathrm{p} t_\mathrm{s} \cos\varphi \sin\varphi \\ 2 f_\mathrm{c} f_\mathrm{s} t_\mathrm{p}^2 \cos^2\varphi \\ 2 f_\mathrm{c} t_\mathrm{p} t_\mathrm{s} \cos\varphi \sin\varphi \end{pmatrix}, \qquad (5.3.80)$$

where $\chi_\mathrm{el}^{(2\omega)}$ is the susceptibility tensor for the respective surface configuration, Φ, φ are the angles of polarization for the reflected and fundamental light, $F_\mathrm{c,s}, f_\mathrm{c,s}$ are the corresponding Fresnel coefficients, and $T_\mathrm{s,p}, t_\mathrm{s,p}$ the linear transmission coefficients. To distinguish between the three phases in our example it is suitable to keep the SHG signal s-polarized while varying the polarization of the fundamental light. Table 5.8 shows the resulting SHG waves for all three phases. Figure 5.10 qualitatively illustrated the dependence of the s-polarized SHG signal on the polarization of the incident light. Although it is just a rough sketch, the differences between the three phases become obvious. Thus, by performing measurements with s-, p- or mixed-polarized incoming light in correct succession, it should be possible to detect the antiferromagnetic state definitely.

In conclusion, our analysis shows that nonlinear optics presents itself as a useful tool to examine antiferromagnetically ordered surfaces, even if the underlying three-dimensional lattice is of very high symmetry. The time inversion symmetry of these surfaces has not to be broken either.[110] This could be important for the observation of oxidized transition-metal surfaces like NiO, which often possess the above-mentioned symmetry features.

5.4 Microscopic theory for nonlinear magneto-optics

In this section of the chapter we develop an electronic theory of optical second harmonic generation, in particular of the *nonlinear magneto-optical Kerr effect* at ferromagnetic transition-metal surfaces and film interfaces. This theory is designed to establish a quantitative connection between the electronic band

Table 5.8 Reflected SHG signal as a function of different polarizations of the fundamental light for the (001) plane of a fcc crystal

Configuration	$P_{\text{incoming light}}$	$P_{\text{SHG light}}$	SHG wave $E^{(2\omega)}$
para	s	s	0
	p	s	0
	mix(45°)	s	$2i\left(\dfrac{\omega}{c}\right)\lvert E_0^{(2\omega)}\rvert^2\, A_s f_s t_p t_s\, xxx$
ferro	s	s	$2i\left(\dfrac{\omega}{c}\right)\lvert E_0^{(2\omega)}d\rvert^2\, A_s t_s^2\, yyy$
$(M\|x)$	p	s	$2i\left(\dfrac{\omega}{c}\right)\lvert E_0^{(2\omega)}\rvert^2\, A_s t_p^2 (f_c\, yxx + f_s^2\, yzz)$
	mix(45°)	s	$2i\left(\dfrac{\omega}{c}\right)\lvert E_0^{(2\omega)}\rvert^2\, A_s(\tfrac{1}{2} f_c^2 f_p^2\, yxx + \tfrac{1}{2} t_s^2\, yyy$ $+ \tfrac{1}{2} f_s^2 t_p^2\, yzz + f_s t_p t_s\, yyz)$
AFM	s	s	0
	p	s	$2i\left(\dfrac{\omega}{c}\right)\lvert E_0^{(2\omega)}\rvert^2\, A_s t_p^2\, 2 f_c f_s\, yzx$
	mix(45°)	s	$2i\left(\dfrac{\omega}{c}\right)\lvert E_p^{(2\omega)}\rvert^2\, A_s(f_s t_p t_s\, yyz + f_c f_s t_p^2\, yzx)$

structures on the one hand, characterizing the microscopic properties of the different materials, and the (frequency-dependent) nonlinear optical and magneto-optical spectra on the other hand. Contrary to 'jellium' theories, which focus on the material-independent features of the nonlinear response of free-electron metals, here we emphasize the different microscopic (geometric, electronic and optical) 'fingerprints' of various materials (mostly ferromagnetic transition-metal surfaces and interfaces) and how these are reflected by the nonlinear (magneto)-optical spectra.

5.4.1 Calculation of the nonlinear Kerr susceptibility

For a convenient derivation of the nonlinear optical response we use the self-consistent field theory in the form first proposed by Ehrenreich and Cohen,[111] and we extend and adapt it to transition metals. As a further step the required nonlinearity is included. Finally ferromagnetic exchange and spin–orbit coupling are incorporated in the theory.

Linear response

In this subsection we calculate the linear optical susceptibility $\chi^{(1)}$, a second-rank tensor, which is related to the linear optical dielectric tensor $\varepsilon^{(1)}$ via

$$\chi^{(1)} = \frac{\varepsilon^{(1)} - 1}{4\pi}. \tag{5.4.1}$$

Here the linear optical dielectric function is defined as

$$\varepsilon(q, \omega) = 1 - \frac{4\pi e^2}{q^2} \langle\langle \rho(q,t); \rho(-q) \rangle\rangle_\omega, \tag{5.4.2}$$

and the linear dielectric polarization is given by

$$P(q, \omega) = \chi^{(1)}(q, \omega) E(q, \omega). \tag{5.4.3}$$

Here, the tensor indices have been removed for notational simplicity. Instead of a diagrammatic formulation requiring a resummation of an infinite series, we use the equation-of-motion method in a self-consistent field. Thus we consider the Liouville equation of motion for the density operator

$$i\hbar \dot\rho = [H, \rho], \tag{5.4.4}$$

where

$$H = H_0 + V(r, t) \tag{5.4.5}$$

is the Hamiltonian of the total system. H_0 is the Hamiltonian of the unperturbed solid and satisfies the Schrödinger equation

$$H_0 |kl\rangle = E_{kl} |kl\rangle. \tag{5.4.6}$$

Here the eigenvalues E_{kl} correspond to the Bloch states $|kl\rangle = (1/\sqrt{\Omega})u_{kl} e^{ikr}$ of wavevector k in band l of the crystal (volume Ω). The Fourier transform of the self-consistent potential reads

$$V(r, t) = \sum_{q'} V(q', t) e^{-iq'r}. \tag{5.4.7}$$

This potential consists of the external potential (e.g. of the external field of the laser photons) and of the screening potential of the electrons in the solid.

Similarly we expand the density operator ρ in the form

$$\rho = \rho^{(0)} + \rho^{(1)}(\omega) + \cdots, \tag{5.4.8}$$

where $\rho^{(0)}$ satisfies the eigenvalue equation

$$\rho^{(0)} |kl\rangle = f(E_{kl}) |kl\rangle, \tag{5.4.9}$$

with the Fermi function f as eigenvalue. Taking into account exclusively first-order terms in the commutator of eqn (5.4.4) and using the eigenvalue equations (5.4.6) and (5.4.9) one gets the relation for the matrix element of the density operator ρ between the states $|kl\rangle$ and $|k+q, l'\rangle$ from the equation of motion

$$\begin{aligned}
i\hbar \frac{d}{dt} \langle kl| \rho^{(1)} |k+q, l'\rangle \\
\equiv (-\hbar\omega + i\hbar\alpha) \langle kl| \rho^{(1)} |k+q, l'\rangle \\
= \langle kl| [H_0, \rho^{(1)}] |k+q, l'\rangle + \langle kl| [V, \rho^{(0)}] |k+q, l'\rangle \\
= (E_{kl} - E_{k+q, l'}) \langle kl| \rho^{(1)} |k+q, l'\rangle \\
+ [f(E_{k+q, l'}) - f(E_{kl})] V(q, t) \langle kl| e^{-iq\cdot r} |k+q, l'\rangle.
\end{aligned} \tag{5.4.10}$$

The time dependence of the matrix elements is assumed as $e^{i\omega t + \alpha t}$, and α corresponds to a finite lifetime of self-energy due to electron–electron interaction or the experimental resolution.

The polarization $P(q,t)$ is related to the induced change of the electron density via

$$\nabla \cdot P^{(1)}(q,t) = en^{(1)}(q,t) \quad \Rightarrow \quad iq \cdot P^{(1)}(q,t) = en^{(1)}(q,t). \quad (5.4.11)$$

Using the electric field

$$E(q,t) = \frac{iq}{e} V(q,t) \quad (5.4.12)$$

and eqns (5.4.10) and (5.4.11), the linear optical susceptibility is given by

$$\chi^{(1)}(q,\omega) = -\frac{e^2}{q^2} \frac{n^{(1)}(q,t)}{V(q,t)}$$

$$= \frac{-\dfrac{e^2}{q^2 \Omega} \sum_{k',l,l'} \langle k'+q,l' | e^{iq \cdot r} | k'l \rangle \langle k'l | \rho^{(1)} | k'+q,l' \rangle}{V(q,t)}.$$

$$(5.4.13)$$

For the external potential and therefore also for the screening potential (thus for the total self-consistent potential) we assume the same time dependence $e^{i\omega t + \alpha t}$ as for the matrix elements $\langle kl | \rho^{(1)} | k+q,l' \rangle$. With this convention we obtain the following expression for the optical dielectric susceptibility:

$$\chi^{(1)}(q,\omega) = -\frac{e^2}{q^2 \Omega} \sum_{k,l,l'} |\langle kl | e^{-iq \cdot r} | k+q,l' \rangle|^2$$

$$\times \frac{f(E_{k+q,l'}) - f(E_{kl})}{E_{k+q,l'} - E_{kl} - \hbar\omega + i\hbar\alpha}. \quad (5.4.14)$$

Here it is important to note that photons have a linear dispersion relation ($\omega = cq$) implying $|q| \ll 1/a$ where a is the lattice constant. Therefore the longitudinal and transverse dielectric susceptibilities are equal for optical photons. Thus the above expression contains both particle–hole excitations (interband excitations) for $l \neq l'$ and the Drude (intraband) term,[112] which is responsible for the plasmons. To express its tensorial nature one just has to multiply $\chi^{(1)}(q,\omega)$ by $q \otimes q/q^2$.

Nonlinear response

In this subsection we extend the self-consistent field method to terms of second order in the electric field in order to calculate the nonlinear optical susceptibility $\chi^{(2)}$, a third-rank tensor. The second-order polarization is given by

$$P^{(2)}(2q, 2\omega) = \chi^{(2)}(2q, 2\omega) E(q,\omega) E(q,\omega). \quad (5.4.15)$$

The only difference compared to the potential V in eqns (5.4.12) and (5.4.13), which varies with frequency ω, is due to the fact that only the screening potential $V_s^{(2)}$ occurs in second order, since the incident light consists only of photons of frequency ω. Therefore the expansions of the Hamiltonian and the density operator up to second order are

$$H = H_0 + V^{(1)}(\boldsymbol{q},t) + V_s^{(2)}(2\boldsymbol{q},t) \tag{5.4.16}$$

and

$$\rho = \rho^{(0)} + \rho^{(1)}(\omega) + \rho^{(2)}(2\omega), \tag{5.4.17}$$

where the static contributions again satisfy the eigenvalue equations (5.4.6) and (5.4.9). Here local field effects are neglected due to the extended nature of metallic Bloch wavefunctions. They will be included if intensities rather than bare susceptibilities are discussed. It then turns out that the dominant contributions to nonlinear optics generally result from the nonlinear susceptibilities rather than the Fresnel coefficients, thus supporting the point of view adopted for simplicity in the present discussion.

In second order the polarization and the induced electron density are connected to each other by the equation

$$\nabla \cdot \boldsymbol{P}^{(2)}(2\boldsymbol{q},t) = en^{(2)}(2\boldsymbol{q},t) \quad \Rightarrow \quad 2i\boldsymbol{q} \cdot \boldsymbol{P}^{(2)}(2\boldsymbol{q},t) = en^{(2)}(2\boldsymbol{q},t). \tag{5.4.18}$$

For the matrix elements in second order

$$\langle \boldsymbol{k}l | \rho^{(2)} | \boldsymbol{k}+2\boldsymbol{q}, l'' \rangle$$

and the nonlinear screening potential $V_s^{(2)}(2\omega)$, we assume the time dependence $e^{2i\omega t + 2\alpha t}$ and collect the second-order terms in the equation of motion (5.4.4) for these matrix elements

$$i\hbar \frac{d}{dt} \langle \boldsymbol{k}l | \rho^{(2)} | \boldsymbol{k}+2\boldsymbol{q}, l'' \rangle$$
$$= \langle \boldsymbol{k}l | [H_0, \rho^{(2)}] | \boldsymbol{k}+2\boldsymbol{q}, l'' \rangle$$
$$+ \langle \boldsymbol{k}l | [V^{(1)}, \rho^{(1)}] | \boldsymbol{k}+2\boldsymbol{q}, l'' \rangle + \langle \boldsymbol{k}l | [V_s^{(2)}, \rho^{(0)}] | \boldsymbol{k}+2\boldsymbol{q}, l'' \rangle,$$
$$\Leftrightarrow \quad (-2\omega + 2i\hbar\alpha) \langle \boldsymbol{k}l | \rho^{(2)} | \boldsymbol{k}+2\boldsymbol{q}, l'' \rangle$$
$$= (E_{\boldsymbol{k}l} - E_{\boldsymbol{k}+2\boldsymbol{q},l''}) \langle \boldsymbol{k}l | \rho^{(2)} | \boldsymbol{k}+2\boldsymbol{q}, l'' \rangle + \langle \boldsymbol{k}l | [V^{(1)}, \rho^{(1)}] | \boldsymbol{k}+2\boldsymbol{q}, l'' \rangle$$
$$+ [f(E_{\boldsymbol{k}+2\boldsymbol{q},l''}) - f(E_{\boldsymbol{k}l})] V_s^{(2)}(2\boldsymbol{q},t) \langle \boldsymbol{k}l | e^{-2i\boldsymbol{q}\cdot\boldsymbol{r}} | \boldsymbol{k}+2\boldsymbol{q}, l'' \rangle. \tag{5.4.19}$$

It follows from Poisson's equation for the nonlinear screening potential

$$\nabla^2 V_s^{(2)} = -4\pi e^2 n^{(2)}, \tag{5.4.20}$$

that

$$V_s^{(2)}(2\boldsymbol{q},t) = \frac{\pi e^2}{q^2 \Omega} \sum_{\boldsymbol{k}',l,l'} \langle \boldsymbol{k}'+2\boldsymbol{q}, l'' | e^{2i\boldsymbol{q}\cdot\boldsymbol{r}} | \boldsymbol{k}'l \rangle \langle \boldsymbol{k}'l | \rho^{(2)} | \boldsymbol{k}'+2\boldsymbol{q}, l'' \rangle.$$

$$\tag{5.4.21}$$

Using eqns (5.4.15), (5.4.18) and (5.4.21), as well as the first-order result for $\langle kl| \rho^{(1)} |k+q,l'\rangle$ and summing over k, l, l' and l'' finally yields the nonlinear optical dielectric function

$$\chi^{(2)}(2q, 2\omega) = \frac{-ie^3}{2q^3\Omega}$$

$$\times \sum_{k,l,l',l''} \Bigg((\langle k+2q,l''|e^{2iq\cdot r}|kl\rangle \langle kl|e^{-iq\cdot r}|k+q,l'\rangle \langle k+q,l'|e^{-iq\cdot r}|k+2q,l''\rangle)$$

$$\times \frac{\left(\dfrac{f(E_{k+2q,l''}) - f(E_{k+q,l'})}{E_{k+2q,l''} - E_{k+q,l'} - \hbar\omega + i\hbar\alpha} - \dfrac{f(E_{k+q,l'}) - f(E_{kl})}{E_{k+q,l'} - E_{kl} - \hbar\omega + i\hbar\alpha} \right)}{E_{k+2q,l''} - E_{kl} - 2\hbar\omega + 2i\hbar\alpha}$$

$$\times \frac{1}{1 - \dfrac{\pi e^2}{q^2\Omega} \sum_{k,l,l''} |\langle kl|e^{-2iq\cdot r}|k+2q,l''\rangle|^2 \dfrac{f(E_{k+2q,l''}) - f(E_{kl})}{E_{k+2q,l''} - E_{kl} - 2\hbar\omega + 2i\hbar\alpha}} \Bigg). \quad (5.4.22)$$

Surface sensitivity of second harmonic generation

Since the product of the three transition matrix elements vanishes in the bulk of inversion-symmetric media within the electric dipole approximation, the dominant contributions to optical second harmonic generation result from the surface. In thin films, the contributions arise from each interface, in particular the buried interfaces, which are therefore uniquely accessible by nonlinear optical probes. Thus in a multilayer configuration bulk contributions are completely negligible while they may contribute via higher-order multipoles at surfaces of bulk matter. The surface and interface sensitivity of optical second harmonic generation, which is not destroyed by the occurrence of a spontaneous magnetization, which is an axial vector, can be seen in different ways:

1. An expansion of the polarization P in terms of the electric field E

$$P(\omega) = \chi^{(1)}(\omega)E(\omega) + \chi^{(2)}(2\omega)E(\omega)E(\omega) + \cdots \quad (5.4.23)$$

requires that P changes its sign in inversion-symmetric media if E does so. Thus all even orders of the susceptibility tensors χ vanish in bulk, i.e. in particular $\chi^{(2)}$. Only at the surface or at film interfaces, where inversion symmetry or translational symmetry is broken due to the incompleteness of the crystal, can second harmonic radiation be generated.

2. In bulk matter each electric dipole transition (described by a dipole matrix element) is linked to a parity change between initial and final states ($\Delta l = \pm 1$). If initial and final states are identical, as is the case in the optical absorption and emission of photons, only processes involving an even number of photons may take place. Therefore in particular *three*-photon processes such as optical second

harmonic generation (absorption of two photons of frequency ω and emission of a third photon of frequency 2ω or the inverse process, respectively) are forbidden in inversion-symmetric media. Higher-order terms, in particular bulk quadrupole contributions, are much smaller in metals. This has also been demonstrated experimentally on pure surfaces[113] and in many adsorption experiments.[85]

3. The above formula for $\chi^{(2)}(2q, 2\omega)$ can be rewritten by splitting the wavevector q into components parallel and perpendicular to the metal surface (no reaction forces parallel to the surface). This calculation essentially yields that one has to replace the three-dimensional Fourier transforms by two-dimensional ones, q by $q_\parallel$, and the volume Ω by the surface area A. This procedure yields the same result as derived by Persson et al.[114-117] for the linear surface dielectric function $g(q_\parallel, \omega)$ by integrating the Poynting vector over the metal surface:

$$\operatorname{Im} g(q_\parallel, \omega) = \frac{2\pi^2}{q_\parallel A} \sum_{\gamma, \gamma'} |\langle \gamma | \Phi_1 | \gamma' \rangle|^2 (f_\gamma - f_{\gamma'}) \delta(\varepsilon_\gamma - \varepsilon_{\gamma'} - \hbar\omega + i\hbar\alpha). \tag{5.4.24}$$

Here $|\gamma\rangle, |\gamma'\rangle$ are the single-particle wavefunctions of the solid and Φ_n is given by

$$\Phi_n = (1 - g(0, n\omega)) e^{-niq_\parallel \cdot r_\parallel - n q_\parallel z}. \tag{5.4.25}$$

In the nonlinear case we find the following result for the nonlinear optical surface dielectric function

$$\chi^{(2)}(2q_\parallel, 2\omega) = \frac{e^3 |q_\parallel| a}{q_\parallel^2 A}$$

$$\times \sum_\sigma \Bigg(\sum_{k,l,l',l''} \langle k + 2q_\parallel, l'' | e^{2iq_\parallel \cdot r} | kl\sigma \rangle \langle kl\sigma | e^{-iq_\parallel \cdot r} | k + q_\parallel, l'\sigma \rangle$$

$$\times \langle k + q_\parallel, l'\sigma | e^{-iq_\parallel \cdot r} | k + 2q_\parallel, l''\sigma \rangle$$

$$\times \frac{\left(\dfrac{f(E_{k+2q_\parallel, l''\sigma}) - f(E_{k+q_\parallel, l'\sigma})}{E_{k+2q_\parallel, l''\sigma} - E_{k+q_\parallel, l'\sigma} - \hbar\omega + i\hbar\alpha} - \dfrac{f(E_{k+q_\parallel, l'\sigma}) - f(E_{kl\sigma})}{E_{k+q_\parallel, l'\sigma} - E_{kl\sigma} - \hbar\omega + i\hbar\alpha} \right)}{E_{k+2q_\parallel, l''\sigma} - E_{kl\sigma} - 2\hbar\omega + 2i\hbar\alpha}$$

$$\times \frac{1}{1 - \dfrac{\pi e^2}{q_\parallel A} \sum_{k,l,l''} |\langle kl\sigma | e^{-2iq_\parallel \cdot r} | k + 2q_\parallel, l''\sigma \rangle|^2 \dfrac{f(E_{k+2q_\parallel, l''\sigma}) - f(E_{kl\sigma})}{E_{k+2q_\parallel, l''\sigma} - E_{kl\sigma} - 2\hbar\omega + 2i\hbar\alpha}} \Bigg).$$

$$\tag{5.4.26}$$

The most important feature of this formula is the prefactor qa, which would be absent if the matrix elements were calculated from first principles, since in that case the wavefunctions would govern the symmetry and response depth of the nonlinear surface response.

Illustration of the factor qa

A scenario of frequency doubling at a metal surface ('jellium') looks as follows:

1. Light of frequency ω hits the surface and penetrates one skin depth ($1/q_\parallel$).
2. It generates a polarization $P(\omega)$ up to a depth of $1/q$.
3. This volume polarization causes an oscillatory charge density $Q(\omega)$ within the interface zone of one lattice constant a in depth were the symmetry is broken (incomplete unit cell; static charge density decays from its bulk value to zero).
4. The charge density $Q(\omega)$ interacts again with the incident light $E(\omega)$ within the surface or interface zone a.
5. Then the resulting frequency-doubled field $E(2\omega)$ is radiated, which exclusively carries information about the geometric, electronic and magnetic properties of the interface zone of depth a.

So the factor $q_\parallel a$ reflects the ratio of linear excitation depth and nonlinear response depth.

5.4.2 Second harmonic generation and magnetism

In order to extend the theory of optical second harmonic generation to ferromagnetic systems, one first has to clarify the microscopic mechanism that causes the linear and nonlinear magneto-optical Kerr effects.

Microscopic mechanism of magneto-optics

In 1876 Kerr was the first to observe the rotation of the polarization plane of linearly polarized light reflected from a ferromagnet. This rotation and the small ellipticity observed at the same time were not proportional to the externally applied field but rather depended exclusively on the sample magnetization. The basic mechanism responsible for this phenomenon was not resolved until 1955 when Argyres[118] showed that the Kerr effect is caused by the combined action of exchange interaction and spin–orbit coupling.

Spin–orbit coupling: Spin–orbit coupling acts on the electrons like an external magnetic field:

$$-i\hbar\dot{p} = [H_{so}, p] = [\lambda_{so} L \cdot S, p] = \lambda_{so} p \times S = \text{force}, \qquad (5.4.27)$$

rotates the polarization plane of linearly polarized light, and introduces an additional small ellipticity.

Exchange interaction In paramagnetic materials the clockwise and counterclockwise rotations induced by spin−orbit coupling cancel for the two directions of the electron spins. Thus no net effect remains. In ferromagnets, however, the exchange interaction splits the electronic band structure (majority and minority bands are shifted with respect to each other), and the densities of states at fixed energy become different for both spin directions. Thus the two Kerr rotations for majority and minority electrons no longer cancel and a net Kerr rotation occurs. This Kerr rotation angle is strongly dependent on geometry and frequency as has been shown in sec. 5.3.3. In nonlinear magneto-optics this rotation $\phi_K^{(2)}$ is enhanced compared to linear magneto-optics by one (for surfaces) to three (for thin films) orders of magnitude, and may become as large as 90° in a multilayer situation for standard materials such as Fe. In linear optics, however, the Kerr rotation $\phi_K^{(1)}$ is typically small and lies in the range of 0.1° (for Ni) to 0.5° (for Fe). To increase the Kerr angle in linear optics one has to resort to Heusler alloys such as PtMnSb or PtBi, exhibiting Kerr rotations of 1°, or to alloys such as NiUSe ($\phi_K^{(1)} = 5°$ in the infrared and $\phi_K^{(1)} = 2°$ in the visible range) or TmS, TmSe, CeSb ($\phi_K^{(1)} = 14°$ at energy $\hbar\omega = 0.5\,\text{eV}$, magnetic field $B = 5\,\text{T}$ and temperature $T = 2\,\text{K}$), $CeSb_{0.75}Te_{0.25}$ ($\phi_K^{(1)} = 3°$ at energy $\hbar\omega = 1.0\,\text{eV}$, magnetic field $B = 6\,\text{T}$ and temperature $T = 2\,\text{K}$) and CeTe ($\phi_K^{(1)} = 3°$ at energy $\hbar\omega = 2.0\,\text{eV}$, magnetic field $B = 5\,\text{T}$ and temperature $T = 2\,\text{K}$).[119] While the uranium compound is radioactive, the other alloys require relatively low temperatures and a precise frequency selection in a narrow window ($\pm\Delta\hbar\omega = \pm 0.1\,\text{eV}$) for the observation of the enhanced Kerr rotation. Thus nonlinear magneto-optics may overcome these difficulties.

In the 1960s theoretical work was mainly concerned with different derivations of the linear magneto-optical response functions and the discussion of the interband and intraband contributions.[120,121] At that time no comparison with experimental spectra was made. For the Ce compounds there was a controversy whether the origin of the enhanced linear Kerr angle results from an effect of the plasma resonance or form a transition of the Ce^{3+} ion from the $4f^1$ to the $4f^0 5d^1$ configuration. Thus it was first of interest to gain a microscopic understanding of the linear Kerr effect in the less complicated 3d transition metals.

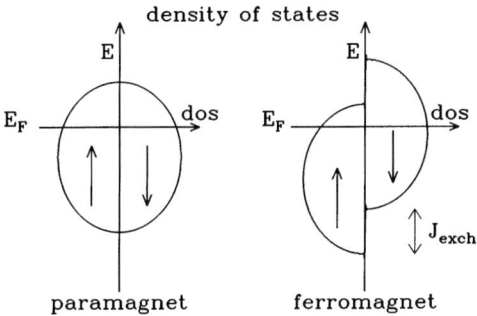

Fig. 5.20. Necessity of the exchange interaction for the occurrence of a Kerr rotation.

Despite the existence of various spin-polarized band-structure calculations for ferromagnetic transition metals,[122-127] only in the 1990s did we start to gain a microscopic understanding of the linear Kerr spectra (besides one early attempt).[128]. Since it is the goal of this work to review the theory of *nonlinear* magneto-optics (where the spectroscopic measurements are still in their infancy), it is of importance to test this theory also by comparison of the *linear* Kerr spectra (which always are calculated as a by-product) with available experimental linear Kerr spectra.

Nonlinear magneto-optical susceptibility

In order to calculate the nonlinear magneto-optical susceptibility $\chi^{(2)}(2q, 2\omega)$, which is the central quantity responsible for the nonlinear Kerr spectrum, we start from eqn (5.4.26). As additional interactions, *spin–orbit coupling* and *exchange interaction* have to be included to describe the additional nonlinear optical effects induced by magnetism. The exchange interaction is reflected by the spin splitting of the band structure $E_{kl\sigma}$ (with additional spin index σ), while spin–orbit coupling is treated to a good approximation within first-order Heitler–Ma perturbation theory in the spin–orbit coupling constant λ_{so}. We proceed similarly to the treatment used by Cohan and Hameka[129] for external magnetic fields in gases and liquids. Although we will later incorporate spin–orbit coupling from first principles, the present approach is justified as a first approximation for the following five reasons:

1. The method contains the correct physics since both the linear and the nonlinear magneto-optical Kerr effects depend linearly on spin–orbit coupling.
2. Our method saves computer time, which can be used more efficiently for other parts of the nonlinear calculations. Note that the computation of a single convergent nonlinear Kerr spectrum was already at the limits of the computational power available around 1990 (even within a tight-binding scheme). Thus any simplification not giving up the decisive physics of nonlinear magneto-optics had to be used in the early stages of the theoretical work.
3. The spin–orbit coupling constants in the 3d transition metals is 50 meV (Fe) or 70 meV (Ni) and thus small compared to typical band separations and optical excitations (some eV). So a perturbation treatment seems justified.
4. Comparisons of perturbation theory with the incorporation of the spin–orbit operator in the diagonal ($L_z S_z$) and off-diagonal elements

$$L_x S_x + L_y S_y = \tfrac{1}{2}(L_+ S_- + L_- S_+) \tag{5.4.28}$$

   of the Hamiltonian matrix[47,130,131] yields no significant deviation of the linear Kerr spectra from the linear dependence on λ_{so}.
5. In contrast to our method, fully relativistic method of band-structure calculation (solution of the Dirac equation) have the disadvantage that besides the

neglect of some of the electron–electron interactions relevant for ferromagnetic transition metals (as usual in density-functional calculations employing the local-density approximation (LDA)[132]), the single-particle band positions are worse than in nonrelativistic band-structure calculations.[133]

If we apply perturbation theory to the product of three dipole matrix elements entering the nonlinear susceptibility (which corresponds to the three photons contributing to each elementary frequency-doubling process)

$$\langle k+2q_\|, l''|e^{2iq\cdot r}|kl\sigma\rangle\langle kl\sigma|e^{-iq\cdot r}|k+q, l'\sigma\rangle\langle k+q, l'\sigma|e^{-iq\cdot r}|k+2q, l''\sigma\rangle,$$

we use first-order perturbation theory for each of the six occurring wavefunctions and each of the three occurring dipole operators. Then we collect all terms of first order in spin–orbit coupling, which are perturbed in *one* wavefunction or in *one* dipole operator and are unperturbed in all other *eight* quantities. So we obtain the following *nonlinear magneto-optical surface response function*:

$$\chi^{(2)}(2q_\|, 2\omega) = \frac{e^2\hbar|q_\||a}{4m^2c^2q_\|^2 A}\sum_\sigma \Bigg\{\sum_{k,l,l',l''}\Bigg[\sum_{k_1} P\Bigg(\frac{V_{k_1;k+2q_\|,l''\sigma}[k_1\times(k+2q_\|)\langle\sigma\rangle]}{E_{k+2q_\|,l''\sigma}-E_{k_1}}$$

$$\times\langle k_1|e^{2iq_\|\cdot r}|kl\sigma\rangle\langle kl\sigma|e^{-iq_\|\cdot r}|k+q_\|, l'\sigma\rangle$$

$$\times\langle k+q_\|, l'\sigma|e^{-iq_\|\cdot r}|k+2q_\|, l''\sigma\rangle\Bigg)$$

$$+Q(V_{kl\sigma;k+2q_\|,l''\sigma}[k\times(k+2q_\|)\langle\sigma\rangle]\langle kl\sigma|e^{-iq_\|\cdot r}|k+q_\|, l'\sigma\rangle$$

$$\times\langle k+q_\|, l'\sigma|e^{-iq_\|\cdot r}|k+2q_\|, l''\sigma\rangle)\Bigg]$$

$$\times\frac{\left(\dfrac{f(E_{k+2q_\|,l''\sigma})-f(E_{k+q_\|,l'\sigma})}{E_{k+2q_\|,l''\sigma}-E_{k+q_\|,l'\sigma}-\hbar\omega+i\hbar\alpha}-\dfrac{f(E_{k+q_\|,l'\sigma})-f(E_{kl\sigma})}{E_{k+q_\|,l'\sigma}-E_{kl\sigma}-\hbar\omega+i\hbar\alpha}\right)}{E_{k+2q_\|,l''\sigma}-E_{kl\sigma}-2\hbar\omega+2i\hbar\alpha}$$

$$\times\frac{1}{1-\dfrac{\pi e^2}{q_\| A}\sum_{k,l,l''}|\langle kl\sigma|e^{-2iq_\|\cdot r}|k+2q_\|, l''\sigma\rangle|^2\dfrac{f(E_{k+2q_\|,l''\sigma})-f(E_{kl\sigma})}{E_{k+2q_\|,l''\sigma}-E_{kl\sigma}-2\hbar\omega+2i\hbar\alpha}}\Bigg\}.$$

(5.4.29)

Here, $P(\ )$ is the sum of the six first-order permutations where the wavefunctions are perturbed, which therefore contain an additional energy denominator, while $Q(\ )$ describes the permutation of the three terms without additional energy denominator, where one of the three dipole operators has been perturbed by spin–orbit coupling. The quantity $V_{k_1;k+2q_\|,l''\sigma}[k_1\times(k+2q_\|)\langle\sigma\rangle]$

describes spin–orbit coupling where $V_{k_1;k+2q_\parallel,l''\sigma}$ is the matrix element of the full crystal potential. It is important to note again that the interface sensitivity of $\chi^{(2)}(2q_\parallel,2\omega)$ is governed by the wavefunctions entering the dipole matrix elements. If these matrix elements are treated as constants (independent of k) the prefactor $q_\parallel a$ guarantees the surface sensitivity. For further calculation it makes sense to simplify the tensor $\chi^{(2)}(2q_\parallel,2\omega)$. We will apply four approximations, which are not very restrictive but give a drastic simplification of the nonlinear magneto-optical susceptibility:

1. We neglect the nonlinear screening term

$$\frac{1}{1 - \frac{\pi e^2}{q_\parallel A} \sum_{k,l,l''} |\langle kl\sigma|e^{-2iq_\parallel \cdot r}|k+2q_\parallel,l''\sigma\rangle|^2 \frac{f(E_{k+2q_\parallel,l''\sigma}) - f(E_{kl\sigma})}{E_{k+2q_\parallel,l''\sigma} - E_{kl\sigma} - 2\hbar\omega + 2i\hbar\alpha}} \quad (5.4.30)$$

in the denominator. The calculation of this term would require the inversion of this linear (magneto-)optical tensor, which needs knowledge of *all* elements:

$$\left[1 - \chi^{(1)}(2q_\parallel,2\omega)\right]^{-1} \to \left[\delta_{ij} - \chi^{(1)}_{ij}(2q_\parallel,2\omega)\right]^{-1}. \quad (5.4.31)$$

Furthermore this term is of the order of magnitude of the linear response function $\chi^{(1)}(2q_\parallel,2\omega)$ and has decayed from a value of order unity at an energy of $\hbar\omega = 0.5\,\text{eV}$ to 0.15 at $\hbar\omega = 2\,\text{eV}$. Thus this contribution becomes negligible at not too low (= optical) frequencies.

2. Kittel[134] (cf. Argyres[118]) showed by a simple estimate of the orders of magnitude that the linear magneto-optical Kerr effect results from the spin–orbit induced change of the wavefunctions. Even if the optical momentum of ferromagnetic transition metals were not completely quenched, the change of the energy levels induced by the diagonal part of spin–orbit coupling would be very small compared to the optical excitation energy and the exchange splitting. This level shift is of second order in spin–orbit coupling while the effect upon the wavefunctions (which involves no spin-flip processes) is of first order and therefore dominates. For completely quenched orbital angular momentum there would be no effect of spin–orbit coupling upon the energy denominators at all. However it is not *a priori* clear that the effect of the spin–orbit perturbation of the dipole operators is negligible compared to the effect upon the wavefunction. But it turns out that for crystals with 90° rotation axes the operator effect vanishes to first order.[135] Note that the conventional argument by Misemir[130] comparing the two terms of the matrix elements

$$\left(\frac{-i\hbar}{E_k - E_{k+q}} r_{k;k+q} + \frac{\hbar}{4mc^2}(\sigma+r)_{k;k+q}\right)\left(\frac{\partial V}{\partial r}\right)_{k;k+q} \quad (5.4.32)$$

is not sufficient since it neglects the fact that the first term yields no magneto-optical effect in the absence of spin–orbit coupling in the wavefunctions. Thus the first matter element vanishes no matter how small its energy denominator may be. Since the second term vanishes only for certain symmetry conditions, it might be interesting in the future to work it out in more detail for interfaces or (111) surfaces which lack 90° rotational symmetry. For cubic systems we neglect the $Q(\ )$ terms.

3. For energies $\hbar\omega \geq 0.5\,\mathrm{eV}$ the main contribution to $\chi^{(2)}$ results from interband transitions while the Drude (intraband) contribution rapidly decays ($\sim 1/\omega^2$) and therefore becomes negligible already at energies of several hundred meV. So we keep only interband transitions in $\chi^{(2)}$ within the electric dipole approximation in the expansion of the exponentials. Furthermore, SHG experiments show clearly that higher multipole orders are usually unimportant at surfaces of metals: the electric quadrupole contribution in linear optics is only of the order of the electric dipole term in SHG.[136] In films and multilayers higher-order multipoles are negligible anyway.

4. The treatment of spin–orbit coupling can be simplified considerably in line with the use of constant matrix elements. Expanding the exponentials the transition matrix elements become of the form

$$\langle \boldsymbol{k}_1|x|kl\sigma\rangle\langle kl\sigma|z|\boldsymbol{k}+\boldsymbol{q}_\|,l'\sigma\rangle\langle \boldsymbol{k}+\boldsymbol{q}_\|,l'\sigma|z|\boldsymbol{k}+2\boldsymbol{q}_\|,l''\sigma\rangle. \quad (5.4.33)$$

Replacing the x matrix element by a z matrix element, which contains the unperturbed wavefunction rather than the state $\boldsymbol{k}_1$, and replacing the prefactor

$$\frac{V_{\boldsymbol{k}_1;\boldsymbol{k}+2\boldsymbol{q}_\|,l''\sigma}[\boldsymbol{k}_1\times(\boldsymbol{k}+2\boldsymbol{q}_\|)\langle\boldsymbol{\sigma}\rangle]}{E_{\boldsymbol{k}+2\boldsymbol{q}_\|,l''\sigma}-E_{\boldsymbol{k}_1}} \quad (5.4.34)$$

by

$$\frac{\lambda_{\mathrm{so}}}{\hbar\omega}, \quad (5.4.35)$$

its average value, where λ_{so} is the atomic spin–orbit coupling parameter, we effectively replace the matrix element

$$\langle \boldsymbol{k}+2\boldsymbol{q}_\|,l''\sigma|z|kl\sigma\rangle \quad (5.4.36)$$

by

$$\frac{\lambda_{\mathrm{so}}}{\hbar\omega}\langle \boldsymbol{k}+2\boldsymbol{q}_\|,l''\sigma|z|kl\sigma\rangle \quad (5.4.37)$$

rather than by the more complicated expression

$$\sum_{\boldsymbol{k}_1}\frac{V_{\boldsymbol{k}_1;\boldsymbol{k}+2\boldsymbol{q}_\|,l''\sigma}[\boldsymbol{k}_1\times(\boldsymbol{k}+2\boldsymbol{q}_\|)\langle\boldsymbol{\sigma}\rangle]}{E_{\boldsymbol{k}+2\boldsymbol{q}_\|,l''\sigma}-E_{\boldsymbol{k}_1}}\langle \boldsymbol{k}_1|x|kl\sigma\rangle. \quad (5.4.38)$$

The prefactor $\lambda_{\mathrm{so}}/\omega$ is a measure of the effective strength of the perturbation.

Since computational methods have improved since the theory was formulated for the first time, it is a goal of present research to give up this approximation. However, up to now no code exists that can handle magnetism and spin–orbit coupling without the spherical approximation for an interface situation in magneto-optics at the same time. Thus such a first-principles code is currently being set up.[19]

Using all four simplifications the nonlinear magneto-optical susceptibility reduces to the considerably shorter and much easier computational expression

$$\chi^{(2)}_{zzz}(2q_\|, 2\omega, (M)y)$$

$$= \frac{e^3 |q_\|| a}{\Omega} \frac{\lambda_{so}}{\hbar \omega} \sum_\sigma \sum_{k,l,l',l''} \left(\langle k + 2q_\|, l'' \sigma | z | kl\sigma \rangle \right.$$

$$\times \langle kl\sigma | z | k + q_\|, l'\sigma \rangle \langle k + q_\|, l'\sigma | z | k + 2q_\|, l''\sigma \rangle$$

$$\times \frac{\left(\dfrac{f(E_{k+2q_\|, l''\sigma}) - f(E_{k+q_\|, l'\sigma})}{E_{k+2q_\|, l''\sigma} - E_{k+q_\|, l'\sigma} - \hbar\omega + i\hbar\alpha} - \dfrac{f(E_{k+q_\|, l'\sigma}) - f(E_{kl\sigma})}{E_{k+q_\|, l'\sigma} - E_{kl\sigma} - \hbar\omega + i\hbar\alpha} \right)}{E_{k+2q_\|, l''\sigma} - E_{kl\sigma} - 2\hbar\omega + 2i\hbar\alpha} \right).$$

(5.4.39)

The indices of this tensor element reflect the choice of the following geometry. Since the nonlinear Kerr effect senses the properties of the surface, the vector of the magnetization is assumed to lie in the surface plane $M \| \hat{y}$ while the light beam hits the surface of grazing incidence. Its polarization is perpendicular to the surface ($E \| \hat{s}$). So the nonlinear Kerr rotation leads to an additional (and not small!) x component of the polarization. This geometry (p polarization for incident and reflected beams) exploits best the broken symmetry at the surface (maximum nonlinearity perpendicular to the surface) and yields the largest experimental yield. Since spin–orbit coupling is treated perturbatively and the exchange interaction is isotropic, we assume that the direction of the magnetization is fixed by a small external magnetic field, which defines the quantization axis accordingly.

The above tensor $\chi^{(2)}$ contains the following properties in an easily accessible way:

1. The *frequency doubling* is reflected by the energy denominator with frequency $2\hbar\omega$, by the terms containing $2q_\|$, and by the three matrix elements each corresponding to the three photons contributing to the frequency-doubling process.
2. The *interface sensitivity* (also in the presence of the axial vector M) is expressed by the product of the three dipole matrix elements, which would vanish in the bulk of inversion-symmetric media. For constant matrix elements the prefactor $|q_\|| a$ guarantees the surface sensitivity.

3. *Spin–orbit coupling* is explicitly taken into account in λ_{so}.
4. *Exchange splitting* and thus magnetism are expressed by the ubiquitous spin indices σ.
5. This is in particular the case for the spin-polarized *band structure* $E_{kl\sigma}$ which guarantees the element specificity and material sensitivity of our theory.

The main problem with the calculation of the nonlinear magneto-optical susceptibility $\chi^{(2)}(2q_\parallel, 2\omega)$ are the treatment of (a) the band structure (for Ni and Fe) and (b) the dipole matrix elements $M_{ij} = \langle ki\sigma | r | kj\sigma \rangle$.

Linear magneto-optical susceptibility

Incorporating spin–orbit coupling and exchange interaction in eqn (5.4.14) for the *linear optical* susceptibility, one may derive the *linear magneto-optical* susceptibility in analogy to the previous subsection for the nonlinear tensors. Similarly the application of the approximations 2–4 (1 is meaningless in the linear case) simplifies the formula without losing the key features of the linear Kerr effect. Then one obtains the result

$$\chi^{(1)}_{xy}(q, \omega, (M)_z) = -\chi^{(1)}_{yx}(q, \omega, (M)_z)$$

$$= \frac{4\pi e^2}{\Omega} \frac{\lambda_{so}}{\hbar \omega} \sum_{k,l,l',\sigma} \left(\langle k+1, l', \sigma | x | kl\sigma \rangle \langle kl\sigma | x | k+q, l'\sigma \rangle \right.$$

$$\left. \times \left(\frac{f(E_{k+q,l'\sigma}) - f(E_{kl\sigma})}{E_{k+q,l'\sigma} - E_{kl\sigma} - \hbar\omega + i\hbar\alpha_1} \right) \right). \quad (5.4.40)$$

In this case the 'polar' Kerr configuration has been chosen where the magnetization M and the beam direction q are assumed to be parallel to the surface normal ($\parallel \hat{z}$) while the photon polarization points in the x direction ($E \parallel \hat{x}$). The Kerr rotation then adds a (small, in linear optics) y component to the polarization reflecting the magnetic bulk properties of the second. $\chi^{(1)}_{xy} = -\chi^{(1)}_{yx}$ is the only nonvanishing off-diagonal element of the second-rank tensor $\chi^{(1)}$. All other off-diagonal elements vanish since in these cases the selection rules for the two dipole matrix elements $\Delta l = \pm 1$, $\Delta m = 0, \pm 1$ are not satisfied.

Plasmon pole

Now we show that our formula, which can also be derived from the current–current correlation function,[112] contains the plasmon pole for intraband transitions.

In the limit of large wavelength (which is always fulfilled in optics) the matrix element can be rewritten

$$\lim_{q \to 0} \langle kl | e^{-iq \cdot r} | k+q, l' \rangle = \delta_{u'} - i \langle kl | q \cdot r | k+q, l' \rangle (1 - \sigma_{u'}). \quad (5.4.41)$$

Here, the first and second terms denote the intraband and interband contributions, respectively. Thus the intraband contribution to the dielectric function is given by

$$1 - \lim_{q \to 0} \frac{4\pi e^2}{q^2} \sum_{k,l} \frac{f(E_{k+q,l}) - f(E_{kl})}{E_{k+q,l} - E_{kl} - \hbar\omega - i\hbar\alpha}$$

$$= 1 - \lim_{q \to 0} \frac{4\pi e^2}{q^2} \sum_{k,l} [f(E_{k+q,l}) - f(E_{kl})]$$

$$\times \left(\frac{-1}{\hbar\omega + i\hbar\alpha} - \frac{E_{k+q,l} - E_{kl}}{(\hbar\omega + i\hbar\alpha)^2} + \cdots \right)$$

$$\approx 1 + \frac{4\pi e^2}{q^2} \sum_{k,l} [\mathbf{q} \cdot \nabla_k f(E_{kl})](\mathbf{q} \cdot \nabla_k E_{kl}) \frac{1}{(\mathbf{b}\omega + i\hbar\alpha)^2}$$

$$= 1 + \frac{4\pi e^2}{q^2} \frac{q^2}{(\mathbf{b}\omega + i\hbar\alpha)^2} \sum_{kl} f(E_{kl}) \frac{\partial^2 E_{kl}}{\partial k^2}$$

$$= 1 - \frac{4\pi e^2}{(\hbar\omega + i\hbar\alpha)^2} \sum_{kl} \frac{f(E_{kl})}{m_{kl}}$$

$$= 1 - \frac{\omega_{pl}^2}{(\omega + i\alpha)^2}, \qquad (5.4.42)$$

where m_{kl} is the effective-mass tensor and ω_{pl} is the plasma frequency.

Finally we show that (besides the perturbative treatment of spin–orbit coupling, which is justified under the mentioned conditions) the linear magneto-optical susceptibility obtained in this way is identical to the result by Wang and Callaway.[128] If we split the independent summation over l and l' in eqn (5.4.40) into sums over occupied and unoccupied levels, and if we consider for simplicity only the optical dielectric function, we obtain for the interband contribution to $\varepsilon(q, \omega)$:

$$\lim_{q \to 0} \frac{4\pi e^2}{q^2} \sum_{kl, \text{occ}} \sum_{kl', \text{unocc}} |\langle kl|\mathbf{q} \cdot \mathbf{r}|k+q, l'\rangle|^2$$

$$\times \left(\frac{1}{E_{k+q,l} - E_{kl} - \hbar\omega - i\hbar\alpha} + \frac{1}{E_{k+q,l} - E_{kl} + \hbar\omega + i\hbar\alpha} \right). \quad (5.4.43)$$

This is exactly the partial fraction decomposition of the result by Wang and Callaway:[128]

$$\lim_{q \to 0} \frac{4\pi e^2}{q^2} \sum_{kl, \text{occ}} \sum_{kl', \text{unocc}} |\langle kl|\mathbf{q} \cdot \mathbf{r}|k+q, l'\rangle|^2$$

$$\times \frac{2(E_{k+q,l} - E_{kl})}{(E_{k+q,l} - E_{kl})^2 - (\hbar\omega + i\hbar\alpha)^2}. \qquad (5.4.44)$$

Thus our result for the optical dielectric function can also be written as

$$\varepsilon(\omega) = 1 - \frac{\omega_{pl}^2}{(\omega + i\alpha)^2}$$

$$- 4\pi e^2 \sum_{kl,\text{occ}} \sum_{kl',\text{unocc}} |\langle kl| \mathbf{q} \cdot \mathbf{r} |k+q,l'\rangle|^2$$

$$\times \frac{2(E_{k+q,l} - E_{kl})}{(E_{k+q,l} - E_{kl})^2 - (\hbar\omega + i\hbar\alpha)^2}, \quad (5.4.45)$$

which leads, after subtracting unity and subsequently multiplying by $[i(\omega + i\alpha)]$, to the formula for the optical conductivity given by Wang and Callaway (see eqn (5.5a) of ref. 128, since

$$|\langle kl| \mathbf{r} |k+q,l^2\rangle|^2 = \frac{|\langle kl| \mathbf{p} |k+q,l^2\rangle|^2}{m^2(E_{k+q,l} - E_{kl})^2}. \quad (5.4.46)$$

5.4.3 Interpretation of the susceptibility tensor

Microscopic picture of the linear and nonlinear Kerr effect

Figure 5.21 shows the level scheme of the optical transitions for the linear and nonlinear Kerr effect. Here we consider for simplicity only the situation of the linear Kerr effect.

Without spin–orbit and exchange interaction The dipole matrix element $\langle f| x |i\rangle$ may describe a maximally allowed transition. Then the orthogonal matrix element $\langle f| y |i\rangle$ vanishes and so does the magneto-optical susceptibility $\chi^{(1)}(\mathbf{q}, \omega)$, which contains the product

$$\langle f| x |i\rangle \langle f| y |i\rangle = 0.$$

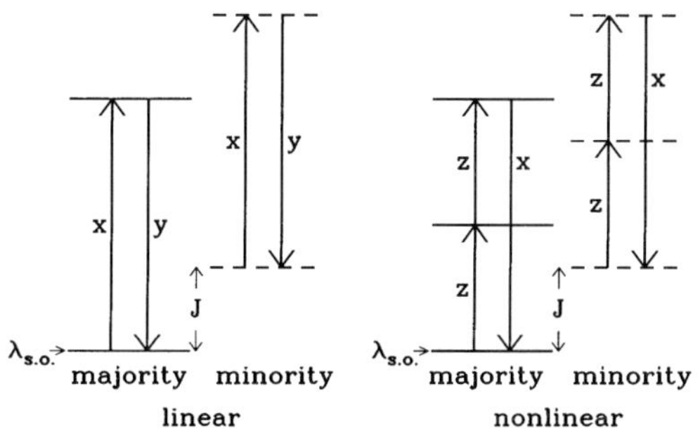

Fig. 5.21. Energy level scheme for the linear and nonlinear Kerr effect.

With spin–orbit interaction only Now we have

$$\langle f|y|i + O(\lambda_{so})\rangle \neq 0 \quad \Rightarrow \quad \langle f|x|i\rangle\langle f|y|i + O(\lambda_{so})\rangle \neq 0.$$

Thus a rotation is generated that is *linear* in spin–orbit coupling also for the nonlinear Kerr effect. However clockwise and counterclockwise rotations are still equally probable and therefore cancel each other.

With spin–orbit and exchange interaction Transitions between up-spins now occur at a different rate than transitions between down-spins. This implies that a net Kerr rotation remains. Note there are no spin-flips in optics in linear order of spin–orbit coupling (see eqn (5.4.32)).

5.4.4 Tight-binding theory for nonlinear surface magneto-optics

In the later 1980s, when the theory of nonlinear magneto-optics was formulated, there was no chance for an *ab initio* description of the nonlinear Kerr effect. In the meantime, however, computational power has increased so drastically that first-principles methods even for thin-film situations are no longer beyond feasibility. Furthermore, the *ab initio* methods have become more refined (e.g. full-potential methods, generalized gradient approximation, etc.) and thus are more suitable for computations in optics. Thus step by step many of the approximations made in the early stages of the theory can now be relaxed. It is necessary in particular to include the momentum-dependent matrix elements and the nonspherical character of spin–orbit coupling in the first-principles theory of nonlinear magneto-optics. Thus the first-principles theory will become able to bridge the gap between the macroscopic symmetry classification via group theory and the electronic theory of the nonlinear Kerr effect using band structures alone and neglecting the ***k*** dependence of the matrix elements. The *ab initio* theory can reveal the symmetry properties of nonlinear magneto-optics via the dipole and spin–orbit matrix elements without any adjustable parameter. For thin films there are no tight-binding parameters available, which imposes an additional restriction on parametrized calculations that is absent in the first-principles methods, since the latter are applicable to any geometry.

Despite these remarkable perspectives for first-principles theories in nonlinear magneto-optics, tight-binding methods will still be used in the foreseeable future for larger systems and nonideal multilayer structures which rapidly may outrange the capabilities of *ab initio* methods. On the other hand, tight-binding methods offer a well defined starting point for the incorporation of many-body effects while first-principles methods do not.

Many-body effects and ab initio methods

For the computation of the nonlinear magneto-optical Kerr effect *ab initio* techniques had and partly still have the following deficiencies:

Table 5.9 Comparison of theory and experiment

Ratio theory/experiment	Fe	Co	Ni
Exchange interaction	1.0	1.2	2.2
Occupied d-band width	1.1	1.2	1.45

1. They take a lot of computer time and thus they can be applied only to relatively small and ideal structures.
2. They have difficulties in correctly predicting the ground state of ferromagnetic transition metals: Fe is found to be an fcc antiferromagnet rather than a bcc ferromagnet. However, note that the recent implementation of the generalized gradient approximation in the full-potential linear augmented plane-wave method remedies this insufficiency of the local-density approximation.
3. Even more frequently deviations occur in the excited states, since density-functional calculations take into account only part of the electronic correlations (local-density approximation, GW approximation, self-interaction correction, generalized gradient approximation).

As an illustration we refer to the comparison of the *ab initio* computed width of the occupied d band and the exchange interaction with results from spin-polarized photoemission (Table 5.9). We notice two points:

1. The *ab initio* theories yield too large values due to an insufficient account of the narrowing caused by correlations.
2. In Ni the difference between theory and experiment is the largest since electron–electron interaction is most important in this metal.[137-142]

The Kerr effects (linear and nonlinear), however, result from spin-dependent magneto-optical absorption. So it is precisely the spin-polarized excitation spectrum of Fe, Co, or Ni that is needed with high accuracy. Therefore *ab initio* theories yield no satisfactory results for the linear[143] Kerr spectrum in particular of Ni. The peaks of the theory are shifted towards higher energies by approximately 1 eV; this shift also occurs at the zero, which is found at approximately 4 eV in experiment. So a range exists where even the signs of the theory and the experiment are different. In addition the shape of the spectrum with two positive peaks (at 1.4 and 3.2 eV) is not correctly reproduced by the theory.

Thus we originally avoided *ab initio* calculations and rather used for surfaces the 'combined interpolation scheme' (CIS). This method is based on a parametrized band structure and allows for the incorporation of many-body effects, the importance of which is reflected in the band width and exchange interaction. This method has the following advantages:

1. One is able to calculate the band structure throughout the complete Brillouin zone and thus one is not confined to high-symmetry directions.

2. Since the computational time for the calculation of the bands at one point of the Brillouin zone is short, many reciprocal lattice points can be taken into account. This is of particular importance for the calculation of the nonlinear Kerr spectrum. Furthermore it becomes possible to calculate also more complicated materials, thin films, or alloys.
3. The method gives access to physically easily interpretable control parameters, the variation of which simplifies the understanding of the different features of the nonlinear (and linear) Kerr spectra and of their physical origin, as we will see later.
4. The technique allows for the incorporation of experimentally accessible quantities that can be derived from several independent measurements, such as photoemission and inverse photoemission, de Haas–van Alphen effect, or different spin-polarization spectroscopies, and from many-body theories in a simple and efficient way.
5. The 'combined interpolation scheme' effectively takes into account the most important many-body effects of the spin-polarized magneto-optically excited states, which are probed very sensitively by the nonlinear and linear magneto-optical Kerr effects.

In order to obtain now the band structure $E_{kl\sigma}$ entering both the linear and the nonlinear magneto-optical susceptibility, we use two types of parametrized band-structure calculations:

1. We adopt band structures from other calculations or experiments. However, since these are displayed in the literature only for high-symmetry directions of the Brillouin zone, we superimpose several of the three-dimensionally continued 'Brillouin spheres' with different weight factors in order to incorporate all relevant features of the band structure in the calculation. The superposition of Brillouin spheres is very similar to the Debye approximation for phonons, with two exceptions: (i) Three Brillouin spheres are used rather than one Debye sphere. (ii) We take the realistic band structure with up to nine bands per spin into account rather than assuming linear dispersion of a single band as the Debye model does. This procedure yields already the dominant features of the Kerr spectra but always relies on externally computed or measured band structures. Thus we will not report on this pathway in this chapter. For more details see ref. 13.

2. Mainly we calculate our own three-dimensional band structures using the 'combined interpolation scheme'. For that a restricted set of parameters is taken from independent experiments or theories in order to set up the Hamiltonian matrix. This Hamiltonian represented in a suitable set of basis functions is diagonalized at some thousands of points (up top 23 000) in the irreducible subspace of $\frac{1}{48}$th of the Brillouin zone.

Nickel
We use the 'combined interpolation scheme' (CIS) essentially in the form first

proposed by Hodges et al.[144] and independently in a somewhat different form (with nonlocal pseudopotentials) by Mueller.[145] A good review is the article by Ehrenreich and Hodges.[146] Since this method has also been used by Weling and Callaway[147] to explain photoemission spectra, our bands should agree in the high-symmetry directions with theirs. However, we will deviate from their scheme and go beyond whenever this is necessary.

The CIS is a method for the calculation of band structures which is particularly well suited for transition metals and allows for the incorporation of ferromagnetism in a simple manner. It may be applied not only to pure elemental crystals but also to alloys, surfaces, clusters, or specific geometries. It creates a set of parameters that are fitted to experimental data, many-body theories, or other theories that are mainly able to treat idealized situations (mostly pure bulk) and have the above-mentioned problems for Ni. The deviations in this case are unacceptable for nonlinear magneto-optics, which reacts very sensitively to the precise band positions at high-symmetry points, and can only in part be compensated for by using improved density-functional theories. In the case of thin films, however, it will turn out that the preceding statements have to be qualified, since the use of bulk parameters turns out to be less adequate than the *ab initio* band structures, which have admittedly improved their quality considerably over the past years.

The philosophy in setting up the Hamiltonian matrix is to describe the electron–core interaction of the sp-type orbitals by pseudopotentials while the d orbitals are treated by the tight-binding method, though using Fletcher–Wohlfarth parameters[148,149] rather than the usual Slater–Koster parameters.[150] The basis set for Ni then consists of nine wavefunctions per spin, namely, five linear combinations of atomic orbitals (LCAO) for the localized 3d electrons and four orthogonalized plane waves (OPW, in fact usually only plane waves) for the delocalized 4s and 4p states. The OPWs have the form

$$\psi_i(\boldsymbol{k},\boldsymbol{r}) = (Nv_a)^{-\frac{1}{2}} e^{i(\boldsymbol{k}+\boldsymbol{K})\cdot\boldsymbol{r}} \qquad (i=1,2,3,4) \qquad (5.4.47)$$

with $\boldsymbol{K}_1 = (0,0,0)$, $\boldsymbol{K}_2 = (0,-16,0)(\pi/4a)$, $\boldsymbol{K}_3 = (-8,-8,-8)(\pi/4a)$, $\boldsymbol{K}_4 = (-8,-8,8)(\pi/\pi a)$. N is the number of atoms in the solid (which is cancelling finally), a is the lattice constant (3.52 Å for Ni), v_a the normalization volume

$$v_a = \frac{a^3}{4Z^3}, \qquad (5.4.48)$$

$Z = 28$ is the nuclear charge of Ni, and the factor of 4 corresponds to the four atoms per conventional cubic unit cell of volume a^3 in fcc Ni. These four plane waves represent the minimum, symmetry-adapted basis set for the description of the low-lying levels of the 4s and 4p electrons.

The vectors $\boldsymbol{K}_i$ are given as the union of sets of all vectors

$$\boldsymbol{G} = n_1\boldsymbol{b}_1 + n_2\boldsymbol{b}_2 + n_3\boldsymbol{b}_3, \qquad (5.4.49)$$

which map equivalent symmetry points of the Brillouin zone into each other.

The vectors G have to be part of the same Brillouin zone, and the b_i are the basis vectors of the reciprocal lattice, which read for fcc Ni

$$b_1 = \frac{4\pi}{a}\frac{1}{2}(\hat{y}+\hat{z}), \qquad b_2 = \frac{4\pi}{a}\frac{1}{2}(\hat{z}+\hat{x}), \qquad b_3 = \frac{4\pi}{a}\frac{1}{2}(\hat{x}+\hat{y}). \qquad (5.4.50)$$

Formally they follow from the general theorem,[151] which says that the Hamiltonian matrix for the plane-waves states is determined by applying four rules:

1. For each high-symmetry point k_j one determines a set $B^{(k_j)}$ of plane waves $\psi_i(k,r)$ according to

$$B^{(k_j)} := \left\{ \psi_i(k,r) \,\Big|\, |k_j + K| = \min_{K'} |k_j + K'| \right\}. \qquad (5.4.51)$$

The corresponding K vectors are those closest to k_j.

2. As a basis B for the s states one takes for each k point the union of the $B^{(k_j)}$:

$$B = \left\{ \psi_i(k,r) \,\Big|\, \psi_i(k,r) \in B^{(k_j)} \text{ for a high-symmetry point } k_j \right\}. \qquad (5.4.52)$$

3. For each basis state $\psi_i(k,r)$ a *symmetry factor* $F_j(k)$ is determined such that for each high-symmetry point k_j

$$F_j(k_j) = \begin{cases} 1, & \text{if } \psi_i(k,r) \in B^{(k_j)}, \\ 0, & \text{if } \psi_i(k,r) \notin B^{(k_j)}. \end{cases}$$

Between the high-symmetry points the $F_j(k)$ are interpolated continuously and monotonically. The symmetry factors 'switch off' those elements of the basis which do not comply with the symmetry of k_j.

3. Concerning the basis set B or rule 3, the Hamiltonian matrix H^s has the form (for plane waves)

$$\begin{aligned} H_{ii}^s &= V_{(0,0)} + \frac{\hbar^2}{2m}|k+K_i|^2, \\ H_{ij}^s &= V_{ij}F_i(k)F_j(k) \qquad \text{for } i \neq j. \end{aligned} \qquad (5.4.53)$$

The proof of this theory is given in ref. 151.

In the case of fcc Ni one has the high-symmetry points shown in Table 5.10 and Fig. 5.22.[152] Each K point is equivalent to two U points. The total number of points of one symmetry type is generally obtained as the product of the total numbers of equivalent and inequivalent points. Since the point W possesses the largest number of equivalent points of the same symmetry type, it yields already the four plane waves mentioned above. For Ni it turns out that the other symmetry points do not lead to further basis functions. This is due to the fact that the point W has the lowest symmetry. This situation is not the general case. For Fe, for example, it will turn out that several high-symmetry points contribute to the set of K_i.

Now we have found the plane waves describing the sp orbitals. The LCAO

Table 5.10 High-symmetry points for fcc Ni

High-symmetry point (position)	Equivalent points	Nonequivalent points	Total number of points
Γ (zone centre)	1	1	1
X (square centre)	2	3	6
K (edge centre)			12
L (hexagon centre)	2	4	8
W (corner)	4	6	24
U (edge centre)			24

contribution to the basis set, which describes the 3d orbitals, reads

$$\psi_i(\mathbf{k},\mathbf{r}) = N^{-\frac{1}{2}} \sum_l e^{i\mathbf{k} \cdot \mathbf{R}_l} \varphi_i(\mathbf{r} - \mathbf{R}_l) \qquad (i = 5,\ldots,9), \qquad (5.4.54)$$

Here, the $\varphi(\mathbf{r} - \mathbf{R}_l)$ are atomic d orbitals centred on site $\mathbf{R}_l$:

$$\varphi_5(\mathbf{r}) = \left(\frac{15}{4\pi}\right)^{\frac{1}{2}} \frac{xy}{r^2} f(|\mathbf{r}|),$$

$$\varphi_6(\mathbf{r}) = \left(\frac{15}{4\pi}\right)^{\frac{1}{2}} \frac{yz}{r^2} f(|\mathbf{r}|),$$

$$\varphi_7(\mathbf{r}) = \left(\frac{15}{4\pi}\right)^{\frac{1}{2}} \frac{zx}{r^2} f(|\mathbf{r}|), \qquad (5.4.55)$$

$$\varphi_8(\mathbf{r}) = \left(\frac{15}{16\pi}\right)^{\frac{1}{2}} \frac{x^2 - y^2}{r^2} f(|\mathbf{r}|),$$

$$\varphi_9(\mathbf{r}) = \left(\frac{5}{16\pi}\right)^{\frac{1}{2}} \frac{3z^2 - r^2}{r^2} f(|\mathbf{r}|).$$

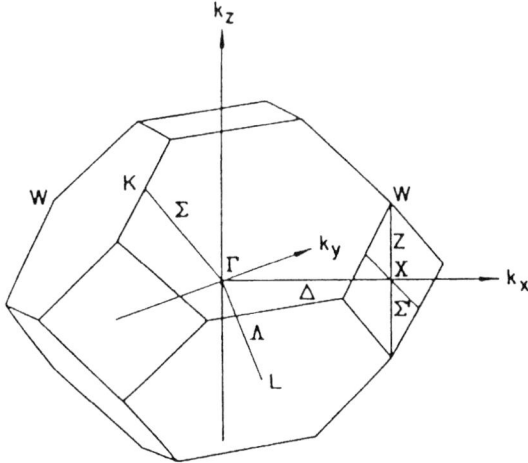

Fig. 5.22. Brillouin zone of the fcc lattice: truncated octahedron.

The function

$$f(|r|) = \frac{2\sqrt{2}}{3\sqrt{5}} \left(\frac{Z}{3a_B}\right)^{\frac{3}{2}} \left(\frac{Z|r|}{3a_B}\right)^2 \exp\left(-\frac{Z|r|}{3a_B}\right) \quad (5.4.56)$$

is the normalized solution of the radial Schrödinger equation for the isolated atom, and $a_B = 5.29 \times 10^{-11}$ m is Bohr's radius. The normalization is due to the condition $\int_0^\infty (rf(r))^2 \, dr = 1$. We neglect the non-orthogonality of the atomic orbitals localized on different sites, as is frequently done also in the Slater–Koster interpolation scheme. Nevertheless, the plane waves for the sp electrons hybridize with the d states causing an indirect d–d interaction.

This particular choice of the nine basis functions has two remarkable features:

1. The number of sp basis functions is sufficiently small and avoids the extension to reciprocal lattice vectors pertinent to several different Brillouin zones.
2. On the other hand, the basis set is sufficiently large to reflect the symmetry of the Brillouin zone and to take into account a considerable indirect d–d interaction. Thus the incorporation of other than nearest-neighbour interactions in the d-band portion of the Hamiltonian matrix is not required (at least not for fcc Ni).

Thus, the respect to the d portion of the Hamiltonian matrix, the combined interpolation scheme in the Fletcher–Wohlfarth parametrization is equivalent to the orthogonal *three*-centre approximation in the tight-binding description (in Slater–Koster parametrization).

The Hamiltonian then is represented as an (18×18) matrix of the form:

$$H = \begin{pmatrix} \uparrow\uparrow & 0 \\ 0 & \downarrow\downarrow \end{pmatrix}. \quad (5.4.57)$$

Since spin–orbit coupling is treated perturbationally, both spin directions decouple in the Hamiltonian. These spin contributions are themselves (9×9) matrices, e.g.

$$\uparrow\uparrow = \begin{pmatrix} s\uparrow s\uparrow & s\uparrow s\uparrow \\ (s\uparrow d\uparrow)^\dagger & d\uparrow d\uparrow \end{pmatrix}. \quad (5.4.58)$$

The three different blocks schematically denoted as $s\uparrow s\uparrow$, $s\uparrow d\uparrow$ and $d\uparrow d\uparrow$ are of the form

$$s\uparrow s\uparrow = \begin{pmatrix} \beta + \alpha |k+K_1|^2 & V_{200} F_2 & V_{111} F_3 & V_{111} F_4 \\ V_{200} F_2 & \beta + \alpha |k+K_2|^2 & V_{111} F_2 F_3 & V_{111} F_2 F_4 \\ V_{111} F_3 & V_{111} F_2 F_3 & \beta + \alpha |k+K_3|^2 & V_{200} F_3 F_4 \\ V_{111} F_4 & V_{111} F_2 F_4 & V_{200} F_3 F_4 & \beta + \alpha |k+K_4|^2 \end{pmatrix}.$$

$$(5.4.59)$$

This block corresponds to Harrisons[153] pseudopotential Hamiltonian for Al. The OPW–LCAO block is given by

$$s\uparrow d\uparrow = H_{ij} = B_2 j_2(|k+K_i| B_1) \left(\frac{(k+K_i)_\mu (k+K_i)_\nu}{|k+K_i|^2} \right) F_i(k),$$

$$(i = 1,2,3,4); \quad (j,\mu,\nu) = (5,x,y), (6,y,z), (7,z,x); \quad (5.4.60)$$

$$H_{i8} = B_2 j_2(|k+K_i| B_1) \left(\frac{(k+K_i)_x^2 - (k+K_i)_y^2}{2|k+K_i|^2} \right) F_i(k); \quad (5.4.61)$$

$$H_{i9} = B_2 j_2(|k+K_i| B_1) \frac{1}{6}\sqrt{3} \left(\frac{3(k+K_i)_z^2}{|k+K_i|^2} - 1 \right) F_i(k). \quad (5.4.62)$$

Here $j_2(x)$ denotes the spherical Bessel function of angular momentum $l = 2$:

$$j_2(x) = \frac{3}{x^3}(\sin x - x \cos x) - \frac{\sin x}{x}. \quad (5.4.63)$$

The LCAO–LCAO block $d\uparrow d\uparrow$ reads ($\xi = k_x a/2$, $\eta = k_y a/2$, $\zeta = k_z a/2$):

$$H_{55} = E_0 - 4A_1 \cos\xi \cos\eta + 4A_2 \cos\zeta(\cos\xi + \cos\eta)$$

$$H_{66} = E_0 - 4A_1 \cos\eta \cos\zeta + 4A_2 \cos\xi(\cos\eta + \cos\zeta)$$

$$H_{77} = E_0 - 4A_1 \cos\xi \cos\zeta + 4A_2 \cos\eta(\cos\xi + \cos\zeta)$$

$$H_{88} = E_0 + \Delta + 4A_4 \cos\xi \cos\eta - 4A_5 \cos\zeta(\cos\xi + \cos\eta)$$

$$H_{99} = E_0 + \Delta - \tfrac{4}{3}(A_4 + 4A_5)\cos\xi \cos\eta + \tfrac{4}{3}(2A_4 - A_5)\cos\zeta(\cos\xi + \cos\eta)$$

$$H_{56} = H_{65} = -4A_3 \sin\xi \sin\zeta$$

$$H_{57} = H_{75} = -4A_3 \sin\eta \sin\zeta$$

$$H_{67} = H_{76} = -4A_3 \sin\xi \sin\eta$$

$$H_{58} = H_{85} = 0$$

$$H_{68} = H_{86} = -4A_6 \sin\eta \sin\zeta \quad (5.4.64)$$

$$H_{78} = H_{87} = 4A_6 \sin\xi \sin\zeta$$

$$H_{59} = H_{95} = -\frac{8}{\sqrt{3}} A_6 \sin\xi \sin\eta$$

$$H_{69} = H_{96} = \frac{4}{\sqrt{3}} A_6 \sin\eta \sin\zeta$$

$$H_{79} = H_{97} = \frac{4}{\sqrt{3}} A_6 \sin\xi \sin\zeta$$

$$H_{89} = H_{98} = \frac{4}{\sqrt{3}} (A_4 + A_5)\cos\zeta(\cos\eta - \cos\xi).$$

This Hamiltonian contains 14 parameters: the two pseudopotential coefficients V_{200} and V_{111}, the value β and curvature α of the free-electron bands at the Γ point, the s–d intermixing parameters B_2 (a kind of hybridization strength) and B_1 (corresponds to the peak position in the radial part $r^2 f(|r|)$ of the wavefunction), the two parameters E_0 and Δ fixing the d-band position (where the latter describes the crystal-field splitting and would vanish within the *two*-centre approximation), and the six Fletcher parameters A_i ($i = 1,\ldots,6$), which correspond to the *three*-centre integrals of the tight-binding approximation. The two-centre approximation would contain only the three Slater–Koster parameters $(dd\sigma)$, $(dd\pi)$ and $(dd\delta)$ rather than the six A parameters. From a technical point of view, the combined interpolation scheme is nearly as accurate as the *non*-orthogonal three-centre approximation, but for Ni it is much better, since it describes the photoemission and de Haas–van Alphen data very well (in contrast to the tabulated non-orthogonal Slater–Koster parameters[49]).

The Fletcher parameters are defined as follows, with the crystal potential V and the atomic potential U:

$$A_1 = -\int \varphi_5^*(x - \tfrac{1}{2}a, y - \tfrac{1}{2}a, z)(V-U)\varphi_5(x,y,z)\,d^3r,$$

$$A_2 = \int \varphi_5^*(x, y - \tfrac{1}{2}a, z - \tfrac{1}{2}a)(V-U)\varphi_5(x,y,z)\,d^3r,$$

$$A_3 = \int \varphi_5^*(x - \tfrac{1}{2}a, y, z - \tfrac{1}{2}a)(V-U)\varphi_6(x,y,z)\,d^3r,$$

$$A_4 = \int \varphi_8^*(x - \tfrac{1}{2}a, y - \tfrac{1}{2}a, z)(V-U)\varphi_8(x,y,z)\,d^3r, \qquad (5.4.65)$$

$$A_5 = -\int \varphi_8^*(x, y - \tfrac{1}{2}a, z - \tfrac{1}{2}a)(V-U)\varphi_8(x,y,z)\,d^3r,$$

$$A_6 = \int \varphi_6^*(x, y - \tfrac{1}{2}a, z - \tfrac{1}{2}a)(V-U)\varphi_8(x,y,z)\,d^3r.$$

These parameters can be linked to the usual two- and three-centre integrals of Slater and Koster by the relations

$$E_0 = -0.95\,\text{eV} = E_{xy,xy}(000) = d_0,$$
$$E_0 + \Delta = -0.890\,640\,\text{eV} = E_{3z^2-r^2, 3z^2-r^2}(000) = d_0,$$
$$A_1 = 0.25\,\text{eV} = -E_{xy,zy}(100) = -\tfrac{3}{4}(dd\sigma) - \tfrac{1}{4}(dd\delta),$$
$$A_2 = 0.106\,250\,\text{eV} = E_{xy,xy}(011) = \tfrac{1}{2}(dd\pi) + \tfrac{1}{2}(dd\delta),$$
$$A_3 = 0.121\,385\,\text{eV} = E_{xy,xz}(011) = \tfrac{1}{2}(dd\pi) - \tfrac{1}{2}(dd\delta), \qquad (5.4.66)$$
$$A_4 = 0.152\,923\,\text{eV} = E_{x^2-y^2, x^2-y^2}(110) = (dd\pi),$$
$$A_5 = 0.015\,131\,\text{eV} = -\tfrac{3}{4}E_{3z^2-r^2,3z^2-r^2}(110) - \tfrac{1}{4}E_{x^2-y^2,x^2-y^2}(110)$$
$$= -\tfrac{3}{16}(dd\sigma) - \tfrac{1}{4}(dd\pi) - \tfrac{9}{16}(dd\delta),$$
$$A_6 = 0.103\,386\,\text{eV} = \tfrac{1}{2}\sqrt{3}\,E_{xy,3z^2-r^2}(110) = -\tfrac{3}{8}(dd\sigma) - \tfrac{3}{8}(dd\delta).$$

For their band-structure calculation, Weling and Callaway[147] fitted these 14 parameters of the Hamiltonian to the photoemission data by Eberhardt and Plummer[154] and to the Fermi-surface measurements by Tsui and Stark[155,156] at those symmetry points where the bands can be calculated analytically in terms of these parameters. Their values for E_0, $E_0 + \Delta$ and the A_i ($i = 1, \ldots, 6$), which we also use, were already given above. The values for the remaining parameters are: $\beta = -8.8$ eV, $\alpha = 0.204\,937$ eV$/C^2$, $B_1 = 0.480\,651/C$, $B_2 = 12.870\,937$ eV, $V_{111} = 2.036\,977$ eV, $V_{200} = -0.387\,444$ eV, with $C = \pi/4a$.

The only modification of this matrix caused by ferromagnetism affects the diagonal elements of the LCAO–LCAO block. The other blocks are identical: $H_{ij} = H_{i+9, j+9}$ ($1 \leq i \leq 9, 1 \leq j \leq 9$). Although the experimentally determined exchange interaction[157] is 310 meV, Weling and Callaway find a better overall agreement between theory and experiment if they use two different exchange splittings, viz. 400 meV for the t_{2g} orbitals and 100 meV for the e_g orbitals. So one has the following energy levels: $E_0 \downarrow = -0.75$ eV, $E_0 \uparrow = -1.15$ eV, $E_0 \downarrow + \Delta \downarrow = -0.84$ eV, $E_0 \uparrow + \Delta \uparrow = -0.94$ eV. Thus, the bands are much closer to experiment than the values of 600 to 800 meV of usual *ab initio* band-structure calculations. In a suitable many-body model this parameter choice also yields a reasonable Curie temperature of the order of 630 K and guarantees the exact reproduction of the magnetic moment of Ni, which is 0.56 μ_B. In order to obtain this value from *ab initio* calculations one has to compensate for the too low energetic position of many occupied states (resulting from a too large band width) by an unreasonably large exchange splitting at the Fermi level.

The neglect of the symmetrization factors $F_i(k)$ ($i = 1, \ldots, 4$) in the Hamiltonian (i.e. $F_i(k) = 1$) would lead to level shifts and splittings of the order of the difference between the band structures obtained in the combined interpolation and those from *ab initio* calculations. We calculate these factors in two different ways: (i) The first option is to use the interpolation formulae given by Ehrenreich and Hodges, which, however, lead to small discontinuities at the point X. (ii) We therefore introduce a smooth interpolation of the $F_i(k)$ values being exactly known at the high-symmetry points ($F_1(k) = 1$ throughout) (Table 5.11). (iii) A third smooth interpolation method using sine functions has been implemented by Moos.[151]

In method (ii), which we will describe here, we subdivide the irreducible $\frac{1}{48}$th of the Brillouin zone, where the Hamiltonian matrix is to be diagonalized, into three disjunct tetrahedra ΓXUW, ΓLUW and ΓLKW with the basis vectors *X*, *U*, *W*, or *L*, *U*, *W*, or *L*, *K*, *W*, respectively. At each *k* point obeying the constraint $0 \leq k_z \leq k_x \leq k_y \leq 2\pi/a$ we diagonalize the three (3 × 3) matrices in order to obtain the coordinates of *k* in the basis of each of the three tetrahedra. Whenever for one of the tetrahedra the sum of the coordinates is in the interval [0, 1], the corresponding *k* point is contained in this tetrahedron. In this way we scan the relevant portion of the Brillouin zone (truncated octahedron for fcc Ni). This procedure automatically yields also the symmetry of each *k* point and thus the weight for the *k* space summation in the linear and nonlinear magneto-optical response functions. Usually each *k* point is found in only one tetrahedron. However, if a *k* point belongs to two or three tetrahedra, we know

Table 5.11 The $F_i(k)$ values at high-symmetry points

Symmetry point	$F_2(k)$	$F_3(k)$	$F_4(k)$
$\Gamma(0,0,0)$	0	0	0
$X\left(0, \dfrac{2\pi}{a}, 0\right)$	1	0	0
$L\left(\dfrac{\pi}{a}, \dfrac{\pi}{a}, \dfrac{\pi}{a}\right)$	0	1	0
$K\left(\dfrac{3\pi}{2a}, \dfrac{3\pi}{2a}, 0\right)$	0	1	1
$W\left(\dfrac{\pi}{a}, \dfrac{2\pi}{a}, 0\right)$	1	1	1
$U\left(\dfrac{\pi}{2a}, \dfrac{2\pi}{a}, \dfrac{\pi}{2a}\right)$	1	1	0

that the corresponding point is part of the interface between neighbouring tetrahedra or lies on the straight line between Γ and W (which belongs to all three of the tetrahedra). Then the coordinates with respect to the tetrahedra are taken as the coefficients of the linear interpolation of the F functions, which are known at the corners of the tetrahedra. This procedure guarantees a smooth interpolation of the $F_i(k)$ and eliminates the discontinuities of method (i) at the point X. Furthermore it takes into account the detailed anisotropy of the Brillouin zone, which has to be included in the k-space summations of the response functions eqns (5.4.39) and (5.4.40). Thus the weights of the k points inside the tetrahedra and at the interfaces and edges between adjacent tetrahedra within the same Brillouin zone are now determined. The weights at the boundary of the Brillouin zone, i.e. at interfaces and edges between *different* Brillouin zones, will be discussed later for the example of Fe.

Thus all quantities are known, and the Hamiltonian matrix can be diagonalized. For that purpose we use standard numerical routines. One of them has been taken from the book by Papaconstantopoulos,[49] the other is a Householder library code. Both yield identical results, which are also, in the high-symmetry directions ΓX, XW, WL, LΓ, ΓK and UX, identical to those of ref. 147 within several significant digits at the high-symmetry points and within plotting accuracy in between. Our spin-polarized band structure is plotted in Fig. 5.23. According to ref. 147, self-consistency could be reached by a small shift of the Fermi level by 0.054 eV. The small discontinuity at the point X vanishes by the application of interpolation method (ii) for the symmetrization factors $F_i(k)$; see above. It is important to note that our calculated band structure agrees topologically exactly with the result of *ab initio* calculations. However, the d-band positions are different by up to 1 eV (at least for the older *ab initio* calculations; this difference reduces to half of its value for our recent *ab initio* calculations). Thus, the d-band width is strongly reduced by electronic correlations. Conse-

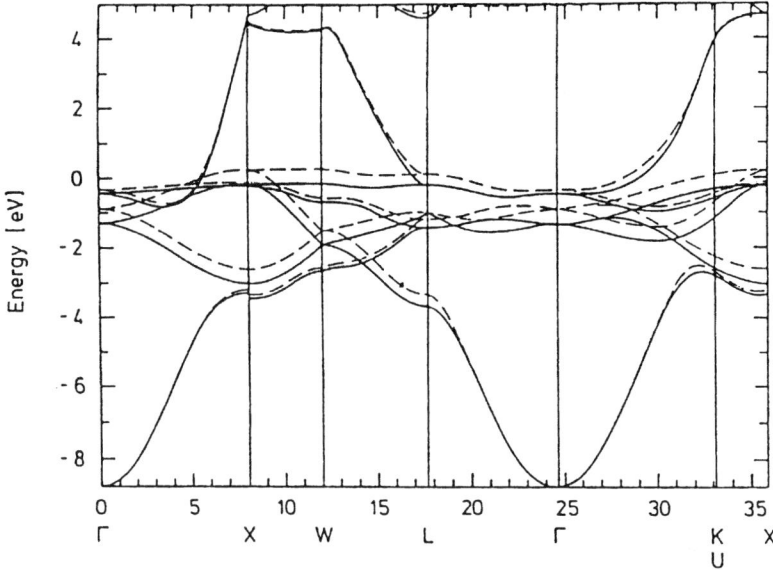

Fig. 5.23. Spin-polarized band structure of Ni; the dashed lines are minority bands while the full curves refer to majority electrons.

quently, our band structure is able to reproduce the magnetic moment of Ni (0.56 μ_B) with such a small exchange interaction of 100 meV and 400 meV for the e_g and t_{2g} orbitals, respectively, while *ab initio* calculations, due to the too large d-band width, require an unphysically large exchange splitting of 600 to 800 meV (in some cases up to 1000 meV) at the Fermi level in order to explain the magnetic moment of Ni. We will later show how these deviations of the band positions and the exchange coupling are reflected by the linear and nonlinear Kerr spectra.

Iron

In view of the results of spin-polarized photoemission one expects that many-body effects are relevant also for Fe, although they might not play a role as important as in the case of Ni. This results in part from the fact that the bcc crystal structure is less densely packed and the coordination number is smaller than in fcc Ni. For comparison we consider the Brillouin zones ('primitive unit cells of reciprocal space') of fcc Ni (truncated octahedron, see Fig. 5.22) and bcc Fe (rhombic dodecahedron, see Fig. 5.24). The primitive unit cells in real space (Wigner–Seitz cells) then are interchanged (rhombic dodecahedron for fcc Ni and truncated octahedron for bcc Fe). There are only two Fe atoms rather than four Ni atoms within the conventional cubic unit cell of volume a^3 (a = lattice constant). Owing to the reduced packing density and the only partially filled 3d shell, the *electronic* many-body effects will be somewhat less important than for Co or Ni. On the other hand, for Fe the influence of the *atomic* coordination

Table 5.12 The Wigner–Seitz cells for Ni and Fe

Material	Ni	Fe
Volume	$a^3/4$ (fcc)	$a^3/2$ (bcc)
Coordination	12	8 ($\Rightarrow$ +next-nearest neighbours)
	smooth surface	open surface
	many-body effects	atomic states

will be more dominant. Thus it is necessary to incorporate the next nearest-neighbour hoppings for Fe. For Ni, one expects a smooth surface, which is dominated by the electronic correlations. These are also increased due to the reduced band width and large band filling. The surface of Fe, however, should be dominated by roughness and by the atomic geometry and possible surface states. Table 5.12 summarizes the different features of the electronic structure of Ni and Fe.

For the calculation of the Fe band structure we use both the combined interpolation scheme as well as two first-principles methods (the full-potential linear muffin-tin orbitals (FP-LMTO) and the full-potential linear augmented plane-wave (FP-LAPW) techniques). While the *ab initio* methods are mostly used for the calculation of nonlinear Kerr spectra in thin magnetic films and will be discussed later, the combined interpolation scheme is used mainly for the nonlinear magneto-optics at surfaces and will be presented here. This method proposed for Ni by Hodges *et al.* has been used by Baker and Smith[158] for paramagnetic Fe and in a model calculation by Misemer. In the following we present our treatment for Fe, which closely follows the calculation of the Ni band structure.

As basis set we use again plane waves for the sp orbitals and a linear combination of atomic orbitals for the d states. Again we use pseudopotentials for the sp contributions and a tight-binding description of the d portion of the Hamiltonian matrix. The basis set consists of 12 orbitals per spin, seven of them being plane waves describing sp orbitals

$$\Phi_i(\mathbf{k},\mathbf{r}) = e^{i(\mathbf{k}+\mathbf{K}_i)\mathbf{r}}, \qquad (i=1,\ldots,7) \qquad (5.4.67)$$

and five LCAOs for the description of the 3d electrons

$$\psi_\mu(\mathbf{k},\mathbf{r}) = \frac{1}{\sqrt{N}} \sum_l e^{i\mathbf{k}\cdot\mathbf{R}_l} \varphi_\mu(\mathbf{r}-\mathbf{R}_l), \qquad (\mu=8,\ldots,12). \qquad (5.4.68)$$

Here, as in Ni, the $\varphi_\mu(\mathbf{r}-\mathbf{R}_l)$ are atomic d orbitals centred around the lattice

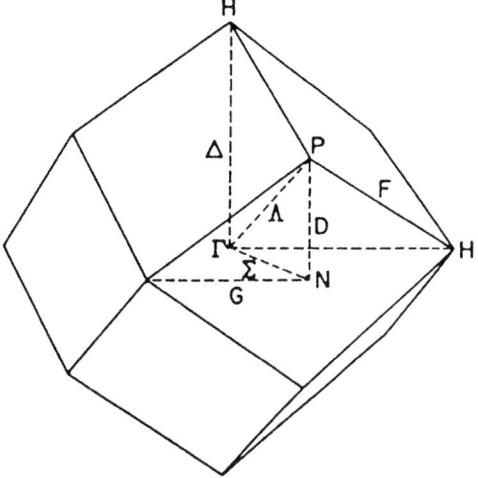

Fig. 5.24. Brillouin zone of the bcc lattice: rhombic dodecahedron.

site R_l:

$$\varphi_1 = \frac{xy}{r^2} f(r),$$

$$\varphi_2 = \frac{yz}{r^2} f(r),$$

$$\varphi_3 = \frac{xz}{r^2} f(r), \qquad (5.4.69)$$

$$\varphi_4 = \frac{1}{2} \frac{x^2 - y^2}{r^2} f(r),$$

$$\varphi_5 = \frac{1}{2\sqrt{3}} \frac{3z^2 - r^2}{r^2} f(r),$$

with

$$f(r) = \frac{2\sqrt{2}}{3\sqrt{5}} \left(\frac{Z}{3a_B}\right)^{3/2} \left(\frac{Zr}{3a_B}\right)^2 e^{-Zr/3a_B}.$$

The choice of the vectors K_i and thus the number and shape of the sp basis states is dictated by the Brillouin-zone symmetry (Fig. 5.24). Bcc Fe has the high-symmetry points shown in Table 5.13.[152]

The vectors K_i are the union of sets of all vectors connecting equivalent symmetry points of the Brillouin zone. In fcc Ni only the points W are corner points and its turned out that the difference vectors of the four equivalent W points were already sufficient to yield all required vectors K_i according to the mentioned theorem. For bcc Fe, however, the Brillouin zone contains two

Table 5.13 High-symmetry points for bcc Fe

High-symmetry point (position)	Equivalent points	Nonequivalent points	Total number of points
Γ (zone centre)	1	1	1
N (face centre)	2	6	12
H (corner)	6	1	6
P (corner)	4	2	8

corner points of different symmetry, H and P. Thus one obtains seven vectors K_i, six of them resulting from the equivalent H points and one in addition, K_7, from the P points, which is not contained in the set of the other six vectors:

$$K_1 = (0,0,0),$$
$$K_2 = (-2,0,0)\frac{2\pi}{a},$$
$$K_3 = (-1,-1,0)\frac{2\pi}{a},$$
$$K_4 = (-1,1,0)\frac{2\pi}{a}, \qquad (5.4.70)$$
$$K_5 = (-1,0,-1)\frac{2\pi}{a},$$
$$K_6 = (-1,0,1)\frac{2\pi}{a},$$
$$K_7 = (0,-1,-1)\frac{2\pi}{a}.$$

All other connection vectors of equivalent symmetry points do not lead to further basis states for bcc Fe. Note that Misemer finds another not further specified eighth state, while our result (seven plane waves) agrees with the calculation by Baker and Smith[158] for paramagnetic Fe.

Since now the 12 basis functions are known, we can set up the Hamiltonian matrix represented in this basis. For the sp bands one finds the following pseudopotential Hamiltonian:

$$H_{k+K_i;k+K_i} = \beta + \alpha |k+K_i|^2 + Sj_2^2(|k+K_i|R)$$
$$H_{k+K_i;k+K_j} = F_i(k)F_j(k) \cdot \{V(|K_i - K_j|) \qquad (5.4.71)$$
$$+ Sj_2(|k+K_i|R)j_2(|k+K_j|R)P_2[(k+K_i)\cdot(k+K_j)]\}.$$

The difference compared to Ni consists of the additional orthogonalization factors ('S terms') introduced by Baker and Smith in order to orthogonalize the sp states with respect to the d orbitals. So the local pseudopotentials V are combined with the S terms into nonlocal pseudopotentials. This somehow corresponds to the method by Mueller.[145] Similarly to the $l = 2$ terms of the 'Augmented plane-wave' (APW) method and the Korringa–Kohn–Rostoker–Ziman (KKRZ) method, the parameter S could be identified with an expression containing phase shifts or logarithmic derivatives while R could be seen as the muffin-tin radius. Thus it is not surprising that the additional S term gives an improved band structure.

The sd-hybridization block has the following form:

$$H_{k+K_i;\mu} = B_\mu F_i(k) j_2(|k + K_i| R) \varphi_\mu(k + K_i). \tag{5.4.72}$$

Here, extending the procedure for Ni, we use different constants $B_\mu = B_{t_{2g}}$ ($\mu = 8, 9, 10$) and $B_\mu = B_{e_g}$ ($\mu = 11, 12$) for orbitals of different symmetry (t_{2g} and e_g orbitals, respectively) to take into account the crystal-field splitting.

The d portion of the Hamiltonian matrix is treated again within the orthogonal three-centre approximation of the tight-binding scheme. As in the case of Ni, the additional crystal-field parameter Δ is introduced. The hoppings are again parametrized in terms of Fletcher–Wohlfarth parameters while also next-nearest-neighbour interactions have to be included, since for bcc lattices the distance a of the six next-nearest neighbours exceeds the distance $a\sqrt{3}/2$ of the eight nearest neighbours only slightly. For comparison note that for fcc lattices the distance difference of the 12 nearest neighbours ($a\sqrt{2}/2$) and the eight next-nearest neighbours(a) is much larger.

So the elements of the d portion of the Hamiltonian matrix read:

$$H_{8,8} = E_0 + 2A_1(\cos 2\xi + \cos 2\eta) + 2A_2 \cos 2\zeta + 8A_3 \cos \xi \cos \eta \cos \zeta$$

$$H_{8,9} = -8A_7 \sin \xi \cos \eta \sin \zeta$$

$$H_{8,10} = -8A_7 \cos \xi \sin \eta \sin \zeta$$

$$H_{8,11} = 0$$

$$H_{8,12} = -8A_6 \sin \xi \sin \eta \cos \zeta$$

$$H_{9,9} = E_0 + 2A_1(\cos 2\eta + \cos 2\zeta) + 2A_2 \cos 2\xi + 8A_3 \cos \xi \cos \eta \cos \zeta$$

$$H_{9,10} = -8A_7 \sin \xi \sin \eta \cos \zeta$$

$$H_{9,11} = -\sqrt{3}\, H_{9,12} = -4\sqrt{3}\, A_6 \cos \xi \sin \eta \sin \zeta \tag{5.4.73}$$

$$H_{10,10} = E_0 + 2A_1(\cos 2\xi + \cos 2\zeta) + 2A_2 \cos 2\eta + 8A_3 \cos \xi \cos \eta \cos \zeta$$

$$H_{10,11} = \sqrt{3}\, H_{10,12} = 4\sqrt{3}\, A_6 \sin \xi \cos \eta \sin \zeta$$

$$H_{11,11} = E_0 + \Delta + \tfrac{3}{2}A_4(\cos 2\xi + \cos 2\eta) + \tfrac{1}{2}A_2(\cos 2\xi + \cos 2\eta + 4\cos 2\zeta)$$
$$\quad + 8A_5 \cos \xi \cos \eta \cos \zeta$$

$$H_{11,12} = \tfrac{1}{2}\sqrt{3}(A_2 - A_4)(\cos 2\xi - \cos 2\eta)$$

$$H_{12,12} = E_0 + \Delta + \tfrac{3}{2}A_2(\cos 2\xi + \cos 2\eta) + \tfrac{1}{2}A_4(\cos 2\xi + \cos 2\eta + 4\cos 2\zeta)$$
$$\quad + 8A_5 \cos \xi \cos \eta \cos \zeta.$$

Weight of boundary points For Fe the Ni theory is extended by carrying out a *weighted* three-dimensional summation over the points within the irreducible $\frac{1}{48}$th of the Brillouin zone in order to compensate for boundary effects due to the finite mesh. The volume around each point on the boundary of the irreducible $\frac{1}{48}$th depends on the number of adjacent $\frac{1}{48}$ths. For example, each P point is weighted by a factor of $\frac{1}{24}$, since 24 irreducible $\frac{1}{48}$ths share this point. A similar weighting procedure has already been performed for the three tetrahedra forming the irreducible $\frac{1}{48}$th of Ni. There, however, only the interior boundaries of the three tetrahedra were considered, while we extend the weighting to the exterior boundaries in the case of Fe, the irreducible $\frac{1}{48}$th of the Brillouin zone of which consists of only one tetrahedron ΓNPH. Thus we reach improvements of the theoretical accuracy ranging typically from some per cent for very dense k-space meshes to 15% for less dense grids.

The ferromagnetism of Fe is again taken into account by the exchange splitting of the diagonal matrix elements of the d bands ($E_{0\uparrow} \neq E_{0\downarrow}$). The s bands see this exchange splitting of the d bands via hybridization effects. Only pure s bands are not exchange-split. The value of the exchange splitting J is taken from the spin-polarized photoemission experiment by Eastman *et al.*,[157] who found an exchange splitting of 1.5 eV for the Fe d bands at the P_4 point. Since this point is sd-hybridized, this value corresponds to a pure d exchange of 1.776 eV.

In order to obtain good agreement with the observed d and s band width, we reduce all hopping parameters by 40% compared to the values of Baker *et al.* and shift the lower sp band edge by 1.03 eV.

Spin–orbit coupling is again treated perturbatively. In agreement with Argyres[118] we use the atomic value of $\lambda_{so} = 0.05$ eV for the spin–orbit coupling constant. Later we also performed *ab initio* calculations taking into account spin–orbit coupling in the wavefunctions of the dipole matrix elements. This, however, makes sense only for the *ab initio* calculations of the matrix elements.

All other parameters are taken from ref. 158.

Dipole matrix elements

Influence of the matrix elements For the calculation of the linear and nonlinear Kerr spectra the matrix elements play an important role. The basic questions are to what extent the spectra are determined by the band structure alone, and whether structures resulting from the bands are suppressed due to the dipole transition matrix elements. In addition, the quality of the wavefunctions entering the matrix elements has to be checked. Thus, we follow two pathways for the treatment of the dipole matrix elements in the combined interpolation scheme:

1. We estimate the matrix elements using atomic wavefunctions and consider them to be k-independent. Thus the loss of symmetry information has to be compensated for by the cutoff factor qa in the nonlinear susceptibility.
2. We diagonalize the Hamiltonian matrix and use not only the eigenvalues

(band structure) but also the eigenvectors. The eigenfunctions of the Hamiltonian obtained in this way are inserted in the dipole matrix elements. Thus, the matrix elements become *k*-dependent. It is to be expected that this procedure may preserve the symmetry and thus the selection rules. However, the absolute numbers for the matrix elements are presumably only correct *on average*.

Momentum-independent matrix elements If we assume the matrix elements to be constant (momentum-independent), we average over the detailed shape of the wavefunction. For Ni the following estimate holds:

$$\langle k_1 | e^{niq_\| \cdot r} | k_2 \rangle \approx niq_\| \times 10^{-11} \text{ m}; \qquad n = 1, 2.$$

Here 10^{-11} m is the spatial extension of a hydrogen-like 3d orbital of Ni (nuclear charge $Z = 28$). The comparison with experimental linear Kerr spectra will show that this estimate is rather reasonable and describes the order of magnitude of the spectra very well.

Momentum-dependent matrix elements In order to investigate the influence of the wavefunction symmetry upon the matrix elements and thus upon the linear and nonlinear Kerr spectra, we now calculate the momentum-dependent matrix elements. In particular, the question arises if peaks that occur in the spectra for momentum-independent matrix elements due to the *k*-dependent convolution of densities of states can vanish due to the selection rules imposed by the dipole matrix elements. To check if and to what extent our basis set and Hamiltonian are sufficient for the description of magneto-optical absorption, we calculate the wavefunctions and the dipole matrix elements, which enter the linear and nonlinear magneto-optical susceptibilities ($\chi^{(1)}$ and $\chi^{(2)}$), from the same Hamiltonian as the band structure. Therefore Haydock's recursion method,[159] which calculates only local (*k*-independent) densities of states but determines neither band structure nor wavefunctions, is not applicable to our problem, at least not in its conventionally used form.

In this subsection we consider the bulk-like matrix elements. Later we will discuss also the surface selection rules, which tend to be less stringent than in bulk due to the absence of three-dimensional inversion symmetry.

The bulk-like matrix elements are of the form $\langle kl\sigma | \hat{O} | k'l'\sigma \rangle$, where the operator $\hat{O}$ denotes the dipole operators x, y, or z. It holds to a good approximation that $k' = k + q_{(\|)} \approx k$ or $k' = k + 2q_{(\|)} \approx k$, since $|q_{(\|)}| \ll |k|$.

The wavefunctions $|k'l'\sigma\rangle$ are superpositions of the basis functions

$$\Psi_j(k, r) = \sum_{i=1}^{9} c_{ij}(k\sigma) \psi_i(k, r); \qquad j = 1, \ldots, 9. \qquad (5.4.74)$$

Thus

$$\langle kl\sigma | \hat{O} | kl'\sigma \rangle = \sum_{i,j} c_{li}(k\sigma) c_{l'j}(k\sigma) \int \psi_i^*(k, r) \hat{O} \psi_j(k, r) \, d^3r, \quad (5.4.75)$$

where the ψ_i are given by eqns (5.4.47) and (5.4.54). The coefficient matrices C then are the diagonalization matrices of the Hamiltonian H:
$$C^\dagger(k)H(k)C(k) = E(k). \tag{5.4.76}$$
So the calculation of the band structure automatically produces the matrix $C(k)$, which contains the complete spin dependence for each point of the Brillouin zone. The remaining task consists of the determination of the spin-independent matrix elements of the basis functions

$$\int \psi_i^*(k,r)\hat{O}\psi_j(k,r)\,d^3r \quad (i=1,\ldots,9 \text{ and } j=1,\ldots,9)$$

$$= \frac{1}{N} \sum_{R_l, R'_l} e^{-ikR_l} e^{ikR'_l} \int \varphi_i(r-R_l)\hat{O}\varphi_j(r-R'_l)\,d^3r$$

$$= \frac{1}{N} \sum_{R_l, R'_l} \delta_{R_l, R'_l}\left(\int \varphi_i(r)\hat{O}\varphi_j(r)\,d^3r + R_l \int \varphi_i(r)\varphi_j(r)\,d^3r\right)$$

$$= \int \varphi_i(r)\hat{O}\varphi_j(r)\,d^3r. \tag{5.4.77}$$

This relation holds for an *atomic* basis set. Therefore, in the case of an atomic basis (i.e. employing hydrogen orbitals φ_i for all wavefunctions ψ_i including the sp functions), the k dependence is completely given by the matrices $C(k)$, and the remaining k-independent r-integrals $\int \varphi_i(r)\hat{O}\varphi_j(r)\,d^3r$ can be computed analytically. We evaluate these three-dimensional integrals for all combinations of i and j (i.e. one 4s orbital, three 4p and five 3d orbitals) and obtain the following results:

$$\begin{aligned}
\langle 4p_{m=0}|z|4s\rangle &= -0.2535\,\text{Å}, \\
\langle 4p_{m=0}|z|3d_{m=0}\rangle &= +0.01271\,\text{Å}, \\
\langle 4p_{m=-1}|z|3d_{m=-1}\rangle &= +0.0110029\,\text{Å}, \\
\langle 4p_{m=+1}|z|3d_{m=+1}\rangle &= +0.0110029\,\text{Å}, \\
\langle 4p_{m=-1}|\binom{x}{y}|4s\rangle &= \binom{1}{i}\frac{1}{\sqrt{6}} \times 0.82993 \times a_B, \\
\langle 4p_{m=+1}|\binom{x}{y}|4s\rangle &= \binom{1}{-i}\frac{1}{\sqrt{6}} \times 0.82993 \times a_B, \\
\langle 4p_{m=-1}|\binom{x}{y}|3d_{m=-2}\rangle &= \binom{1}{-i}\frac{1}{\sqrt{5}} \times 0.046509 \times a_B, \\
\langle 4p_{m=-1}|\binom{x}{y}|3d_{m=0}\rangle &= \binom{-1}{-i}\frac{1}{\sqrt{30}} \times 0.046509 \times a_B, \\
\langle 4p_{m=+1}|\binom{x}{y}|3d_{m=0}\rangle &= \binom{-1}{i}\frac{1}{\sqrt{30}} \times 0.046509 \times a_B, \\
\langle 4p_{m=+1}|\binom{x}{y}|3d_{m=+2}\rangle &= \binom{1}{i}\frac{1}{\sqrt{5}} \times 0.046509 \times a_B, \\
\langle 4p_{m=0}|\binom{x}{y}|3d_{m=-1}\rangle &= \binom{1}{-i}\frac{1}{\sqrt{10}} \times 0.046509 \times a_B, \\
\langle 4p_{m=0}|\binom{x}{y}|3d_{m=+1}\rangle &= \binom{1}{i}\frac{1}{\sqrt{10}} \times 0.046509 \times a_B.
\end{aligned} \tag{5.4.78}$$

The interchange of the wavefunctions leads to the complex conjugate value. The other 203 matrix elements vanish (in bulk). The result corresponds to the atomic dipole selection rules $\Delta l = \pm 1$, $\Delta m = 0, \pm 1$ and gives a rough estimate of the order of magnitude of the matrix elements in the solid (see above subsection on momentum-independent matrix elements).

Unfortunately the combined interpolation scheme shows an additional complication. The s and p orbitals have, in contrast to the d orbitals, no atomic character but rather are represented by plane waves. Therefore decoupling between k and r space is no longer possible and Harrison's approximation[160] for the matrix elements is also not applicable. So one has to calculate the integrals of the basis functions for each k point separately. Nevertheless it will turn out that this task can be considerably simplified, since the number of independent nonvanishing matrix elements reduces from 162 to 40.

1. First we show that all sp–sp dipole matrix elements vanish (dipole operator $\hat{O} = r = (x, y, z)$; $i = 1, \ldots, 4$; $j = 1, \ldots, 4$):

$$\langle \text{sp} | r | \text{sp}' \rangle = \int \psi_i^*(k, r) r \psi_j(k, r) \, d^3r,$$

$$= \frac{1}{N v_a} \int e^{-i(k + K_i) \cdot r} r \, e^{i(k + K_j) \cdot r} \, d^3r$$

(5.4.79)

$$= \begin{cases} \int r \, d^3r = 0, & \text{if } K_i = K_j \\ \int r \{ \cos[(K_j - K_i) \cdot r] + i \sin[(K_j - K_i) \cdot] \} \, d^3r \\ \text{otherwise.} \end{cases}$$

The integral over the three-dimensional space vanishes (as well as in 1D and 2D) for $K_i = K_j$, because the r and $(-r)$ contributions cancel.

The same argument also holds for the cosine term in the integral if $K_i \neq K_j$. For the remaining sine term (for $K_i \neq K_j$) one can show using the formula for the Γ-function[161]

$$\int_u^\infty x^{\mu - 1} \sin x \, dx = \frac{i}{2} [e^{-(i\pi\mu/2)} \Gamma(\mu, iu) - e^{i\pi\mu/2} \Gamma(\mu, -iu)], \qquad \text{Re } \mu > -1$$

(5.4.80)

that the integral over the three-dimensional space vanishes although its absolute value increases and finally diverges. Thus any finite integration would yield an incorrect result. Furthermore the sp wavefunctions do not contribute to magnetism.

2. Next we prove that the d–d dipole matrix elements do not contribute either ($5 \leq i \leq 9$, $5 \leq j \leq 9$):

$$\langle d | \mathbf{r} | d' \rangle = \frac{1}{N} \sum_{R_l, R_l'} \int e^{-i k R_l} \varphi_i(\mathbf{r} - \mathbf{R}_l) \mathbf{r}\, e^{i k R_l'} \varphi_j(\mathbf{r} - \mathbf{R}_l')\, d^3r$$

$$= \frac{1}{N} \sum_{R_l, R_l'} \delta_{R_l, R_l'} \int \varphi_i(\mathbf{r} - \mathbf{R}_l) \mathbf{r} \varphi_j(\mathbf{r} - \mathbf{R}_l)\, d^3r$$

$$= \int \varphi_i(\mathbf{r}) \mathbf{r} \varphi_j(\mathbf{r})\, d^3r + \frac{1}{N} \sum_{R_l} \mathbf{R}_l \int \varphi_i(\mathbf{r}) \varphi_j(\mathbf{r})\, d^3r = 0. \quad (5.4.81)$$

The second term contributes only for $i = j$, since the atomic orbitals are orthonormal. But then the sum over all reciprocal lattice vectors vanishes. The first term yields zero due to the angular momentum selection rule for atomic wavefunctions, which is not fulfilled for two d orbitals ($l = 2$ for both states). The integrand factorized into radial and angular parts, and the replacement $\mathbf{r} \to -\mathbf{r}$ shows that the integral over the angular part vanishes while the radial part depends only on the absolute value $|\mathbf{r}|$.

3. Next we show that the matrix elements $\langle d_i | \mathbf{r} | sp_j \rangle$ can be considerably simplified ($5 \leq i \leq 9$, $1 \leq j \leq 4$):

$$\langle d_i | \mathbf{r} | sp_j \rangle = \frac{1}{N \sqrt{v_a}} \sum_{R_l} e^{-i k R_l} \int \varphi_i(\mathbf{r} - \mathbf{R}_l) \mathbf{r}\, e^{i(k + K_j)r}\, d^3r$$

$$= \frac{1}{N \sqrt{v_a}} \left(\sum_{R_l} e^{i K_j R_l} \int \varphi_i(\mathbf{r}) \mathbf{r}\, e^{i(k + K_j)r}\, d^3r \right.$$

$$\left. + \sum_{R_l} e^{i K_j R_l} \mathbf{R}_l \int \varphi_i(\mathbf{r}) e^{i(k + K_j)r}\, d^3r \right). \quad (5.4.82)$$

For fcc Ni the lattice vectors $\mathbf{R}_l$ are given by

$$\mathbf{R}_l = \mathbf{a} = n_1 \mathbf{a}_1 + n_2 \mathbf{a}_2 + n_3 \mathbf{a}_3 = \frac{a}{2} \begin{pmatrix} n_2 + n_3 \\ n_1 + n_3 \\ n_1 + n_2 \end{pmatrix}, \quad (5.4.83)$$

where

$$\mathbf{a}_1 = \frac{a}{2} \begin{pmatrix} 0 \\ 1 \\ 1 \end{pmatrix}, \quad \mathbf{a}_2 = \frac{a}{2} \begin{pmatrix} 1 \\ 0 \\ 1 \end{pmatrix}, \quad \mathbf{a}_3 = \frac{a}{2} \begin{pmatrix} 1 \\ 1 \\ 0 \end{pmatrix},$$

are the fcc basis vectors, and the n_i ($i = 1, 2, 3$) are positive or negative integers ($n_i = 0, \pm 1, \pm 2, \ldots$). So factors $e^{i K_j R_l}$ read as follows:

$$e^{i K_2 R_l} = 1; \qquad e^{i K_2 R_l} = e^{-2\pi i(n_1 + n_2)} = 1;$$

$$e^{i K_3 R_l} = e^{-2\pi i(n_1 + n_2 + n_3)} = 1; \qquad e^{i K_4 R_l} = e^{-2\pi i n_3} = 1.$$

Therefore the second sum in eqn (5.4.82) vanishes due to the factor R_1, while the first sum produces N times the same contribution

$$\langle d_i| r |sp_j\rangle = \frac{1}{N\sqrt{v_a}} \int d^3r\, \varphi_i(r) r\, e^{i(k+K_j)\cdot r}, \tag{5.4.84}$$

where N cancels out versus the normalization constant. Thus it is sufficient to calculate 40 complex matrix elements, viz. $5 \times 4 = 20$, for the dipole operators x and y or x and z in the linear and nonlinear response functions $\chi_{xy}^{(1)}(q,\omega,(M)_z)$ and $\chi_{xzz}^{(2)}(2q_\|, 2\omega, (M)_y)$, respectively. The exchange of d states by sp wavefunctions again simply leads to the complex conjugate of the matrix element. The remaining matrix elements are well defined, since the atomic d orbitals are localized in a very small range, each Wigner–Seitz cell yields the same contribution and the sum is normalized to the number of cells. Furthermore, here intraband contributions are neglected, which would be of importance only for small frequencies below the optical range ($\hbar\omega \leq 0.5\,\text{eV}$). Thus the singularities mentioned by Aspnes[162] and Moss et al.[163] cancel due to this analytical treatment, which explicitly takes into account the particular symmetry, periodicity and localization properties of the basis set. On the other hand, one can take advantage of the localization of the d electrons, since for that reason the integrand of the matrix elements gives contributions only very close to the position of its maximum, which lies at approximately 10^{-11} m. Therefore we need only few mesh-points close to this distance for the integration, but in all three spatial directions.

The shape of the Wigner–Seitz cell of Ni (a rhombic dodecahedron) can be taken into account in detail by the disjunct union of sets of 24 tetrahedra. These tetrahedra are spanned by the corner and each possible set of three elements of the following 14 corners of the Wigner–Seitz cell:

$$\frac{a}{2}(\pm 1, 0, 0),\quad \frac{a}{2}(0, \pm 1, 0),\quad \frac{a}{2}(0, 0, \pm 1),\quad \frac{a}{4}(1, 1, \pm 1),$$

$$\frac{a}{4}(1, -1, \pm 1),\quad \frac{a}{4}(-1, 1, \pm 1),\quad \frac{a}{4}(-1, -1, \pm 1).$$

Similarly as in the case of Brillouin zone (for the symmetrization factors of the Hamiltonian matrix) each space point can be represented in the 24 basis sets of the different tetrahedra in order to decide if it is contained in the Wigner–Seitz cell or not. After carrying out this procedure it turns out *a posteriori* that it is not worth the effort, since all contributions lie to a very good approximation automatically within the Wigner–Seitz cell due to the localization of the d orbitals. Thus we can renounce this refinement of the theory and save computational time at this point.

5.4.5 Results: nonlinear Kerr spectra for Ni and Fe surfaces

Magnitude of the nonlinear Kerr effect

Before we start to present results for the linear and nonlinear Kerr spectra from our microscopic theory we give, in this section, an estimate of the magnitude of the nonlinear magneto-optical Kerr effect. For that purpose we consider the quantity $|\omega^2\chi^{(2)}(2q_\|, 2\omega, M)|$ for a typical laser wavelength of 1.06 μm ($\simeq 1.17$ eV). From our microscopic theory, we will find for that quantity a value of 4×10^{15} s^{-2} m V^{-1}, while the experimental detection limit[84] is about 5×10^9 to 5×10^{10} s^{-2} m V^{-1}. On the other hand, Murphy et al.[164] found for Al(111) a maximum SHG conversion efficiency in p polarization of

$$R = \frac{I_{2\omega}}{(I_\omega)^2} = 1.7 \times 10^{-19} \text{ cm}^2 \text{ W}^{-1} \qquad (5.4.85)$$

We use this experimental result for our estimate of the nonlinear magneto-optical susceptibility of Ni.

Inserting the nonlinear polarization for $M = 0$ in the wave equation ($c =$ speed of light)

$$\nabla \times \nabla \times E(2q_\|, 2\omega) - \frac{4\omega^2}{c^2} \varepsilon^{(1)}(2q_\|, 2\omega) E(2q_\|, 2\omega) = 16\pi \frac{\omega^2}{c^2} P(2q_\|, 2\omega) \qquad (5.4.86)$$

yields

$$E(2q_\|, 2\omega) = \frac{4\pi}{\chi^{(1)}(2q_\|, 2\omega)} \chi^{(2)}(2q_\|, 2\omega, 0)\big(E(q_\|, \omega)\big)^2, \qquad (5.4.87)$$

where

$$\chi^{(1)}(2q_\|, 2\omega) = \varepsilon^{(1)}(2q_\|, 2\omega) - 1. \qquad (5.4.88)$$

Since the intensity I_ω (and similarly $I_{2\omega}$) is given by the absolute value of the Poynting vector S,

$$I_\omega = |S| = \varepsilon_0 c |E(q_\|, \omega)|^2, \qquad (5.4.89)$$

$\varepsilon_0 =$ vacuum permittivity, we find for the nonlinear optical susceptibility of Al(111)

$$\left[\omega^2\chi^{(2)}(2q_\|, 2\omega, 0)\right]_{\text{SHG, Al}} = (\omega^2\varepsilon_0 cR)^{\frac{1}{2}} \frac{|\chi^{(1)}(2q_\|, 2\omega)|}{4\pi}$$

$$= 6.7 \times 10^{17} \text{ s}^{-2} \text{ m V}^{-1} \cdot \frac{|\chi^{(1)}(2q_\|, 2\omega)|}{4\pi}$$

$$= 2.9 \times 10^{18} \text{ s}^{-2} \text{ m V}^{-1}, \qquad (5.4.90)$$

where the linear optical susceptibility for bulk Al (a free-electron metal) has been calculated from the measured plasma frequency ($\hbar\omega = 15.8$ eV). Since $\chi^{(2)}$

is proportional to $\varepsilon_1(\boldsymbol{q}_\|, \omega)\varepsilon_1^{1/2}(2\boldsymbol{q}_\|, 2\omega)$ and the ratio of nonlinear optical to nonlinear magneto-optical susceptibility[13,14] is given by

$$\left|\frac{\chi^{(2)}(2\boldsymbol{q}_\|, 2\omega, \boldsymbol{M})}{\chi^{(2)}(2\boldsymbol{q}_\|, 2\omega, 0)}\right| \approx 0.07, \qquad (5.4.91)$$

the experiment by Murphy *et al.* implies a maximum nonlinear magneto-optical susceptibility in Ni of

$$\left[\omega^2\chi^{(2)}(2\boldsymbol{q}_\|, 2\omega, \boldsymbol{M})\right]_{\text{Kerr, Ni}} \approx 4 \times 10^{15} \text{ to } 1.3 \times 10^{16} \text{ s}^{-2}\,\text{m}\,\text{V}^{-1}, \quad (5.4.92)$$

depending on the numbers for the optical constants of Ni.

This result agrees well with our microscopic theory and demonstrates again that and why the nonlinear magneto-optical Kerr effect is observable. It clearly shows the pronounced enhancement of optical second harmonic generation at metallic surfaces and supports our assumption[13] that the nonlinear response results mainly from the topmost one or two layers (of some ångströms in thickness), where the electronic charge density decreases from its bulk value to zero. Therefore an estimate based on the *atomic* polarizability of the surface atoms does not apply to metals.

In Ni it is clear that the main contributions to the linear and nonlinear susceptibilities result from the interband transitions of the 'magnetic' d electrons, while in Al the 'free' sp electrons are responsible for optical second harmonic generation. Thus, as indicated by our estimate, the SHG intensity should be smaller at a Ni surface, since s–d hybridization reduces the delocalized behaviour of the electrons. The high density of d electrons[49] cannot compensate for their more localized behaviour.

Kerr spectra of nickel

Next we discuss the linear and nonlinear Kerr spectra of fcc Ni calculated by the theoretical methods previously discussed. It is the goal of this subsection to investigate the influence of the band structure and the matrix elements on the linear and nonlinear susceptibilities $\omega^2 \operatorname{Im} \chi_{xy}^{(1)}$ and $\omega^2 \operatorname{Im} \chi_{xzz}^{(2)}$ by comparing $\boldsymbol{k}$-independent with $\boldsymbol{k}$-dependent spectra.

Linear Kerr spectrum for momentum-independent matrix elements Our results for the linear magneto-optical spectrum $\omega^2 \operatorname{Im} \chi_{xy}^{(1)}(0, \omega, \boldsymbol{M})$ are shown in Fig. 5.25. For the spin–orbit interaction in Ni we use a value of 70 meV throughout as was calculated by different groups (Callaway, Freeman). In this subsection we assume a value for the Lorentzian broadening of $\hbar\alpha = 0.35 \times 10^{-19}\,\text{J} = 0.218\,\text{eV}$ in the spectral range of 'quartz optics' (excitation energy below 5 eV) and of $\hbar\alpha = 0.7 \times 10^{-19}\,\text{J} = 0.437\,\text{eV}$ otherwise. We neglect that $\boldsymbol{q}$ dependence in the linear theory, since for typical optical experiments $|\boldsymbol{q}| \ll |\boldsymbol{k}|$. For the dipole matrix elements we use the 'atomic value' $\langle|r|\rangle = 10^{-11}$ m as in ref. 13. The $\boldsymbol{k}$ summation goes over 6206 $\boldsymbol{k}$ points in the irreducible $\frac{1}{48}$th of the Brillouin zone

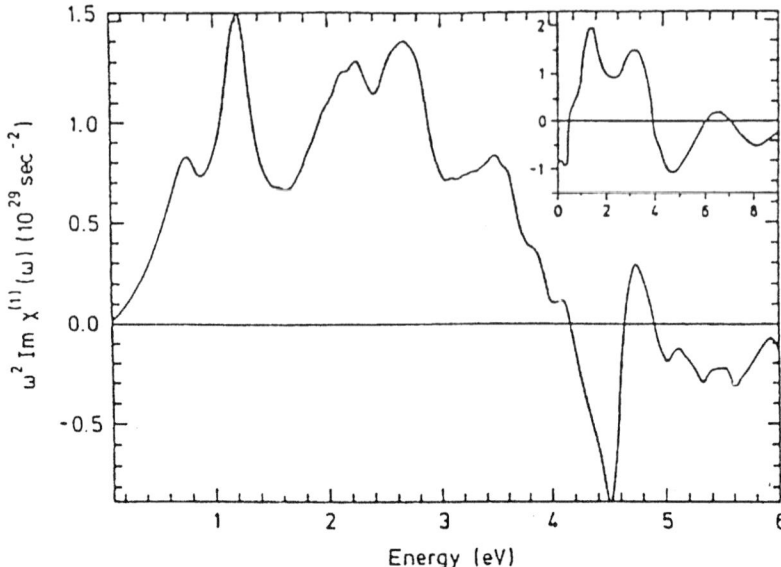

Fig. 5.25. Linear magneto-optical Kerr susceptibility $\omega^2 \operatorname{Im} \chi^{(1)}_{xy}(\omega)$ as a function of frequency ω for bulk Ni. The complete three-dimensional band structure is included, while the dipole matrix elements are treated as constants. The inset shows the experimental results by Erskine.

$(0 \le k_z \le k_x \le k_y \le 2\pi/a)$. The tetrahedra decomposition of the truncated octahedron (fcc Brillouin zone) is used in the form described earlier. In order to check the convergence behaviour of the k summation we compare also with the summations over 891 and 148 points. It turns out that the main features of the spectrum already show up clearly for the 148 k-point sum despite the poor convergence due to this small k-point number. In order to obtain good accuracy we diagonalize the Hamiltonian at each k point and use no interpolation at all. The value of the energy interval (distance between two abscissa points in Fig. 5.15) is 0.02 eV. The complete three-dimensional band structure calculated within the combined interpolation scheme is included, while the dipole matrix elements are treated as constants. For comparison, the inset shows a typical experimental result for the same quantity by Erskine.[165] The axes are scaled as in the main figure.

The calculated spectrum (in Fig. 5.25 for 6206 k points clearly exhibits the same main structures as the experimental spectrum. There are three peaks at 1.2 eV, 2.7 eV (both of positive sign) and 4.5 eV (of negative sign). The experimental peak positions are (for each experiment somewhat different) 1.15 to 1.4 eV, 2.7 to 3.2 eV (positive sign) and 4.3 to 4.6 eV (negative sign), where we compare with the magneto-optical experiments by Erskine[165] (performed with synchrotron radiation, see inset), Yoshino et al.[166] and Krinchik et al.[167] (using a conventional light source). We find good agreement with experiment also concerning the absolute magnitude of the spectra. For example, we obtain for the first peak a height of $1.55 \times 10^{29}\,\text{s}^{-2}$ in comparison with the experimental value of $(1.2 \text{ to } 2.0) \times 10^{29}\,\text{s}^{-2}$. Furthermore, the peak height ratio of the first three calculated peaks of $2:1.8:(-1.1)$ agrees well with the experimental value of $2:1.5:(-1)$ (see refs 165 and 167). Similarly we find very good agreement of the peak width ratio of the first two peaks with the measurement while the third peak is narrower than the corresponding experimental structure. Nevertheless the zero of the experimental spectrum at 3.9 eV is well reproduced by our calculation (4.15 eV). In general, our spectra show, as is typical for theories, more details than the experiments, the resolution of which is restricted.

This good overall agreement of our theory with experiment is in pronounced contrast to *ab initio* calculations exhibiting deviations of the peak positions by more than 1 eV compared to experiment and line shapes too different to be compared with any experiment. Thus it becomes clear again that the band structure parameters by Weling and Callaway derived from photoemission data, de Haas–van Alphen measurements and spin-polarization measurements are well suited for the electronic calculation of magneto-optical Kerr spectra.

Choosing a parametrized description of the Hamiltonian matrix we incorporate the relevant many-body effects (such as self-energy corrections), which play an important role for the interpretation of the magneto-optical spectra, in the calculation. For Ni, the single-particle picture of usual density-functional theory (e.g. 'linear augmented plane-wave' calculations) is—despite recent improvements of the latter—not yet sufficient to describe satisfactorily the magneto-optical excitation spectra. Thus, for Ni surfaces (a particularly strong correlated electron system), our semiempirical band structure calculation (with the parameters mostly taken from ref. 147 gives much better results than the usual *ab initio* calculations[128,168,169] or the fit of the parameters to these *ab initio* band structures.[170] Only recently improved full-potential linear augmented plane-wave methods are able to reduce this discrepancy.

Linear Kerr spectrum for momentum-dependent matrix elements Next we include the nonconstant (k-dependent, but bulk-like) matrix elements in our theory as previously described in order to investigate the hereby induced modifications of the linear Kerr spectrum. The wavefunctions for the calculation of the dipole matrix elements are taken from the diagonalization of the same Hamiltonian whose eigenvalues yield the band structure. As in all eigenvalue problems it is not *a priori* clear that the eigenvalues (wavefunctions), and the optical transition matrix elements calculated from them and reflecting the off-diagonal elements of the Hamiltonian matrix, give results as reasonable as the eigenvalues (energy bands) resulting from the diagonal elements.

For the calculation we again use 6206 k points, and the integration grid consists of 216 mesh points. We calculate the matrix elements between the *basis functions* only once and store them for the rest of the calculation to save computer time. In order to keep a remainder of the selection rules, which are lost in replacing the atomic p orbitals by plane waves in the combined interpola-

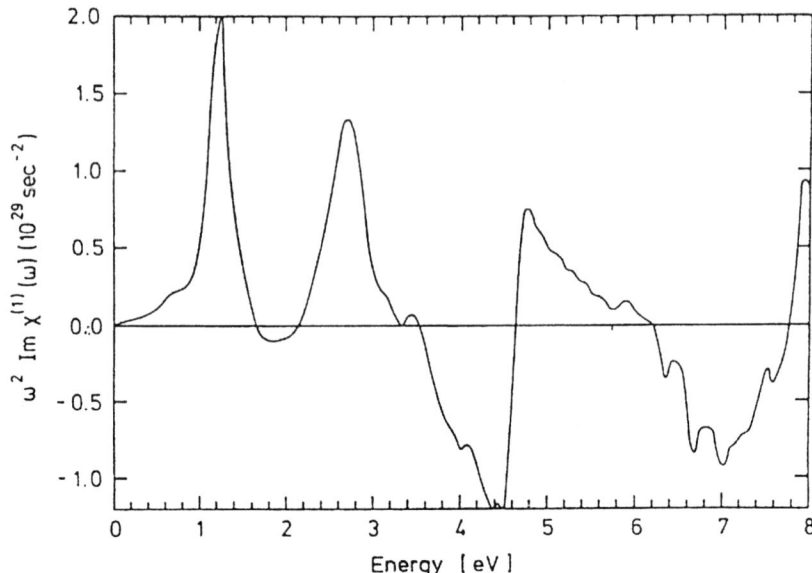

Fig. 5.26. Linear magneto-optical Kerr susceptibility $\omega^2 \, \mathrm{Im} \, \chi_{xy}^{(1)}(\omega)$ for bulk Ni as a function of frequency ω with three-dimensional band structure and $\mathbf{k}$-dependent dipole matrix elements.

tion scheme, we choose the matrix elements (as is also done later in the nonlinear case) as being invariant under the operation $x \cdot y \to -x \cdot y$ and in addition keep the exact invariance $z \to -z$. The energy interval of the abscissa is now 0.07 eV. Figure 5.26 shows the calculated magneto-optical susceptibility $\omega^2 \chi_{xy}^{(1)}(0, \omega, (M)_z)$, which corresponds to the usual *polar* Kerr geometry (magnetization $\mathbf{M} \parallel \hat{z}$, electric field of incident photons $\mathbf{E} \parallel \hat{x}$, incident beam perpendicular to the surface, polarization of reflected beam acquires small y component) in the energy range 0–8 eV. The comparison with Fig. 5.25 shows that the inclusion of the $\mathbf{k}$-dependent matrix elements does not change the overall structure of the spectrum. The main difference consists of the absence of the constant background between the first two peaks of the spectrum: the structures have narrowed. The pronounced minimum in between is typical for spectra with $\mathbf{k}$-dependent matrix elements (see ref 128, 169). This minimum does not occur in experiments due to the instrumental broadening of the spectrum.

The peak positions remain unchanged. Two further broad peaks centred around 5.5 eV (positive sign) and around 7 eV (negative sign) can possibly be identified with the structures at 6.5 eV and 8 eV in Erskine's experimental spectrum (inset of Fig. 5.25). The peak height ratio of $2:1.5:(-1.18)$ for the first three peaks agrees even better than before. The absolute value of the first maximum is $2.0 \times 10^{29} \, \mathrm{s}^{-2}$ and its deviation from its previous value $1.5 \times 10^{29} \, \mathrm{s}^{-2}$ measures the difference between the true average value of the calculated dipole matrix elements and the value of 10^{-11} m assumed in the calculations for constant matrix-elements and being obviously quite realistic.

The width of the peak at 4.5 eV, however, has considerably increased and now agrees better with experiment. The zero between the peaks at 2.7 and 4.5 eV lies at 3.55 eV (rather than 4.15 eV for constant matrix elements) and thus 0.35 eV below the experimental value. For several spectral features the inclusion of the dipole matrix elements on the same footing as the band structure yields somewhat improved agreement with linear Kerr experiments. However, none of the peaks in the spectrum obtained for constant matrix elements vanishes due to the inclusion of the k-dependent matrix elements.

In general the following 'distribution of tasks' between the band structure and the matrix elements can be inferred. The band structure determines the *peak positions* and *peak height ratios* while the k-dependent matrix elements are responsible for the *peak shape and width* and the *absolute magnitude* of the spectrum. This observation is physically plausible, since it again emphasizes the importance of the interband transitions in high-symmetry directions for the occurrence of the peaks in the Kerr spectra. The momentum dependence of the matrix elements averages out by the hybridization of states with different symmetry. Therefore the atomic selection rules appear to be somewhat relaxed, and the matrix elements are only *on average* responsible for the magnitude of the spectra, since they occur as k-dependent prefactors in the susceptibilities.

The narrowing of the peaks for k-dependent matrix elements can be understood in the following way: For energies outside the peak maximum ('off-resonance') the influence of certain high-symmetry transitions is no longer prevailing, and the 'incoherent' behaviour of many other matrix elements (also for the different spins) causes a compensation effect, which rapidly suppresses the pronounced peaks and consequently causes the narrowing (in contrast to the 'coherent' behaviour of the matrix elements).

Owing to some bands that reach quite high energies, the calculated spectra exhibit structures up to 46 eV. However, since we did not include *all* bands up to these large energies, the susceptibilities are reliable only up to approximately 8 eV. Consequently, the abscissae of Figs 5.25 and 5.26 are restricted to that energy range.

Finally we would like to list some possible reasons for the occurrence of the negative part observed in some of the measured linear Kerr spectra but absent in our calculation. Note that this feature looks very different in the various experiments and is less pronounced than the three main peaks.

1. The most important reason might be the important role of many-body effects close to the Fermi level of Ni[171-173] such as counter-spin admixtures. These effects show up clearly in theory and experiments on spin-polarized photo-emission and spin-polarized electron-capture spectroscopy and exceed our incorporation of many-body corrections as self-energy shifts of the bands.
2. Our theory treats the influence of spin–orbit coupling on the wavefunctions perturbatively and, in this way, describes the magneto-optical spectra very well. Therefore no real coupling exists between the electrons of opposite spin which could affect the low-frequency behaviour in the infrared (via the listing of degeneracies due to spin–orbit coupling; see ref. 174, in particular the

work by Halilov and Uspenskii). In optical range, it is reasonable to exclude light-induced spin flips.
3. Experimentalists extract the quantity $\omega^2 \operatorname{Im} \chi_{xy}^{(1)}(0, \omega, (M)_z)$ from the measured Kerr angles or intensity differences with the help of formulae containing optical constants and dielectric functions. In many cases the frequency dependence of these quantities has been neglected although it is as strong as that of the Kerr rotation itself (in linear optics). Both diagonal and off-diagonal elements of the tensor $\chi^{(1)}(\omega)$ depend on frequency, of course.
4. The Drude term, which has so far been neglected in our theory, would cause a vertical upward shift of the spectrum.
5. Possible inaccuracies of the matrix elements might affect the energy range below 0.5 eV more strongly than band-structure effects. However, the interesting range for laser experiments in metals is clearly at higher energies where these effects are unimportant.

Nonlinear Kerr spectrum of Ni for constant matrix elements We now calculate the nonlinear magneto-optical susceptibility

$$\omega^2 \operatorname{Im} \chi_{xzz}^{(2)}(2q_\|, 2\omega, (M)_y) \qquad (5.4.93)$$

on the basis of the same semiempirical band structure as in the linear case. If the matrix elements are treated as constants, a prefactor $|q_\||a$ is inserted in the susceptibility which guarantees that the SHG signal results exclusively from the topmost atomic layer. This factor (ratio of linear excitation and nonlinear response depth) was already discussed in detail together with the scenario of optical SHG and results from the *two*-dimensional Fourier transform. It is important to note that the details of the electronic charge-density profile at the surface are fairly unimportant for the *frequency* dependence of the nonlinear *magneto*-optical Kerr effect, since the latter mainly results from the 'magnetic' d electrons, which are localized in a relatively narrow range (in real space). Figure 5.27 shows the nonlinear Kerr spectrum for constant matrix elements and abscissa intervals of 0.1 eV. The spectrum has the order of magnitude $10^{16}\,\text{s}^{-2}\,\text{m}\,\text{V}^{-1}$. Thus the nonlinear Kerr effect is observable according to the estimate at the beginning of the previous subsection. The structure of the nonlinear Kerr spectrum is much richer than that of the linear magneto-optical susceptibility, since the three-photon processes carry more information than linear optics does. The nonlinear Kerr effect measures the difference of majority- and minority-state differences ('double-difference differences') and therefore very sensitively probes the details of surface electronic structure and magnetic surface properties such as magnetization strength, magnetic easy axis, exchange splitting and spin−orbit coupling.

Although these spectra exhibit structures up to 46 eV (as in the linear case, due to the ω resonances rather than the 2ω denominators) they should not be taken too seriously above 5 to 6 eV, because not all bands were included up to 46 eV and the convergence decreases for higher energies, since at that time only

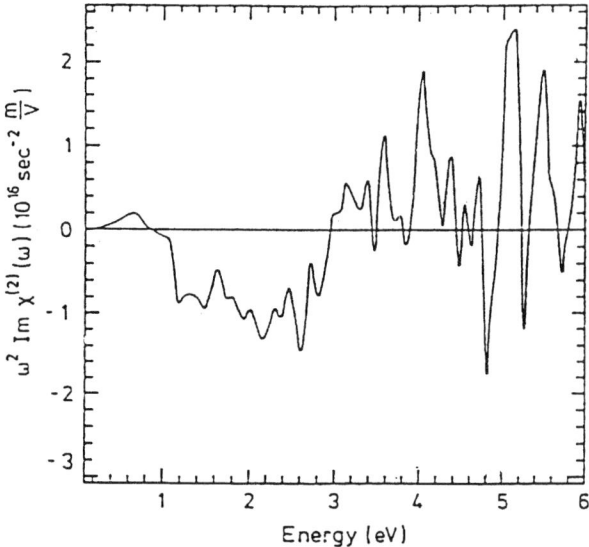

Fig. 5.27. Nonlinear magneto-optical Kerr susceptibility $\omega^2 \, \text{Im} \, \chi^{(2)}_{xzz}(\omega)$ as a function of frequency ω for the surface of ferromagnetic Ni with in-plane magnetization M_y for constant matrix elements. Incident and frequency-doubled reflected beams are assumed to be p-polarized.

891 k points within $\frac{1}{46}$th of the irreducible portion of the Brillouin zone could be used for the nonlinear calculation. In the optical range our spectrum shows structures at 0.6 eV (positive sign), in a broad range around 2 to 2.5 eV (negative sign) and at 4 eV (positive sign). These peaks will be analysed in more detail in the case of Fe.

Nonlinear Kerr spectrum for momentum-dependent matrix elements If we use k-dependent matrix elements for the calculation of the nonlinear Kerr spectrum we have to drop the prefactor $|q_\parallel| a$, since the surface sensitively is already contained in the product of the three dipole matrix elements, which, in inversion-symmetric media, gives only a contribution at the surface (where inversion symmetry is broken). The inclusion of the momentum-dependent matrix elements in the nonlinear magneto-optical susceptibility (see Fig. 5.28 for 891 k points and abscissa intervals of 0.05 eV) has essentially the same consequences as for the linear Kerr spectra. The basic structure of the spectrum remains unchanged: the peaks occur at 0.6 eV (positive sign), in a broad range around 2.6 eV (negative sign) and around 4 eV (positive sign). As in the linear case the peaks become narrower and therefore more pronounced for k-dependent matrix elements.

The most interesting difference induced by the momentum dependence of the matrix elements appears in the absolute value of the susceptibility, which is now of the order $10^{17} \, \text{s}^{-2} \, \text{m} \, \text{V}^{-1}$ (rather than $10^{16} \, \text{s}^{-2} \, \text{m} \, \text{V}^{-1}$ for constant matrix

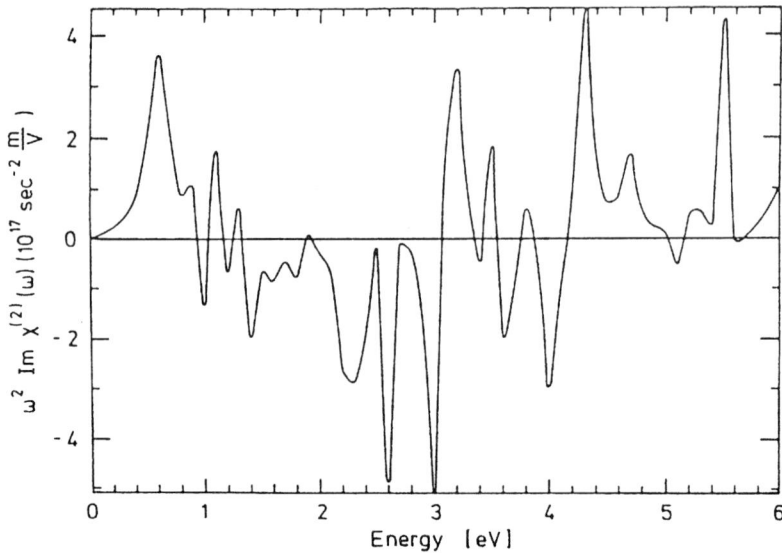

Fig. 5.28. Nonlinear magneto-optical Kerr susceptibility $\omega^2 \operatorname{Im} \chi^{(2)}_{xzz}(\omega)$ as a function of frequency ω for the surface of ferromagnetic Ni with in-plane magnetization M_y for k-dependent matrix elements. Incident and frequency-doubled reflected beams are assumed to be p-polarized.

elements). Although this change is not dramatic its origin is worth while being considered. Of course, the value of 10^{-11} m for the dipole matrix elements results from an atomic estimate and deviations may have a large effect, since the matrix elements enter the nonlinear susceptibility to the third power. A similar argumentation also applies to the numerical computation of the product of the k-dependent matrix elements, which consists of three triple (nine single) integrals. Thus small inaccuracies could cause large changes in the magnitude of the nonlinear magneto-optical susceptibilities due to the limited number of mesh points.

However, the most probable explanation for this difference of the nonlinear Kerr spectra for constant and momentum-dependent matrix elements might be the different surface treatment:

For *constant* matrix elements the influence of the surface is taken into account by the prefactor $|q_\parallel| a$ while we use of a modified 'truncated bulk approximation' for *k-dependent* matrix elements. So the total nonlinear (SHG and NOLIMOKE) yield, which exactly vanishes in bulk due to the invariance of the product of the three matrix elements under the operation $z \to -z$, results from the incompleteness of the topmost layer of Wigner–Seitz cells. This cut is performed in the middle of the cell, thus maximizing the nonlinear output, and expresses the broken electronic symmetry at the surface, which is responsible for the frequency doubling. A small shift of the cut at the crystal face by 0.1 Å could reduce the yield already by one order of magnitude due to compensation effects from different portions of the Wigner–Seitz cell and therefore could explain the discrepancy. In real crystals the surface cut adjusts upon crystal cleavage or surface preparation as the minimum of free energy at a certain temperature or locks in as a metastable configuration under fixed external conditions.

On the other hand, one could replace the wavevector $q_\parallel$ describing the inverse penetration depth in the prefactor $|q_\parallel|a$, which occurs for constant matrix elements, by the plasma wavevector $q_{\parallel,\text{plasma}}$. This vector is larger and connects the skin depth with the onset of transparency in metals. This replacement would not only remove the discrepancy but also cancel the suppression of the low-energy peaks for constant matrix elements.

By and large, the order of magnitude of our calculated nonlinear Kerr spectra agrees well with recent SHG[164,175] and NOLIMOKE experiments and SHG spectra.[48] This confirmed our prediction of the observability of the nonlinear Kerr effect. In particular the SHG experiments by Murphy *et al.* on Al(111) in ultrahigh vacuum (UHV) and by Guyot-Sionnest on Ag(111) in electrolyte solution as well as the jellium calculation by Liebsch[164] support our prediction of an enhanced SHG and NOLIMOKE yield in metals which exceeds that of an atomic estimate[84,85] by four orders of magnitude. Thus, SHG and especially NOLIMOKE might experimentally be realized even with a simple semiconductor (diode) laser and consequently are even technologically competitive for applications in nondestruction remote sensing as well as nonlinear magneto-optical layer-by-layer recording.

Therefore our microscopic theory explains the mechanism, the order of magnitude and the frequency dependence of the nonlinear Kerr effect and clearly demonstrates that NOLIMOKE is a promising probe for the optical investigation of magnetic phenomena at surfaces, interfaces, multilayers and in thin films.

Summary of the results for nickel

1. In order to develop a detailed theory of nonlinear magneto-optics and to investigate the influence of the complete anisotropic band structure and of the three-dimensional k-dependent dipole transition matrix elements, we calculated the linear and nonlinear magneto-optical Kerr spectra of ferromagnetic Ni from an electronic theory as a function of frequency. We computed the three-dimensional band structure semiempirically and treated the matrix elements on the same footing.
2. The good agreement between theory and experiment shows that the parameters of the band-structure calculation derived from spin-polarized photoemission data and de Haas–van Alphen measurements also describe magneto-optical absorption very well.
3. In general, most of the peaks in the spectra become narrower for k-dependent matrix elements. Although the incorporation of momentum-dependent matrix elements improves some of the details of the spectra (e.g. the

peak height ratios and the width of structures at higher energies), the most important features are already obtained for constant matrix elements. The 'averaging out' of 'artificial' symmetry effects by constant matrix elements describes the effects in real metallic solids very well. On the other hand, the incorporation of the matrix elements (preferably on the *ab initio* level) is required for the description of geometric effects in particular experimental configurations.
4. This detailed electronic theory confirms our previous statements that the nonlinear Kerr effect is surface- and interface-sensitive (in particular for p-polarized photons) and is of course easily observable.
5. We find that the peak height ratios are determined by the band structure (interband transitions) while the *k*-dependent matrix elements are responsible for the absolute magnitude and the peak widths.
6. The nonlinear magneto-optical Kerr effect is very sensitive to the detail of the band structure and yields a lot of information on the magnetic (exchange interaction, spin–orbit coupling) and electronic properties of ferromagnetic transition-metal surfaces.

Kerr spectra of iron

Motivated by the microscopic calculations of NOLIMOKE spectra for a nickel surface[12,13] first experiments on Fe were performed in 1991 by Reif *et al.*[16] The (110) surface of single-crystal Fe in a single-domain state was studied under ultrahigh-vacuum conditions. Upon reversing the direction of the in-plane magnetization with the help of a weak external magnetic field defining the quantization axis, a drastic change of the SHG yield was observed. For the clean Fe (110) surface the ratio of the nonlinear magnetic to nonmagnetic susceptibility amounts to 25%. This value is in agreement with the theoretical prediction, which is 17.5%. The latter results not only from scaling with the magnetic moments, as was previously assumed for the analysis of the experiment.[16,176] It is remarkable that our estimate for $\chi^{(2)}_{\text{mag.}}/\chi^{(2)}_{\text{nonmag.}} \approx \delta/I_{\text{nonmag.}}$ is in good agreement with the experimental value.

If the Fe surface is deliberately contaminated by chemisorbing CO, an exponential decay of the *magnetic* signal is observed, whereas the *nonmagnetic* contribution to the signal remains constant. This dramatic influence of contamination clearly demonstrates the sensitivity of this technique for surface *magnetism*. Further observations show that the SHG signal vanishes if the surface is intentionally damaged by a too strong laser pulse.

Besides comparing with the experimental results by Reif *et al.*, it is of general interest to analyse theoretically the differences between the Kerr effect spectra of iron and nickel. The differences regarding the magnetic properties like the exchange splitting and magnetic moments and also the crystal structure require an extension of our previous theory. For Fe one expects a reduction of electronic many-body effects and at the same time an increase of geometric structure effects. The surfaces of Fe are not as smooth as those of Ni, where

Table 5.14 Parameters used for the calculation of the magneto-optical susceptibilities

$\beta = 2.5\,\text{eV}$	$A_1 = 0.0774\,\text{eV}$	$B_1 = 20.57\,\text{eV}$
$\alpha = 0.315\,\text{eV}$	$A_2 = -0.00816\,\text{eV}$	$B_2 = 18.28\,\text{eV}$
$V_1 = 1.40\,\text{eV}$	$A_3 = -0.1074\,\text{eV}$	$E_0 = 9.646\,\text{eV}$
$V_2 = 2.15\,\text{eV}$	$A_4 = -0.2016\,\text{eV}$	$\Delta = 0.068\,\text{eV}$
$S = 15.54\,\text{eV}$	$A_5 = 0.1272\,\text{eV}$	$J_0 = 1.78\,\text{eV}$
$R = 2.48 \times 10^{-9}$	$A_6 = -0.0756\,\text{eV}$	$\lambda_{\text{so}} = 0.05\,\text{eV}$
	$A_7 = A_3 - A_5 - \dfrac{\sqrt{3}}{2} A_6$	$\hbar\alpha_1 = 0.1\text{ or }0.4\,\text{eV}$

many-body effects play a dominant role.[171–173] Thus, a larger wavefunction basis set for the electronic band-structure calculation is necessary.

While for Ni the formulation of the nonlinear magneto-optical theory and the demonstration of the observability of the nonlinear Kerr effect were the main theoretical issues, for Fe (after the previous questions had been settled) the main interest is focused on the information on the electronic and magnetic surface properties that can be obtained from the nonlinear Kerr spectra.

We perform the calculation for the nonlinear magneto-optical Kerr susceptibility $\chi^{(2)}(\omega)$ in iron in order to compare directly with experiment. The nonlinear and linear susceptibilities are determined via eqns (5.4.39) and (5.4.40). For the calculations we use the parameters listed in Table 5.14. To demonstrate the validity of the electronic structure used in our calculations of $\chi^{(1)}(\omega)$ and $\chi^{(2)}(\omega)$ (calculated with the parameters given in Table 5.14), we show in Fig. 5.29 the resulting ferromagnetic band structure of Fe and compare with the results obtained from the spin polarized version of the paramagnetic band structure in ref. 158. The position of the Fermi level was calculated self-consistently for both cases by fixing the total number of conduction electrons: $N_{\text{el}} = 8$. To judge the quality of our band structure note that we obtain good agreement with *ab initio* calculations by Wang and Callaway[123] and by Hathaway *et al.*[177] if we use the *un*-modified parameters (of ref. 158). To obtain the observed d-band width we reduce, as mentioned already, the tight-binding hopping parameters A_i ($i = 1, \ldots, 7$). The points a, b and c indicate the shift of the respective pure d-like or sd-hybridized high-symmetry points. Similarly, the modification of the energies β of the Γ point leads to a shift of the s-like symmetry points. This becomes most pronounced for the shift of the Γ point indicated by the point d in Fig. 5.29. Note that we include the effects of ferromagnetism only via the spin splitting the d bands. Thus, our calculation yields a spin splitting of the various bands according to their different sd hybridization. Note that as aimed, we have very good agreement with the experimentally observed d-band width (3.1 eV for the occupied portion at point P, see ref. 157).

Figure 5.30 shows the nonlinear magneto-optical Kerr susceptibility $\omega^2\,\text{Im}\,\chi^{(2)}_{xzz}(\omega)$ for Fe as a function of frequency ω. Note that $\omega^2\,\text{Im}\,\chi^{(2)}_{xzz}(\omega)$ for

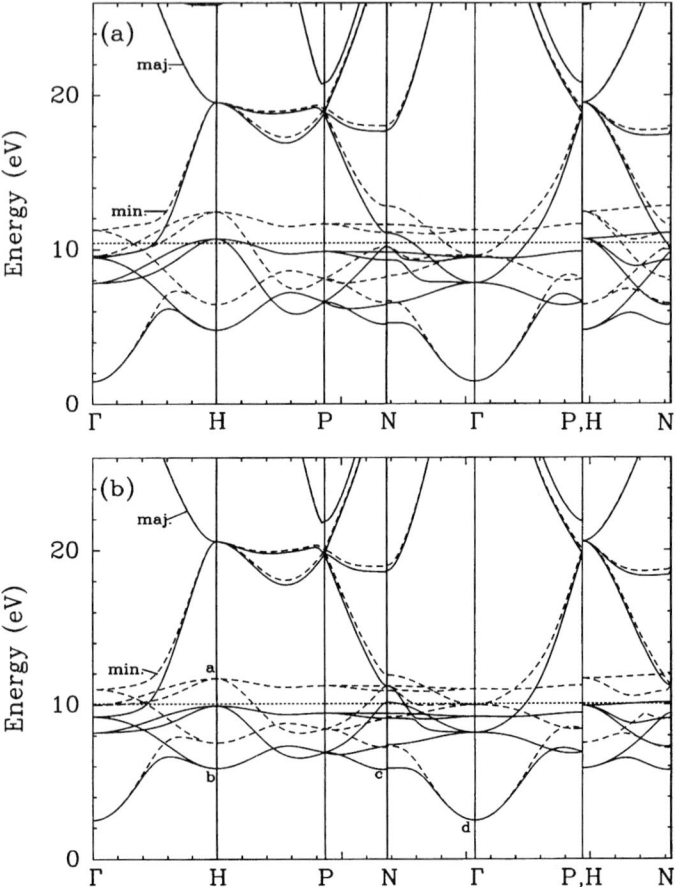

Fig. 5.29. Spin-polarized band structure or iron calculated within the combined interpolation scheme using parameters (a) given by Baker and Smith, and (b) given in Table 5.14. Solid curves refer to the majority-spin bands and dashed curves to the minority-spin bands. Note the energy shifts in particular at a, b, c and d.

Fe is of the order of $10^{16}\,\text{s}^{-2}\,\text{mV}^{-1}$, which is of similar magnitude as for nickel. The low-energy part of the spectrum (energy below 10 eV) consists of a minimum at 2.8 eV with $\omega^2\,\text{Im}\,\chi^{(2)}_{xzz}(\omega) < 0$ and two peaks at 4.5–6.0 eV and at 7.8 eV. A reduction of the damping constant $\hbar\alpha_1$ from 0.4 eV (dashed curve) to 0.1 eV (solid curve) reveals a large variety of secondary structure. In particular, the minimum is split into two subminima. The shoulder at approximately 1.3 eV, which becomes more pronounced for the smaller value of $\hbar\alpha_1$, corresponds to the first peak at 0.6 eV in the nonlinear Kerr spectrum of nickel. To understand the physical origin of this structure, we analyse its dependence on the spin-dependent electron density of states, sd hybridization and band width in the following paragraph. Note that minority spins dominate the spectra over the entire range of visible light in agreement with Krinchik's observation.[167]

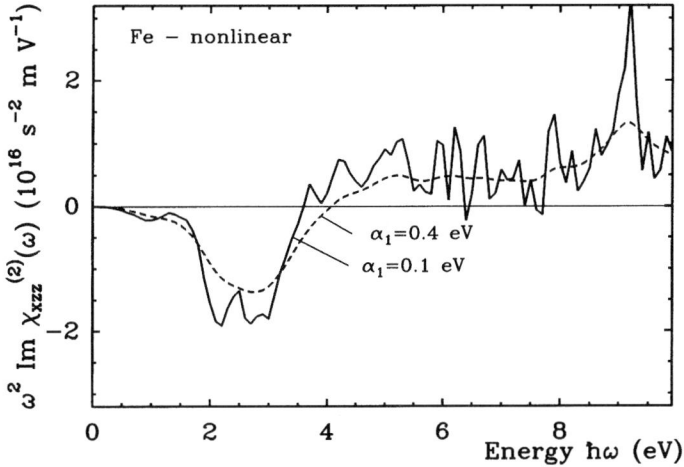

Fig. 5.30. Nonlinear magneto-optical Kerr susceptibility $\omega^2 \operatorname{Im} \chi^{(2)}_{xzz}(\omega)$ for the surface of ferromagnetic iron with in-plane magnetization M_y as a function of frequency ω. The dashed curves shows results using a resolution factor $\hbar \alpha_1 = 0.4\,\text{eV}$, and the solid line refers to $\hbar \alpha_1 = 0.1\,\text{eV}$.

In Fig. 5.31 we show the nonlinear (a) and linear (b) magneto-optical Kerr spectra for different values of the Fletcher–Wohlfarth parameters. For the dotted curves, we use the value of the hopping parameters A_i ($i = 1, \ldots, 7$) given in ref. 158. The dashed and solid curves correspond to a reduction of all A_i to 80% and 60% of these values, respectively. This reduction of the d–d hopping leads to a narrowing of the d band (see Fig. 5.29), which may partly simulate the non-negligible effects of electron–electron interactions. Owing to such correlations the probability of transitions with lower energies is increased. This essentially leads to a shift of the first structures in the linear and nonlinear Kerr spectra to lower energies (by 0.5 and 0.3 eV for $0.8 A_i$, and by 0.9 and 0.4 eV for $0.6 A_i$, respectively) and to a simultaneous reduction of their heights. The second structure is affected less drastically: in the nonlinear spectrum, the shift towards lower energies is 0.2 eV for $0.8 A_i$ and 0.3 eV for $0.6 A_i$ and the peak height increases, whereas the second peak of the linear spectrum decreases and is shifted to lower energies by 0.3 and 0.4 eV for $0.8 A_i$ and $0.6 A_i$, respectively. The shift of the second structures is due to transitions between the d band and s-like or sd-hybridized bands and is therefore always smaller than the shift of the first structures, which predominantly result from transitions between d-like bands close to the Fermi level. The modifications of the third structure in the nonlinear and linear spectra increasingly correspond to bands far from the Fermi level and do not show a clear dependence on the d–d hopping elements. Thus in summary we may conclude that changes in the d-band width affect different parts of the spectra, both linear and nonlinear ones, differently. We then expect such behaviour also with respect to changes in the position of the s band as can be seen in Fig. 5.32.

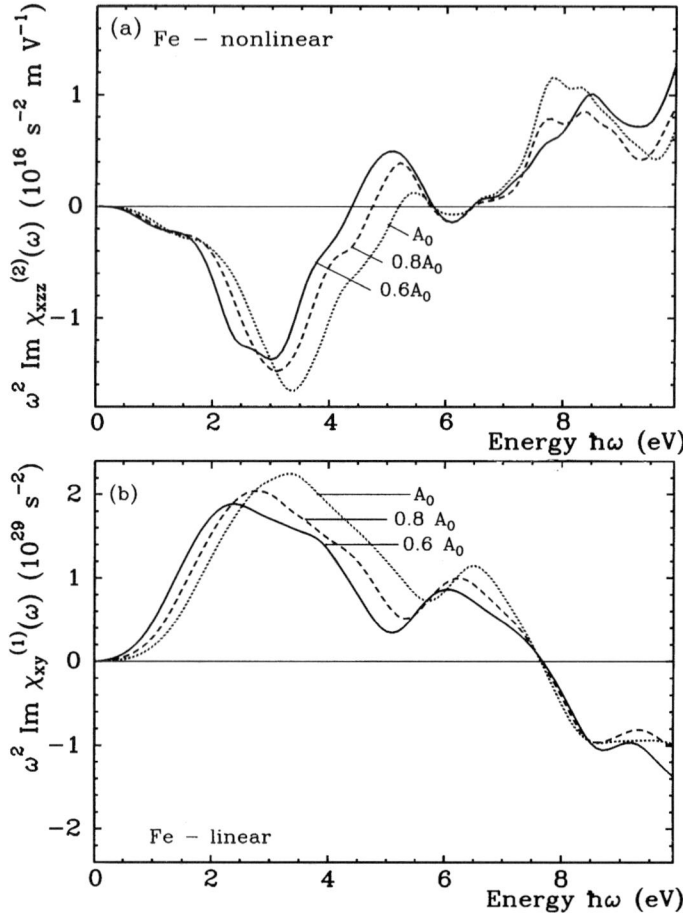

Fig. 5.31. (a) Nonlinear and (b) linear magneto-optical Kerr susceptibilities of Fe for different d-band widths as functions of frequency ω. The hopping parameters A_i ($i = 1,\ldots,7$) are those by Baker and Smith (dotted line). The dashed and solid curves correspond to reduced values of 80% and 60% of the band width used by Baker and Smith, respectively.

Figure 5.32 shows the nonlinear and linear magneto-optical spectrum for different values of the bottom of the s band (Γ point). The position β of the energy at the Γ point is increased from 1.47 eV via 2.0 eV and 2.5 eV to 3.0 eV. In the low-energy part of the spectra (below 8 eV), this shift accounts for the decreasing s-band width due to electron–electron interactions. Changing β affects in particular the first peak of the nonlinear spectrum and the second peak of the linear spectrum. These are shifted to smaller energies. The peaks at large energies (above 8 and 8.5 eV in the nonlinear and linear spectrum, respectively), however, are shifted to larger frequencies since the top of the s band lies at increasingly higher energies (high above the Fermi level). Thus,

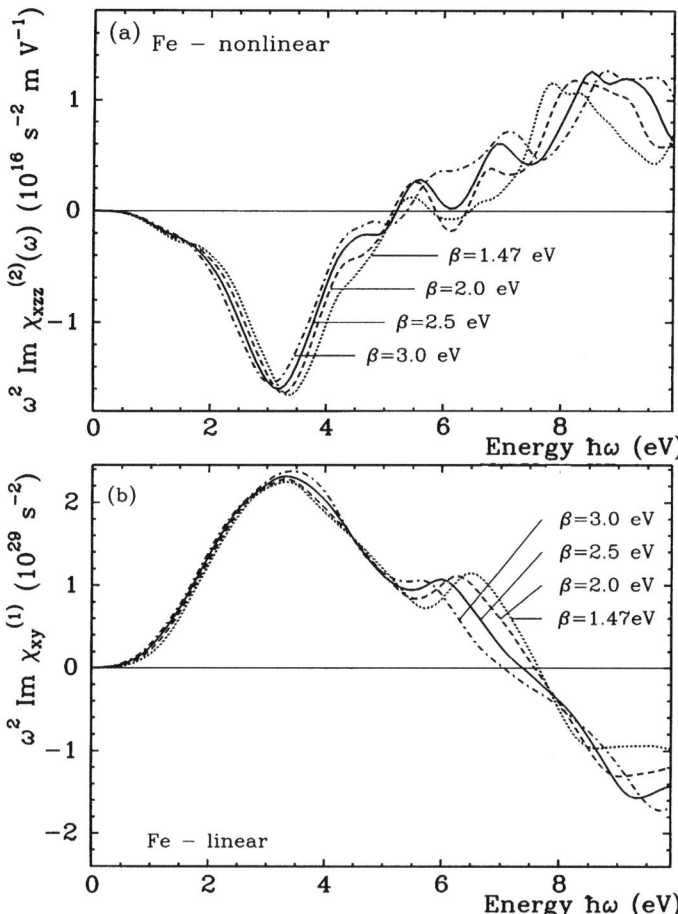

Fig. 5.32. Frequency dependence of (a) nonlinear and (b) linear magneto-optical Kerr susceptibilities of Fe for shifted s bands (β is the energy of the Γ point).

transitions from occupied levels into these unoccupied states occur at higher energies. Note that the first peak of the linear Kerr spectrum is mainly independent of β, which is in agreement with our previous observation (see Fig. 5.31) that this peak involves mostly d–d transitions. Concerning the dependence of the spectra on spin–orbit coupling one should note the following: the nonlinear magneto-optical Kerr spectrum depends linearly on spin–orbit coupling. This not only holds for our perturbational treatment of spin–orbit interaction, but is also true if this interaction is included in the diagonal ($L_z S_z$) and off-diagonal matrix elements ($L_x S_x + L_y S_y = \frac{1}{2}(L_+ S_- + L_- S_+)$) of the Hamiltonian.[130,131]

These general features of the theory are of interest for a systematic understanding of the Kerr effect in the transition-metal series and will also be useful for the comparison of the Kerr spectra of Ni and Fe.

Next, we compare the calculated nonlinear magneto-optical Kerr spectra of Fe with those obtained previously for Ni (see ref. 5). Of course, it matters that the electronic structures of bcc Fe and fcc Ni are different. For the Fe spectra, a detailed calculation of the weight of boundary points of the Brillouin zone is performed for the k-space summation yielding the Kerr spectra. Such weights were not included in the Ni spectra calculation. Thus, for the comparison of Fe and Ni spectra we determined for Ni a weight factor $F = 1.33$ for the boundary points. Thus we obtain for Fe and Ni for the structure at 2 to 3 eV a height ratio $(P_{Fe}/P_{Ni})F$ of 2.2 to 2.5 (rather than 1.7 to 1.9 for $F = 1$). Note that the height ratio is not simply given by the ratio of the exchange couplings, or that of the magnetic moments, but also by the difference of spin–orbit coupling, which enhances relatively the height of the Ni spectra ($\lambda_{so}(Fe) = 50$ meV and $\lambda_{so}(Ni) = 70$ meV). The larger minima and maxima in the nonlinear magneto-optical Kerr spectrum of Fe compared with Ni (and similarly for the linear spectra) result mainly for two reasons: (i) The exchange splitting is larger in Fe by a factor of 4.5 (compared to 400 meV for the t_{2g} orbitals of Ni). Here, however, it plays a role that the J dependence of the Kerr effect tends to saturate for large values of J. (ii) The filling of the 3d bands is different. Fe is much closer to half-filling, which leads to a larger number of d–d transitions. Thus, it is clear that the maxima and minima in the nonlinear (and linear) magneto-optical spectrum are more pronounced for Fe than for Ni. Furthermore, the peaks and minima in the optical region of Fe are shifted to higher energies compared to Ni because of the larger width of the Fe d band. This discussion shows how the Kerr spectra change from transition metal to transition metal.

Dependence on the surface magnetization After the discussion of the electronic surface properties that are reflected in the nonlinear Kerr spectra we now focus on the *magnetic* features for the determination of which the nonlinear Kerr effect has been designed mostly.

In the inset of Fig. 5.34 we show the dependence of the nonlinear Kerr spectra on the exchange coupling constant J. We find a monotonic increase of the peak heights with increasing J. As is shown in the inset, this increase starts linearly in J. For larger values of the exchange interaction, the dependence tends to saturate. In view of our results presented in sec. 5.3 from symmetry arguments and in sec. 5.4 based on electronic theory, it is possible to understand more about the differences in the Kerr spectrum expected for different ferromagnetic metals.

Below we will demonstrate that the nonlinear and linear magneto-optical susceptibilities are linear in the magnetization M. This holds for itinerant as well as for an atomic ferromagnetic behaviour. Here, we show that a simple and very transparent atomic picture can already account for most of this behaviour.

First, we need the dependence on the exchange coupling J, which may be understood as follows. The increase of J has basically three effects:

1. Starting with small values of J, the difference of allowed majority- and

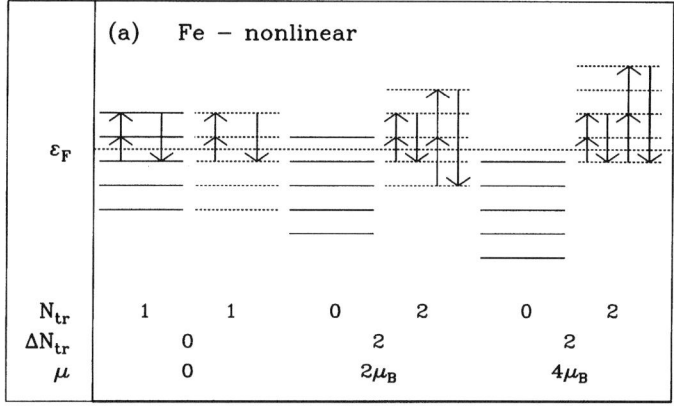

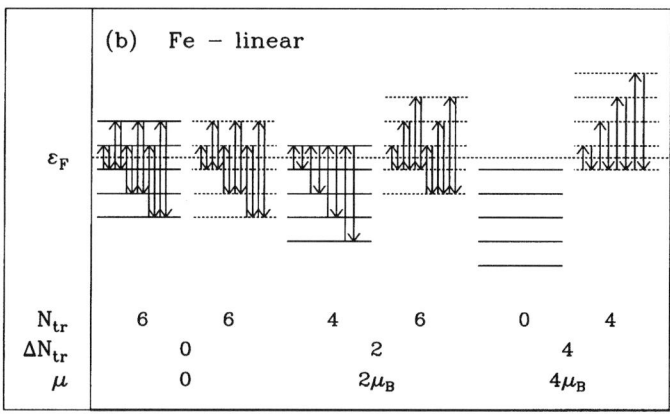

Fig. 5.33. Illustration of the d-band transitions in Fe in atomic picture: (a) the nonlinear case and (b) the linear case. Here ε_F, μ, N_{tr}, ΔN_{tr} denote the Fermi energy, magnetic moment, number of transitions and difference of transitions of minority- and majority-spin bands, respectively.

minority-electron dipole transitions at a certain frequency increases linearly in J; see Fig. 5.33 for an atomic model. For large J, however, this difference saturates, since the self-consistently determined Fermi level becomes pinned and moves rigidly upon further increase of J.

2. As long as the increase of J and the subsequent self-consistent readjustment of the Fermi level do not cause bands to cross the Fermi energy, the overall shape and peak structure of the spectrum remain unaltered. As soon as Fermi-level crossings occur (either gradually for relatively steep bands or more abruptly for rather flat bands), the occupation character of the bands is changed, thus opening up new previously forbidden channels for electronic dipole transitions or closing previously allowed channels (see Fig. 5.33). This effect becomes

noticeable if more than one electron per atom is transferred from one band to another. As a consequence, one may get drastic changes in the structure of the nonlinear and linear Kerr spectra. This is in particular the case if all transitions contributing to a pronounced structure change. Although the value of J in ferromagnetic Fe is 1.78 eV, which is quite large, this possibility occurs only for energies larger than 7 eV and for a value $J = 2J_0$. Otherwise the features of the spectra appear to be very robust.

3. As will be shown below, the increase of J leads also to a small shift of the peak energies to larger values. This is so in particular for the first structure of the nonlinear and linear Kerr spectra. The reason is that the exchange splitting of the d-like bands affects differently the s-like and sd-hybridized states via sd-hybridization. This explains also the larger shift of the second structure which is due to s–d transitions, in the linear (and to some extent also in the nonlinear) spectrum compared to the shift of the first structure. Note, in general, that these shifts are relatively small and thus do not change the overall features of the spectra.

The J dependence of the spectra: some analytical results First we calculate the J dependence of the peak position in the presence of sd-hybridization. The exchange splitting of one d band and one sd band yields a different shift of majority- and minority-spin bands. The distances ΔE between the split bands depend on the exchange splitting value J:

$$\Delta E(\uparrow) = \sqrt{\left(E_1 - E_2 - \frac{J}{2}\right)^2 + 4V^2},$$
$$\Delta E(\downarrow) = \sqrt{\left(E_1 - E_2 + \frac{J}{2}\right)^2 + 4V^2}, \qquad (5.4.94)$$

where E_1 and E_2 are the original paramagnetic band positions and V describes the band hybridization.

Secondly, we calculate the J dependence of the linear magneto-optical Kerr spectrum consisting of one hypothetical peak of Lorentzian shape. Assuming different J dependences for the majority- and minority-spin peaks, i.e. the majority- and minority-spin electrons undergo transitions between bands with different J dependence, the resulting linear Kerr susceptibility may be written in the following form:

$$\chi^{(1)} = \frac{AJ^2 + BJ}{CJ^2 + DJ + E'}, \qquad (5.4.95)$$

where A, B, C, D, E are J-independent constants. For $J \to 0$ the susceptibility depends linearly on J; for $J \to \infty$ it tends to saturate.

The corresponding nonlinear spectrum consists of peaks whose shape is that of the derivative of a Lorentzian. The result for the J dependence of this nonlinear Kerr spectrum is

$$\chi^{(2)} \sim \frac{J^3 + aJ^2 + bJ}{J^3 + cJ^2 + dJ + e}, \quad (5.4.96)$$

which again shows linear J dependence for small J and saturation behaviour for large values of J.

Thirdly, we calculate the J dependence of the linear and nonlinear magneto-optical Kerr spectra for Fe *atoms*. In Fig. 5.33 the atomic 3d bands of iron ($N_{Fe} = 6$) are shown schematically. They cross the Fermi level with increasing exchange interaction. We obtain the number of possible transitions in majority- and minority-spin bands, the difference of which is a measure for the Kerr susceptibility. For sufficiently large exchange interaction the bands no longer cross the Fermi level, which again demonstrates the case of saturation, which occurs for smaller J in the nonlinear case compared to the linear one.

Taking now the information on the J dependence of the magnetization and of the Kerr spectra (see inset of Fig. 5.34) together, one is able to calculate the nonlinear and (for comparison) linear Kerr spectra as a function of the magnetization M. In Fig. 5.34 we show the dependence of the nonlinear and linear susceptibilities on M. The triangles and squares are the calculated values, the lines serve as a guide to the eye. For magnetization values up to 2.45 μ_B we find for both cases a linear dependence. Thus, the nonlinear Kerr effect is *linear* in the magnetization and can consequently sense two-dimensional magnetism. This result is therefore of particular importance.

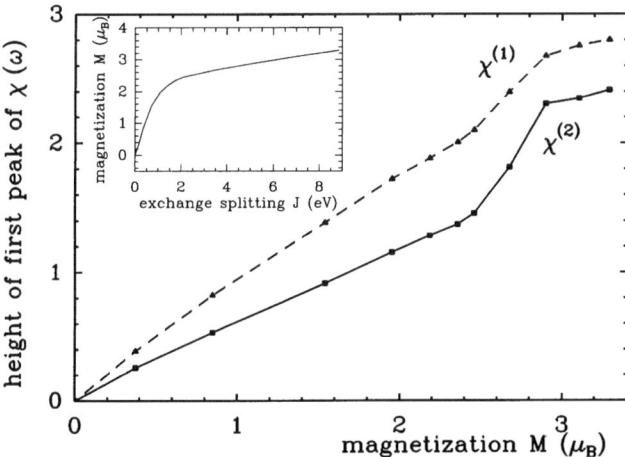

Fig. 5.34. Magnetization dependence of the linear and nonlinear susceptibilities. The inset shows the dependence of the magnetization on the exchange coupling constant J. A curve similar to that of the inset would result for the depth of the first minimum as a function of the exchange coupling constant J.

The susceptibilities increase more rapidly for larger M (corresponding to surfaces) since pronounced level crossing of the Fermi surface occur. The inset shows the dependence of the magnetization on the exchange coupling constant J. We find a behaviour similar to the $\chi(J)$ dependence derived from the atomic as well as the itinerant picture.

Summary of the results for iron

1. We calculated the band structure of Fe using the 'combined interpolation scheme' and determined the nonlinear and linear magneto-optical Kerr spectra with the appropriate weighting of zone boundary points in $\boldsymbol{k}$ space.
2. For small frequencies ($\hbar\omega \leq 7\,\text{eV}$ in the linear and $\hbar\omega \leq 4\,\text{eV}$ in the nonlinear case) the minority electrons dominate.
3. The peak positions depend sensitively on d-band width and s-band position. The low-energy part of the spectrum is dominated by d–d transitions while d–s transitions govern the higher-energy range.
4. The nonlinear Kerr effect of Fe surfaces is about 2.5 times as large as that of Ni surfaces.
5. The nonlinear (as well as the linear) Kerr effect is *linear* in the magnetization, thus being an ideal probe for two-dimensional magnetism.
6. The nonlinear (as well as the linear) Kerr effect is linear in spin–orbit coupling.
7. The nonlinear (as well as the linear) Kerr effect is linear in the exchange splitting J for small J and saturates for larger J.

5.4.6 *First-principles theory for thin films*

The magnetism of low-dimensional metallic structures such as surfaces, thin films and multilayer sandwiches has recently become an exciting new field of research and applications.[61] In particular, artificially grown thin magnetic films and multilayers exhibit a rich variety of properties not previous found in bulk magnetism such as enhanced or reduced moments,[178] oscillatory exchange coupling through nonmagnetic spacers,[179–181] giant magnetoresistance,[182,183] and reorientation of the magnetic easy axis upon thickness and temperature variation[184–188] and the observation of spin-polarized quantum-well states.[189–191] These are responsible for the oscillatory behaviour of the exchange coupling of ferromagnetic thin films via nonmagnetic spacers. NOLIMOKE as a sensitive probe for two-dimensional magnetism is therefore particularly well suited also for the nonlinear optical investigation of thin films. Striking examples are the giant quantum-well oscillation amplitudes and the arbitrarily large Kerr rotations in thin magnetic films.

Density-function theory

Density-functional theory uses the particle density $n(\boldsymbol{r})$ as basic variable. Hohenberg and Kohn proved that also the external potential is a unique functional $n(\boldsymbol{r})$. Thus the density determines all physical properties of the system. Kohn and Sham split the total energy functional into noninteracting and interacting parts

$$E_v[n(r)] = T_s[n(r)] + \int v(r)n(r)\,d^3(r) + \int\int \frac{n(r)n(r')}{|r-r'|}\,d^3r\,d^3r'$$
$$+ E_{xc}[n(r)], \qquad (5.4.97)$$

with kinetic energy of noninteracting electrons $T_s[n(r)]$ and exchange–correlation term $E_{xc}[n(r)]$. Variation of eqn (5.4.97) together with particle conservation yields Euler's condition

$$\frac{\delta T_s[n(r)]}{\delta n(r)} + \underbrace{v^{\text{ext}}(r) + \int \frac{n(r')}{|r-r'|}\,d^3r' + V_{xc}[n(r)]}_{v^{\text{eff}}} - \mu = 0, \qquad (5.4.98)$$

where $\mu = \varepsilon_F$ is the Lagrange parameter of the Fermi energy. The functional derivative

$$V_{xc}[n(r)] = \frac{\delta E_{xc}[n(r)]}{\delta n(r)} \qquad (5.4.99)$$

is the exchange–correlation potential. The variational problem (5.4.98) corresponds to the Schrödinger equation

$$[-\nabla^2 + v^{\text{eff}}(r)]\psi_i(r) = \varepsilon_i \psi_i(r), \qquad (5.4.100)$$

a single-particle equation of N particles in the effective potential v^{eff}. The particle density is then given by

$$n(r) = \sum_{i=1}^{N} n_i^{\text{occ}}(\varepsilon_i, T^{\text{el}})|\psi_i(r)|^2. \qquad (5.4.101)$$

In the *local-density approximation* (LDA) the exchange–correlation energy

$$E_{xc}^{\text{LDA}}[n(r)] = \int n(r)\varepsilon_{xc}^{\text{LDA}}(n(r))\,d^3r \qquad (5.4.102)$$

does not contain terms $\sim \nabla n(r)$ or of higher order.

This formalism can be extended to magnetic systems. Within the local spin-density approximation (LSDA), the particle density and spin polarization are represented as

$$n(r) = n_\uparrow(r) + n_\downarrow(r) = \sum_i^{\text{occ}} |\psi_{i\uparrow}(r)|^2 + \sum_i^{\text{occ}} |\psi_{i\downarrow}(r)|^2 \qquad (5.4.103)$$

and

$$\zeta(r) = \frac{n_\uparrow(r) - n_\downarrow(r)}{n(r)}. \qquad (5.4.104)$$

In this approximation the single-particle equations, which are generally coupled by the spin-dependent potential, decouple and become

$$[-\nabla^2 + v^{\text{eff},\sigma}(r)]\psi_{i\sigma}(r) = \varepsilon_{i\sigma}\psi_{i\sigma}. \qquad (5.4.105)$$

We use the exchange–correlation functional in the Vosko–Wilk–Nusair parametrization[192] of the Ceperley-Alder results.[193]

Linear muffin-tin orbitals (LMTO) method

The LMTO method is an approximation to the Korringa–Kohn–Rostoker method, which treats the many-body problem as a multiple-scattering problem in a periodic potential. Using the muffin-tin potential

$$v_{\text{MT}}(r) = \begin{cases} v(|r|) - v_{\text{MTZ}} & r \leq R_{\text{MT}}, \\ 0 & r \leq R_{\text{MT}}, \end{cases} \qquad (5.4.106)$$

the Schrödinger equation separates into radial and angular parts. The radial part in the sphere is solved by Hankel and Bessel functions. In the interstitial region, the outgoing waves, which are singular at the centre of the muffin-tin sphere, are determined at a fixed energy. Since the scattering phases for the outgoing waves tend to zero for angular momentum l larger than a cutoff (like in the Mie scattering problem, see sec. 5.2), only the incident waves survive for large l.

Thus the spherical-wave basis exhibits two advantages: (i) the small and efficient basis set and (ii) the applicability of the method also for non-periodic systems. Finally the linearization of the basis set in the energy is carried out and allows for a determination of all eigenvalues for *arbitrary* energies at the same time.[194]

Full-potential extension

The *full-potential* extension of the LMTO method by Methfessel[195] combines efficiency with accuracy. The potential matrix elements contain three-centre integrals which are reduced to a sum over two-centre integrals. Finally the value and derivative of the wavefunctions at the muffin-tin radius are used to express the physical quantities in terms of structure constants given by Clebsch–Gordan coefficients.

Since these are calculated only once at the beginning, the method is very efficient. On the other hand, the atomic position is fixed throughout the iteration cycle, which hampers the incorporation of forces (contrary to the full-potential linear augmented plane-wave (FLAPW) method).

On the other hand, FLAPW will become more and more attractive, since it contains the forces in a natural way and it becomes accessible due to the increasing availability of powerful workstations. Thus, we started to implement spin–orbit coupling in the wide spread 'WIEN95' FLAPW code, since it is the key quantity for many applications in magnetism such as linear and nonlinear magneto-optics, magnetocrystalline anisotropy, giant magnetoresistance, magnetic dichroism and spin–lattice relaxation as some time limit for magnetic recording. In particular in nanostructures spin–orbit coupling might be different from its atomic value.

Spin–orbit coupling and WIEN95

The implementation of spin–orbit coupling in WIEN95 is performed in the following way. After the self-consistent determination of the charge, the potential and the basis functions, we set up a Hamiltonian matrix again including spin–orbit coupling and solve the corresponding secular equation

$$\sum_{ij} \langle \sigma_{k_i}^{sc} | H^{sc} + H_{so} | \sigma_{k_j}^{sc} \rangle = \sum_{ij} \varepsilon(q) \rho_i(q) \langle \sigma_{k_i}^{sc} | \sigma_{k_j}^{sc} \rangle$$

to obtain the eigenfunctions

$$\Phi(q) = \sum_n \rho_n(q) \phi_{k_n}, \qquad q = 1, 2, \ldots$$

and the corresponding eigenenergies $\varepsilon(q)$ shifted by spin–orbit coupling. Here, the quantities denoted by 'sc' are calculated self-consistently. The spin–orbit operator

$$H_{so} = \frac{\alpha^2}{2} s \cdot (\nabla V \times p)$$

is treated in the spherical approximation

$$\nabla V \simeq \frac{r}{r} \frac{\partial V}{\partial r},$$

since spin–orbit coupling heavily weights the region near the nucleus while contributions outside the muffin-tin sphere are negligible ($\partial V/\partial r \approx 0$). This yields

$$H_{so} = \frac{\alpha^2}{2} s \cdot (r \times p) \frac{1}{r} \frac{\partial V}{\partial r} = \frac{\alpha^2}{2} s \cdot L \frac{1}{r} \frac{\partial V}{\partial r}.$$

The spin–orbit matrix elements are calculated using the self-consistently determined basis functions

$$\phi_{k_i} = \begin{cases} \sum_{lm} [A_{lm}(k_i) u_l(r, E_l) + B_{lm}(k_i) \dot{u}_l(r, E_l)] Y_{lm} & t < R_{mt} \\ (1/\sqrt{\omega}) e^{i k_i r} & r > R_{mt} \end{cases}$$

and computing the matrix elements

$$\langle \sigma^{sc}_{k_i} | H_{so} | \sigma^{sc}_{k_j} \rangle = \int_{r<R_{mt}} d\mathbf{r}\, \phi^{sc}_{k_i} \left(\frac{\alpha^2}{2} \mathbf{s} \cdot \mathbf{L} \frac{1}{r} \frac{\partial V}{\partial r} \right) \phi^{sc*}_{k_j},$$

since

$$\int_{r>R_{mt}} d\mathbf{r}\, \phi^{sc}_{k_i} H_{so} \phi^{sc*}_{k_j} \equiv 0.$$

The separation into angular and radial parts

$$\langle \sigma^{sc}_{k_i} | H_{so} | \sigma^{sc}_{k_j} \rangle = \frac{\alpha^2}{2} \sum_\sigma \sum_{ij} \Bigg\{ \int_{r<R_{mt}} [A_{lm}(\mathbf{k}_i)u_l(r,E_l) + B_{lm}(\mathbf{k}_i)\dot{u}_l(r,E_l)] $$
$$\times r \frac{\partial V}{\partial r} [A_{lm}(\mathbf{k}_i)u_l(r,E_l) + B_{lm}(\mathbf{k}_i)\dot{u}_l(r,E_l)] \Bigg\}$$
$$\times \left\langle \sigma \Big| \int d\Omega\, Y_{lm}(\hat{r}) \mathbf{s} \cdot \mathbf{L} Y^*_{lm'}(\hat{r}) \Big| \sigma' \right\rangle$$

then yields

$$\langle \sigma^{sc}_{k_i} | H_{so} | \sigma^{sc}_{k_j} \rangle = \sum_{lmm'} \Big\{ \lambda^l_{uu} A_{lm}(\mathbf{k}_i) A^*_{lm'}(\mathbf{k}_j) + \lambda^l_{\dot{u}\dot{u}} B_{lm}(\mathbf{k}_i) B^*_{lm'}(\mathbf{k}_j) $$
$$+ \lambda^l_{u\dot{u}} [A_{lm}(\mathbf{k}_i) B^*_{lm'}(\mathbf{k}_j) + B_{lm}(\mathbf{k}_i) A^*_{lm'}(\mathbf{k}_j)] \Big\}$$
$$\times \left\langle \sigma \Big| \int d\Omega\, Y_{lm}(\hat{r}) \mathbf{s} \cdot \mathbf{L} Y^*_{lm'}(\hat{r}) \Big| \sigma' \right\rangle,$$

which couples minority and majority spins, thus leading to a doubling of the size of the Hamiltonian matrix.

As a result we obtain several spin–orbit coupling constants in the solid

$$\lambda^l_{uu} \equiv \frac{\alpha^2}{2} \int_{r<R_{mt}} dr\, u_l(r) r \frac{\partial V}{\partial r} u_l(r),$$

$$\lambda^l_{u\dot{u}} \equiv \frac{\alpha^2}{2} \int_{r<R_{mt}} dr\, u_l(r) r \frac{\partial V}{\partial r} \dot{u}_l(r),$$

$$\lambda^l_{\dot{u}\dot{u}} \equiv \frac{\alpha^2}{2} \int_{r<R_{mt}} dr\, \dot{u}_l(r) r \frac{\partial V}{\partial r} \dot{u}_l(r),$$

rather than only one in the atom. The maximum λ might be considerably larger than in the atom, since chemical bonding smears out the electronic charge density. Thus, its tails come closer to the nucleus.

In addition, the off-diagonal terms of

$$s \cdot L = \begin{pmatrix} L_z & L_x + iL_y \\ L_x - iL_y & -L_z \end{pmatrix}$$

describe spin-flip terms, which are a second-order effect for magneto-optics and thus negligible, but yield an essential contribution to the magnetic anisotropy energy. The first-order effect of spin–orbit coupling in linear and nonlinear magneto-optics results from the change of the electronic wavefunctions in $\chi^{(i)}(M)$, while the change of the eigenenergies is of second order.

Band structure and nonlinear magneto-optics in thin films

Motivation So far no NOLIMOKE theory exists for thin films. We will show that it is necessary for the determination of important characteristic film properties to perform *ab initio* electronic calculations. As is the case for linear surface MOKE experiments on Au/Fe(bcc)/Au(001),[196] we expect also film-characteristic new structures in the frequency-dependent *nonlinear* Kerr spectrum Im $\chi^{(2)}(\omega)$ for thin Fe films. Besides structural and magnetic properties, possibly, even quantum-well states are reflected in recent experiments on this system.[197]

Most important for the calculation of the nonlinear susceptibility is the electronic band structure $E_{kl\sigma}$. For thin Fe films, we calculate the band structure within the full-potential linear muffin-tin orbital (FP-LMTO) method developed by Methfessel *et al*.[195] in the spin-polarized version by M. van Schilfgaarde, adapted for fcc and bcc Fe films on several substrates.[198] Thus, we combine the advantages of a parameter-free density-functional method with the possibility of consideration of many k points in the Brillouin zone, which are required to describe optical processes (transitions).

For the calculation of the nonlinear Kerr spectrum we extract after the self-consistency cycle the eigenvalues and the eigenfunctions at all k points. Here, we perform calculations of free-standing bcc iron films within a slab geometry.[132,199] For simplicity we have used slabs with an odd number of layers.[200] However, the features of the calculated Kerr spectra show a physical thickness dependence.

To simulate structural film effects we perform calculations for different film lattice constants and find that (i) nonlinear Kerr spectra reflect the film structure more sensitively than linear ones and (ii) reveal special features of films and monolayers which are different from those of bulk surfaces. Our results indicate a magnetic-moment enhancement at surfaces, interfaces and in thin films, which is clearly visible in the NOLIMOKE spectra.

From studies of the Kerr spectra Im $\chi^{(2)}_{ijl}(\omega)$ one may learn about the atomic, electronic and magnetic structure of the film, the occurrence of bcc → fcc structural changes and accompanying magnetic changes. Further extension of our theory will include quantum-well states, interface effects due to different interfaces, and magnetic effects induced by a magnetic field.

Theory Using our microscopic theory we take for the calculations of the nonlinear magneto-optical Kerr spectra of Ni[46] and Fe[47] surfaces the following expression:

$$\chi^{(2)}_{xzz}(2q_\parallel, 2\omega, M) = \frac{e^3 C}{\Omega} \frac{\lambda_{so}}{\hbar \omega}$$

$$\times \sum_\sigma \sum_{k,l,l',l''} \Biggl\{ \langle k+2q_\parallel, l'' \sigma | z | kl\sigma \rangle \langle kl\sigma | z | k+q_\parallel, l'\sigma \rangle$$

$$\times \langle k+q_\parallel, l'\sigma | z | k+2q_\parallel, l''\sigma \rangle$$

$$\times \frac{\dfrac{f(E_{k+2q_\parallel,l''\sigma}) - f(E_{k+q_\parallel,l'\sigma})}{E_{k+2q_\parallel,l''\sigma} - E_{k+q_\parallel,l'\sigma} - \hbar\omega + i\hbar\alpha_1} - \dfrac{f(E_{k+q_\parallel,l'\sigma}) - f(E_{kl\sigma})}{E_{k+q_\parallel,l'\sigma} - E_{kl\sigma} - \hbar\omega + i\hbar\alpha_1}}{E_{k+2q_\parallel,l''\sigma} - E_{kl\sigma} - 2\hbar\omega + i2\hbar\alpha_1} \Biggr\},$$

(5.4.107)

where λ_{so} is the spin–orbit coupling constant, $E_{kl\sigma}$ are the electronic energy levels resulting from the band-structure calculations, and the factor C determines the surface response as will be discussed below. The other symbols are the same as in previous studies (e.g. ref. 47). Here, we present the susceptibility tensor element χ_{xzz} as a typical odd element in the longitudinal Kerr geometry (M parallel to the y axis, p-polarized incident electric field). Other magnetic (odd) and nonmagnetic (even) tensor elements can be written similarly. Obviously, $\chi^{(2)}$ vanishes in the case of inversion symmetry within the electric dipole approximation. Equation (5.4.107) shows clearly that $\chi^{(2)}$ is determined by the electronic structure of the material.

The energies $E_{kl\sigma}$ and the transition matrix elements are calculated *ab initio* for films. We use wavefunctions in the transition matrix elements that depend on the position of the appropriate atom in the film. This is necessary for determining layer-dependent contributions.

The breakdown of inversion symmetry at surfaces or interfaces is essentially determined by the product of the three transition matrix elements. Previous calculations of nonlinear and linear Kerr spectra of Ni surfaces[46] have shown that the position of the main structures in $\chi^{(2)}$ and the peak height ratio of $\chi^{(2)}$ do not depend on the k dependence of the transition matrix elements. Mainly, the energy eigenvalues define the spectral structure. This was also checked by comparison of MOKE calculations and experimentally obtained spectra. Thus, we use in our calculations constant transition matrix elements. Since this destroys the symmetry sensitivity of NOLIMOKE we must recover the effects due to the breakdown of inversion symmetry by introducing the cutoff factor C. For films on a substrate, C guarantees that the nonlinear response results exclusively from the topmost film layers as would be the case for properly calculated matrix elements. Here, we assume that atoms in the other layers experience inversion symmetry. Actually, atoms in the second layer should contribute slightly to $\chi^{(2)}$, since they still feel a small deviation from inversion symmetry. Furthermore, since the contributions from the surface and the substrate interface are very different, we take into account only the surface contribution to $\chi^{(2)}$ and determine $\chi^{(2)}$ accordingly. In the case (i) of the *surface of bulk material* we perform a three-dimensional electronic structure calculation and use a cutoff for the surface response $C = q_\| a$, where $q_\|$ is the component of the photon wavevector parallel to the surface and a is the lattice constant. This factor gives the ratio of the response depth a to the excitation depth in the crystal ($\approx 1/q_\|$). In the case (ii) of *films* we take into account the surface layer response using (a) a fraction of the film-averaged $\chi^{(2)}$ corresponding to *one* atomic layer and (b) the projection of the six wavefunctions in the three matrix elements to atoms in the *first* layer amounting to a nonlinear response only from this first layer. The corresponding surface cutoff factors are in case (a) $C = 1/n^3$, with n atomic layers in the film, and in case (b) $C = W_{k+2q_\|,l''\sigma} W_{k+q_\|,l'\sigma} W_{kl\sigma}$, where W_α denotes the weight of the density of states $|kl\sigma\rangle$ in the Wigner–Seitz cell of the first monolayer. Note that this approximation (b) is better with respect to the film-thickness dependence of the response, since nonlinear optics senses the interface while in linear optics the whole film is involved in agreement with experiment.[201-204]

5.4.7 Results: nonlinear Kerr spectra for thin Fe films

Thickness dependence of the nonlinear Kerr effect

In Figs 5.35–5.37 we present results for the film-thickness dependence of the spectra Im $\chi^{(2)}(\omega)$. For comparison also results for Im $\chi^{(1)}(\omega)$ are given.

First, in Figs 5.35 and 5.36 results are shown for the thickness dependence of the magneto-optical Kerr spectra of bcc Fe films obtained from *ab initio* calculations. The film-specific features are recognized by comparing the results for the surface of corresponding bulk materials. In Fig. 5.35(a) the surface sensitivity of the optical response was simulated by using the factor $1/n^3$. Note the differences between the nonlinear and linear spectra. In the nonlinear monolayer spectrum the first minimum is shifted compared to the surface of the bulk spectrum by about 2 eV to lower energies. This results from the reduced d-band width of a film. Furthermore, the tiny maximum at 5.6 eV in the surface of the bulk spectrum gets enhanced and shifted to 4.5 eV, having then a width of nearly 3 eV in the case of a monolayer spectrum. In the nonlinear case the

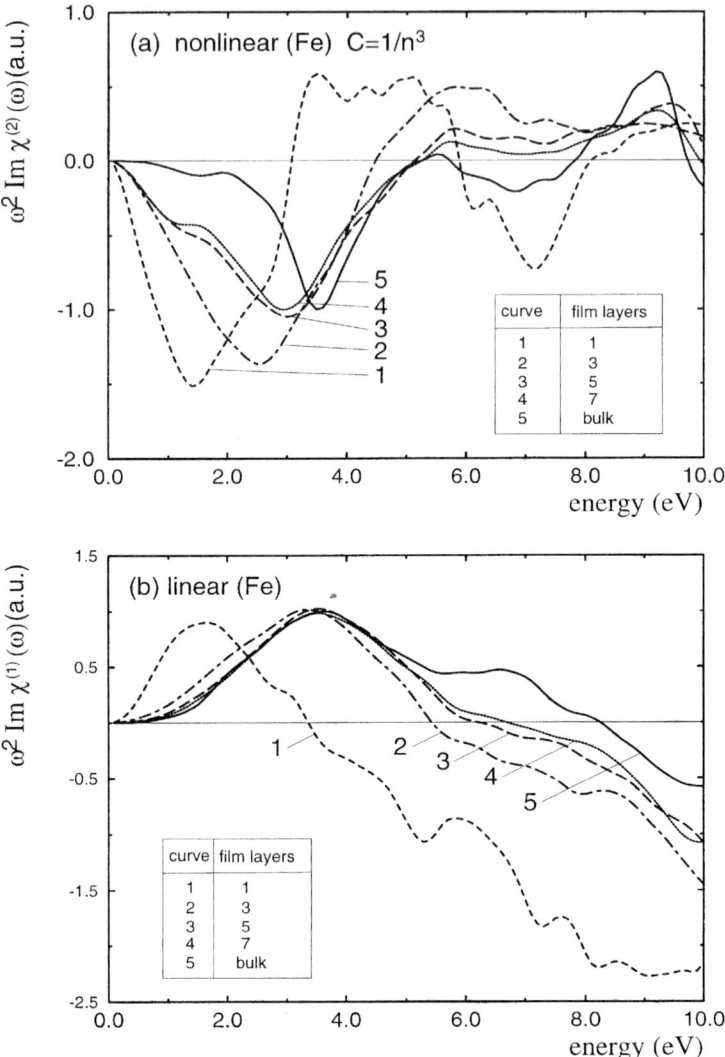

Fig. 5.35. *Ab initio* calculated (a) nonlinear and (b) linear magneto-optical Kerr spectra of Fe for a truncated bulk surface (solid line), a monolayer (dashed curve) and films having three layers (dashed-dotted), five layers (long-dashed) and seven atomic layers (dotted curve). The second harmonic response of the surface layer results by averaging the electronic input structure over the whole film. Note, again, that we use for the surface layer of the film $C = 1/n^3$ and energies $E_{kl\sigma}$ obtained from film calculations, while for the surface of bulk we use $C = q_\| a$ and $E_{kl\sigma}$ obtained from bulk calculations.

position of the minimum between 2 and 4 eV and the slope up to 5.5 eV in the seven-layer spectrum lie close to those of the surface of the bulk spectrum. Note, consistent with our approximation regarding the surface sensitivity, that

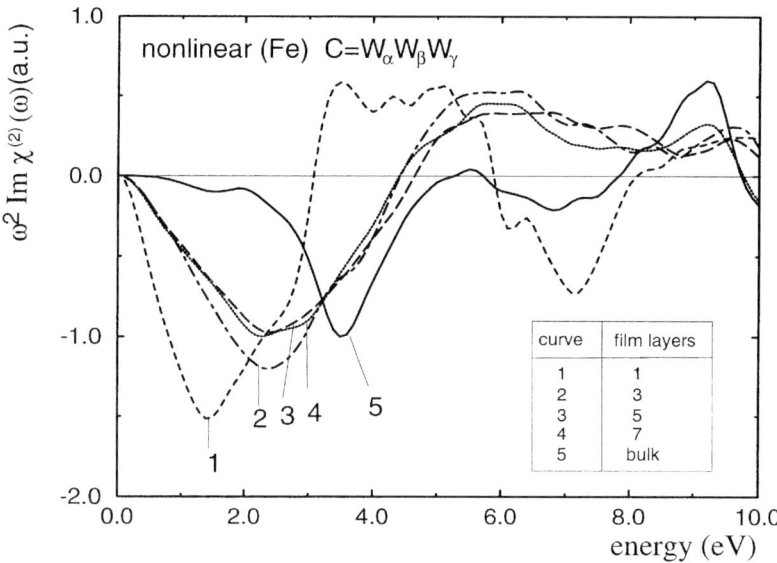

Fig. 5.36. *Ab initio* calculated nonlinear magneto-optical Kerr spectra of Fe for a truncated bulk surface (solid line), a monolayer (dashed curve) and films with three layers (dashed-dotted), five layers (long-dashed) and seven atomic layers (dotted curve). The second harmonic response results from the first atomic layer. This SH response is obtained by projecting the wavefunctions to the first atomic layer yielding the factor $C = W_{k+2q_\|,l''\sigma} W_{k+q_\|,l'\sigma} W_{kl\sigma}$.

results for both the film and the surface of bulk spectrum converge. The first minimum around 1 to 2 eV is deepest for a monolayer.

In Fig. 5.36 the nonlinear spectra result from the topmost layer as follows from the approximation (b) $C = W_{k+2q_\|,l''\sigma} W_{k+q_\|,l'\sigma} W_{kl\sigma}$, which is the more physical approximation for the nonlinear case. Consistent with this approximation the structure changes not too much going from three to five to seven film layers. Thus, the essential contribution to NOLIMOKE results from the surface layer. However, the electronic structure of this layer is affected by the other layers due to hybridization and next-nearest-neighbour interaction. This point is corroborated by the results for the monolayer. Note that the results are different from those for the surface of bulk spectrum, where we have used the truncated bulk approximation. One should also note the layer-dependent change of the depth of the first minimum, which is related to the value of the magnetization in the top layer.

In Fig. 5.35(b) we also present results for the linear Kerr effect in order to compare with the nonlinear case and with recent experiments. For the comparison the amplitudes in the linear spectrum are scaled with the film thickness, e.g. by $1/n^2$, since only two matrix elements are involved in $\chi^{(1)}$. Without such scaling the spectra will grow with increasing film thickness as is physically expected. Note that the film spectra and surface of bulk spectra converge as

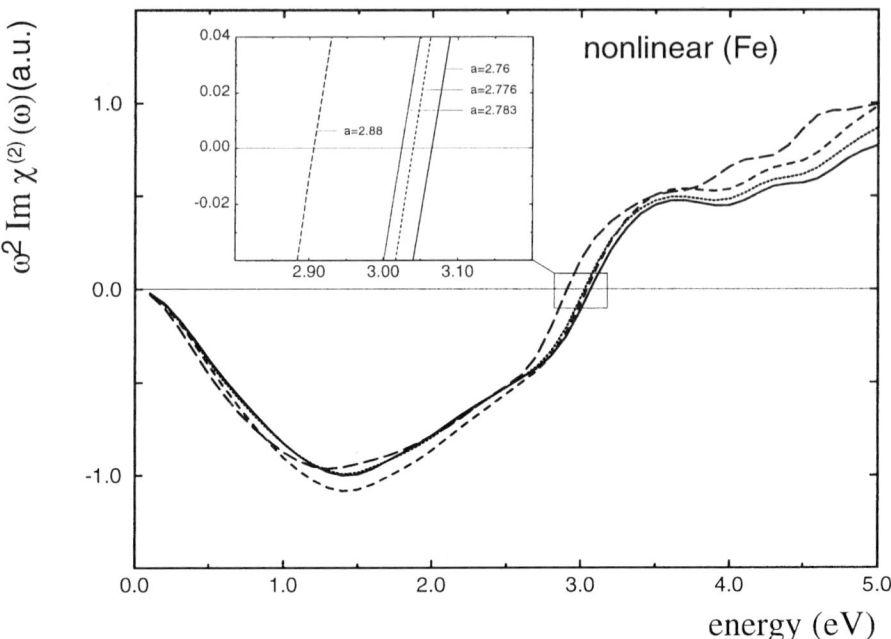

Fig. 5.37. Film-lattice-constant dependence of *ab initio* calculated nonlinear Kerr spectra of an Fe monolayer. The solid curve refers to the bulk bcc Fe lattice constant, the dashed curve to $a = 2.776\,\text{Å}$ (bulk Au), and the dotted curve to $a = 2.783\,\text{Å}$ (bulk Ag). The long-dashed curve refers to the experimental a for Fe. The inset shows at an enhanced scale the effects of different lattice constants for the zero of Im $\chi^{(2)}$ at $\hbar\omega \approx 3\,\text{eV}$.

should be expected. The linear spectra of the seven-layer film and of the truncated bulk do not differ up to 5 eV. In the linear monolayer spectrum the transitions between majority-spin electron bands begin to dominate at energies higher 3.3 eV (sign change). In the optical range we find a new structure at 5.2 eV.

In agreement with experiment[196, 197] we observe the shift of the first peak with respect to the one- and three-monolayer spectra. It is interesting that we do not obtain the extra peak in the linear spectrum of thin films. This supports the interpretation of this peak as resulting from quantum-well state transitions.

In Fig. 5.37 we show results for the nonlinear magneto-optical Kerr spectra for an Fe monolayer assuming different lattice constants to simulate film-structure (or substrate) effects. We compare spectra obtained using the Fe bulk lattice constant $a = 2.76\,\text{Å}$ (determined by total energy minimization), using $a = 2.776\,\text{Å}$ corresponding to an Fe monolayer on Au, and $a = 2.783\,\text{Å}$ corresponding to Fe on Ag, and using the experimental Fe lattice constant $a = 2.88\,\text{Å}$. The values for Au and Ag follow from the theoretical lattice constant applying the experimental lattice mismatch. Whereas the amplitude of the general minimum at about 1.5 eV shows no clear dependence on the lattice constant, the energy for which Im $\chi^{(2)}$ changes sign corresponds quantitatively to the ratios of the lattice constants.

Thus, we demonstrate that the nonlinear Kerr spectra can reveal the structural changes induced by different substrates, temperatures, or magnetostriction. Furthermore, the *ab initio* calculated NOLIMOKE spectra confirm the theoretical assumption concerning the response depth of about one atomic layer and show clearly the enhanced magnetization in thin films. Note that these results are not restricted to bcc Fe films, but they reflect general properties of all transition-metal thin films.

Discussion We learn from our results for Im $\chi^{(1)}$ and Im $\chi^{(2)}$ obtained by an electronic theory that NOLIMOKE reflects sensitively the electronic and magnetic structure of thin films. In contrast to the MOKE signal, the NOLIMOKE signal originates for flat surfaces essentially from the surface layer.[205] However, that the electronic structure of this layer depends on the film thickness can be clearly seen in our results. Very interesting also are the results for Fe on Ag and Au substrates, simulated by the lattice constant of the Fe monolayer, since they demonstrate that even structural effects are reflected in the NOLIMOKE and SHG spectra. This is of general interest regarding film growth and structural changes occurring as a function of film thickness.

We emphasize that our results show that characteristic features of magnetism in thin films, like changes of the magnitude of the magnetic moments and of the magnetization, can be seen in the NOLIMOKE spectra. Particularly, this can be seen from the depth of the minimum in $\chi^{(2)}$ around 3 eV. The enhanced minimum results from an increase of the magnetic moments. Actually, we obtain an enhanced moment of 2.8 μ_B in the first film layer. Since the magnitude of the magnetic moment is a fingerprint of the geometric structure, this can be used to determine the film geometry. While already MOKE reflects characteristic film-averaged features, NOLIMOKE does this even more sensitively and clearly exhibits further interesting details. While both MOKE and NOLIMOKE are suitable to study the structure of thin films in a material-specific way, only NOLIMOKE can be used to study interface effects for which we expect similar contributions to $\chi^{(2)}(\omega)$ with features as for the surface.

Concerning the comparison with experiment it has been observed in agreement with our theory that the MOKE signal increases linearly with the film thickness up to 20 monolayers whereas the NOLIMOKE signal remains nearly constant. Our theory suggests that the experimental enhancement of the NOLIMOKE signal for three- and four-layer films at a frequency of 1.55 eV (corresponding to a wavelength of 800 nm; see Fig. 5 in ref. 203) results from the electronic structure and in particular from the enhancement of the surface magnetization. Furthermore, the experiments describing the thickness dependence of MOKE or NOLIMOKE signals are usually performed at one fixed frequency. Our calculations have shown that for very thin films structures of the Kerr spectra are shifted due to changes in the electronic structure of the films. Thus, the Kerr susceptibility at fixed frequency does not show a monotonic increase with increasing film thickness depending on the choice of the frequency with respect to the spectrum structure.

The onset of the NOLIMOKE signal due to the appearance of magnetism will reflect the structure of the surface. For flat surfaces, the slope of the signal should be much larger than for corrugated surfaces. However, the signal itself might be enhanced by the corrugation. For film thickness larger than a few atomic layers the NOLIMOKE signal will remain nearly constant, while the MOKE signal increases with film thickness. An enhanced magnetization in thin films is reflected by an enhanced NOLIMOKE signal.

Regarding the enhancement of magnetic moments in thin films we deduce from our NOLIMOKE spectra for a three-layer film $\mu_{\text{surface}} = 2.64\,\mu_B$.

From our electronic band structure we obtain directly a film-averaged magnetic moment of $2.62\,\mu_B$.

From polarized-neutron reflection for a 5.5 monolayer Fe film on Ag(001) covered by Ag, Bland et al.[206] deduced an averaged magnetic moment of $2.58\,\mu_B$. Here again the comparison of MOKE and NOLIMOKE spectra can clear up whether the magnetic moments are enhanced (at the film–substrate interface or at the film–vacuum interface, or at both).

Since we present also the first theoretical results of the thickness dependence of the linear magneto-optical Kerr spectra we compare these with recent experiments by Suzuki et al.[196] and Geerts et al.[197] In agreement with these experimental results we obtain the shift of the first minimum in $\chi^{(1)}$ dependent on the film thickness and a quick convergence towards bulk results for more than four layers. Note that we have not included the confinement of the substrate electrons and thus quantum-well state effects are not included in the calculation. Quantum-well states would exhibit a larger periodicity in nonlinear magneto-optics.

The comparison of the *ab initio* results with semiempirical tight-binding calculations shown in Fig. 5.38 may demonstrate that it is necessary to use *ab initio* calculations for determination of film-specific structure. Note that the *ab initio* and the semiempirical tight-binding spectra of the bulk surface agree rather well in the optical range. However, the calculations for the monolayer using *ab initio* or different tight-binding methods differ drastically. Curve 5 in Fig. 5.38(a) is obtained by including in the hopping parameters correlations effects and the reduced number of nearest neighbours of a monolayer. The situation is similar in the linear case; see Fig. 5.38(b). This comparison illustrates that bulk tight-binding parameters are acceptable for the Kerr spectra at the bulk surface, but care should be taken for films.

For future studies on NOLIMOKE it will be interesting to extend our calculations to multilayer systems and to analyse in more detail interface contributions and lateral resolution. In particular, we will extend our NOLIMOKE calculations to analyse the bcc vs. fcc Fe structure on a substrate during film growth. A first-principles evaluation of SHG should support our assumption that essentially only surface and interface layers contribute and that second-layer contributions are of less importance, since the breakdown of inversion symmetry is not as strongly felt as in the interface layer.

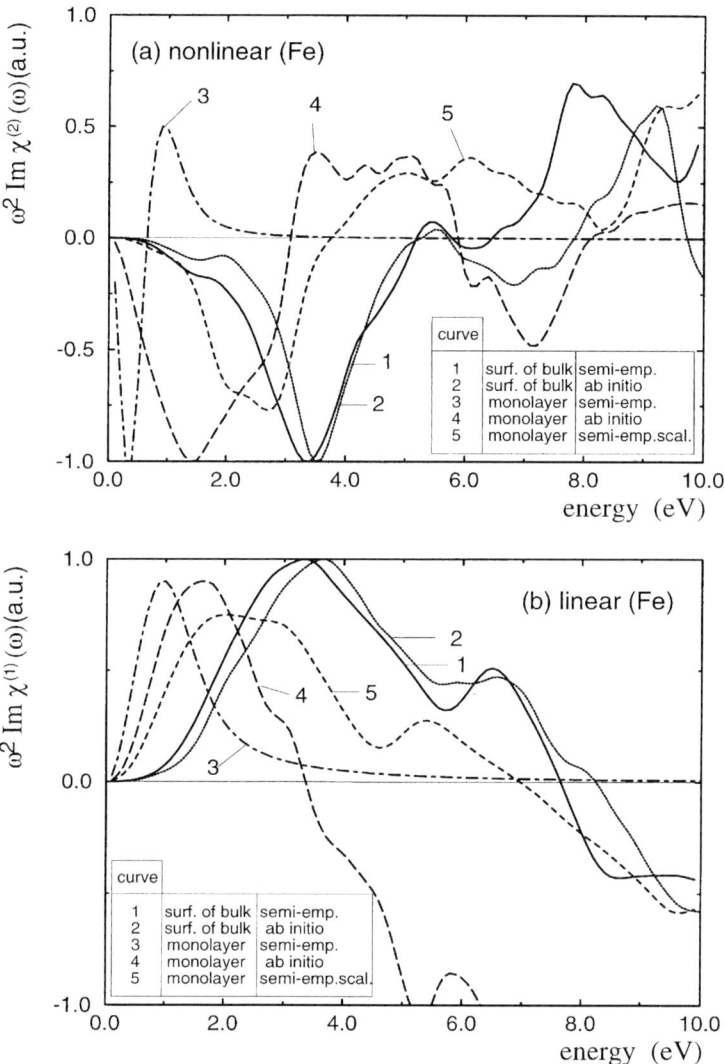

Fig. 5.38. Comparison of *ab initio* and semi-empirical calculations of (a) nonlinear and (b) linear magneto-optical susceptibilities of Fe. The bulk surface results from truncated bulk and the semiempirical calculations for a monolayer are performed with reduced hopping parameters A_0. The dashed-dotted curve results from the diagonalization of a square-lattice Hamiltonian in the tight-binding approximation.

Magnetic quantum-well states

Thin magnetic films and multilayers exhibit a rich variety of properties not previously found in bulk magnetism, such as enhanced or reduced moments,[178] oscillatory exchange coupling through nonmagnetic spacers,[179-181] giant magnetoresistance,[182,183] and the reorientation of the magnetic easy axis upon thickness and temperature variation.[184-188] Especially the observation of spin-polarized quantum-well states (QWS)[189-191,207] in Cu/Co(001) has attracted a great deal of attention. It has become clear that quantum-well states are indeed responsible for the important oscillatory behaviour of the exchange coupling of ferromagnetic thin films via nonmagnetic spacers.[208,209] Presently mainly photoemission (PE) and inverse photoemission (IPE)[189-191,207] have been used to identify QWS effects. Very recently a possible connection between thickness-dependent changes in NOLIMOKE and QWS[204] has been proposed.

It is the goal of this section of the chapter to show also that *nonlinear* optics, in particular NOLIMOKE, is a new sensitive tool for studying QWS. We find very interesting structure in the NOLIMOKE signal due to particular transitions in k space. This is very remarkable since it indicates that NOLIMOKE is able to detect very sensitively k-dependent structures. This new effect seems to be of general interest for the physics of nonlinear optics and its relationship to the underlying electronic structure. Note that this is not the case for linear optics, since there the contribution of the Drude term of the dielectric function creates a strong background of transitions from all k directions. Nonlinear optics, in contrast to linear optics, is able to give angle-resolved information about the underlying electronic structure. We demonstrate this by extending previous work on the Fe/Cu(001) bilayer system[210] to the sandwich system x Cu/Fe/Cu(001) where the layer number x is varied between 3 and 25. Thus we calculate the magnetic intensity contrast

$$\Delta I_{2\omega} = \frac{I_{2\omega}(M) - I_{2\omega}(-M)}{I_{2\omega}(M) + I_{2\omega}(-M)}$$

of NOLIMOKE for these systems and find very large quantum-well oscillations, originating from particular transitions in k space. In Fig. 5.39 we show the result for the nonlinear magneto-optical susceptibility for an Fe/Cu(001) bilayer and compare it to that of the Fe(001) monolayer. Both calculations have been done within the combined interpolation scheme including only Fe d and Cu s and d orbitals. There are two differences between the two spectra: (i) The peak due to the Fe d electrons in the low-energy range around 1 eV (which is too low due to the tight-binding description of the monolayer) is shifted and its sign is reversed. This results from a phase change due to the Fe/Cu interface. (ii) More striking, however, is the second effect, namely the occurrence of a completely new and very pronounced structure in the spectrum around 3.5 eV. This new structure is a purely nonlinear optical effect and results from the Cu d electrons, which become visible in the nonlinear *magneto*-optical Kerr spectrum via the spin polarization of the Fe d electrons. The d electrons of Cu are strongly excited and the Fe d electrons act as a spin-polarized resonant intermediate state. This effect is a very clear-cut example for the additional degree of freedom, which makes nonlinear magneto-optics a novel and a powerful spectroscopic probe of magnetic interfaces. Note that this additional Cu structure would be present in the spectrum also in the absence of any possible small Cu spin polarization and

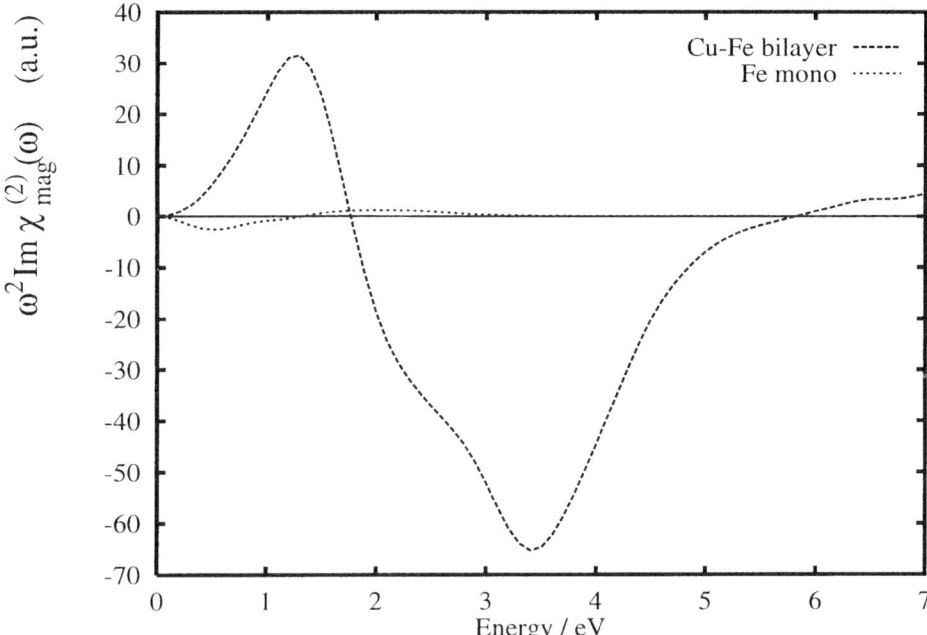

Fig. 5.39. Nonlinear Kerr spectrum of the Fe/Cu(001) bilayer in comparison with an Fe(001) monolayer.

does not require any hybridization of the electronic states of Cu and Fe. These two effects would just lead to a small modification of the striking Cu structure in NOLIMOKE. Thus, this new effect in the Fe/Cu bilayer demonstrates already that NOLIMOKE benefits from a large density of states close to the interface even if it is magnetically inert. Therefore this effect can be regarded as a precursor of the quantum-well behaviour which we are now going to discuss.

In view of the NOLIMOKE spectrum for the Fe/Cu(001) bilayer and of electronic structure presented in Fig. 5.40, a simple physical picture already explains the occurrence of quantum-well oscillations in NOLIMOKE from the x Cu/1 Fe/Cu(001) system. One gets the main peak of the multilayer system, since for 11 layers and multiples of this the marked transitions (a) between Cu d bands and the quantum-well states as final states become resonant at $2\hbar\omega$. Obviously this causes an oscillation with a period of 11 monolayers (ML). Correspondingly the period of 6–7 ML results from the marked transitions (b) in Fig. 5.40. Also it becomes clear that the spin polarization of the intermediate Fe states will cause a magnetization dependence and in particular a shifting of the peak for the magnetization direction M to lower periods. For the situation sketched in Fig. 5.40 the k selectivity becomes immediately obvious since unoccupied final states are necessary for a contribution to the SHG yield.

To verify these physical expectations we performed calculations using our previous theory[13,44] to evaluate the SHG intensity $I_{2\omega}(\omega)$ for opposite magneti-

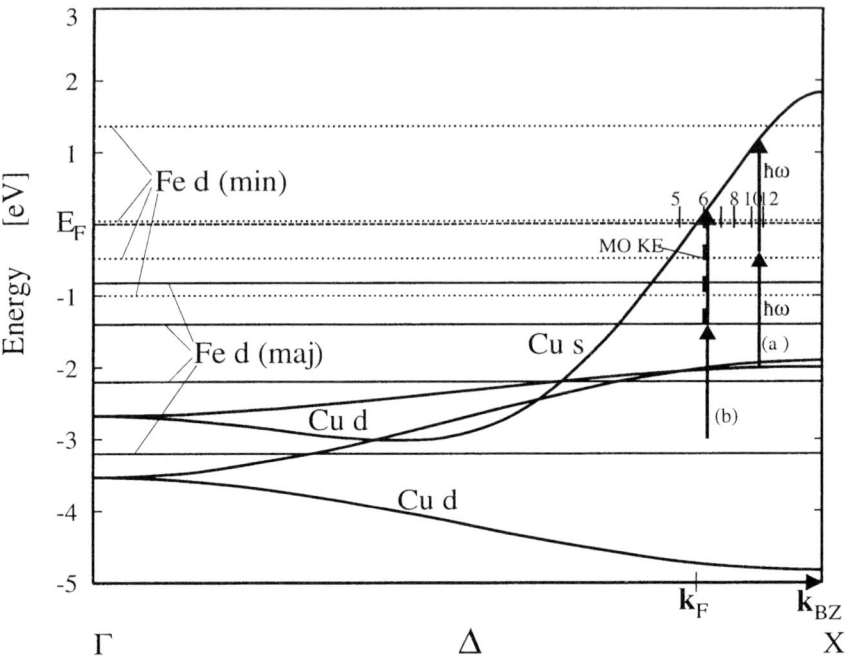

Fig. 5.40. Band structure of the x Cu/1 Fe/Cu(001) sandwich along $\mathbf{k}^\perp$. Bands with energy less than $-5\,\text{eV}$ are not drawn. The numbers 5, 6, 8, 10, 12 on the Fermi energy level indicate the $\mathbf{k}$ position of unoccupied QWS as they occur at the corresponding layer number. The marked transition (a) has a resonance at $2\hbar\omega$ and is responsible for the main peak at approximately 11 ML. The nonlinear transition (b) is responsible for the oscillation period of 6–7 ML. Also the dominating transition for MOKE making the 6–7 ML oscillation is indicated (dashed arrow).

zation directions.[211] Employing an electronic theory for both the nonlinear susceptibility and the dielectric function and separating $\chi^{(2)}(\omega)$ into even and odd parts under magnetization reversal $\chi^{(2)}_{\text{even}}(\omega)$ and $\chi^{(2)}_{\text{odd}}(\omega)$, we get for the SHG yield within the electric dipole approximation for the polar geometry (i.e. $\mathbf{M}$ normal to the surface)[51]

$$I_{2\omega}(\pm M) = \left| 2i\,|E_0^{(\omega)}|^2 \left\{ \chi^{(2)}_{\text{even}}(\omega) A_p \left[2F_c f_c f_s + N^2 F_s (f_c^2 + f_s^2) \right] t_p^2 \cos^2\varphi \cos\Phi \right\} \right.$$
$$\left. + \left\{ A_s \left[\chi^{(2)}_{\text{even}}(\omega) 2 f_s t_p t_s \cos\varphi \sin\varphi \pm \chi^{(2)}_{\text{odd}}(\omega) 2 f_c f_s t_p^2 \cos^2\varphi \right] \right\} \sin\Phi \right|^2.$$
(5.4.108)

Here φ and Φ denote the angle of polarization of the incident light and the outgoing second harmonic light and were chosen as $\varphi = 0°$ (p-polarized) and $\Phi = 75°$. The linear amplitudes A_p and the transmission and Fresnel factors $t_{p,s}$, $f_{c,s}$ and $F_{c,s}$ are derived from the dielectric function $\varepsilon(\omega)$. Note the nonlinear susceptibility tensor $\chi^{(2)}(\omega)$ is material-specific via the electronic band structure, and so are the linear dielectric function $\varepsilon(\omega)$ and the indices of refraction n and N. To simplify our calculation, we assume constant matrix elements, which are fitted to the linear dielectric function $\varepsilon(\omega)$.[46,211] This approximation is reasonable because the $\mathbf{k}$ dependence of the matrix elements is expected to become less important in two dimensions due to the shrinking of the d-band width for the reduced coordination number and also due to the occurrence of additional allowed optical transitions.[211] Selection rules excluding dipole transitions of the type $\langle \Psi_{m=\pm 2} | r | \Psi_{m=0} \rangle$ were taken into account.[44] To compare with experiment,[212,213] we choose 1.61 eV as incident photon energy. A normalization with respect to the Cu layer number has to be performed to take the interface sensitivity of SHG into account in order to make the nonlinear response comparable for various films thicknesses. Therefore, we divided $I_{2\omega}(\omega)$ by the layer number, ensuring that for a band structure without dispersion the response is identical for all layer numbers. To calculate $\varepsilon(\omega)$ and $\chi^{(2)}(\omega)$ from the electronic band structure of the x Cu/1 Fe/(001)Cu system we use a Cu bulk Hamiltonian (thus depending on k_x, k_y, k_z; $k_z = \mathbf{k}^\perp$ is perpendicular to the layers) combined with an Fe monolayer. The Hamiltonian is calculated within the combined interpolation scheme,[146] and the parametrization is according to Fletcher and Wohlfarth.[148,149] The parameters for the Cu bulk band structure are taken from ref. 214, and for the Fe monolayer they have been achieved from a fit to an *ab initio* calculation.[47] Of course, in Γ–X direction there is no dispersion of the Fe monolayer band structure. We evaluate the SHG response at $(k_x, k_y) = (0,0)$, since for the (001) direction the high density of states due to the extremal Fermi-surface diameter (calliper) at $\mathbf{k}_\| = (0,0)$, which gives the QW period from Ruderman–Kittel–Kasuya–Yoshida (RKKY) calculations,[215] dominates the output.[216,217]. The $\mathbf{k}$ summation is performed over k points along the $\mathbf{k}^\perp$ direction.

Note that, due to the two resonance denominators of $\chi^{(2)}(\omega)$,[46] it is not necessary for the intermediate state to be unoccupied to give a contribution to $\chi^{(2)}(\omega)$. It is sufficient if the final state (usually a Cu s state) is unoccupied. As a consequence, a high density of intermediate states leads to a large number of terms contributing to $\chi^{(2)}(\omega)$, thus enhancing the SHG response. In our electronic structure, this amplification is caused by the spin-polarized Fe d states. Since at least one of the three states involved in a nonlinear transition must be unoccupied to give a contribution to the SHG yield, the QWS above E_F are of great importance as final states for the NOLIMOKE signal. The QWS result from the confinement of the electrons in thin films, causing an equally spaced discretization in $\mathbf{k}^\perp$ direction, whereby the number of k points equals the number of layers. Clearly this discretization of the k values affects the SHG intensity since photon transitions are limited to these distinct $\mathbf{k}^\perp$ points.

For the fundamental period Λ of intensity oscillations as observed in photoemission, transitions at $\mathbf{k}^\perp$ vectors are decisive, for which (for Cu bulk) the s band crosses the Fermi surface. If the layer number increases, such unoccupied $\mathbf{k}^\perp$ states at the Fermi surface occur if the layer number equals $m k_{BZ}^\perp /$

($k_{BZ}^{\perp} - k_{F}^{\perp}$), with $m = 1, 2, 3 \ldots$. Then these new unoccupied s states permit additional transitions, and the optical response increases. The ratio $k_{BZ}^{\perp}/(k_{BZ}^{\perp} - k_{F}^{\perp})$ gives the fundamental period Λ. Obviously, since optical transitions may occur to all states above E_F, this period only marks a lower limit of possible oscillation periods and is not as strict as for (I)PE experiments, and depends on the photon energy and the position of the initial bands. In particular for the nonlinear response, due to the additional degree of freedom and due to $2\hbar\omega$ resonances, the SHG intensity increases and new (larger) oscillation periods occur. Although every period longer than the fundamental one may occur in the SHG spectrum if d states allow for resonances with unoccupied QWS at a $k^{\perp}$ vector between $k_F^{\perp}$ and $k_{BZ}^{\perp}$, the period Λ obtained from photoemission experiments and the doubled period 2Λ have an outstanding importance. If a QWS allows for a SHG signal with period 2Λ, there is a QWS at k_F too, and both resonant transitions (a) and (b) indicated in Fig. 5.40 contribute to the SHG signal at layer thickness $n2\Lambda$, thus enlarging the SHG amplitude at $m2\Lambda$. Owing to interferences of the various transitions, this enhancement is not compensated by the performed normalization. This effect is only present for multiples of 2Λ and is completely absent in linear optics. Our calculation shows that the spin-polarized Fe d bands are responsible for the occurrence and amplification of the observed oscillations. This becomes apparent if we compare the resulting SHG intensity of the x Cu/1 Fe/Cu(001) system for opposite magnetization directions with the system without Fe interlayer (but keeping the confinement for the Cu overlayer), which is 50 times weaker, in good agreement with experiment.[212] This strong enhancement is caused by the additional terms to be summed for the calculation of $\chi^{(2)}(\omega)$ when more bands are present, even if they are not resonant with the QWS. This amplification mechanism is not possible in linear optics, in agreement with experimental observations.[196]

In Fig. 5.41 we show results of our calculation of the NOLIMOKE signal demonstrating the pronounced QWS oscillations and their strong spin dependence. The large peak at approximately 11 ML (and a corresponding peak at 22 ML) results from the amplification due to the Fe bands, while the resonant $2\hbar\omega$ transition is between Cu d and Cu s bands (transition (a) in Fig. 5.40). Since the position of the Fe bands is less important for such a constellation, this peak dominates the SHG spectrum for both magnetization directions. At 6–7 ML the Cu d band edge is too far below E_F to give resonances with the QWS for the photon energy of 1.61 eV, hence a much reduced intensity results. While for the majority spins both $\hbar\omega$ resonances with Fe d bands as intermediate states and $2\hbar\omega$ resonances with Fe in initial state are important, the minority transitions involve mainly $\hbar\omega$ resonances with intermediate Fe d states and Cu d states as initial states. Since these resonances are not well matched by the photon energy, the short period of the SHG yield from the minority electrons is less pronounced. This can be traced back to the $I_{2\omega}(-^{(2)}M)$ yield, which is (in the geometry under consideration) influenced mainly by the minority transitions. The observed slight difference of the corresponding periods between the two magnetization directions is caused by the exchange splitting of the Fe d bands,

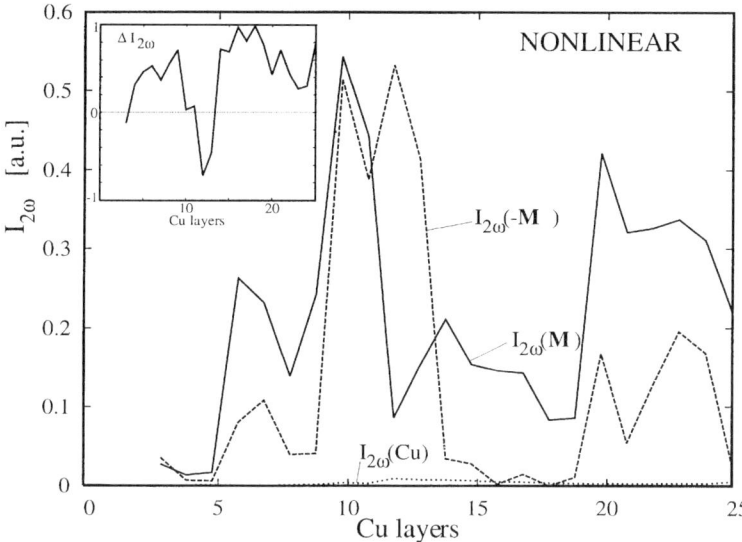

Fig. 5.41. SHG yield for opposite magnetization directions M and $-M$. The dominating 11 ML period is due to a $2\hbar\omega$ resonance between Cu d states and quantum-well states, drastically enhanced by the Fe d bands and thus demonstrating the k selectivity of NOLIMOKE. The signal for neglecting Fe bands is nearly vanishing on this intensity scale. The peak shift between the M and $-M$ signals is due to the spin polarization of the Fe d bands. $I_{2\omega}$ refers to the case where Fe is absent, but the confinement of the Cu layers is kept. The inset shows the magnetic contrast

$$\Delta I_{2\omega} = \frac{I_{2\omega}(M) - I_{2\omega}(-M)}{I_{2\omega}(M) + I_{2\omega}(-M)}.$$

allowing for resonances with the QWS at different layer numbers. The inset of Fig. 5.41 showing results for the magnetic contrast $\Delta I_{2\omega}$ gives further evidence for the importance of the Fe d bands. The result indicates clearly that the exchange splitting of the Fe interlayer is involved. The contrast varies between 100% and −80% and changes sign several times, due to the same magnitude of the SHG intensity for both magnetization directions. This coincidence of the two intensities at fixed frequency could not be explained if the SHG yield were generated solely by transitions between three spin-polarized quantum-well states, since then one signal should be much more pronounced than the other one. Furthermore, then the signal of a pure Cu surface should be of the same magnitude as that of the sandwich. Calculations for different exchange splittings showed that the occurring oscillation periods and the phase shift between the SHG yield for opposite magnetization directions are strongly influenced by the strength of the spin splitting. These results indicate a suppression of periods at particular exchange energies in sandwich structures.

In Fig. 5.42 we show the dielectric function $\varepsilon(\omega)$, its M dependence and the

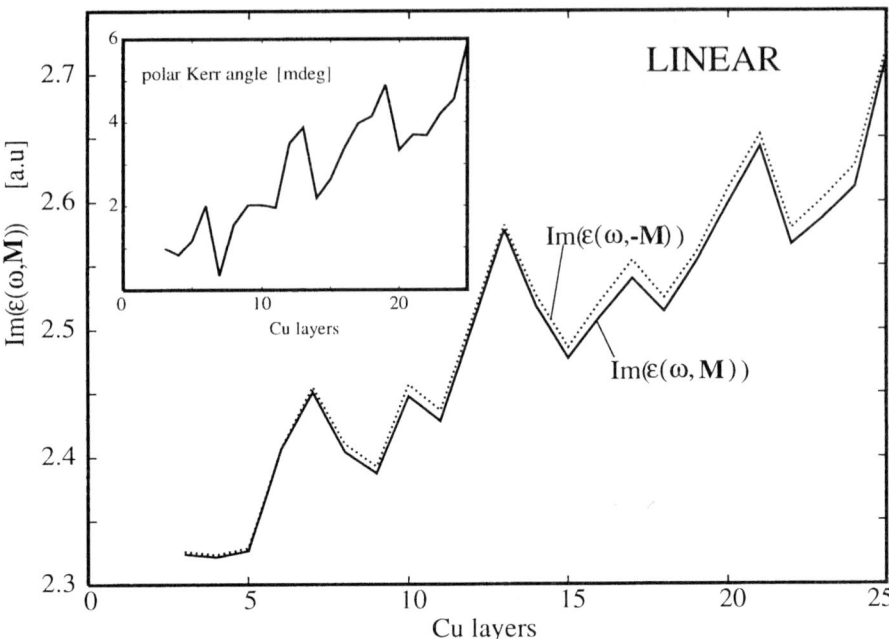

Fig. 5.42. Linear dielectric function of the x Cu/1 Fe/Cu(001) sandwich for opposite magnetizations as a function of the Cu layer thickness. Note that the 6–7 ML period is visible, while the 11 ML period is completely absent. The magnetic contrast is much smaller than for the NOLIMOKE signal. The inset shows the linear Kerr angle as a function of the Cu layer thickness.

linear Kerr angle. These results demonstrate the enhanced sensitivity of NOLIMOKE regarding oscillations due to QWS as compared with the linear optical response. Two oscillation periods for the imaginary part of the dielectric function $\varepsilon(\omega)$ for both magnetization directions can be seen, a dominant one with period 7 ML and a less pronounced oscillation with a period of 3–4 ML. From Fig. 5.40 the origin of these oscillation periods becomes clear, since for about 6–7 ML (and multiples of these thicknesses) there are QWS above E_F resonant with an Fe minority d band. Similarly, for the other peaks there are respective resonances with Fe d bands. In contrast to nonlinear optics, an overall increase of the linear signal with Cu thickness is observed, since it results not only from the interface, but from all layers, so that the normalization with respect to the layer thickness has not to be performed. Of course, the doubled period (11 ML) is absent, since $2\hbar\omega$ resonances do not contribute to the linear signal. Note that the magnetic effect is three orders of magnitude smaller than for the nonlinear signal, due to the strong influence of the nonmagnetic intraband transitions on the linear signal. Thus for the linear susceptibility, $\chi^{(1)}_{\text{even}}(\omega, M) \gg \chi^{(1)}_{\text{odd}}(\omega, M)$. This small magnetic effect becomes obvious from the linear Kerr angle Φ_{Kerr} shown in the inset of Fig. 5.42. Φ_{Kerr} is of the order of millidegrees, whereas the nonlinear Kerr angle is two to three orders of magnitude larger.[47] The overall increase with increasing layer thickness is again due to the long range of MOKE. The period of 6–7 ML is due to $\hbar\omega$ resonances between QWS above E_F and Fe majority d band at -1.4 eV. If this transition is resonant, the majority contribution of $\varepsilon(\omega)$ increases, resulting in an increase of Φ_{Kerr} at the corresponding layer thickness.

In conclusion, we showed that QWS give rise to strongly enhanced oscillation. The electronic origin of this strong enhancement is analysed. We find that NOLIMOKE is able to probe particular transitions in k space. Our results demonstrate that, although caused by the s QWS, the amplitude of the oscillation is due to the high density of Fe d states. Periods different from the fundamental period found in PE experiments are possible, depending on the position of resonant d bands below E_F. In contrast to linear optics, in NOLIMOKE even $2\hbar\omega$ resonances strongly influence the oscillation. In the considered sandwich structure, this causes the doubled period to dominate the spectrum.

5.5 Hot-electron dynamics in nonlinear optics

Owing to its surface sensitivity, the nonlinear optical response has become a powerful probe for investigating the electronic structure of surfaces, interfaces, thin films and multilayers. Recently, a combination of linear and nonlinear experiments[218] has been performed exploiting the time dependence of SHG. Thus, effects of hot electrons not at equilibrium with the lattice and their changes in time can be analysed. This opens a new route to investigate the dynamics of the system during relaxation to the equilibrium state. Note that the different electronic temperatures of hot electrons at equilibrium not with the lattice but among themselves result from varying the light irradiation. If intense and short laser pulses in the range of 100 fs to 1 ps are used for the SHG experiment, only the electrons will quickly thermalize (even in the approximation of Fermi liquid theory), since the slow electron–photon coupling does not come into play.[219,220] Only later, at times of the order of several picoseconds, will the electrons heat up the lattice. Thus, at short times the temperature dependence of SHG is essentially due to the varying nonequilibrium electronic temperature, which may be considerably different from the equilibrium temperature at later times, when electrons and lattice are at equilibrium.

Here, we present results for the electron temperature dependence of the SHG yield at noble-metal surfaces using an electronic theory for the nonlinear response.[221] The model allows already the identification of characteristic features of this time-dependent nonlinear optical response to be expected for the surface of bulk Cu and for other metals like Ag and Au and recent experimental results for Cu and Au[222] to be explained. Our analysis allows identification of the essential origin of the electron-temperature dependence of the nonlinear optical response. Furthermore our studies explain the different behaviour of linear and nonlinear optics on nonequilibrium electronic temperatures. In general, our theory is of interest for the dynamics of nonequilibrium electronic systems.

5.5.1 Theory

We calculate the SHG intensity $I^{(2)}(\omega, T_{el})$ using our electronic theory. Then the SHG yield for p polarization within the electric dipole approximation is, in the usual notation,[44]

$$I(\text{p-SH}) = \left|\left(2i\frac{\omega}{c}\right)\right|^2 |E_0(\omega)|^4 \left|A_p\{[F_c \chi^{(2)}_{xzx} 2f_c f_s \right.$$
$$\left. + N^2 F_s (\chi^{(2)}_{zxx} f_c^2 + \chi^{(2)}_{zzz} f_s^2)] t_p^2 \cos^2 \varphi + N_2 F_s \chi^{(2)}_{zxx} t_s^2 \sin^2 \varphi\}\right|^2. \quad (5.5.1)$$

The contributions to $I^{(2)}(\omega, T_{el})$ due to the Fresnel factors F_c, F_s, f_c and f_s and the transmission coefficients T_p, t_p and t_s and $\chi^{(2)}(\omega, T_{el})$ are all temperature-dependent. In both cases the temperature dependence arises from the Fermi functions $f(E,T)$ which, due to the many hot electrons resulting from the light irradiation, have to be taken at considerably elevated electron temperatures. Note that we limit ourselves to the time regime where the electrons have already thermalized but electron–phonon coupling has not become really effective. By taking both temperature dependences into account it becomes possible to decide theoretically whether $\chi^{(2)}(\omega, T_{el})$ or the Fresnel and transmission coefficients cause the essential temperature dependence of the SHG intensities. Since it is known from theory and also experiment that the $\chi^{(2)}_{zzz}$ tensor element dominates over $\chi^{(2)}_{xzx}$ and $\chi^{(2)}_{zxx}$,[44] we restrict our calculation to this single element of the nonlinear susceptibility for the SHG yield. Then approximately

$$I(\text{p-SH}) = \left|\left(2i\frac{\omega}{c}\right)\right|^2 |E_0(\omega)|^4 \left|A_p N^2 F_s \chi^{(2)}_{zzz} f_s^2 t_p^2 \cos^2 \varphi\right|^2. \quad (5.5.2)$$

The band structure involves five d bands and four plane waves and has been obtained[146] within the combined interpolation scheme (CIS). The d bands are parametrized in terms of Fletcher–Wohlfarth parameters,[148,149] and their values are given in Table 5.15. The parameters of the d bands were evaluated by a fit to an *ab initio* LAPW band-structure calculation by Krakauer et al.[223] Furthermore, to get also the correct onset of interband transitions we shifted the d states correspondingly.[224] The energy eigenvalues were calculated for 1861 **k** points in the irreducible part of the Brillouin zone ($\frac{1}{8}$th of the whole Brillouin zone). The resulting band structure shown later in Fig. 5.46 should thus be representative for the surface layer of bulk Cu.

The calculation of the dielectric function $\varepsilon(\omega, T_{el})$ is performed by including

Table 5.15 Fletcher–Wohlfarth parameters fitted to *ab initio* calculations for the Cu monolayer (E_0 = on-site energy, A_i = overlap integrals of the d orbitals)

Parameter	Value (eV)
E_0	−3.645
A_1	0.196
A_2	0.163
A_3	0.272
A_4	0.407
A_5	0.314
A_6	0.370

both the intra- and interband electronic transitions. Neglecting the q dependence of ε, we take[47,112]

$$\varepsilon(\omega,T) = 1 - \frac{\omega_{\text{pl}}^2}{\omega(\omega + \text{i}/\tau_{\text{pl}})} - \frac{4\pi e^2}{\Omega} \sum_{k,l,l',\sigma} M^2 \frac{f(E_{kl'\sigma},T) - f(E_{kl\sigma},T)}{E_{kl'\sigma} - E_{kl\sigma} - \hbar\omega + \text{i}\hbar\alpha_1},$$

(5.5.3)

where M are the dipole matrix elements, $f(E,T)$ are the Fermi functions, ω is the fundamental frequency, Ω is the unit area, $\alpha_1 = 0.1\,\text{eV}$ is the Lorentzian broadening, τ_{pl} is the Drude relaxation time, and $\omega_{\text{pl}} = 4\pi n_c e^2/m_c^*$ is the plasma frequency with n_c the electron density and m_c^* the effective electron mass. Both ω_{pl} and τ_{pl} are fitted to a dielectric function using literature values[225,226] at low fundamental energies, where no interband transitions occur. We use $\hbar\omega_{\text{pl}} = 9\,\text{eV}$ and $\tau_{\text{pl}}/\hbar = 8.5\,\text{eV}^{-1}$.

To simplify our calculation, we assume constant matrix elements M, which fit to the dielectric function $\varepsilon(\omega, T = 0\,\text{K})$.[47] This approximation is reasonable because the k dependence of the matrix elements is expected to become less important at the surface due to the shrinking of the d-band width for the reduced coordination number and also due to the occurrence of additional allowed optical transitions. Additionally, as will be shown later, the main contribution of the SHG intensity has its origin in d → d → s transitions, so the matrix elements give a simple prefactor to the sum and cancel when only intensity differences are considered. Once $\varepsilon(\omega, T_{\text{el}})$ is calculated, it is straightforward to get the Fresnel factors and transmission factors from eqn (5.5.3).

For evaluating the second-order susceptibility, we employ the microscopic theory developed in ref. 46. Note, for experiments not detecting absolute intensities (as is usually the case), that the prefactor in this formula is of no further importance for comparison with experimental data. As in the calculation of $\varepsilon(\omega)$, we neglect the k dependence of the matrix elements, taking the values for M_z from the $\varepsilon(\omega)$ fit. We only take into account the dominant $\chi_{zzz}^{(2)}$ tensor element and neglect the contributions of the other tensor elements, as pointed

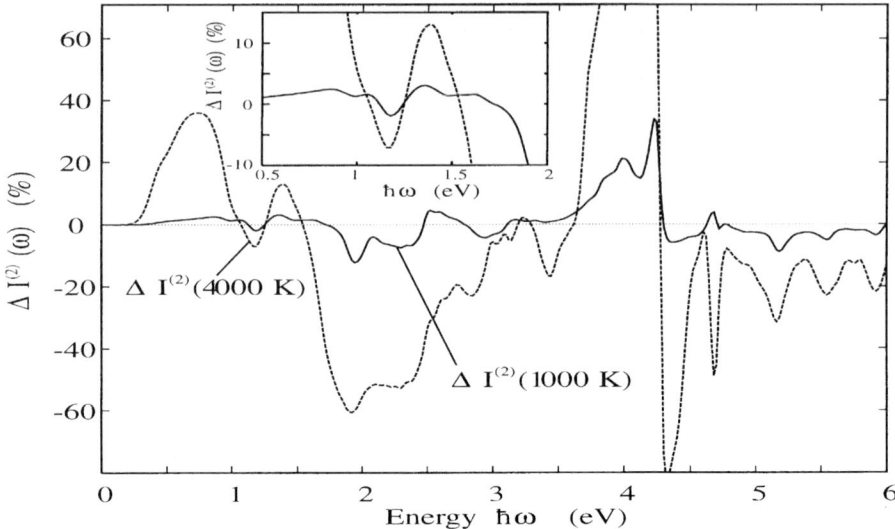

Fig. 5.43. Change of the SHG intensity $\Delta I^{(2)}(\omega)$ given by eqn (5.5.4) due to a rise of electron temperature as a function of the incident photon energy $\hbar\omega$ in percent for $T_{el} = 1000\,\mathrm{K}$ and $T_{el} = 4000\,\mathrm{K}$. The inset shows the energy range between 0.5 and 2 eV at an enlarged abscissa scale.

out earlier. This completes then the theory for the determination of the SHG intensity $I^{(2)}(\omega, T_{el})$ with the two essential inputs $\varepsilon(\omega, T_{el})$ and $\chi^{(2)}(\omega, T_{el})$.

5.5.2 Results

The most important results of our calculations are presented in Figs 5.43–5.45. For these results we used an electronic structure that is shown in Fig. 5.46. In Fig. 5.43 we present results for the change

$$\Delta I^{(2)}(\omega) = \frac{I^{(2)}(\omega, T_{el}) - I^{(2)}(\omega, 300\,\mathrm{K})}{I^{(2)}(\omega, 300\,\mathrm{K})} \tag{5.5.4}$$

of the SHG yield as a function of frequency for different nonequilibrium electronic temperatures T_{el}. In particular, in the inset of fig. 5.43 we show for the frequency range from 0.5 to 2 eV that $I^{(2)}(\omega)$ may be reduced or enhanced due to increasing nonequilibrium temperature for the electrons. In order to demonstrate what essentially causes the dependence on the electronic temperature, namely $\chi^{(2)}(\omega, T_{el})$ or the Fresnel coefficients, we present in Fig. 5.44 results for

$$\Delta\chi^{(2)}(\omega, T) = \frac{\chi^{(2)}(\omega, T_{el}) - \chi^{(2)}(\omega, 300\,\mathrm{K})}{\chi^{(2)}(\omega, 300\,\mathrm{K})}, \tag{5.5.5}$$

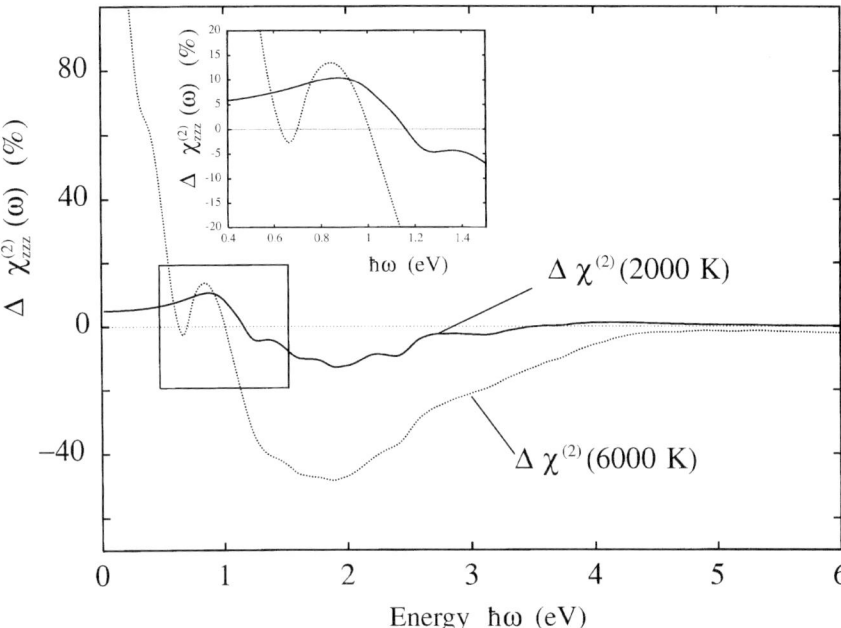

Fig. 5.44. Change of the second-order susceptibility $\Delta\chi(\omega,T)$ given by eqn (5.5.5) due to a rise of electron temperature as a function of the incident photon energy $\hbar\omega$ in per cent. $\chi^{(2)}_{zzz}$ is very similar to the SHG yield, indicating that the temperature effects for these fundamental energies on the SHG intensity are caused by $\chi^{(2)}_{zzz}$. The inset displays this similarity for incident energies between 0.5 and 2 eV.

and in Fig. 5.45 results for the difference

$$\Delta\frac{I^{(2)}}{|\chi^{(2)}|^2} = \frac{I^{(2)}(\omega,T_{el})/|\chi^{(2)}(\omega,T_{el})|^2 - I^{(2)}(\omega,300\,\text{K})/|\chi^{(2)}(\omega,300\,\text{K})|^2}{I^{(2)}(\omega,300\,\text{K})/|\chi^{(2)}(\omega,300\,\text{K})|^2}.$$

(5.5.6)

Note that in Fig. 5.45 we present results for $I^{(2)}(\omega,T)$ using a frequency- and temperature-independent susceptibility $\chi^{(2)}$. The comparison of the results in Figs 5.44 and 5.45 shows clearly that $I^{(2)}(\omega,T_{el})$ results essentially from $\chi^{(2)}(\omega,T_{el})$. Furthermore, from comparing the results in Figs 5.43 and 5.45 together with those in Figs 5.44 and 5.45, we note that the nonmonotonic behaviour with respect to the temperature dependence at 1.1 and 1.7 eV results from the resonances at $2\hbar\omega$ and $\hbar\omega$, respectively. The first one is only present in the nonlinear response $I^{(2)}(\omega,T_{el})$, but not in the Fresnel and transmission coefficients. The second one, however, occurs in both contributions due to the nonlinear susceptibility and the Fresnel coefficients but is still dominated by the nonlinear susceptibilities.

For a detailed analysis of the results presented in Figs 5.43–5.45 we would like

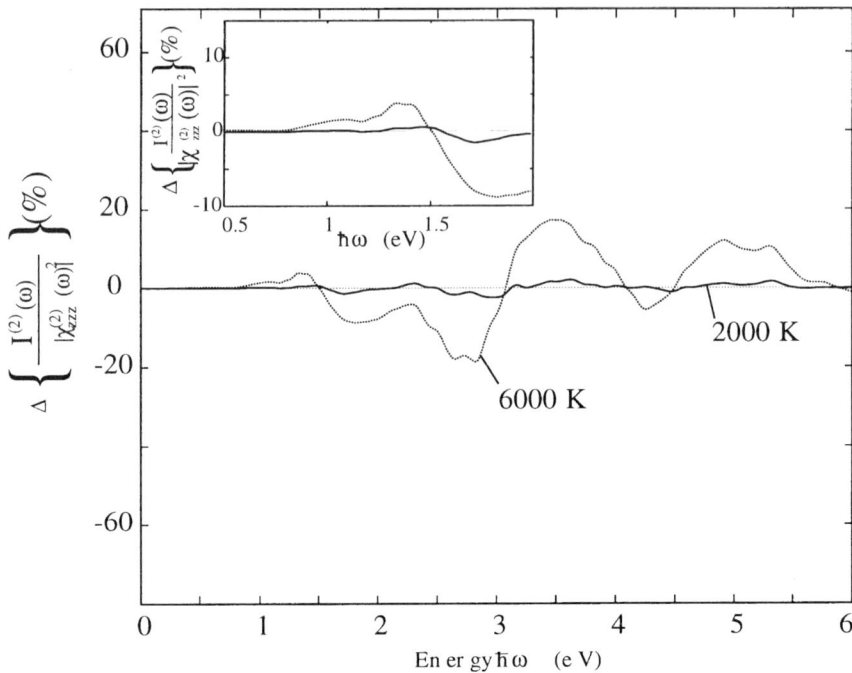

Fig. 5.45. Results for the relative change of the SHG yield $\Delta(I^{(2)}/|\chi^{(2)}|^2$ given by eqn (5.5.6) for $T_{el} = 1000$ K and $T_{el} = 4000$ K. In contrast to the results shown in Fig. 5.43, the frequency and temperature dependence of $\chi^{(2)}(\omega, T_{el})$ is neglected. The inset shows at an enlarged abscissa scale the energy range between 0.5 and 2 eV. These results demonstrate the importance of the temperature dependence of $\chi^{(2)}(\omega, T_{el})$.

to make the following remarks. It turns out that the largest contributions to the electron-temperature dependence result from transitions where the lowest two states are in the Cu d band. Investigating the relative changes of SHG intensity with temperature and its contributions due to the Fresnel factors and the nonlinear susceptibility $\chi^{(2)}(\omega, T_{el})$, we observe in Figs 5.43–5.45 that the shape of the dependence on the fundamental frequency is essentially given by the second-order susceptibility $\chi^{(2)}(\omega, T_{el})$. In the range between 1.8 and 3 eV, the SHG intensity decreases with increasing temperature. Most important is the result that indeed two energy windows at fundamental energy of 1.1 eV and 1.7 eV exist, where a small temperature increase (300 K to 1000 K) results in an SHG intensity increase, but a stronger temperature increase (300 K to 4000 K) results in a rapid SHG intensity decrease. From Figs 5.44 and 5.45 we conclude that $\chi^{(2)}(\omega, T_{el})$ is responsible for this. This clarifies then the physical origin of the temperature dependence of $I^{(2)}(\omega, T_{el})$ observed on polycrystalline Cu surfaces,[222] where in 2 eV photon energy pump–probe SHG experiments on Cu surfaces for weak pump pulses an increase of the SHG intensity has been observed, but a decrease when the pump pulse becomes stronger and thus the

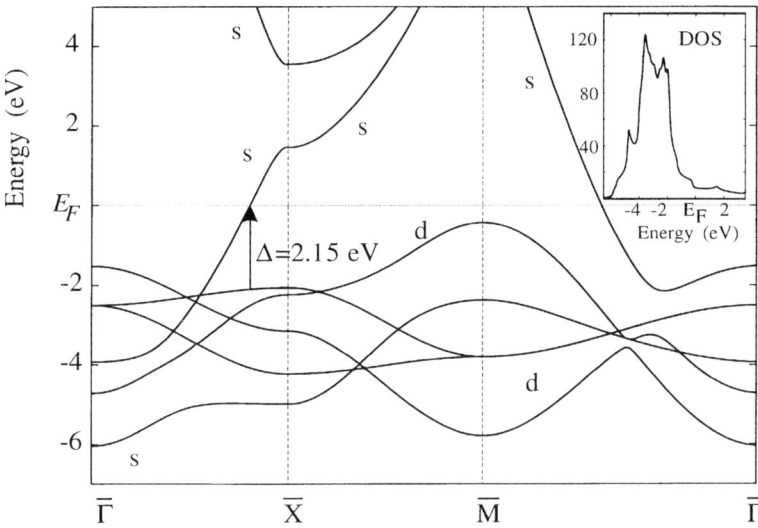

Fig. 5.46. Calculated band structure of a Cu surface using the combined interpolation scheme. Δ is the interband transition onset energy. The symmetry of the bands is indicated by the s and d labels. The arrow indicates the transition with lowest energy. The inset shows the corresponding density of states.

electron temperature higher. For our results, we expect such behaviour also for fundamental photon energies close to 1 eV.

Note it is remarkable that our model band structure (Fig. 5.46) yields already such a fair agreement with the experimental results. This is so, since the temperature dependence of $\chi^{(2)}(\omega, T_{el})$ and $I^{(2)}(\omega, T_{el})$ is strongly governed by the correct positon of the upper d band edge and also because the important features of the band structure are already simulated correctly by our model. Furthermore, SHG probes the surface layer only, so our band structure is a fair approximation to the experimental situation.

Since the 'mismatch' between $(E_d - E_F)$ and the photon energy $\hbar\omega$ in Au is larger by 230 meV than in Cu (Cu bulk: $E_d - E_F = 2.15$ eV), one may simulate the Au surface in our calculation by using approximately the SHG spectrum of the Cu monolayer, but considering a photon energy that is 230 meV below the respective energy value for the Cu monolayer. While for Cu we found the crossover behaviour at 1.1 eV and 1.7 eV, for Au we have to take fundamental photon energies of 0.87 eV and 1.47 eV. Then our results presented in Fig. 5.43 yield an increase of SHG for all temperatures. The increase for 4000 K is higher than that for 1000 K. These results seem in good agreement with experiments for a polycrystalline surface, where a monotonic increase of the SHG yield for both temperatures has been observed.[222] Thus, our electronic theory is able to explain the ultrafast electronic relaxation process on both Cu and Au surfaces. From our direct calculation of the dielectric functions we find that the temperature dependence of $I^{(2)}$ is mainly caused by $\chi^{(2)}(\omega, T_{el})$. At a frequency

$\hbar\omega = 0.9\,\text{eV}$ we find that the $\chi^{(2)}(\omega, T_{el})$ dependence on the electron temperature begins to saturate for higher temperature, as observed in the experiment.[222,227] Note that the SHG response of Ag may be modelled similarly as has been described for Au.

In summary, our results for $I^{(2)}(\omega, T_{el})$ suggest that the temperature dependence is essentially due to $\chi^{(2)}(\omega, T_{el})$. In particular, $\chi^{(2)}(\omega, T_{el})$ is responsible for the nonmonotonic temperature dependence $I^{(2)}(\omega, T_{el})$.

5.5.3 Conclusions

We calculated the dependence of the SHG yield on electron temperatures for electrons not at equilibrium with the lattice, caused by light irradiation for a Cu surface and for Ag and Au surfaces. We find that the temperature effects result mainly from $\chi^{(2)}(\omega, T_{el})$ and not from the Fresnel coefficients. In particular, our calculation yields a nonmonotonic electron-temperature dependence near 1.1 eV and 1.7 eV, which explains both the nonmonotonic intensity dependence for Cu and the monotonic increase of the SHG yield for Au. Note that for Au we use the same band structure and smaller light frequencies in order to study the same electronic transitions as for Cu. Clearly, $I^{(2)}(\omega, T_{el})$ depends on the position of the d band with respect to the Fermi energy. The interesting effect that mainly $\chi^{(2)}(\omega, T_{el})$ causes the temperature dependence comes about due to extra two-photon absorption processes making use of the high d-band density of states and occurring only at elevated electronic temperatures. Note, however, that only due to the interference of these transitions with others can a considerable temperature dependence result. The situation is illustrated in Fig. 5.47. Note

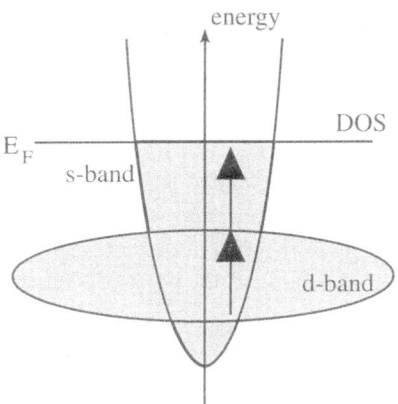

Fig. 5.47. Illustration of important electronic transitions generating SH from noble metals. Note both initial and intermediate states belong to the d band. Actually, by neglecting such transitions in our calculation, we find a drastically reduced $I^{(2)}(\omega, T_{el})$. Since in the case of the linear optical response, at best only the initial state can be a d state, nonequilibrium temperature effects are drastically reduced and less interesting with regards to probing the electronic structure.

that, without the effects due to the indicated transitions, the nonmonotonic temperature dependence would be absent. The figure illustrates also the difference between the temperature dependence in nonlinear and linear response, where for the latter only d states are possible as initial states. In addition, one expects a nonmonotonic temperature dependence also at fundamental photon energies close to 1.1 eV where the $2\hbar\omega$ transitions become resonant.

Regarding the time dependence of our results, for larger times the electron–phonon coupling will become more effective and cause the decrease of the electronic temperature via an energy transfer from the hot electrons to the lattice. The resulting equilibrium temperature will be only a few hundred kelvins higher than before the light irradiation. Then, $I^{(2)}(\omega, T_{el})$ will be similar to $I^{(2)}(\omega, 300\,\text{K})$. Such temperature dependences should be compared with those obtained for systems due to the usual equilibrium thermodynamic effects.

As a résumé, our studies show that SHG can be used to study the dynamics of excited hot electrons in crystals. Note that here we considered the time window where the electrons are far from equilibrium and have their own temperature different from the lattice temperature. Obviously, using SHG for studying the time evolution of the electronic system far from equilibrium offers new perspectives, in particular for studying magnetism. For transition metals, we expect in general a faster dynamics due to the increase of electron–phonon coupling.

It is of considerable interest to search for faster spin-switching mechanisms using intense femtosecond laser pulses which may lead directly to a breakdown of magnetism via electron–electron correlations and may therefore bypass the lattice, thus reducing lattice heating. It is to be expected that more interesting results will be found on the *femtosecond* timescale, which is now also accessible using Ti:sapphire lasers. Upon intense laser excitation, the magnetic state may break down already within some femtoseconds without the influence of the lattice and it is recovered within τ_{SL}, which involves coupling of the spins to the lattice via *anisotropic* crystal-field fluctuations. In this case, the spins are cooled by the lattice rather than heated as in the experiment by Vaterlaus *et al.*, which requires a theoretical explanation. These timescales should be optically accessible in metallic thin-film media in the near future.

5.6 Summary and outlook

The presented review discusses the new field of nonlinear magneto-optics at surfaces and interfaces, in particular the electronic theory of the nonlinear magneto-optical Kerr effect.

5.6.1 Summary

Our main results can be summarized as follows:

1. Already the extension of classical electrodynamics to the case of higher

harmonic generation yields an enhanced sensitivity of nonlinear optics to the size, shape, curvature and magnetism of metallic nanostructures. The angular profile of nonlinear Mie scattering is a particularly useful tool for the characterization of nano- and microparticles such as fullerene-based structures, blood cells, or droplets polluting the atmosphere. For larger particles, this profile exhibits a magic angle such as in rainbow scattering rather than getting more and more complicated.

2. Performing the symmetry analysis of nonlinear optics at surfaces one finds a characteristically different input polarization dependence of second harmonic generation from noble-metal surfaces than from transition-metal surfaces, since s and d electrons feel the symmetry breaking very differently.

3. For magnetic interfaces the nonlinear Kerr effect allows for the determination of the magnetic easy axis due to its characteristic polarization dependence. The Kerr rotation becomes as large as 90° for p-polarized perpendicularly incoming light in the longitudinal configuration and is always enhanced by several orders of magnitude for thin films compared to linear magneto-optics.

4. The symmetry analysis of nonlinear magneto-optics predicts new nonlinear effects probing the symmetry of the superconducting order parameter for high-temperature superconductors as well as uniquely determining the spin configuration at antiferromagnetically ordered surfaces even of cubic materials with magnetic unit-cell doubling.

5. An electronic theory for the nonlinear magneto-optical Kerr effect is formulated, which includes the element-specific effects of the band structure, the exchange and spin–orbit coupling and the symmetry breaking at surfaces and interfaces.

6. Using a parametrized band-structure calculation we compute the nonlinear Kerr spectra of Fe and Ni surfaces. We clearly demonstrate that the nonlinear Kerr effect sensitively probes the geometric, electronic and magnetic surface properties. It is linear in the magnetization and spin–orbit coupling.

7. For thin films, nonlinear Kerr spectra using *ab initio* calculated band structures exhibit a clear film-thickness dependence of the interface-layer response and clearly reveal the lattice parameter imposed by the substrate as well as its electronic structure in addition to the electronic film structure even in the absence of any substrate spin polarization. This is due to an additional degree of freedom in nonlinear optics.

8. We find strong spin-polarized quantum-well oscillations of amplitude unity and single and double periods in the nonlinear Kerr effect from x Cu/Fe(001) sandwiches. These oscillations are much more pronounced than in linear magneto-optics.

9. The dependence of the SHG yield on electron temperature for hot elec-

trons on Cu surfaces exhibits a nonmonotonic behaviour close to the onset of d-electron excitations by one or two photons: the SHG yield increases for weak excitation intensity while it increases for strong excitation intensities. The SHG intensity changes result from the nonlinear susceptibilities rather than the Fresnel coefficients and are of the order of 30% to 50%, while the linear reflectivity at the same time changes only by a few per cent. Thus nonlinear optics probes hot electrons much more sensitively than linear optics.

5.6.2 Outlook

It is clear that the new and promising field of nonlinear magneto-optics requires further development and cultivation. In particular the first-principles theory of the nonlinear Kerr effect, refined by the inclusion of the dipole matrix elements and spin−orbit coupling on the same footing, has to be thoroughly tested and compared with forthcoming frequency-dependent measurements. This electronic theory then contains all elements of the symmetry analysis on the microscopic level. This allows NOLIMOKE to become a routine tool for the optical investigation of surfaces and, in particular, buried interfaces. One should keep in mind that NOLIMOKE not only has the potential to separate the response from different interfaces present vertically on the scale of the skin depth but also offers the unique opportunity of studying laterally nanostructured films with a nanometre resolution due to the local-field enhancements and the in-plane symmetry breaking. The nonlinear Kerr effect may also provide an alternative to Stern−Gerlach apparatus for the investigation of magnetic clusters on surfaces.

The most promising extension of nonlinear magneto-optics in our view, however, is its application to femtosecond spectroscopy and coherent control in magnetic films and multilayers. This part of the field is just beginning. However, it is to be expected that ultrafast switching processes can be triggered by nonlinear magneto-optical pump−probe spectroscopy. Regarding applications, two remarks have to be made.

1. Certainly nonlinear (magneto-)optics can be applied to the remote sensing and characterization of pollutants in the atmosphere, to the optical measurements of blood cell sizes and shapes, and to fullerene materials, using higher harmonic Mie scattering.

2. Moreover, NOLIMOKE is right now at the verge of being detectable upon excitation with laser diodes. If this is possible, NOLIMOKE might even be technologically interesting, since it enables much higher lateral resolution than linear magneto-optics with the option of layer-by-layer recording, thus strongly increasing the bit density per unit volume in magneto-optical recording.

Acknowledgements

I would like to thank Professor K. H. Benneman for many stimulating discussions, his continued interest and strong support of this research.

I am indebted to my colleagues, coworkers and students A. Dähn, J. P. Dewitz, R. Družinić, Dr M. E. Garcia, Dr P. J. Jensen, R. Knorren, T. A. Luce, A. Lesard, T. H. Moos, Dr U. Pustogowa and Dr J. Schmalian for many discussions and their collaboration.

Furthermore I gratefully acknowledge the fruitful and successful collaboration with the (mostly) experimental groups headed by Dr J. Ferré (Orsay), Professor D. Fröhlich (Dortmund), Professor J. Kirschner (Halle), Dr I. L. Lyubchanskii (Donetsk), Professor E. Matthias (Berlin), Professor R. Pisarev (St. Petersburg), Dr T. Rasing (Nijmegen) and Professor J. Reif (Cottbus).

References

1. M. Göppert-Mayer, *Ann. Phys. (Leipzig)* **9**, 273 (1931).
2. N. Bloembergen and P. S. Pershan, *Phys. Rev.* **128**, 606 (1962).
3. P. L. Kelley, *J. Phys. Chem. Solids* **24**, 607 and 1113 (1963).
4. S. S. Jha, *Phys. Rev.* **140**, A 2020 (1965).
5. P. S. Pershan, in: E. Wolf (Ed.), *Progress in Optics* **5**, 85 (1966).
6. P. N. Butcher, *Nonlinear Optical Phenomena*, pp. 43 ff. (Ohio State University Press, Columbus, 1965).
7. S. Kielich and R. Zawodny, *Optica Acta* **20**, 867 (1973).
8. M. Corvi and W. L. Schaich, *Phys. Rev.* B **33**, 3688 (1986).
9. W. L. Schaich and A. Liebsch, *Phys. Rev.* B **37**, 6187 (1988).
10. A. Liebsch, *Phys. Rev. Lett.* **61**, 1233 (1988).
11. S. B. Borisov and I. L. Lyubchanskii, *Opt. Spectrosc. (USSR)* **61**, 801 (1986).
12. W. Hübner and K. H. Bennemann, *Europhys. Conf. Abstr.* **13A**, A42 (1989).
13. W. Hübner and K. H. Bennemann, *Phys. Rev.* B **40**, 5973 (1989).
14. R. P. Pan, H. D. Wei, and Y. R. Shen, *Phys. Rev.* B **39**, 1229 (1989).
15. A. Vaterlaus, T. Beutler, and F. Meier, *Phys. Rev. Lett.* **679**, 3314 (1991).
16. J. Reif, J. C. Zink, C.-M. Schneider, and J. Kirschner, *Phys. Rev. Lett.* **67**, 2878 (1991).
17. U. Pustogowa, W. Hübner and K. H. Bennemann, *Phys. Rev.* B **49**, 10031 (1994).
18. R. Kubo, *J. Phys. Soc. Jap.* **12**, 570 (1957).
19. J. Dewitz, private communication (1996).
20. X. M. Hua and J. I. Gersten, *Phys. Rev.* B **33**, 3756 (1986).
21. K. Hayata and M. Koshiba, *Phys. Rev.* A **46**, 6104 (1992).
22. D. Östling, P. Stampfli, and K. H. Bennemann, *Z. Phys.* D **28**, 169 (1993).
23. J. P. Dewitz, W. Hübner, and K. H. Bennemann, *Z. Phys.* D **37**, 75 (1996).
24. G. Mie, *Ann. Phys.* **25**, 377 (1908).
25. T. Götz, M. Buck, C. Dressler, F. Eisert, and F. Träger, *Appl. Phys.* A **60**, 607 (1995).
26. J. D. Jackson, *Classical Electrodynamics* (Wiley, New York, 1975).
27. M. Born and E. Wolf, *Principles of Optics* (Pergamon Press, Oxford, 1975).
28. P. B. Johnson and R. W. Christy, *Phys. Rev.* B **9**, 5056 (1974).
29. C. F. Bohren and D. R. Huffman, *Absorption and Scattering of Light by Small Particles* (Wiley, New York, 1983).
30. J. R. Probert-Jones, *J. Opt. Soc. Am.* A **1**, 822 (1984).
31. J. P. Dewitz, W. Hübner, and K. H. Bennemann (unpublished).

REFERENCES

32. P. Barber and C. Yeh, *Appl. Opt.* **14**, 12 (1975).
33. P. A. Waterman, *Proc. IEEE* **53**, 805 (1965).
34. P. A. Waterman, *Phys. Rev.* D **3**, 825 (1971).
35. S. A. Schelkunoff, *Electromagnetic Waves* (Van Nostrand, New York, 1943).
36. S. E. Apsel, J. Deng, and L. A. Bloomfield, *Phys. Rev. Lett.* **76**, 1441 (1996).
37. J. Kasparian, B. Krämer, J. P. Dewitz, S. Wajda, P. Rairoux, B. Vezin, V. Boutou, T. Leisner, W. Hübner, J. P. Wolf, L. Wöste, and K. H. Bennemann, *Phys. Rev. Lett.* **78**, 2952 (1997).
38. P. Chýlek, *J. Opt. Soc. Am.* **66**, 285 (1976).
39. N. Bloembergen, R. K. Chang, S. S. Jha, and C. H. Lee, *Phys. Rev.* **174**, 813 (1968); P. Guyot-Sionnest, W. Chen, and Y. R. Shen, *Phys. Rev.* B **33**, 254 (1986).
40. A. Leibsch and W. L. Schaich, *Phys. Rev.* B **40**, 5401 (1989).
41. J. E. Sipe, D. J. Moss, and H. M. van Driel, *Phys. Rev.* B **35**, 1129 (1987).
42. K. Böhmer, J. Hohlfeld, and E. Matthias, *Appl. Phys.* A **60**, 203 (1995).
43. K. Bömer, Thesis (FU Berlin, 1994).
44. W. Hübner, K. H. Bennemann, and K. Böhmer, *Phys. Rev.* B **50**, 17597 (1994).
45. D. A. Kleinman, *Phys. Rev.* **126**, 1977 (1962).
46. W. Hübner, *Phys. Rev.* B **42**, 11553 (1990).
47. U. Pustogowa, W. Hübner, and K. H. Bennemann, *Phys. Rev.* B **48**, 8607 (1993).
48. G. Petrocelli, S. Martellucci, and R. Francini, *Appl. Phys.* A **56**, 263 (1993).
49. D. A. Papaconstantopoulos, *Handbook of the Bandstructure of Elemental Solids* (Plenum, New York, 1986).
50. Ni forms in air a thin antiferromagnetic and passivating oxide on its surface, probably two monolayers in thickness.
51. W. Hübner, and K. H. Bennemann, *Phys. Rev.* B **52**, 13411 (1995).
52. At the surface and at interfaces also transitions with $\Delta m = 0, \pm 1$ and $\Delta l = 0, \pm 2$ are allowed since the perpendicular inversion symmetry is broken.
53. H. A. Wierenga, M. W. J. Prins, D. L. A. Abraham, and Th. Rasing, *Phys. Rev.* B **50**, 1282 (1994).
54. Nevertheless, in linear optics the Kerr rotation in the polar configuration is usually larger than in the longitudinal configuration, since (i) the polar configuration always implies perpendicular incidence and thus large trigonometric prefactors (which tend to zero for perpendicular incidence in the longitudinal configuration[17]), and (ii) a somewhat enhanced surface contribution might be present for perpendicular magnetization, which is not averaged out over the skin depth of a metal.
55. B. Koopmans, M. Groot Koerkamp, Th. Rasing, and H. van den Berg, *Phys. Rev. Lett.* **74**, 3692 (1995).
56. R. Vollmer, A. Kirilyuk, H. Schwabe, J. Kirschner, H. A. Wierenga, W. de Jong, and Th. Rasing, *J. Magn. Magn. Mater.* **148**, 295 (1995).
57. I. A. Campbell and A. Fert, in: E. P. Wohlfarth (Ed.), *Ferromagnetic Materials*, Vol. 3, p. 747 (North Holland, Amsterdam, 1982).
58. P. Nozières and C. Lewiner, *J. Physique* **34**, 901 (1973).
59. O. A. Aktsipetrov, P. V. Elyutin, A. A. Fedyanin, A. A. Nikulin, and A. N. Rubtsov, *Surf. Sci.* **325**, 343 (1995).
60. J. Reif, C. Rau, and E. Matthias, *Phys. Rev. Lett.* **71**, 1931 (1993).
61. L. M. Falicov, D. T. Pierce, S. D. Bader, R. Gronsky, K. B. Hathaway, H. J. Hopster, D. N. Lambeth, S. S. P. Parkin, G. A. Prinz, M. B. Salamon, I. K. Schuller, and R. H. Victora, *J. Mater. Res.* **5**, 1299 (1990).
62. P. M. Oppeneer, T. Maurer, J. Sticht, and J. Kübler, *Phys. Rev.*B **45**, 10924 (1992).
63. R. A. de Groot and F. M. Mueller, *Phys. Rev. Lett.* **50**, 2024 (1983); P. A. M. van der

Heide et al., *J. Phys. F: Met. Phys.* **15**, L75 (1985); W. Reim et al., *J. Magn. Magn. Mater.* **54–57**, 1401 (1986); G. H. O. Daalderop et al., *J. Magn. Magn. Mater.* **74**, 211 (1988); J. H. Wijngaard, C. Haas, and R. A. de Groot, *Phys. Rev. B* **40**, 9318 (1989).

64. Note that this introduction of the Kerr angle in the nonlinear case essentially makes sense only for p-polarized incident light, since SH yield is predominantly p-polarized.
65. Note that this corresponds in ref. 2 to putting $\alpha = \pi - \theta_S$.
66. Note that ε_S in the denominator of the second term of eqn (4.12) in ref. 2 has to be replaced by ε_R.
67. O. A. Aktsipetrov, O. V. Braginskii, and D. A. Esikov, *Sov. J. Quantum Electron.* **20**, 259 (1990).
68. G. S. Krinchik, and V. A. Artem'ev, *Zh. Eksp. Teor. Fiz.* **53**, 1901 (1967) [*Sov. Phys. JETP* **26**, 1080 (1968)]; *J. Appl. Phys.* **39**, 1276 (1968).
69. H. Feil and C. Haas, *Phys. Rev. Lett.* **58**, 65 (1987).
70. However, since the interband contributions should disappear for $\omega \to 0$, one must replace the factor $\lambda_{so}/\hbar\omega$ by $\lambda_{so}/\hbar\omega_1$ for $\omega \to 0$, where $\hbar\omega_1$ refers to the minimum interband transition energy.
71. K. Bömer and E. Matthias, (private communication).
72. G. Spierings, V. Koutsos, H. A. Wierenga, M. W. J. Prins, D. Abraham, and Th. Rasing, *Surf. Sci.* **287/288**, 747 (1993); *J. Magn. Magn. Mater.* **121**, 109 (1993).
73. J. A. Martindale, S. E. Barrett, C. A. Klug, K. E. O'Hara, S. M. DeSoto, C. P. Slichter, T. A. Friedmann, and D. M. Ginsberg, *Phys. Rev. Lett.* **68**, 702 (1992).
74. Z.-X. Shen, D. S. Dessau, B. O. Wells, D. M. King, W. E. Spicer, A. J. Arko, D. Marshall, L. W. Lombardo, A. Kapitulnik, P. Dickinson, S. Doniach, J. DiCarlo, T. Loeser, and C. H. Park, *Phys. Rev. Lett.* **70**, 1553 (1993).
75. W. N. Hardy, D. A. Bonn, D. C. Morgan, R. Liang, and K. Zhang, *Phys. Rev. Lett.* **70**, 3999 (1993).
76. J. Kane, Q. Chen, K.-W. Ng, and H.-J. Tao, *Phys. Rev. Lett.* **72**, 128 (1994).
77. T. P. Devereaux, D. Einzel, B. Stadlober, R. Hackl, D. H. Leach, and J. J. Neumeier, *Phys. Rev. Lett.* **72**, 396 (1994).
78. D. A. Wollman, D. J. Van Halingen, W. C. Lee, D. M. Ginsberg, and A. J. Leggett, *Phys. Rev. Lett.* **71**, 2134 (1993).
79. P. Chaudhari and Shawn-Yu Lin, *Phys. Rev. Lett.* **72**, 2134 (1994).
80. D. A. Brawner and H. R. Ott, *Phys. Rev. B* **50**, 6530 (1994).
81. C. C. Tsuei, J. R. Kirtley, C. C. Chi, Lock See Yu-Jahnes, A. Gupta, T. Shaw, J. Z. Sun, and M. B. Ketchen, *Phys. Rev. Lett.* **73**, 593 (1994).
82. D. A. Gajewski, M. B. Maple, and R. C. Dynes, *Phys. Rev. Lett.* **72**, 2267 (1994).
83. For a detailed and recent discussion see: D. J. Van Harlingen, *Rev. Mod. Phys.* **67**, 523 (1995).
84. Y. R. Shen, *Annu. Rev. Mater. Sci.* **16**, 69 (1986).
85. G. L. Richmond, J. M. Robinson, and V. L. Shannon, *Prog. Surf. Sci.* **28**, 1 (1988).
86. J. Schmalian and W. Hübner, *Phys. Rev. B* **53**, 11860 (1996).
87. J. Bardeen, L. N. Cooper, and J. R. Schrieffer, *Phys. Rev.* **108**, 1175 (1957).
88. M. Sigrist and K. Ueda, *Rev. Mod. Phys.* **63**, 239 (1991).
89. P. W. Anderson, *Phys. Rev. B* **32**, 2935 (1985).
90. G. E. Volovik and L. P. Gorkov, *Sov. Phys. JETP* **61**, 843 (1985).
91. E. I. Blount, *Phys. Rev. B* **32**, 2935 (1985).
92. M. Sigrist and T. M. Rice, *Phys. Rev. B* **39**, 2200 (1985).
93. M. Sigrist and T. M. Rice, *Z. Phys. B* **68**, 9 (1987).
94. M. Fiebig, D. Fröhlich, B. B. Krichevtsov, and R. V. Pisarev, *Phys. Rev. Lett.* **73**, 2127 (1994).

95. B. Koopmans, A. M. Janner, H. A. Wierenga, Th. Rasing, G. A. Sawatzky, and F. van der Woude, *Appl. Phys.* A **60**, 103 (1995).
96. V.V. Eremenko and N. F. Kharchenko, *Phys. Rep.* **155**, 379 (1987).
97. R. Zawodny and S. Kielich, *Phys. Rev.* A **38**, 3504 (1988).
98. N. N. Akhmedeiev, S. B. Borisov, A. K. Zvezdin, I. L. Lyubchanskii, and Yu. V. Melikhov, *Fiz. Tverd. Tela* **27**, 1075 (1985) [*Sov. Phys. Solid State* **27**, 650 (1985)].
99. S. B. Borisov, I. L. Lyubchanskii, A. D. Petrenko, and G. I. Trush, *JETP* **78**, 279 (1994).
100. R. V. Pisarev, I. G. Siny, and G. A. Smolensky, *Solid State Commun.* **7**, 23 (1969).
101. B. B. Krichetsov, V. V. Pavlov, R. V. Pisarev, and V. N. Gridnev, *J. Phys. Condens. Matter* **5**, 8233 (1993).
102. M. Fiebig, D. Fröhlich, G. Sluyterman v. L., and R. V. Pisarev, *Appl. Phys. Lett.* **66**, 2906 (1995).
103. V. N. Muthukumar, R. Valentí, and C. Gros, *Phys. Rev. Lett.* **95**, 2766 (1995).
104. V. N. Muthukumar, R. Valentí, and C. Gros, *Phys. Rev.* B **54**, 433 (1996).
105. I. Dzyaloshinskii and E. V. Papamichail, *Phys. Rev. Lett.* **75**, 3004 (1995).
106. A. Dähn, W. Hübner, and K. H. Bennemann, *Phys. Rev. Lett.* **77**, 3929 (1996).
107. R. R. Birss, *Symmetry and Magnetism* (North Holland, Amsterdam, 1964).
108. S. Bhagavantam, *Crystal Symmetry and Physical Properties* (Academic Press, London, 1966).
109. Nevertheless, classifying ferromagnetic surfaces Pan *et al.* used time inversion in combination with space symmetry operation. This is a contradiction which has not become clear so far. Since, however, for our configurations R never appears in combination with any space transformation but always as an independent symmetry element, this problem does not affect our symmetry analysis. If R is an allowed operation for examining $\chi_{el}^{(2\omega)}$, according to formula (5.3.79) all the nonzero tensor elements must be invariant under time inversion. (That is, they have to be even in the antiferromagnetic parameter.) Whereas leaving R out of the symmetry analysis, nothing can be said about the behaviour of the tensor elements under time reversal.
110. Domain imaging in the way Fiebig *et al.*[94,102] did it on Cr_2O_3 is impossible for this symmetry, though. To get the necessary effect of interference there had to exist 180° spin domains in the surface which coud not be brought to identity by a translation. In consequence, time inversion symmetry R of the surface had to be broken. Although demonstrating above that this is not the case for our regular lattices, we want to point out that small distortions, caused by magnetostriction, for example, could perhaps induce an appropriate domain structure.
111. H. Ehrenreich and M. H. Cohen, *Phys. Rev.* **115**, 786 (1959).
112. H. Haug and S. Schmitt-Rink, *Prog. Quantum Electron* **9**, 3 (1984).
113. R. Vollmer, M. Straub, and J. Kirschner, *Surf. Sci.* **352–354**, 684 (1996); *ibid.*, 937 (1996).
114. B. N. J. Persson and N. D. Lang, *Phys. Rev.* B **26**, 5409 (1982).
115. B. N. J. Persson and S. Andersson, *Phys. Rev.* B **29**, 4382 (1984).
116. B. N. J. Persson and E. Zaremba, *Phys. Rev.* B **31**, 1863 (1985).
117. J. F. Annett, R. E. Palmer, and R. F. Willis, *Phys. Rev.* B **37**, 2408 (1988).
118. P. N. Argyres, *Phys. Rev.* **97**, 334 (1955).
119. W. Reim, J. Schoenes, F. Hulliger, and O. Vogt, *J. Magn. Magn. Mater.* **54–57**, 1401 (1986).
120. L. M. Roth, *Phys. Rev.* **133**, A 542 (1964).
121. H. S. Bennett and E. A. Stern, *Phys. Rev.* **137**, A 448 (1965).
122. J. Langlinais and J. Callaway, *Phys. Rev.* B **5**, 124 (1972).

123. C. S. Wang and J. Callaway, *Phys. Rev.* B **15**, 298 (1977).
124. H. Ebert, *Phys. Rev.* B **38**, 9390 (1988).
125. H. Eckardt and L. Fritsche, *J. Phys.* F **17**, 925 (1987).
126. J. Tersoff and L. M. Falicov, *Phys. Rev.* B **26**, 6186 (1982).
127. E. Wimmer, A. J. Freeman, and H. Krakauer, *Phys. Rev.* B **30**, 3113 (1984).
128. C. S. Wang and J. Callaway, *Phys. Rev.* B **9**, 4897 (1974).
129. N. V. Cohan and H. F. Hameka, *Physica* **38**, 320 (1967).
130. D. K. Misemer, *J. Magn. Magn. Mater.* **72**, 267 (1988).
131. P. M. Oppeneer, J. Sticht, T. Maurer, and J. Kübler, *Z. Phys.* B **88**, 309 (1992).
132. U. Pustogowa, W. Hübner, K. H. Bennemann, and T. Kraft, *Z. Phys.*, B **702**, 109 (1997).
133. P. M. Marcus, private communication (1992).
134. C. Kittel, *Phys. Rev.* **83**, 208 (A), (1951).
135. J. Chen, private communication (1995).
136. P. Tepper and J. Reif, private communication (1989).
137. A. Liebsch, *Phys. Rev. Lett.* **43**, 1431 (1979).
138. A. Liebsch, *Phys. Rev.* B **23**, 5203 (1981).
139. W. Nolting, W. Borgiel, V. Dose, and Th. Fauster, *Phys. Rev.* B **40**, 5015 (1989).
140. W. Borgiel and W. Nolting, *Z. Phys.* B **78**, 241 (1990).
141. O. Gunnarsson, P. Gies, W. Hanke, and O. K. Andersen, *Phys. Rev.* B **40**, 12140 (1989); P. Gies, private communication (1988).
142. O. Jepsen, J. Madsen, and O. K. Andersen, *Phys. Rev.* B **26**, 2790 (1982).
143. Corresponding nonlinear spectra do not yet exist.
144. L. Hodges, H. Ehrenreich, and N. D. Lang, *Phys. Rev.* **152**, 505 (1966).
145. F. M. Mueller, *Phys. Rev.* **153**, 659 (1967).
146. H. Ehrenreich and L. Hodges, *Meth. Comput. Phys.* **8**, 149 (1968).
147. F. Weling and J. Callaway, *Phys. Rev.* B **26**, 710 (1982).
148. G. C. Fletcher and E. P. Wohlfarth, *Phil. Mag.* **42**, 106 (1951).
149. G. C. Fletcher, *Proc. Phys. Soc. (London)* A **65**, 192 (1952).
150. J. C. Slater and G. F. Koster, *Phys. Rev.* **94**, 1498 (1954).
151. T. H. Moos, Dipolma Thesis, Freie Universität Berlin (1995); T. H. Moos, W. Hübner, and K. H. Bennemann, *Solid State Commun.* **98**, 639 (1996).
152. J. Callaway, *Quantum Theory of the Solid State* (Academic Press, New York, 1974).
153. W. A. Harrison, *Phys. Rev.* **118**, 1182 (1960).
154. W. Eberhardt and E. W. Plummer, *Phys. Rev.* B **21**, 3245 (1980).
155. D. C. Tsui, *Phys. Rev.* **164**, 669 (1967).
156. R. W. Stark and D. C. Tsui, *J. Appl. Phys.* **39**, 1056 (1968).
157. D. E. Eastman, F. J. Himpsel, and J. A. Knapp, *Phys. Rev. Lett.* **44**, 95 (1980).
158. S. K. Baker and P. V. Smith, *J. Phys.* F **7**, 781 (1977).
159. R. Haydock, *Solid State Phys.* **35**, 215 (1980).
160. W. A. Harrison, *Electronic Structure and Properties of Solids* (W. H. Freeman, San Francisco, 1980), p. 118 ff.
161. I. S. Gradshteyn and I. M. Ryzhik, *Table of Integrals, Series, and Products* (Academic Press, New York, 1965), p. 420.
162. D. E. Aspnes, *Phys. Rev.* B **6**, 4648 (1972).
163. D. J. Moss, E. Ghahramani, J. E. Sipe, and H. M. van Driel, *Phys. Rev.* B **41**, 1542 (1990).
164. R. Murphy, M. Yeganeh, K. J. Song and E. W. Plummer, *Phys. Rev. Lett.* **63**, 236 (1989).

165. J. L. Erskine and E. A. Stern, *Phys. Rev. Lett.* **30**, 1329 (1973); J. L. Erskine, *Physica* **89B**, 83 (1977).
166. T. Yoshino and S. Tanaka, *Opt. Commun.* **1**, 149 (1969).
167. G. S. Krinchik and V. A. Artem'ev, *Zh. Eksp. Teor. Fiz.* **53**, 1901 (1967) [*Sov. Phys. JETP* **26**, 1080 (1968)]; *J. Appl. Phys.* **39**, 1276 (1968).
168. N. V. Smith, R. Lässer, and S. Chiang, *Phys. Rev.* B **25**, 793 (1982).
169. H. Ebert, Habilitation Thesis, Universität München (1990).
170. W. Hübner (unpublished, 1988).
171. R. H. Victora and L. M. Falicov, *Phys. Rev. Lett.* **55**, 1140 (1985).
172. W. Hübner and L. M. Falicov, *Solid State Commun.* **85**, 385 (1993).
173. W. Hübner and L. M. Falicov, *Phys. Rev.* B **47**, 8783 (1993).
174. The negative sign of the experimental susceptibility for $\hbar\omega < 0.5$ eV is frequently assigned to interband transitions close to the L point of the Brillouin zone, which are impossible in the band structure by Weling and Callaway (e.g. B. R. Cooper, *Phys. Rev.* **139**, A 1504 (1965) or J. L. Erskine and E. A. Stern, *Phys. Rev. Lett.* **30**, 1329 (1973)). Halilov and Uspenskii link this behaviour to a resonance at the spin–orbit coupling energy [see S. V. Halilov and Yu. A. Uspenskii, *J. Phys.: Condens. Matter* **2**, 6137 (1990)]. In this low-energy range, however, also intraband transitions and magnons can contribute to the signal.
175. P. Guyot-Sionnest, A. Tadjeddine, and A. Liebsch, *Phys. Rev. Lett.* **64**, 1678 (1990).
176. A more careful consideration should include the ω dependence. Obviously, for determining $(\chi^{(2)}_{\text{mag.}}/\chi^{(2)}_{\text{nonmag.}})_{\text{Fe}}$ exactly, one must calculate $\chi^{(2)}_{zzz}$ for Fe and include the detailed geometry of the experiment as well as the ellipticities of (i) the nonlinear Kerr-effect, (ii) second harmonic generation, and (iii) the polarizer in the reflected frequency-doubled beam.
177. K. B. Hathaway, H. J. F. Jansen, and A. J. Freeman, *Phys. Rev.* B **31**, 7603 (1985).
178. D. Weller, S. F. Alvarado, W. Gudat, K. Schröder, and M. Campagna, *Phys. Rev. Lett.* **54**, 1555 (1985).
179. P. Grünberg, R. Schreiber, Y. Pang, M. B. Brodsky, and H. Sowers, *Phys. Rev. Lett.* **57**, 2442 (1986).
180. J. Unguris, R. J. Celotta, and D. T. Pierce, *Phys. Rev. Lett.* **64**, 140 (1991).
181. S. S. P. Parkin, N. More, and K. P. Roche, *Phys. Rev. Lett.* **64**, 2304 (1990).
182. M. N. Baibich, J. M. Broto, A. Fert, F. Nguyen Van Dau, F. Petroff, P. Etienne, G. Creuzet, A. Friederich, and J. Chazelas, *Phys. Rev. Lett.* **61**, 2472 (1988).
183. P. Zhan, I. Mertig, M. Richter, and H. Eschrig, *Phys. Rev. Lett.* **75**, 2996 (1995).
184. C. Liu, E. R. Moog, and S. D. Bader, *Phys. Rev. Lett.* **60**, 2422 (1988).
185. A. Carl and D. Weller, *Phys. Rev. Lett.* **74**, 190 (1995).
186. R. Allenspach and A. Bischof, *Phys. Rev. Lett.* **69**, 3385 (1995).
187. D. P. Pappas, K. P. Kamper, and H. Hopster, *Phys. Rev. Lett.* **64**, 3179 (1990).
188. B. Schulz and K. Baberschke, *Phys. Rev.* B **50**, 13467 (1994).
189. J. E. Ortega and F. J. Himpsel, *Phys. Rev. Lett.* **69**, 844 (1992).
190. F. J. Himpsel, *Phys. Rev.* B **44**, 5966 (1991).
191. C. Carbone, E. Vescovo, O. Rader, W. Gudat, W. Eberhardt, *Phys. Rev. Lett.* **71**, 2805 (1993).
192. S. H. Vosko, L. Wilk, and M. Nusair, *Can. J. Phys.* **58**, 1200 (1980).
193. D. M. Ceperley and B. J. Alder, *Phys. Rev. Lett.* **45**, 566 (1980).
194. O. K. Andersen, *Solid. State Commun.* **13**, 133 (1973); *Phys. Rev.* B **12**, 3060 (1975).
195. M. Methfessel, *Phys. Rev.* B **38**, 1537 (1988); M. Methfessel, C. O. Rodriguez, and O. K. Andersen, *Phys. Rev.* B **40**, 2009 (1989); M. Methfessel and M. Scheffler,

Physica B **172**, 175 (1991).
196. Y. Suzuki, T. Katayama, S. Yoshida, and K. Tanaka, *Phys. Rev. Lett.* **68**, 3355 (1992).
197. W. Geerts, Y. Suzuki, T. Katayama, K. Tanaka, K. Ando, and S. Yoschida, *Phys. Rev.* B **50**, 12581 (1994).
198. T. Kraft, P. M. Marcus, and M. Scheffler, *Phys. Rev.* B **49**, 11511 (1994).
199. U. Pustogowa, W. Hübner, K. H. Bennemann, *J. Magn. Magn. Mater.* **148**, 269 (1995).
200. The cut-off factor C also guarantees the breakdown of inversion symmetry in odd-layer slabs which in a real film results from the substrate.
201. J. Thomassen, F. May, B. Feldmann, M. Wuttig, and H. Ibach, *Phys. Rev. Lett.* **69**, 3831 (1992).
202. Dongqi, Li, M. Freitig, J. Pearson, Z. Q. Qui, and S. D. Bader, *Phys. Rev. Lett.* **72**, 3112 (1994).
203. T. Rasing, *Appl. Phys.* A **59**, 531 (1994).
204. H. A. Wierenga, W. de Jong, M. W. J. Prins, Th. Rasing, R. Vollmer, A. Kirilyuk, H. Schwabe, and J. Kirschner, *Phys. Rev. Lett.* **74**, 1462 (1995).
205. This can also be seen from first-principles calculations treating properly the matrix elements in $\chi^{(2)}$ which sense the breakdown of inversion symmetry required for an SH signal. Such calculations are in progress; J. Dervitr *et al.*, to be published in *Phys. Rev. B*.
206. J. A. C. Bland, C. Daboo, B. Heinrich, Z. Celinski, and R. D. Bateson, *Phys. Rev.* B **51**, 258 (1995).
207. K. Garrison, Y. Chang, and P. D. Johnson, *Phys. Rev. Lett.* **71**, 2801 (1993).
208. D. M. Edwards, J. Mathon, R. B. Muniz, and M. S. Phan, *Phys. Rev. Lett.* **67**, 493 (1991).
209. M. v. Schilgaarde and W. A. Harrison, *Phys. Rev. Lett.* **71**, 3870 (1993).
210. U. Pustogowa, T. A. Luce, W. Hübner, and K. H. Bennemann, *J. Appl. Phys.* **79**, 6177 (1996).
211. T. A. Luce, W. Hübner, and K. H. Bennemann, unpublished.
212. M. Straub, R. Vollmer, and J. Kirschner, private communication.
213. A. Kirilyuk and Th. Rasing, private communication.
214. N. V. Smith, *Phys. Rev.* B **19**, 5019 (1978).
215. P. Bruno and C. Chappert, *Phys. Rev. Lett.* **67**, 1602 (1991).
216. It remains to be shown whether contributions from other k points, which in principle could give rise to additional oscillations, play a significant role.
217. J. E. Ortega, F. J. Himpsel, G. J. Mankey, and R. F. Willis, *Phys. Rev.* B **47**, 1540 (1993).
218. J. Hohlfeld, D. Grosenick, U. Conrad, and E. Matthias, *Appl. Phys.* A **60**, 137 (1995).
219. P. Stampfli and K. H. Bennemann, *Phys. Rev.* B **42**, 7163 (1990).
220. P. Stampfli and K. H. Bennemann, *Phys. Rev.* B **46**, 10686 (1992).
221. T. A. Luce, W. Hübner, and K. H. Bennemann, *Z. Phys.* B **102**, 223 (1997).
222. J. Hohlfeld, U. Conrad, and E. Matthias, *Appl. Phys.* B **63**, 541 (1996).
223. H. Krakauer, M. Posternak, and A. J. Freeman, *Phys. Rev.* B **19**, 1706 (1979).
224. V. Drchal, J. Kudrnovsky, and P. Weinberger, *Phys. Rev.* B **50**, 7903 (1994).
225. P. B. Johnson and R. W. Christy, *Phys. Rev.* B **6**, 4370 (1972).
226. τ_{pl} cannot be calculated within RPA, contrary to ω_{pl}.
227. J. Hohlfeld and E. Matthias, private communication.

6
THEORY FOR NONLINEAR OPTICS AT INTERFACES AND IN THIN FILMS OF METALS: SELECTED PROBLEMS

T. A. Luce and K. H. Bennemann
Freie Universität Berlin, Berlin, Germany

6.1 Introduction

We focus here on some important problems to supplement and extend the theory outlined by Hübner,[1] to demonstrate the potential of nonlinear optics regarding the analysis of the electronic and magnetic structure of interfaces and thin films, and to sketch some interesting problems to be solved. First, we summarize the most important general properties of second harmonic generation (SHG) from metals. Then, we discuss some problems reflecting the interplay of effects due to the optical transition matrix elements and the electronic energy spectrum available for these transitions. A problem that we use as an example is the SHG oscillation due to quantum-well states (QWS) in thin films. We present a general theory[2] describing the various experimental results reported in Chapters 2 and 3 by Vollmer and Rasing, respectively. This demonstrates then the richness of information contained in SHG. Another important problem is the analysis of magnetic structure like ferromagnetic order or antiferromagnetic order, of magnetic domain structure and of magnetic anisotropy.[3] We show that the polarization dependence of SHG can be used rather successfully for determining magnetic structures. Finally we discuss how to use SHG and pump-and-probe spectroscopy to study nonequilibrium states and the resultant dynamics in excited metals.

6.2 General properties of SHG

As well known SHG results only if inversion symmetry is absent. This follows immediately from the expansion

$$\boldsymbol{P} = \chi^{(1)}\boldsymbol{E} + \chi^{(2)}\boldsymbol{E}\boldsymbol{E} + \cdots, \qquad (6.2.1)$$

where $\boldsymbol{P}$ is the polarization caused by the electric field $\boldsymbol{E}$ and $\chi^{(1)}$ and $\chi^{(2)}$ are the linear and nonlinear susceptibilities. Hence, in metals that are inversion-symmetric in bulk, $\chi^{(2)}$ results only from surfaces and interfaces. On an atomic scale one expects that only atomic layers and their electrons close to the interface contribute to SHG. In magnetic metals the susceptibilities $\chi^{(1)}$ and $\chi^{(2)}$ depend on the magnetization $\mathbf{M}$. In particular, $\chi^{(2)}(\mathbf{M})$ can be split into[4]

$$\chi^{(2)}(\mathbf{M}) = \chi^{o}(\mathbf{M}) + \chi^{e}(\mathbf{M}), \qquad (6.2.2)$$

where $\chi^{e}(\mathbf{M})$ is even in $\mathbf{M}$ and $\chi^{o}(\mathbf{M})$ is odd in $\mathbf{M}$, $\chi^{o}(\mathbf{M}) \simeq \chi' \mathbf{M} + \cdots$. Therefore, eqn (6.2.1) can be rewritten as ($\chi_{ijl} = \chi_{ijl}(\mathbf{M}=0)$)

$$\begin{aligned} P_i(2\omega) = {} & \chi_{ij} E_j + \chi_{ijl} E_j E_l \\ & + \chi_{ijlm} E_j E_l M_m + \chi_{ijlmn} E_j E_l M_m M_n \\ & + \chi'_{ijlkm} E_j E_l \nabla_k M_m + \chi'_{ijlkmn} E_j E_l M_k \nabla_m M_n + \cdots . \end{aligned} \qquad (6.2.3)$$

Such a Ginzburg–Landau type of expansion has been used by Rasing *et al.*[5] for the analysis of magnetic domain structures and can be derived from the Maxwell equations following Pershan.[6] Obviously, eqn (6.2.3) is particularly useful for analysing inhomogeneous magnetic structures, magnetic domains and antiferromagnetic-like structures.[3] The nonvanishing tensor elements in the expansion (6.2.3) are determined by symmetry. Using different light polarizations one may determine the various tensor elements. Note that the SHG intensity is given by

$$I(2\omega) = |\mathbf{E}(2\omega)|^2, \qquad (6.2.4)$$

where the radiated SH electric field $\mathbf{E}(2\omega)$ follows from $\mathbf{P}(2\omega)$, involving Fresnel coefficients; for example see ref. 7. In the case of cubic crystals one has for (001) surfaces or interfaces

$$E_j(2\omega; \phi, \varphi) = 2\mathrm{i}\,\frac{\omega}{c} |E_l(\omega) E_m(\omega)|$$

$$\times \begin{pmatrix} A_p F_c \cos\phi \\ A_s \sin\phi \\ A_p N^2 F_s \cos\phi \end{pmatrix} \chi_{jlm} \begin{pmatrix} f_c^2 t_p^2 \cos^2\varphi \\ t_s^2 \sin^2\varphi \\ f_s^2 t_p^2 \cos^2\varphi \\ 2 f_c^2 t_p t_s \sin\varphi \cos\varphi \\ 2 f_c f_s t_p^2 \cos^2\varphi \\ 2 f_c t_p t_s \cos\varphi \sin\varphi \end{pmatrix} \qquad (6.2.5)$$

with χ_{jlm} written as

$$\begin{pmatrix} \chi_{xxx} & \chi_{xyy} & \chi_{xzz} & | & \chi_{xyz} & \chi_{xzx} & \chi_{xxy} \\ \chi_{yxx} & \chi_{yyy} & \chi_{yzz} & | & \chi_{yyz} & \chi_{yzx} & \chi_{yxy} \\ \chi_{zxx} & \chi_{zyy} & \chi_{zzz} & | & \chi_{zyz} & \chi_{zzx} & \chi_{zxy} \end{pmatrix}.$$

Here, appropriate boundary conditions are used for the wave equation and the

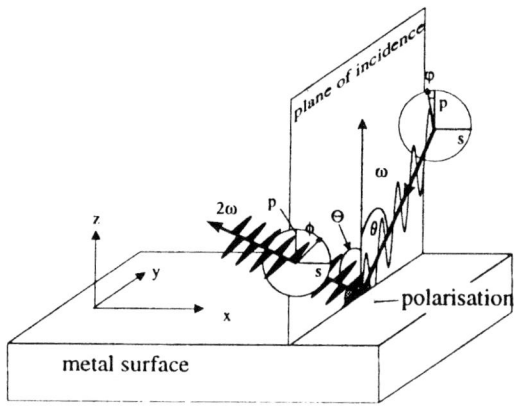

Fig. 6.1. Illustration of the optical configuration. φ and ϕ characterize the polarization of the incident and reflected SH light. θ is the angle of incidence (x, y) is the surface plane, (y, z) the optical plane.

electromagnetic fields, ϕ and φ are the angles of polarization for the SH and fundamental light, θ is the angle of incidence (see Fig. 6.1), and $n(\omega)$ and $N(2\omega)$ are the indices of refraction of the fundamental and SHG light, respectively. $F_{c,s}$ and $f_{c,s}$ are the corresponding Fresnel coefficients and $A_{p,s}$ the amplitudes of the incoming p,s-polarized light.

$$t_p = \frac{2\cos\theta}{n\cos\theta + f_c} \quad \text{and} \quad t_s = \frac{2\cos\theta}{\cos\theta + nf_c}$$

are the linear transmission coefficients. Formula (6.2.5) is important for determining the polarization dependence of SHG.

Following Shen et al.[8] one gets from a symmetry analysis for the tensor elements $\chi_{ijm}(M)$ the results shown in Tables 6.1–6.3. Note that the tensor χ_{ijm} is in dipole approximation a tensor of rank 3 and transforms according to

$$\chi_{ijm} = \sum_{i',j',m'} T_{ii'} T_{jj'} T_{mm'} \chi_{i'j'm'}, \qquad (6.2.6)$$

where $T_{ii'}$ are matrices describing reflections and rotations. Furthermore, χ_{ijm}

Table 6.1 Characterization of nonvanishing even (e) and odd (o) tensor elements $\chi_{ijm} \equiv ijm$ with respect to M for a (001) surface of an fcc crystal. The symmetry plane is given by (xy), z is normal to the surface ($\chi_{ijm} = \chi_{imj}$)

M	χ_{ijm}^e	χ_{ijm}^o
$M \parallel [100]$	zxx, zyy, zzz, yyz, xxz	xyx, yxx, yyy, yzz, zyz
$M \parallel [010]$	zxx, zyy, zzz, yyz, xxz	yxy, xyy, xxx, xzz, zxz
$M \parallel [001]$	xzx, zxx, zzz	xyz = −yxz
$M = 0$	zxx = zyy, zzz, xxz = yyz	

Table 6.2 $\chi_{ijm}(M)$ for a (110) surface (x, y) of an fcc crystal, z is normal to the surface ($\chi_{ijm} = \chi_{imj}$)

M	χ^e_{ijm}	χ^o_{ijm}
$M \parallel [001]$	xzx, yzy, zxx, zyy, zzz	yxy, zxz, xxx, xyy, xzz
$M \parallel [110]$	xzx, yzy, zxx, zyy, zzz	xyz, yzx, zxy
$M = 0$	xzx, yzy, zxx, zyy, zzz	

must also be invariant with respect to the crystal point group. This yields immediately the nonvanishing tensor elements shown in the tables. Obviously, $\chi_{ijm} = \chi_{imj}$ and using a rectangular coordinate system x, y, z with z normal to the surface, tensor elements involving two z will vanish if $M = 0$.

In the case of $M \neq 0$ one understands immediately with the help of Lorentz force that for a (001) surface and $M \parallel x$, for example, $\chi_{xzx}(M)$ is even in M and $\chi_{xyx}(M)$ is odd in M, etc. Regarding the nonvanishing tensor elements, one expects that χ_{zzz} is large, since it feels most effectively the broken inversion symmetry in the z direction. However, note that the χ_{ijm} vary with frequency ω.

While the symmetry analysis is very powerful and elegant in determining the structure of χ_{ijm} and also of the other tensors of higher rank in the expansion (6.2.3), one needs an electronic theory for calculating the relative magnitude of $\chi_{ijm}(M)$, the Kerr spectra $\chi(2\omega, M)$, the Kerr rotation, its polarization dependence and the region at the interface where SHG occurs. Using response theory, one gets for the linear susceptibility

$$\chi^{(1)}(q, \omega) = \frac{-e^2}{q^2 \Omega}$$

$$\times \sum_{k,l,l'} \left\{ |\langle k, l'| e^{-iq \cdot r} |k+q, l\rangle|^2 \frac{f(E_{k+q,l'}) - f(E_{k,l})}{E_{k+q,l'} - E_{k,l} - \hbar\omega + i\hbar\alpha} \right\}$$
(6.2.7)

where $E_{k,l}$ is the band energy with index l and momentum k, q is the photon

Table 6.3 $\chi_{ijm}(M)$ for a (111) surface (x, y) of an fcc crystal, z is normal to the surface, x is in $[2\bar{1}\bar{1}]$ direction ($\chi_{ijm} = \chi_{imj}$)

M	χ^e_{ijm}	χ^o_{ijm}
$M \parallel [2\bar{1}\bar{1}]$	xxx, xyy, xzz, xzx, yzy, yxy, zxx, zyy, zzz, zxz	xyz, xxy, yxx, yyy, yzz, yzx, zyz, zxy
$M \parallel [111]$	xxx = −xyy = −yxy, xxz = yyz, zxx, zyy, zzz	xyz = −yxz, xxy = yxx = −yyy
$M = 0$	xxx = −xyy = −yxy, xxz = yyz, zxx = zyy, zzz	

momentum, $f(E)$ is the Fermi function, $\hbar\omega$ the incident photon energy and Ω is the polarized volume. For the nonlinear susceptibility one gets

$$\chi_{ijm}(2\boldsymbol{q},2\omega) = \frac{-ie^3}{2q^3\Omega} \sum_{k,l,l',l''} M^i_{k,k+2q} M^j_{k+2q,k+q} M^m_{k+q,k}$$

$$\times \left\{ \frac{\dfrac{f(E_{k+2q,l''}) - f(E_{k+q,l'})}{E_{k+2q,l''} - E_{k+q,l'} - \hbar\omega + i\hbar\alpha} - \dfrac{f(E_{k+q,l'}) - f(E_{k,l})}{E_{k+q,l'} - E_{k,l} - \hbar\omega + i\hbar\alpha}}{E_{k+2q,l''} - E_{k,l} - 2\hbar\omega + i2\hbar\alpha} \right\}, \quad (6.2.8)$$

with dipole matrix elements $M^i_{k',k} = \langle \Psi_{k,l'} | r_i | \Psi_{k,l} \rangle$, $r_i = x, y, z$. Comparison of eqns (6.2.7) and (6.2.8) shows clearly that χ_{ijm} contains more information, since it involves three matrix elements according to the optical transitions $\boldsymbol{k} \to \boldsymbol{k} + \boldsymbol{q}$, $\boldsymbol{k} + \boldsymbol{q} \to \boldsymbol{k} + 2\boldsymbol{q}$, and $\boldsymbol{k} + 2\boldsymbol{q} \to \boldsymbol{k}$.

Of course, $\chi_{ijm} \neq 0$ only if inversion is absent, must also follow from the electronic theory for χ_{ijm}. To see this one may rewrite eqn (6.2.8) as

$$\chi_{ijm} \propto \sum_{k,l,l',l''} F_{k,l,l',l''} \sum_{k'} \langle k',l'' | r_i | k',l' \rangle \langle k',l' | r_j | k',l \rangle \langle k',l | r_m | k',l'' \rangle,$$
(6.2.9)

where $F_{k,l,l',l''}$ contains the dependence of the energy levels and $\boldsymbol{k}'$ are generated from $\boldsymbol{k}$ by possible symmetry operations within the Brillouin zone. $\boldsymbol{k}$ runs only over values within the irreducible part of the Brillouin zone. The $\boldsymbol{k}'$-sum can be rewritten

$$\chi_{ijm} \propto \sum_{k,l,l',l''} F_{k,l,l',l''} \sum_{R \in G} \langle k,l'' | R^{-1}(r_i) | k,l' \rangle \langle k,l' | R^{-1}(r_j) | k,l \rangle$$
$$\times \langle k,l | R^{-1}(r_m) | k,l'' \rangle, \quad (6.2.10)$$

where $\boldsymbol{R}$ is an element of the symmetry group G of the crystal and the states are denoted in Dirac form $\Psi_{k,l}(r) = |k,l\rangle$. The resulting performance of the sums leads to

$$D(\boldsymbol{R}) \chi_{ijm} = \chi_{ijm}, \quad (6.2.11)$$

where D is a representation of G acting on χ_{ijm}, since $\boldsymbol{R}$ is a symmetry of the crystal. In the case of inversion symmetry $\boldsymbol{I} \in G$ one gets $I(r_i) = -r_i$ and

$$D(\boldsymbol{I}) \chi_{ijm} = -\chi_{ijm} \quad (6.2.12)$$

resulting in the vanishing of χ_{ijm}. For a system lacking inversion symmetry, $\boldsymbol{I} \notin G$ and this conclusion is not valid.

If there are other symmetry elements remaining, they impose additional restrictions on the tensor elements. Investigating these remaining symmetries, one may also get the vanishing tensor elements from the electronic theory. Regarding an fcc (001) surface, the symmetry elements are the identity, 90°

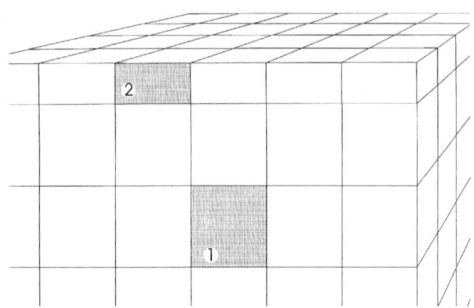

Fig. 6.2. Schematic picture of a surface, divided into Wigner–Seitz cells. Whereas inner cells (1) have an inversion-symmetric environment, for the surface cells (2) the inversion symmetry is broken. Thus, the symmetry group of such a cell (2) is smaller than for cells of type (1).

rotation around the z axis 4_z, 180° rotation around the z axis 2_z and reflections m_y and m_x at the xz and the yz plane. Since $m_y(x) \rightarrow x$, $m_y(y) \rightarrow -y$, and $m_y(z) \rightarrow z$, it follows from eqn (6.2.10) that the tensor elements with an odd number of y fulfil $D(m_y)\chi_{ijm} = -\chi_{ijm}$. Since eqn (6.2.11) is valid for all tensor elements, tensor elements with an odd number of y are zero. A similar conclusion holds for the tensor elements with an odd number of x from inspection of the m_x symmetry. The examination of the rotations gives no additional restrictions. Thus, all but the seven tensor elements

$$\chi_{zxx}, \quad \chi_{xzx}, \quad \chi_{zyy}, \quad \chi_{yzy}, \quad \chi_{xxz}, \quad \chi_{yyz} \quad \text{and} \quad \chi_{zzz}$$

have to vanish. This result, already known from the macroscopic symmetry analysis, demonstrates again the richness of information contained in the nonlinear susceptibility expression eqn (6.2.8).

Of course, it follows also from eqn (6.2.8) that in terms of the atomic structure at surfaces or interfaces only one or two atomic layers next the the interface, where inversion symmetry is broken, contribute to SHG.[9] Considering a system as sketched in Fig. 6.2 that is translation-invariant only in the x and y directions, eqn (6.2.8) has the form

$$\chi(2\omega) \propto \sum_{\sigma} \sum_{k_\parallel, l, l', l''} F_{k_\parallel, l, l', l'', \sigma}$$

$$\times \langle k_\parallel, l'', \sigma | r | k_\parallel, l, \sigma \rangle \langle k_\parallel, l, \sigma | r | k_\parallel, l', \sigma \rangle \langle k_\parallel, l', \sigma | r | k_\parallel, l'', \sigma \rangle, \quad (6.2.13)$$

where the k summation is performed only along k_x and k_y and the matrix element $\langle k_\parallel, n, \sigma | r | k_\parallel, n', \sigma \rangle$ is expressed as

$$\langle k_\parallel, l\sigma | r | k_\parallel, l'\sigma \rangle = \int_{-\infty}^{\infty} dz \int_{WS} dx \int_{WS} dy \, \Psi^*_{l, k_\parallel}(r) r \Psi_{l', k_\parallel}(r). \quad (6.2.14)$$

By rewriting in eqn (6.2.14) the z integration as integration over Wigner–Seitz cells (WS), one gets contributions to the matrix elements from complete Wigner–Seitz cells and from incomplete ones. From expanding the wavefunctions in spherical harmonics

$$\Psi_{l,k_\parallel}(\boldsymbol{r}) = \sum_m C_{k_\parallel} R(r,l) Y_{lm}\left(\frac{\boldsymbol{r}}{r}\right),$$

it is obviously that Wigner–Seitz cells with spherical symmetry (and thus inversion symmetry) yield for the matrix elements the selection rules $\Delta l = \pm 1$ and hence no contribution to χ_{ijm}, since in the product of the three matrix elements in eqn (6.2.8) at least one matrix element must have an even Δl. For an integration over only half a Wigner–Seitz cell, the selection rules are weakened and additional transitions are possible if Δl is an even integer. Thus, a contribution to χ_{ijm} occurs.

Different nonvanishing tensor elements depend sensitively on the photon frequency. This becomes obvious from eqn (6.2.8), since the interplay between symmetry via the matrix elements and the energy spectrum via the energy denominators in eqn (6.2.8) weights differently transition combinations for different photon energies. Consequently, the tensor elements vary with photon energy and no tensor element dominates the others for all frequencies. This shows again the importance of the matrix elements, which cause the symmetry sensitivity in eqn (6.2.8), for calculating χ_{ijm}.

The remaining symmetry felt by the surface cells can be related to the dependence of χ_{ijm} on (001), (111) surfaces, etc., and to differences observed for bcc and fcc crystals, for example. This is also of relevance regarding the dependence of SHG on surface reconstruction and on magnetorestriction. Furthermore, this is of interest regarding the dependence of SHG on the roughness of the surface. Owing to the surface roughness, surface steps, etc., breakdown of inversion symmetry is felt to a different degree by the various atoms at the surface.

The electronic expression for χ_{ijm} (eqn. (6.2.8)), reflects clearly the interplay of the effect due to the matrix elements and the energy denominators. The matrix elements may give particular strong weight to some optical transitions and then SHG is even **k**-selective. The phase factors due to the matrix elements may cause interference effects in thin films, for example, when SHG contributions due to the surface and interface film/substrate superimpose. The matrix elements are also different for s, p, d states and thus for simple and transition metals, etc. The energy denominators describe the dependence of χ_{ijm} on the electronic band structure, on the E_k dispersions and on quantum-well states. Obviously, such states (QWS) in the energy spectrum yield extra optical transitions.

In magnetic systems the optical transitions are spin-dependent. It is important to note, that light couples to the magnetization only via the spin–orbit coupling λ_{so}^o. Hence, to lowest order in λ_{so}, one gets for χ_{ijm}^o the expression

$$\chi_{ijm}^o(2\omega, 2q, M)$$

$$= -\frac{e^3}{\Omega} \sum_{\sigma,k,l,l',l''} P\Bigg[\Bigg(\langle k,l,\sigma|r_k|k+2q,l'\sigma\rangle$$

$$+ \sum_{l''' \neq l'} \frac{\langle k,l''',\sigma|\lambda_{so} L \cdot S|k+2q,l',\sigma\rangle}{E_{k+2q,l',\sigma} - E_{k,l''',\sigma}} \langle k,l,\sigma|r_i|k,l''',\sigma\rangle\Bigg)$$

$$\times \langle k+2q,l',\sigma|r_i|k+q,l'',\sigma\rangle\langle k+q,l'',\sigma|r_m|k,l,\sigma\rangle\Bigg]$$

$$\times \left\{ \frac{\dfrac{f(E_{k+2q,l'',\sigma}) - f(E_{k+q,l',\sigma})}{E_{k+2q,l'',\sigma} - E_{k+q,l',\sigma} - \hbar\omega + i\hbar\alpha} - \dfrac{f(E_{k+q,l',\sigma}) - f(E_{k,l,\sigma})}{E_{k+q,l',\sigma} - E_{k,l,\sigma} - \hbar\omega + i\hbar\alpha}}{E_{k+2q,l'',\sigma} - E_{k,l,\sigma} - 2\hbar\omega + i2\hbar\alpha} \right\}.$$

(6.2.15)

Note, upon performing the summation the term in eqn (6.2.15) due to $\langle k,l,\sigma|r_i|k+2q,l',\sigma\rangle$ vanishes. $P[\cdots]$ represents the sum of six permutations of the product of three matrix elements with one wavefunction disturbed. As is obvious, also from eqn (6.2.3), the nonlinear susceptibility depends on the direction of M and hence reflects magnetic anisotropy and easy axis e ($\chi_{ijl}^o \propto e \cdot M$).

The tensor elements $\chi_{ijl}(2\omega)$ determine essentially the nonlinear optics—see Chapters 3 and 5 by Rasing and Hübner in particular. The Kerr rotation and its polarization dependence follows from (ε = ellipticity)

$$\tan(\phi_K + i\varepsilon) = i\frac{E^+(2\omega) - E^-(2\omega)}{E^+(2\omega) + E^-(2\omega)}. \quad (6.2.16)$$

Here, $E^{+,-}$ refers to right- and left-handed circularly polarized fields, respectively. If the Kerr angle ϕ_K is not too large one has

$$\phi_K \simeq \text{Re}\,\frac{E_\varphi(\text{s-SH})}{E_\varphi(\text{p-SH})}, \quad (6.2.17)$$

where φ is the polarization angle of the incident light and $E_\varphi(\text{s-SH})$ is the reflected SH field polarization perpendicular (s) to the optical plane. Note that in Fig. 6.1 (yz) is the optical plane; p refers to a polarization in the optical plane.

In the case of the longitudinal Kerr configuration ($M \parallel x, (x,z)$ optical plane) one gets for example

$$\phi_{K,p} \simeq \text{Re} \frac{A_s}{A_p} \frac{f_c^2 \chi_{yxx} + f_s^2 \chi_{yzz}}{2 f_c f_s F_c \chi_{xzx} + N^2 F_s (f_c^2 \chi_{zxx} + f_s^2 \chi_{zzz})} \quad (6.2.18)$$

for the Kerr rotation of p-polarized incoming light, and

$$\phi_{K,p} \simeq \text{Re} \frac{A_s}{A_p} \frac{\chi_{yyy}}{N^2 F_s \chi_{zyy}} \quad (6.2.19)$$

for the Kerr rotation of s-polarized incoming light. Thus, note that with the help of the electronic theory, one may express ϕ_K in terms of χ_{ijl}. Similarly, the linear Kerr rotation $\phi_K^{(1)}$ can be expressed in terms of χ_{ij}.[10]

To illustrate that the *ab initio* calculation of the matrix elements yields the relative magnitude of the tensor elements and thus results in a strong dependence of the SHG yield on the polarization, we present results for the SHG calculation of a Cu(001) surface; see Fig. 6.3.[11] The difference between $p_{in} \to P_{out}$ and $s_{in} \to P_{out}$ results from different contributing tensor elements (χ_{zzz}, χ_{zxx}, and χ_{xzx} for $p_{in} \to P_{out}$, χ_{zyy} for $s_{in} \to P_{out}$, with (yz) optical plane).

6.3 General theory for SHG oscillations due to quantum-well states in thin films

The magnetism of metallic surfaces, thin films and multilayer sandwiches has attracted much attention recently due to the discovery of new effects that also have a large potential for applications. The observation of antiferromagnetic coupling of magnetic films separated by a nonmagnetic spacer layer and the subsequent discovery that this coupling could oscillate between ferromagnetic and antiferromagnetic has stimulated intensive efforts to understand these phenomena. It was shown that in ultrathin films electronic potential discontinuities at interfaces lead to reflection of the electronic wavefunctions and consequently to confinement of the electronic states. The resulting quantization of the discrete components of the wavevector $k_\perp$ normal to the film surface gives rise to quantum-well states (QWS) and to resonances in the density of states. Note that these quantum-well states may also act as a mediator for magnetic interlayer coupling.

Photoemission experiments directly demonstrated oscillations of the electron density of states and proved the existence of spin-polarized quantum-well states.[12] However, owing to the short mean free path of electrons, this method is difficult to apply for the investigation of buried interfaces and multilayers. This disadvantage is bypassed by optical techniques, since interfaces of thin metallic films are accessible by light. The linear magneto-optical Kerr effect (MOKE), which is frequently used for studying magnetic properties, is sensitive to changes

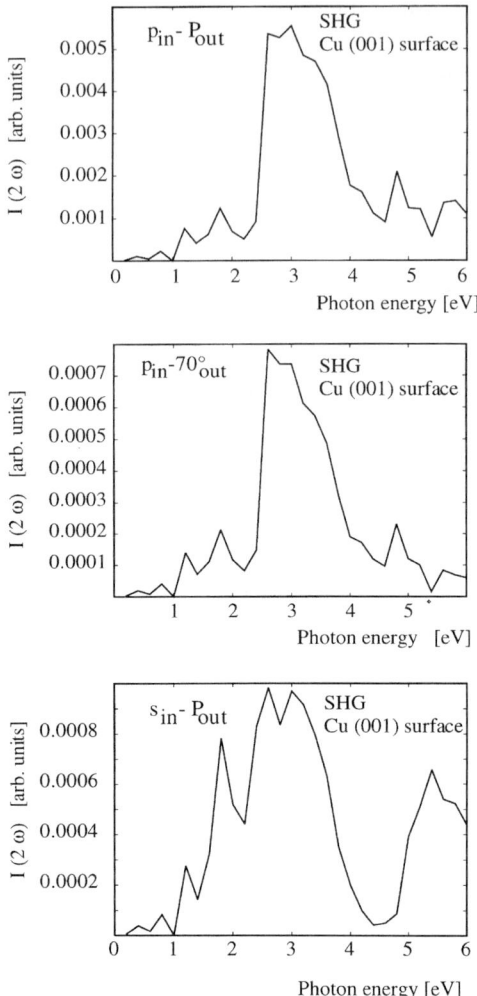

Fig. 6.3. SHG intensity for a Cu (001) surface for three different polarizations ($p_{in} \to P_{out}$, $s_{in} \to P_{out}$ and $p_{in} \to 70°_{out}$) as a function of frequency of the incoming light. The difference between $p_{in} \to P_{out}$ and $s_{in} \to P_{out}$ results from different contributing tensor elements.

of the linear susceptibility $\chi^{(1)}$ as a function of the applied magnetic field and is not interface-specific. Hence, the effects due to QWS in the linear signal are relatively small. In contrast, the nonlinear optical technique of second harmonic generation (SHG) combines a large penetration depth with a strong interface sensitivity, which is derived from the breaking of inversion symmetry at interfaces and surfaces. Therefore, one expects that QWS can be sensitively studied using SHG, since substrate contributions do not disturb the signal.

Recent experiments on xCu/Fe/Cu(001), xCu/Co/Cu(001) and xAu/Co(0001)/Au(111) films undoubtedly proved the huge effects in SHG due to

QWS.[13] Since the occurrence of QWS is thickness-dependent, resonant optical transitions involving QWS cause characteristic oscillations of the SHG as a function of the film thickness. However, the observed oscillations for different film systems exhibited quite different features concerning the oscillation period, the magnitude of the magnetic SHG signal and the dependence of the period on the incident laser frequency.

We will extend here our theory of the nonlinear optical response to show how the SHG from ultrathin films depends on its electronic and magnetic structure and that of its substrate. As a result, we can explain within the same theoretical approach the quite different experimental SHG observations for different thin films with quantum-well states. The theory shows the rich information contained in the various characteristic features of the SHG oscillations. In particular, it demonstrates the significance of the interplay between dipole-matrix-element effects, band-structure effects and the role of the optical interference of SHG contributions from the two film interfaces. If the interference of SHG from the surface and the interface is important, then the SHG response appears to be sensitive to the parity of the QWS, with the result that only a doubled period $2\Lambda_M$ appears. Here, Λ_M is the period as observed in the linear Kerr effect.[14] If this interference of the SHG contributions is not so important or if the SHG signal is dominated by only one of the interfaces, then the SHG oscillations will typically reveal two periods, namely the period Λ_M and a larger period. See Fig. 6.4 for illustration.

In the dipole approximation the nonlinear optical response yielding the SHG

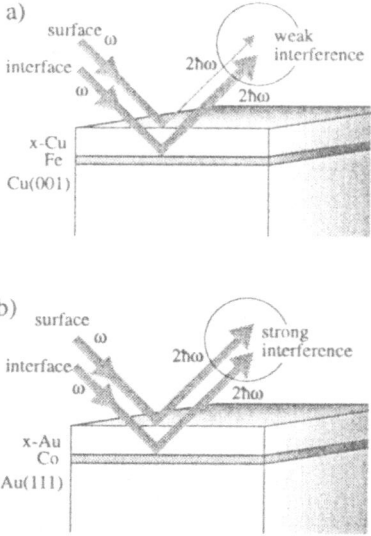

Fig. 6.4. Illustration of the two contributions from the surface and interface to SHG from thin films with QWS. (a) Weak interference, since either the interface (as indicated) or the surface contribution dominates. (b) Both contributions are nearly equal, resulting in strong interference effects.

intensity $I(2\omega) \propto |P(2\omega)|^2$ is given by the polarization $P_i(2\omega) = \chi_{ijm}(2\omega)E_j E_m$. Assuming for simplicity that a single tensor element χ_{ijl} dominates the SHG, one may use $\chi = \chi^e + \chi^o$ and $\boldsymbol{P}(2\omega) = (\chi^e + \chi^o)E^2(\omega)$, where the average effect of the Fresnel factors has been incorporated in an effective χ^e and χ^o, respectively. Approximately, one has $\chi^o_{ijl} = \chi_{ijlm}M_m + \cdots$ and $\chi^e_{ijl} = \chi^{(0)}_{ijl} + \chi_{ijlmn}M_m M_n + \cdots$. In comparison with linear MOKE, where the magnetization induces only very small off-diagonal tensor elements and is suppressed by nonmagnetic excitations, in SHG χ^o is of the same order of magnitude as the nonmagnetic tensor element χ^e, thus causing strong magnetic effects in SHG.

Owing to the periodic appearance of QWS at certain energies with increasing film thickness d, transitions involving these QWS are resonantly enhanced and thus result in oscillations in SHG (and in MOKE). If the states mainly contributing to the optical transitions are spin-polarized, the oscillations in the SHG response $I(2\omega)$ cause corresponding oscillations in the magnetic contrast defined as

$$\Delta I = \frac{I(2\omega, \boldsymbol{M}) - I(2\omega, -\boldsymbol{M})}{I(2\omega, \boldsymbol{M}) + I(2\omega, -\boldsymbol{M})}. \tag{6.3.1}$$

Here, of particular interest are contributions to ΔI due to spin-polarized QWS. Obviously, the oscillation in ΔI should be particularly strong if the electronic confinement, which leads to QWS, acts very differently for electronic spins σ and $-\sigma$. Owing to the spin dependence of the resonant transitions, the oscillation of $I(\boldsymbol{M})$ and of $I(-\boldsymbol{M})$ might be shifted somewhat with respect to each other. Then, of course, the peaks in $I(\boldsymbol{M})$ and $I(-\boldsymbol{M})$ should occur at different film thicknesses.

Also generally the behaviour of ΔI might be rather different if spin polarization is only induced by the substrate or if the film itself is magnetic. For example, ΔI could weight rather differently the surface and the interface as compared to $I(2\omega)$.

If we split the susceptibility $\chi(2\omega)$ into a surface and an interface contribution, $\chi^s(2\omega)$ and $\chi^i(2\omega)$ respectively, then

$$\chi_{ijm}(2\omega) = \chi^s_{ijm}(2\omega) + \chi^i_{ijm}(2\omega). \tag{6.3.2}$$

Hence one gets, depending on the ratio of the interface vs. the surface contribution, that the SHG oscillations resulting from χ_{ijm} are dominated by the symmetry of the dipole matrix elements and thus the optical interference between the two contributions χ^s and χ^i. If the SH intensity has equal contributions from the surface and the interface, then according to eqn (6.3.2) one gets no SHG signal for $\chi^s\chi^i = -(\chi^s)^2$. If the SH intensity is essentially determined by either one of the two contributions to χ_{ijm}, then the relative phase of $\chi^s_{ijm}(2\omega)$ and $\chi^i_{ijm}(2\omega)$ and their interference is not important,[15] see Fig. 6.4 (a). Strong interference effects result if the situation shown in Fig. 6.4 (b) occurs.

The SHG intensity and period of the SHG oscillations are determined by the

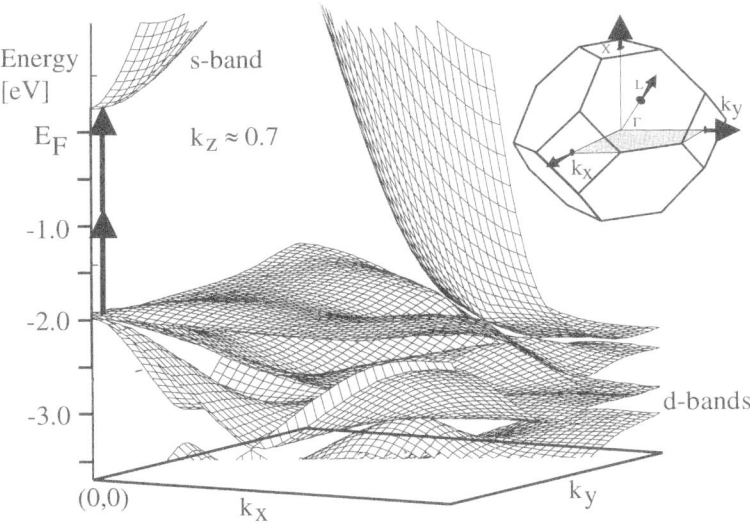

Fig. 6.5. The energy bands for Cu for a plane parallel to the indicated one in the fcc Brillouin zone. The normal of the plane is the Γ-X direction, corresponding to a (001) surface. The energy bands for the (001) surface at $k_\perp \approx 0.7$ (in units of the Brillouin zone length) are drawn. The bands at $(k_x, k_y) = (0,0)$ dominate the optical spectrum due to their flatness there. The dominant SHG transitions are indicated by the arrow.

region of k space contributing to the SHG signal, as is obvious from the expression for χ_{ijl}. For electronic structures whose optical spectra are dominated by a small region in k space, SHG is highly k selective. The structure of the SHG oscillation as a function of the film thickness d is spiky and pronounced, since a few resonant transitions strongly determine the SHG due to the high density of states at these k points. Figure 6.5 illustrates this situation for the case of the (001) surface of Cu, where contributions of the k points in the neighbourhood of the Γ point mostly dominate the SHG. A possible nonlinear transition is indicated by the arrow. On the other hand, if a larger k-space volume is contributing equally to the SHG signal, the SHG spectrum of the film system has a less sharp structure. An example of this case is given in Fig. 6.6 where the electronic structure for the (111) surface of Cu is sketched. Here, a larger region of k points contributes uniformly to the optical response.

To simplify the discussion, we treat separately film systems where interference effects due to χ^s and χ^i are negligible and those where interference effects dominate. In the former case we analyse again separately the situation when the QWS are dominantly involved as final states, as intermediate states and finally when both final and intermediate states together cause SHG oscillations.[16]

First, we analyse the situation where no destructive interference of the surface and interface contributions to SHG occurs or where dominantly χ^s or χ^i contribute to SHG. Then, one gets two oscillation periods Λ_1 and Λ_2 in SHG due to QWS. If QWS above the Fermi energy are mainly available for

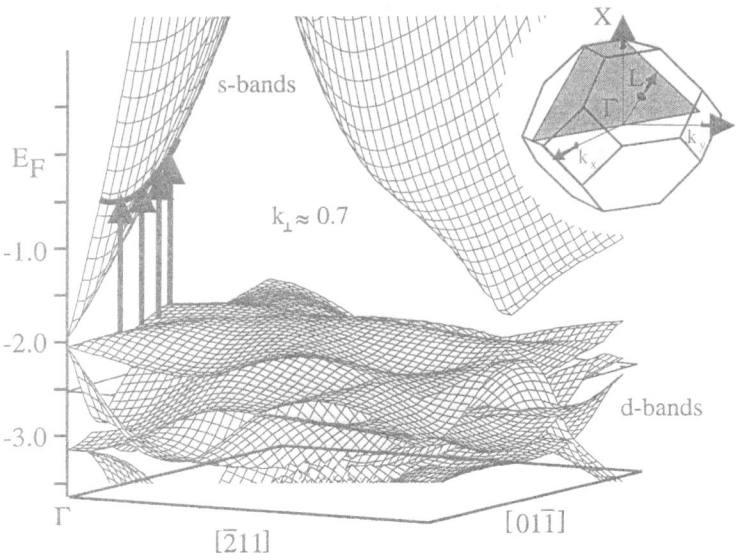

Fig. 6.6. The calculated energy bands for Cu for a plane parallel to the indicated one in the fcc Brillouin zone. The plane corresponds to a (111) surface of Cu. Its normal is the Γ-L direction. The energy bands for the (111) surface at $k_\perp \approx 0.7$ (in units of the Brillouin zone length) are sketched. Apparently, no region with comparable weight as in Fig. 6.5 is present and and consequently k selectivity is less pronounced. Note that various k points contribute equally to the optical response. Some possible optical transitions are indicated by arrows.

optical transitions, then a first peak in SHG appears for a film thickness d_1 at which a QWS has moved across E_F and becomes available as final state. Then, again a SHG peak results for a film thickness $2d_1$ at which the previous situation is repeated and so on—see Fig. 6.7 for illustration. The resulting oscillation period Λ_1 is the same as that calculated for MOKE (Λ_M) at $\hbar\omega = 1.6\,\text{eV}$ and must be longer than or equal to the period observed in photoemission, when only QWS at E_F are investigated. Note that in the MOKE and SHG periods the QWS above the Fermi level may play a role if corresponding optical transitions occur.[17] The result $\Lambda_1 = \Lambda_M$ can be deduced from the electronic structure; see Fig. 6.7 and from using

$$\Lambda_M = \frac{k_{BZ}}{k_{BZ} - k_1}, \tag{6.3.3}$$

where k_1 is the wavevector at which the most important resonances in MOKE occur.

If now for increasing film thickness $d > d_1$ not only the first QWS above E_F, but also the next higher one become available for an optical transition (see Fig. 6.7), then a second peak in the SHG response occurs and consequently a second period appears. This second period Λ_2 depends of course also on the band

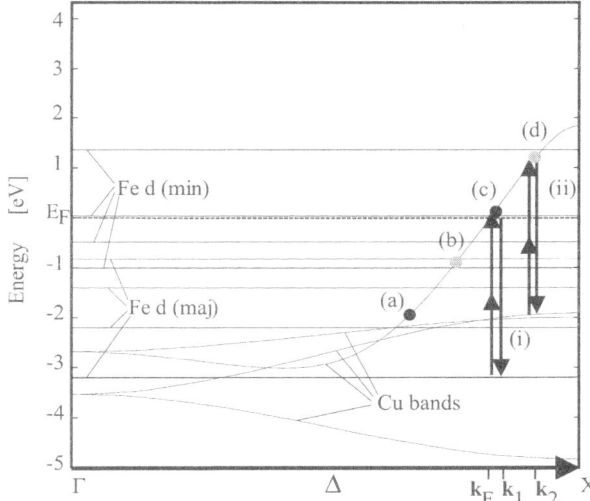

Fig. 6.7. Illustration of the generation of SHG oscillations due to QWS is interference of SHG from surface and interface is negligible. The electronic structure refers to the case of an x Cu/Fe(001) system. Since the Γ-X direction dominates the SHG generation (see Fig. 6.5), only this section of the Brillouin zone is considered. Only the rightmost QWS of a 12 ML Cu film are drawn (dots). Note that for $x = 6$ ML the QWS (b) and (d) are not present. Possible dominant transitions involving QWS as final states are indicated. Whereas transition (i) is possible for films with 6 and 12 ML leading to an oscillation period of 6 ML, transition (ii) is possible only for 12 ML. k_1 and k_2 denote the $k_\perp$-values which determine the oscillation period Λ_1 and Λ_3, respectively.

structure and thus gives additional information about the available final and intermediate states. Note that Λ_2 is generally not related to Λ_M and may only by accident be equal to $2\Lambda_M$. Of course, the occurrence of additional periods depends on the band structure. For the photon energy of 1.6 eV there are no intermediate states for resonant nonlinear transitions, and thus no additional oscillation periods occur.

As is obvious from Fig. 6.7, if only a few dominant optical transitions occur, the k selectivity of the optical response is optimal and the SHG oscillations will be most pronounced. Of source, the strength of the absolute signal depends further on the joint density of states for optical transitions.

Furthermore, the SHG oscillation will change strongly with the incident laser frequency ω, since $\hbar\omega$ must fit the optical transitions. According to the expression for $\chi_{ijl}(2\omega)$ this frequency dependence is strongest if the matrix elements are less dominant than the energy denominators in the susceptibility χ. Note that for different incident laser frequencies, different k points dominate the nonlinear optical response. Then, it is of course necessary that confinement is present at these k points, to obtain oscillations in the SHG signal.

In the case of a ferromagnetic spacer layer between the substrate and the overlayer film, resonant transitions for minority and majority electrons are

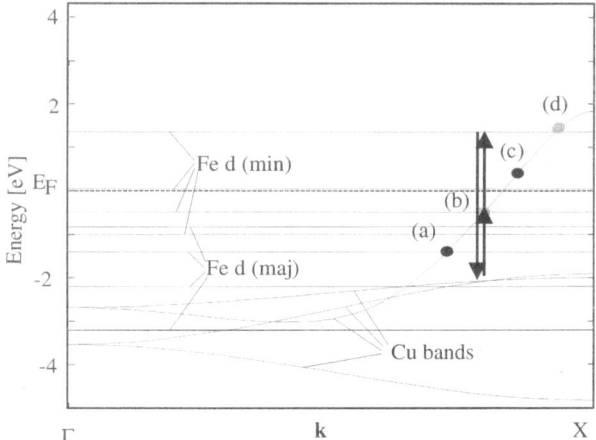

Fig. 6.8. Illustration of optical transitions involving occupied QWS as intermediate states and d states as final states for the same system as in Fig. 6.7. Possible dominant transitions are indicated. Resultant SHG oscillations due to occupied QWS have generally a different period in comparison with those illustrated in Fig. 6.7.

possible at different overlayer thicknesses due to the spin-split d-states. Thus, a phase shift for the $I(M)$- and $I(-M)$-signals occurs. Consequently one may observe an enhanced magnetic contrast of the SHG response.

Figure 6.7 describes the situation when matrix elements and optical interference effects do not dominate. Obviously, one expects a sensitive dependence of the QWS oscillations in SHG on the frequency of the incoming light. For the situation sketched in Fig. 6.7, where SHG oscillations are mainly due to QWS above E_F, one expects that the periods Λ_1 and Λ_2 increase as $\hbar\omega$ increases, since for increasing frequency the resonant transition indicated by (ii) occurs for a QWS appearing at a higher energy and thus at a larger $k_\perp$. This corresponds to an increasing period for increasing $\hbar\omega$. Obviously, if $(k_{BZ} + k_1)/2 \simeq k_2$ one gets

$$2\Lambda_1 \simeq \Lambda_2. \tag{6.3.4}$$

If QWS are formed in an electronic band below E_F and contribute to SHG (see Fig. 6.8 for illustration), then again SHG oscillations result, because of the changing DOS in the initial or intermediate state, if the QWS is at resonance with a final state. In contrast to the previous case (Fig. 6.7) the oscillation period may increase for decreasing photon frequency.

If now QWS below and above E_F together cause oscillations, a period results from a superposition of their two periods. Depending on the weight of these two oscillations, it is possible that the resulting single period increases as the photon frequency decreases, for example. This may explain the behaviour observed for x Cu/Co/Cu(001).

Secondly, we analyse the case where optical interference effects play a dominant role. Owing to $I(2\omega) \propto |\chi_{ijm}(2\omega)|^2$ the relative phase of the two terms $\chi_{ijm}^s(2\omega)$ and $\chi_{ijm}^i(2\omega)$ is important and may cause a cancellation of these two contributions. This is the situation illustrated in Fig. 6.4 (b). The symmetry of the dipole matrix elements is now essential if interference of the two terms in eqn (6.3.2) occurs. Obviously, the interference is strongest if the two SHG contributions at the surface and the interface have nearly the same weight. Using for simplicity $|\chi_{ijm}^s(2\omega)| \approx |\chi_{ijm}^i(2\omega)|$, one gets

$$I(2\omega) \approx 2|\chi_{ijm}^s(2\omega)|^2 + 2\chi_{ijm}^s(2\omega)\chi_{ijm}^i(2\omega). \qquad (6.3.5)$$

For equally weighted contributions one obviously gets perfect cancellation of the thickness-dependent $\chi(2\omega)$, if the resultant phase of the product of $\chi_{ijm}^s(2\omega)\chi_{ijm}^i(2\omega)$ is -1, like for a phase shift π at the interface or for inversion-symmetric films (due to the transformation property of $\chi_{ijm}^s(2\omega)$ under inversion, $\chi_{ijm}^s(2\omega) \to -\chi_{ijm}^i(2\omega)$).

Obviously, also the parity of the QWS plays an important role for $I(2\omega)$, if the surface contribution to SHG is nearly equal to the interface contribution. For example, if for increasing film thickness d (at $d = d_1$) the first unoccupied QWS close to the Fermi energy E_F, which sets the oscillation period, has even parity and if at the interface no phase shift of π occurs, then no SHG results, since the product of the three dipole-transition-matrix elements is small; see Fig. 6.9.

However, QWS having odd parity contribute to SHG. Thus, in the case where the symmetry of the QWS regulates $I(2\omega, d)$, one should observe in SHG a pronounced period doubling of the oscillation period compared with the MOKE period. Only the QWS with odd parity cause oscillations. This is also true for a possible larger second period. However, it will be difficult to detect this experimentally.

In detail this period doubling of the SHG oscillations due to QWS can be understood if one investigates the matrix element product in eqn (6.2.8). Assuming for example that only one QWS is involved, then one has for a typical SHG transition for the product of the matrix elements

$$\langle d| \, \boldsymbol{p} \, |d\rangle \langle d| \, \boldsymbol{p} \, |QWS\rangle \langle QWS| \, \boldsymbol{p} \, |d\rangle.$$

Using for simplicity for the d bands the even symmetry of the corresponding atomic wavefunctions and further taking into account the odd symmetry of the momentum operator $\boldsymbol{p}$, then one gets, for an odd QWS, that only the first matrix element is small. In contrast, for an even QWS all three matrix elements are small.[18] Thus, if surface and interface contribute then the signal from an even QWS is negligible compared to the one from an odd QWS. If using Bloch functions for the d states, one has to keep in mind that the transitions involving QWS take place at specific $\boldsymbol{k}$ values. For these $\boldsymbol{k}$ values the Bloch functions in the film also have definite symmetry properties and the same argument as above is valid.[19]

The minimum between two peaks of $I(2\omega, d)$ is much deeper due to the

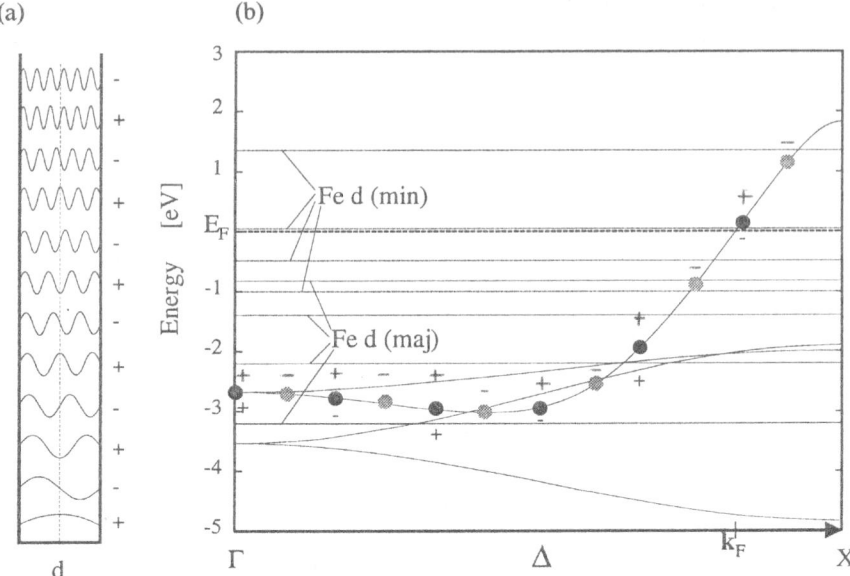

Fig. 6.9. (a) Wavefunctions for a particle in a box of thickness d with infinite walls as a model for QWS wavefunctions. The symmetry of the QWS is alternating with increasing quantum number. (b) Using this model, the symmetry of the QWS is indicated for 6 ML (only full black dots, parity sign of QWS is indicated) and 12 ML (full and grey dots, parity sign of QWS is indicated), for the same system as in Fig. 6.7. Obviously the parity of the QWS at E_F changes sign. Thus, if the plus sign causes constructive interference between surface and interface layer at 12 ML, the minus sign for 6 ML causes destructive interference. For surface and interface contributions of approximately equal weight no SHG will be detectable. Thus, a doubling of the MOKE period Λ_M occurs.

destructive interference (the phase difference due to the thickness increase is negligible compared to the wavelength of light). If the spin polarization of the QWS is not very pronounced, one expects only a relatively small magnetic contrast ΔI in SHG, since the d states of a ferromagnetic spacer are not strongly involved in the transitions contributing to SHG. Otherwise there would be no equally weighted contribution from surface and interface.

Generally, one notes that if the system under consideration shows only a weak k selectivity, the SHG response will display smooth oscillations without strong peaks and almost no frequency dependence, since the wavelength dependence of the period is washed out by the slightly different periods generated from the different k points contributing to the optical transitions. Of course, when both the symmetry of the dipole matrix elements and also the electronic structure of the (magnetic transition-metal) substrate have to be taken into account, the SHG signal will be more complex. However, a strong k selectivity due to the influence of the substrate makes additional states available for resonant transitions. Then, there can be no equally weighted contribution from interface and surface, and interference effects should be small.

Table 6.4 Characteristics of SHG response and its dependence on the electronic structure for thin films

	$|\chi^i| \gg |\chi^s|$	$|\chi^i| \approx |\chi^s|$
k selectivity	(x Cu/Fe/Cu(001) for example)	
	Strong magnetic signal due to strong (magnetic) interface contributions	Weak SHG signal, from only few **k** points and without strong interface contributions
	Sharp SHG peaks due to few contributing **k** points resulting in strong resonances	Doubled period and additional periods are frequency dependent
	Strong frequency dependence of the SHG oscillation period due to the dispersion of the QWS in the $k_\perp$ direction	MOKE period absent; doubled and additional SHG period visible
	MOKE period and larger periods visible; no exact doubling of the MOKE period	
No **k** selectivity		(x Au/Co(0001)/Au(111) for example)
	Strong magnetic signal, since strong interface contribution Broad SHG peaks, since contributions come from many **k** points	Smaller magnetic contribution, since interface and (nonmagnetic) surface contributions are of the same magnitude
	Weak frequency dependence of the oscillation period	Broad, smooth peaks, since interference effects do not change abruptly
	MOKE period and larger periods present	SHG oscillation periods rather independent of the frequency, since the SHG signal is caused by the QWS near E_F
		MOKE period absent, doubled period present

Note that only the QWS show a strong dependence on the layer thickness, in contrast to the other band-like states of the film. Thus, it is sufficient for the thickness dependence of SHG to consider transitions, where QWS are involved. However, there are also other contributions to SHG, e.g. involving only d states. These transitions create only a background contribution to SHG. The strong SHG oscillations result from transitions involving QWS.

In Table 6.4 we have listed characteristic features present if optical interference effects are dominant or negligible, respectively, and for more or less dominant **k** selectivity. Of course, due to the electronic structure of the film, which may be different at its surface and the substrate interface, the above two limiting situations of no interference and perfect interference may not be realized. Then the QWS features in SHG are less pronounced.

As an example we discuss first x Cu/Fe/Cu(001).[13] Here, we find that the essential contribution comes just from one interface and has also a strong k selectivity. For this system it is sufficient to restrict k to $(k_x, k_y) = (0,0)$, since the Fermi surface has a calliper there and a large DOS at E_F. Furthermore, from an inspection of the band structure in the $k_\perp$ direction, one finds that three different types of optical transitions contribute essentially to $\chi_{ijm}(2\omega)$ (p = dipole operator):

(i) $\langle\text{Fe}|\,p\,|\text{Fe}\rangle\langle\text{Fe}|\,p\,|\text{QWS}\rangle\langle\text{QWS}|\,p\,|\text{Fe}\rangle$,

(ii) $\langle\text{Cu}|\,p\,|\text{Fe}\rangle\langle\text{Fe}|\,p\,|\text{QWS}\rangle\langle\text{QWS}|\,p\,|\text{Cu}\rangle$,

(iii) $\langle\text{Cu}|\,p\,|\text{Cu}\rangle\langle\text{Cu}|\,p\,|\text{QWS}\rangle\langle\text{QWS}|\,p\,|\text{Cu}\rangle$.

The optical transitions corresponding to (i) and (ii) are responsible for the dominant Fe/Cu interface contribution to SHG due to QWS, since the important d states and large local DOS of Fe are not present at the Cu film surface. The surface contribution to SHG only involves the matrix element combination (iii). Thus, the Fe/Cu interface dominates the SHG and essentially $\chi_{ijm}(2\omega) \approx \chi^i_{ijm}(2\omega)$. As a consequence, one expects a strong magnetic contrast ΔI due to the spin-split Fe d bands at the magnetic interface. This is in perfect agreement with experiment, which shows a strong magnetic contrast and two wavelength-dependent oscillation periods $\Lambda_1 = \Lambda_M$ and a larger one Λ_2. The latter dominates the SHG signal.

As an example for systems with dominant contributions of both surface and interface we consider x Au/Co(0001)/Au(111).[20] From the band structure one can deduce that the Co states cause confinement for the majority spin states above approximately 1 eV below E_F. Since the 2.38 eV onset of d-band transitions comes from states around the neck of the Fermi surface near the L point, these directions have to be considered. However, since there is no calliper of the Fermi surface, contributions from many k points have to be taken into account. The area of contributing k points is indicated in Fig. 6.10 (a) (region A). This causes only a weak k selectivity and consequently the SHG peak with respect to the Au layer thickness is relatively broad. Although not only the direction Γ-L but also parallel ones matter, one may estimate again approximately the average MOKE period $\Lambda_M \approx 7$ ML from the Au band structure.[21]

From inspection of the Co band structure sketched in Fig. 6.10(b) it can be seen that there are no Co d states serving as intermediate states. Thus, the interface and surface contributions should be very similar, which leads to a strong influence of the QWS parity on the SHG contribution and consequently to a doubling of the Λ_M period, $\Lambda_1 = 2\Lambda_M$. This is in excellent agreement with experiment, where a period of 13 ML was observed due to the period doubling.

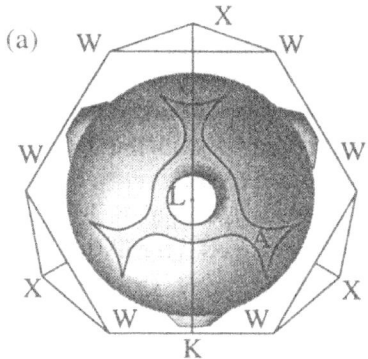

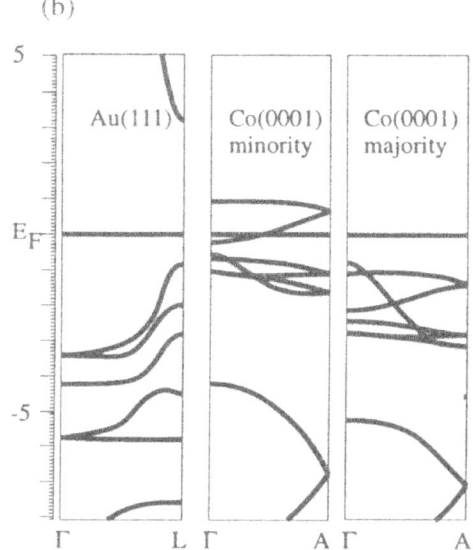

Fig. 6.10. (a) Regions of the irreducible Brillouin zone with s states contributing considerably to the formation of SHG oscillations due to QWS. The Γ point is at the centre of the Brillouin zone (hidden by the L point). Along the indicated k directions (region A) there are states acting as final states. If QWS are present in these regions, they give rise to a strong SHG output. QWS occurring in other regions (e.g. in Γ-L direction) do not contribute significantly since the QWS are then too far above E_F. (b) Illustration of the Au and Co band structures for $k_\perp$ corresponding to the direction perpendicular to the surface for the x Au/Co(001)/Au(111) system. Confinement for the electrons of Au is possible for energies $> -1\,\text{eV}$ due to the lacking overlap between the Au and Co bands. However, the possible Au QWS there are too high above E_F to be reached by photons of $\hbar\omega$ less than 1.7 eV.

An even larger period Λ_2 is expected to be in the range of more than 20 ML and therefore hardly detectable.

The magnetic SHG signal can be understood as follows. Since mainly QWS with electrons having the same spin as the majority electrons in Co occur, the magnetic SHG signal should show the same oscillation period as the nonmagnetic SHG signal. This is in excellent agreement with experiment too. The oscillation period for the total SHG response and for the magnetic contrast equals 13 ML for p polarization of both the incident ω and the outgoing reflected 2ω light beam. The linear MOKE period is found to be 7 ML.[22] Thus, the period doubling indicates that in the xAu/Co/Au(111) system the Co states are not strongly involved as initial states, but are necessary for the formation of the confinement of the Au states. A further indication for the minor influence of the Co d states is the relatively small magnetic signal of -5% to $+20\%$ as compared to the one in the x Cu/Fe/Cu(001) system of -50% to $+50\%$ and the absence of any dependence of the observed period on the incident laser frequency. This shows that no k points with resonances involving Co states are dominating the spectrum.

In summary, theoretical arguments and calculations supported by experimental results have shown the rich information that can be obtained from SHG of ultrathin films. Our analysis shows that rather different experimental observations of oscillating SHG signals due to the presence of QWS[13,20] can be understood within the same theoretical framework. Clearly, since more optical transitions are involved, (M)SHG can reveal more information than MOKE. Generally, one may get oscillatory behaviour of the SHG intensity for increasing film thickness d, which is dominantly controlled either by the interference and the symmetry of the dipole matrix elements or, if interference effects are negligible, by the energy spectrum and strong optical transitions controlled by the light energy $\hbar\omega$. The period of the SHG oscillations due to QWS may change characteristically with photon frequency, thus revealing prominent optical transitions and k selectivity. If besides s bands only flat d bands are present, SHG oscillations should largely result from s-d transitions. Further analysis is needed to understand whether QWS in flat d bands cause changes in the period Λ of the SHG oscillations and the damping of these compared to the QWS originating from s bands.

For other film structures one may easily derive corresponding conclusions using the arguments presented in this chapter. For layered film structures with QWS in several films one expects interesting interference effects, which may also reflect the magnetic structure of the multilayer structure.

6.4 SHG analysis of magnetic structures

The SHG theory can be used in particular to determine the magnetic structure at surfaces and interfaces. Changes in the magnetic moments and their order and in the direction of the magnetization are reflected by the Kerr rotation nd the polarization dependence of the SHG signal. Thus, ferromagnetism, magnetic domain structures, antiferromagnetism and magnetic configuration of multifilm

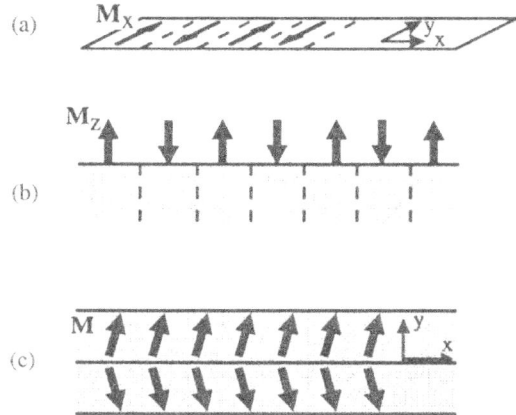

Fig. 6.11. (a) Illustration of in-plane (x, y) antiferromagnetic domain pattern $(M \parallel x)$, (b) out-of-plane $(M \parallel z)$ antiferromagnetic domain pattern (surface normal $\alpha\, z$), and (c) canted domain pattern with $M = (M_x, M_y, 0)$.

structures can be analysed.[23] It follows from the symmetry analysis[8] that the different tensor elements χ_{ijl} characterize various magnetic structures like ferromagnetism, and antiferromagnetism, etc. Note that, by varying the light polarization, the various elements χ_{ijl} can be analysed.[3,24] Thus, antiferromagnetism has been studied using SHG in the garnets.[5]

The basis of our theory for the analysis of magnetic structure are eqns (6.2.1)–(6.2.8). We study some simple magnetic structures to characterize this analysis.

For simplicity we assume simple domain structures as shown in Fig. 6.11. The system of ferromagnetic domains forms an antiferromagnetic array, e.g. neighbouring domains are antiferromagnetically ordered. The light spot (size in the range of micrometres) is assumed to cover many domains and for simplicity we assume first $\lambda \gg l$, where λ is the light wavelength and the distance l characterizes the size of the domain. Then, one should obtain approximately a superposition of light without geometrical phase shift from different domains.

First, we assume for the magnetization M and $M \perp z$, where z is the coordinate along the surface normal, and that the domain walls are very thin—see Fig. 6.11 (a). Then the spatial change from $M \to (-M)$ within the domain walls can be neglected.[25] We consider the longitudinal configuration with $M \parallel x$ and where (x, z) is the optical plane. The s-polarized SHG light generated from s-polarized incoming light results then from the electric field[24]

$$E_s(2\omega; \text{s-SH}) \sim \chi_{yyy}(M), \qquad (6.4.1)$$

where χ_{yyy} is odd in M (see Tables 6.1–6.3 for χ_{ijl} and results by Shen et al.[8]). Here, E_s(s-SH) is the s-polarized SH electric field resulting from the incoming s-polarized linear electric field. Consequently, for domains of equal size the

SHG intensity vanishes ($I \propto I(M) + I(-M) + \cdots \to 0$, and $I(M)$ refers to the light due to domains with magnetization M), e.g.

$$I_{s \to s} = 0. \tag{6.4.2}$$

$I_{s \to s}$ refers to the intensity of s-polarized SH light resulting from incoming s-polarized light with frequency ω. Similarly, one gets for p-polarized incident light and s-polarized SH light

$$E_p(2\omega; \text{s-SH}) \sim \left(\chi_{yzz}(M) f_s^2 + \chi_{yxx}(M) f_c^2 \right) + \cdots . \tag{6.4.3}$$

Note, χ_{yzz} and χ_{yxx} are odd in M.[26] Consequently, one gets for equally sized domains

$$I_{p \to s}(2\omega) \approx 0. \tag{6.4.4}$$

For incoming p-polarized light one gets p-polarized SH light due to

$$E_p(\text{p-SH}) \sim \{ 2 F_c f_s f_s \chi_{xzx} + N^2 F_s (f_c^2 \chi_{zxx} + f_s^2 \chi_{zzz}) \}. \tag{6.4.5}$$

Note that all tensor elements in this equation are even in M. Thus, for the intensity of p-polarized SH light due to the incoming p-polarized ω light one has

$$I_{p \to p} \neq 0. \tag{6.4.6}$$

Furthermore, one gets

$$E_s(\text{p-SH}) \sim \chi_{zyy}. \tag{6.4.7}$$

Hence, one finds also

$$I_{s \to p} \neq 0. \tag{6.4.8}$$

These results are in contrast to those for a ferromagnetic structure and which may be induced for an ensemble of domains by applying a magnetic field. Then, according to eqns (6.4.1), (6.4.3), (6.4.5) and (6.4.7) one has ($i = $ s, p)

$$E_i(\text{p-SH}) \neq 0, \quad E_i(\text{s-SH}) \neq 0, \tag{6.4.9}$$

and in general $I_{i \to p} \neq I_{i \to s}$.

Clearly, this analysis shows how the polarization dependence of SHG can be used to determine domain structure. The situation is illustrated in Fig. 6.12. Of course, from the change in the SH intensity upon applying a magnetic field one may estimate the fraction of $(-M)$ domains to M domains and whether the domains are of equal size.

If the domains with M and $-M$ are not of equal size then possibly also $I_{s \to s} \neq 0$ and $I_{p \to s} \neq 0$. This might be used to characterize the size distribution of the domains. The transverse configuration $M \parallel y$ may be analysed similarly by using Tables 6.1–6.3 for χ_{ijl}. Note that for a 4-fold symmetry of the (x, y) plane the transverse configuration gets equivalent to the longitudinal one with $M \parallel x$ by changing the optical plane from (x, z) to (y, z). One gets no Kerr rotation

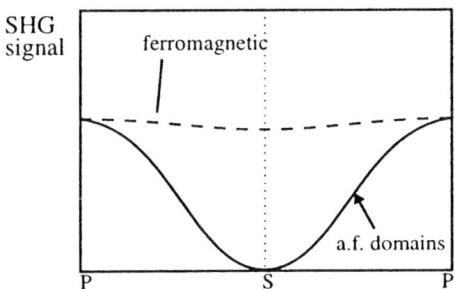

Fig. 6.12. Polarization dependence of SHG for antiferromagnetic domain pattern and for $M \parallel z$, respectively. $I_{p \to \alpha}$ is the intensity of α ($\alpha \equiv \phi$) polarized SH light resulting from p-polarized linear light. ($\phi = 90° \leftrightarrow s$). For comparison, we show results for a ferromagnetic structure, induced by an externally applied magnetic field, for example. Note that if the neighbouring domains with M and $(-M)$ are not of equal size, then $I_{p \to s}$ varies between zero and $I_{p \to s}$ (ferromagnetic).

(MSHG signal) for the transverse configuration. However, this changes if an external magnetic field is applied aligning M such that $M \parallel x$.

Assuming a magnetization $M \parallel z$, e.g. the polar configuration (see Fig. 6.11 (b)), and again antiferromagnetic ordering of neighbouring magnetic domains (of size $l \ll \lambda \ll d$, where d is the diameter of the light spot), one gets for s-polarized incident light for the SHG light

$$E_s(2\omega; \text{s-SH}) \sim 0, \quad (6.4.10)$$

and hence

$$I_{s \to s}(2\omega) \sim 0. \quad (6.4.11)$$

For p-polarized incident light one gets

$$E_p(2\omega; \text{s-SH}) \sim i\chi_{yxz}. \quad (6.4.12)$$

Note that $\chi_{yxz}(M)$ is odd in M. Hence

$$I_{p \to s}(2\omega) = 0. \quad (6.4.13)$$

Furthermore,

$$E_s(\text{p-SH}) \sim \chi_{zyy}. \quad (6.4.14)$$

and then

$$I_{s \to p} \sim |\chi_{zyy}|^2 \neq 0, \quad (6.4.15)$$

since $\chi_{zyy}(M)$ is even in M. For incident p-polarized light one gets

$$E_p(\text{p-SH}) \sim \{2 F_c f_c f_s \chi_{xxz} + N^2 F_s (f_c^2 \chi_{zxx} + f_s^2 \chi_{zzz})\} \quad (6.4.16)$$

and thus ($I \sim |\cdots|^2$)

$$I_{p \to p} \neq 0. \quad (6.4.17)$$

Note that applying a magnetic field yielding ferromagnetically aligned domains ($M \parallel z$), one generates approximately only light intensities $I_{p \to s}(2\omega)$ and $I_{s \to p}(2\omega)$ with mainly $I_{p \to s} \sim |\chi_{xyz}|^2$. Of course, the results due to eqn (6.4.13) might change characteristically if the domains are not of equal size. This also can be used to learn about the domain structure.

Again, one notices that domains can be identified with the help of the polarization dependence of SHG. Using a magnetic field one may also align M such that $M \parallel x$, for example, and compare SHG with the one expected for the longitudinal configuration. Also note that $\chi^e(M) \simeq \chi^e(0)$ and there is inversion symmetry in the surface (x, y) plane. Consequently, for nearly normal incidence yielding only $E_x(2\omega)$ and $E_y(2\omega)$ the nonmagnetic SHG signal vanishes.[27]

In conclusion, assuming equally sized domains and antiferromagnetic ordering of the domains, one gets in both cases of in-plane (xy plane) magnetization and perpendicular magnetization ($M \parallel z$) for thin films for the SH intensity $I_{p \to p} \neq 0$, $I_{s \to p} \neq 0$, but $I_{p \to s} = 0$ and $I_{s \to s} = 0$. Note that this is in contrast to what is expected for ferromagnetic order and paramagnetism. This shows that the polarization dependence of the SH light may be useful for identifying magnetic domain structures.

For a striped domain pattern (see Fig. 6.11 (c)), one has to analyse $\chi_{ijl}(M)$, with $M = (M_x, M_y, 0)$, where $M_y > 0$ everywhere, while M_x varies antiferromagnetically from domain to domain. Then, for incident s-polarized light ((x, z) optical plane)

$$E_s(2\omega; \text{s-SH}) \sim \chi_{yyy}. \qquad (6.4.18)$$

Since M_x varies antiferromagnetically and χ_{yyy} is odd in M_x, one gets for the MSHG

$$I_{s \to s}(2\omega) \approx 0, \qquad (6.4.19)$$

and

$$I_{s \to p} \sim |\chi_{zyy}|^2 \neq 0, \qquad (6.4.20)$$

with $\chi_{zyy}(M_x, M_y, 0)$. For p-polarized incident light one has

$$E_p(2\omega; \text{s-SH}) \simeq 2\mathrm{i} |E_0(\omega)|^2 A_s t_p^2 \left(f_c^2 \chi_{yxx} + f_s^2 \chi_{yzz} \right) + \cdots . \qquad (6.4.21)$$

One has approximately

$$\chi_{yxx}(M_x, M_y, 0) = \chi_{yxx}(M_x, 0, 0) + \chi_{yxx}(0, M_y, 0) + \delta\chi.$$

Here, $\delta\chi$ refers to higher expansion terms in M. The tensor element $\chi_{yzz}(M_x, M_y, 0)$ is similarly given. Note that χ_{yxx} and χ_{yzz} are odd in M_x (see Tables 6.1–6.3 for χ_{ijl}). Hence,

$$I_{p \to s}(2\omega) \propto |E|^2 \neq 0 \qquad (6.4.22)$$

due to higher expansion terms in M. Also, similarly $I_{p \to p} \neq 0$. The reflected light differs characteristically when p-polarized incident light changes to s-

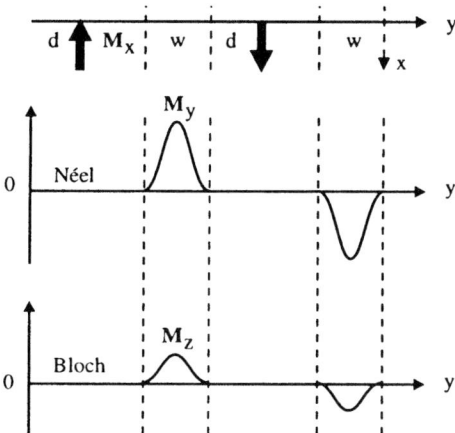

Fig. 6.13. Magnetic characterization of domain walls (w) ($M \parallel x$ is parallel to the domain (d) wall): Néel-type wall, where M rotates within the wall in the (x, y) plane with $M_y \neq 0$; and Bloch-type, where M rotates out of the (x, y) plane with $M_z \neq 0$. (Note that antiferromagnetic variation of M_y or M_z from wall to wall seems energetically slightly favourable. However, this depends on a delicate interplay of magnetic anisotropic and dipolar interactions. MSHG could discriminate between antiferromagnetic and ferromagnetic variation.)

polarized incident light. Also using (y, z) as optical plane one gets characteristic changes. Then,

$$E_p(2\omega; \text{s-SH}) \sim \chi_{xyy}(M_x, M_y, 0) + \gamma \chi_{xzz}(M_x, M_y, 0), \qquad (6.4.23)$$

where for example χ_{xyy} is odd in M_y, etc. Since $M_y > 0$ one gets $I_{p \to s}(2\omega) \neq 0$, and so on. Obviously, by changing the optical plane from (x, z) to (y, z), one will obtain quite different SH signals which characterize the striped domain structure.

Obviously, the polarization dependence of SHG can also identify a striped domain structure. In a magnetic field aligning M in the (x, y) plane one gets also a magnetic signal $I_{s \to s}$, for example. These results suggest immediately how domain walls contribute to MSHG; see Fig. 6.13 for illustration. If within the domains $M \parallel x$, then in the case of Bloch walls M rotates in the (z, x) plane and acquires within the domain wall approximately an M_z component. If this component dominates, then locally for the walls χ_{ijl} has approximately the form[24]

$$\begin{pmatrix} 0 & 0 & 0 & \chi_{xyz} & \chi_{xzx} & 0 \\ 0 & 0 & 0 & \chi_{xzx} & \chi_{xyz} & 0 \\ \chi_{zxx} & \chi_{zxx} & \chi_{zzz} & 0 & 0 & 0 \end{pmatrix}, \qquad (6.4.24)$$

which corresponds to the polar configuration and where χ_{xyz} and χ_{zxy} are odd in M. This determines then the polarization dependence of the MSHG from Bloch walls. Choosing as optical plane (x, z), the strongest MSHG signal from a single Bloch wall results for $E_\alpha(2\omega\text{: s-SH})$, while for a Néel wall this would be for $E_\alpha(2\omega\text{; p-SH})$.

The symmetry of χ_{ijl} for a Bloch wall, where one has approximately a polar configuration, and that for the domain with a longitudinal configuration are different. Thus, a different polarization-dependent contribution to SHG results from a wall vs. a domain. However, writing for SHG

$$I(2\omega) \sim \left| \sum_i \chi_i^d + \sum_i \chi_i^w \right|^2, \qquad (6.4.25)$$

where χ_i^d and χ_i^w are the contributions for the ith domain and wall, respectively, one finds immediately that there is for $\lambda \gg d$ and $\lambda \gg w$ no resultant MSHG signal (linear in M). This is so, since M of neighbouring domains and walls varies antiferromagnetically. In the presence of a magnetic field aligning M_i of the i-domains and walls one gets in contrast a magnetic signal $I(2\omega; M)$.

In the case of a Néel wall the magnetization with $M \| x$ for the domains rotates within the wall in the (y, x) plane and acquires a M_y component. Thus, within the wall the longitudinal configuration of the domains changes to a transverse one. Hence, for a single wall the MSHG field $E_p(2\omega; \text{p-SH})$ is larger than $E_s(2\omega; \text{s-SH})$. Note that for the single domain one gets that $E_s(2\omega; \text{p-SH})$ is strongest. However, again the resultant MSHG from an ensemble of domains and walls vanishes (see eqn (6.4.24)).

For Néel walls the $M \| z$ in the domains one has a change from a polar configuration to a longitudinal one for the walls, if (x, z) is the optical plane and $M \perp x$ in the wall. A strong MSHG signal for a single wall is then observed for $E_p(2\omega; \text{s-SH})$ with $E_p(2\omega) \sim \chi_{yzz}$. Here, χ_{yzz} is odd in M. According to eqn (6.4.24) the MSHG for such a configuration and an ensemble of domains and walls vanishes.

There are other combinations of light polarization and M configurations that can be used to characterize domain walls. Generally, note that a phase-shift of $90°$ occurs between tensor elements that are even and odd in M, respectively.[24]

While the case $\lambda \gg d$ and $\lambda \gg w$ one gets cancellations of SH light from different domains or domain walls, respectively, this is different when the light wavelength is of the order of the domain size, $\lambda \simeq d$, but $\lambda \gg w$. Here, w refers to the width of the domain wall and λ to the wavelength of the light. Then

$$I(2\omega) = \sum_i (I_i^d + I_{i,\text{interf.}}) + \sum_i I_i^w, \qquad (6.4.26)$$

where I_i^w is the SHG intensity from the ith wall and I_i^d from the ith domain and $I_{i,\text{interf.}}$ results from the interference due to χ from neighbouring domains and the wall between these domains. One may choose the incidence angle of the incoming light and the polarizations such that the domain contribution to SHG vanishes and then[28]

$$I(2\omega) \simeq \sum_i I_i^w(M). \qquad (6.4.27)$$

Owing to the polarization dependence of $\chi_{ijl}(M)$ one finds that the SHG wall contribution is different for Néel and Bloch walls. For example, for $M \| x$ in the domains and for an optical plane (x, z) one gets approximately in the case of (a) Néel walls

$$p_{in} \to SHG_p, \qquad s_{in} \to SHG_s, \qquad (6.4.28)$$

e.g. incoming p-polarized (s-polarized) light yields mainly p-polarized (s-polarized) SH light and in the case of (b) Bloch-walls

$$p_{in} \to SHG_s, \qquad s_{in} \to SHG_p. \qquad (6.4.29)$$

Furthermore, for normal incidence and $M \| x$ one gets no SHG from the domains and for a Néel wall (transverse configuration) $E_x(2\omega) \sim \chi_{xyy}$ for s-polarized incident light and $E_x(2\omega) \sim \chi_{xxx}$ for p-polarized incident light. Note that both χ_{xyy} and χ_{xxx} are odd in M. For a Bloch wall with polar configuration one gets no SHG contribution.

In conclusion, for $\lambda \sim d$ and $\lambda \sim w$ the polarization dependence of SHG can straightforwardly characterize domain walls and distinguish between Bloch and Néel walls.

The size of the domains and of the domain wall may be comparable to λ. Then the phase shift of the light reflected from neighbouring domains needs to be taken into account. For a regular array of domains one expects, in analogy to a diffraction pattern, characteristic interference behaviour of the reflected light:

$$I(2\omega) \sim |E_1(2\omega; M) + E'_2(2\omega; -M) + \cdots|^2.$$

Here, E'_2 is the reflected SH field from domain 2 with $(-M)$ neighbouring domain 1 with M and which has a phase difference given by the projection of l on the direction of the outgoing light. The situation is illustrated in Fig. 6.14. Then one gets for the SHG intensity

$$I(2\omega) = I_1 + I_2 + I_{int}, \qquad (6.4.30)$$

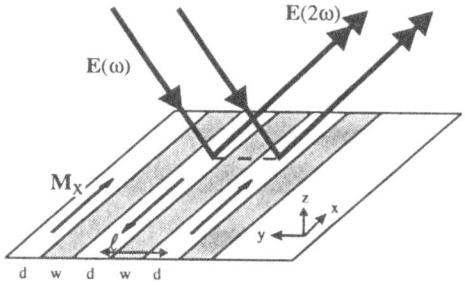

Fig. 6.14. Illustration of SHG for $\lambda \sim l$ when interference of reflected light from neighbouring domains is expected (λ = wavelength of light, $l \sim$ size of domains). For a regular array of domains one expects a characteristic interference pattern with a sensitive polarization dependence. $M_x \| x$ lies in the surface (x, y) or $M \| z$ (surface normal) is also possible.

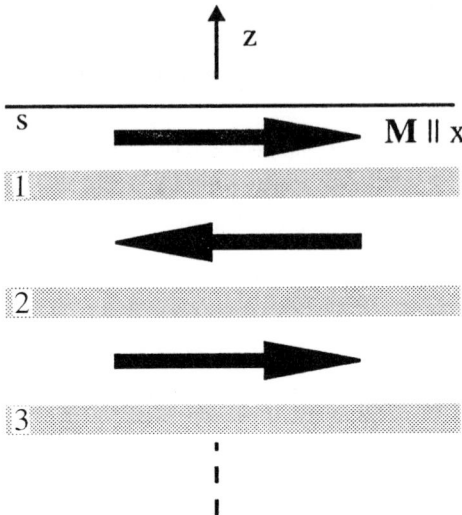

Fig. 6.15. Succession of thin films with antiferromagnetic magnetization of neighbouring films. The shaded regions indicate the interface area of SHG. SHG results also at the surface. Approximately, the first two atomic layers at the surface s will contribute to SHG.

where I_i refers to the contribution from the ith domain. Of particular interest is the interference described by $I_{\text{int}} \sim (E_1 E_2' + \cdots)$. The polarization dependence of $I(2\omega)$ again reveals characteristically the magnetic structure. Generally, if (x, z) is the optical plane, one has as discussed for the domains with $\boldsymbol{M} \parallel x$

$$p_{\text{in}} \rightarrow \text{SHG}_s. \tag{6.4.31}$$

And for the walls one gets in the case of Néel and Bloch walls respectively, the results given by eqns (6.4.28) and (6.4.29).

This completes then our theoretical analysis of domain structures using nonlinear magneto-optics. Clearly, the polarization dependence of the nonlinear susceptibility χ_{ijl} and its symmetry dependence on the magnetization permit a general SHG study of domains, including crystals with no inversion symmetry. Further experimental studies are necessary to check details of the theory and to compare with previous experimental studies of domain structure.[5] Note that usually $\boldsymbol{M} \perp z$ at surfaces. In thin films also $\boldsymbol{M} \parallel z$ occurs and possibly a reorientation transition appears for increasing film thickness and temperature. Domain walls may be of similar size as the domain, in particular near the reorientation transition $\boldsymbol{M}_\perp \rightarrow \boldsymbol{M}_\parallel$. Note that the transition $\boldsymbol{M}_\perp \rightarrow \boldsymbol{M}_\parallel$ could be observed by SHG.

A multifilm structure with antiparallel magnetization in neighbouring films (see Fig. 6.15), may be viewed as a particular antiferromagnetic domain structure. We neglect for simplicity quantum-well states (QWS). Then, we have for SHG

$$I(2\omega) \sim |\chi_s(\boldsymbol{M}) + \chi_{\text{int.1}}(\boldsymbol{M}) + \chi_{\text{int.2}}(\boldsymbol{M}) + \cdots|^2. \tag{6.4.32}$$

Writing $\chi = \chi^e + \chi^o$, $\chi^o \sim M$, one finds

$$I(2\omega) \sim \left| \chi_s^e(M) + \chi_s^o(M) + \sum_i \chi_{\text{int.}i}^e(M) \right|^2. \qquad (6.4.33)$$

Here, $\chi_{\text{int.}i}(M)$ refers to the ith interface of two films. In contrast, for a ferromagnetic structure of the films one has that not only $\chi_{\text{int.}i}^e$ contributes, but also $\chi_{\text{int.}i}^o$, and hence

$$I(2\omega) \sim \left| \chi_s(M) + \sum_i \chi_{\text{int.}i}(M) \right|^2. \qquad (6.4.34)$$

By applying a magnetic field aligning the antiferromagnetic magnetization of the films, one is able to identify the magnetic structure. Such a determination is of particular interest when the magnetization of neighbouring films is not collinear, corresponding somewhat to a striped domain structure. Clearly also this can be determined by using MSHG. The analysis simplifies if only the first few films are probed optically. Interesting cases are when phase shifts occur at the interface and spin-polarized quantum-well states are present.

We can summarize the theoretical basis for the SHG characterization of magnetic domain structures and magnetization of multifilms as follows: The polarization dependence of χ_{ijl} and the different symmetry of $\chi^e(M)$ and $\chi^o(M)$ as well as the dependence on the angle of light incidence and on temperature via $M = M(T)$ permit a structure analysis. For the analysis, in particular if $M = (M_x, M_y, M_z)$, one may also use eqn (6.2.3). This is very useful for describing SHG from various domain structures, in particular for describing striped domain structures with $\chi_{ijl}(M_x, M_y, 0)$. Obviously, the third term in eqn (6.2.3) is of particular significance if the first two terms vanish. Also, note that terms of the form of $(\cdots)\nabla_k M_m$ should be added to the expansion to treat better the spatial variation of M in the domain walls.[3,5]

Finally, it is of interest to point out that SHG pump-and-probe spectroscopy could be used to study the dynamics of magnetic structures. Thus, in particular one may attempt to analyse optically motion of domain walls as well as a reorientation transition of the magnetization.

6.5 Application of SHG to nonequilibrium physics

Recently, SHG has also been used to study the magnetic dynamics of metals like Ni, Fe, Co, etc.[29,30] Clearly, it follows immediately from the electronic theory for χ_{ijl} (see eqn (6.2.8)) that SHG responds to electronic excitations. By laser irradiation induced non-equilibrium one has to use the nonequilibrium Fermi distribution functions $f_{k,l,\sigma}$ in the expression for χ_{ijl} and one has also to take into account the time-dependent changes of the electronic temperature, $T_{el} = T_{el}(t)$. At nonequilibrium, the Fermi distribution functions must be determined from a Boltzmann–Master type of equation. The excited electrons due to the pump laser irradiation are first not in equilibrium with the other electrons remaining in the Fermi sea and with the lattice, but then due to electron–electron interactions thermalization of the nonequilibrium electrons begins. This causes an increase of the electronic temperature $T_{el}(t)$ as time t progresses, but finally when heat is transferred out of the electron bath into the lattice via electron–lattice coupling $T_{el}(t)$ stops increasing and then begins to decrease again. Regarding the magnetic nonlinear response to light irradiation, one has to use $M(t)$ in χ_{ijl}. Thus, after thermalization of the excited electrons occurs, one may use $\chi_{ijl}(M(T_{el}), T_{el})$ in the nonlinear susceptibility.

With a probe laser at time t, the pump laser is assumed to stop irradiation at $t = 0$, one analyses the electronic state for electronic temperatures $T_{el} = T_{el}(t)$. This is the general scenario for the pump-and-probe spectroscopy using SHG.

The magnetic SH signal is given by

$$\Delta I_- \equiv I(M, t) - I(-M, t). \quad (6.5.1)$$

The SHG intensity $I(M, t)$ is calculated by $I(2\omega) \propto |P(2\omega)|^2$ and the polarization is given in terms of the electric field by $P_i(2\omega) = \chi_{ijl} E_j E_l$. Since the second harmonic generation intensity depends on M and T_{el} one has $I(T_{el}, M)$. Expanding the nonlinear susceptibility $\chi_{ijl}(M)$ into even and odd terms in M, namely $\chi = \chi^e + \chi^o$, $\chi^o \simeq \chi' M$, one gets ($I \propto |\chi|^2$) neglecting phase differences[31]

$$\Delta I_- \propto 4\chi^e \chi' M. \quad (6.5.2)$$

Here, we assumed that mainly one tensor element of χ_{ijl} dominates. Similarly, we find for the quantity

$$\Delta I_+ \equiv I(M, t) + I(-M, t) \quad (6.5.3)$$

the result

$$\Delta I_+ \propto \left(|\chi^e|^2 + |\chi'|^2 M^2\right). \quad (6.5.4)$$

For thermalized electrons and spins one may use in eqns (6.5.2) and (6.5.4) for the magnetization $M = M(T_{el})$. Then, for example,

$$\Delta I_- \propto A(T_{el}) M(T_{el}), \quad (6.5.5)$$

and approximately $M \sim \sqrt{1 - (T/T_c)}$, $M^2 \sim (1 - T/T_c)$, $T \equiv T_{el}$. The coefficient $A(T_{el})$ includes Fresnel factors and magnetic anisotropy energy, which also depend on T_{el} (besides on T_{el} via $M(T_{el})$). Thus, ΔI_- exhibits more the magnetic dynamics and ΔI_+ reflects directly the electronic temperature T_{el}.[32] This has been discussed in details by Hohlfeld et al.[30]

We calculate ΔI_- and ΔI_+ using eqns (6.5.1)–(6.5.5) and assuming for simplicity that the dominant dependence on T_{el} is due to $M(T_{el})$. The results are compared with experiments on Ni by Hohlfeld et al.[30] The electronic structure of Ni is sketched in Fig. 6.16. In Fig. 6.17 results are shown for the SHG signals

$$\delta_\pm(t) = \{\Delta I_\pm(t) - \Delta I_\pm(t_0)\}/\Delta I_\pm(t_0),$$

where t_0 refers to the time at which T_{el} starts to increase initially above T_{latt} due

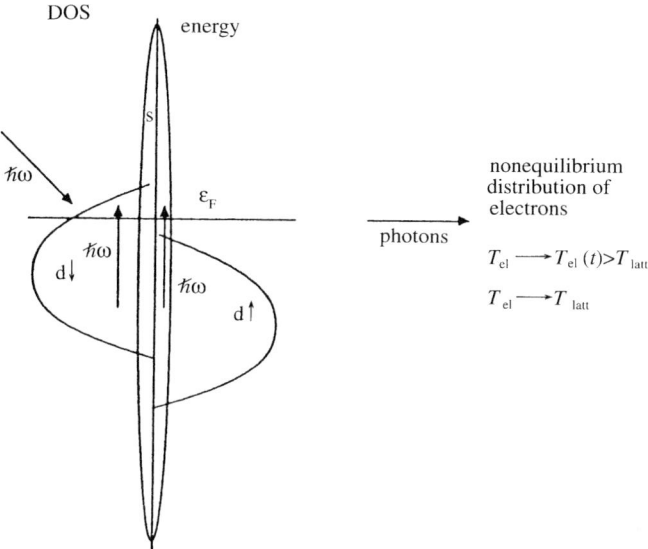

Fig. 6.16. Spin-split density of states for d electrons. Photons excite s and d electrons into states above ε_F, no spin flips occur. Since states are s-d hybridized, electric dipole selection rules have to include this. Using the Hubbard Hamiltonian for the dynamics of the itinerant electrons, magnetism responds due to the interplay of the intra-atomic Couloumb interaction U, the exchange coupling $J < U$, the hopping integral t_{ij} and spin–orbit coupling. Note that, for given photon frequency $\hbar\omega$, one gets different numbers of spin-up and spin-down hot electron due to differences in the initial DOS $N_\sigma(\epsilon)$.

to the probe laser irradiation. Note that one gets approximately $(\delta_-(t) = -[1 - \Delta I_-(t)/\Delta I_-(t_0)])$

$$\delta_-(t) \simeq -[1 - \alpha(t)m(t)], \qquad (6.5.6)$$

with

$$\alpha(t) = \frac{\chi^e(t)}{\chi^e(t_0)} \frac{\chi'(t)}{\chi'(t_0)}$$

and $m(t) = M(t)/M(t_0)$. Similarly, we find $(\delta_+(t) = -[1 - \Delta I_+(t)/\Delta I_+(t_0)])$ the expression

$$\delta_+(t) \cong -\left(1 - \frac{|\chi^e(t)|^2 + |\chi'(t)|^2 M^2(t)}{|\chi^e(t_0)|^2 + |\chi'(t_0)|^2 M^2(t_0)}\right). \qquad (6.5.7)$$

We model for simplicity the electronic temperature $T_{el}(t)$ by the curves shown in the inset of Fig. 6.17. Note that such behaviour is expected from the solutions of the Boltzmann equation.[33] For simplicity the temperature dependence of χ^e and χ' is first neglected and then $\chi^e(t) = \chi^e(t_0)$ and $\alpha = 1$.[32]

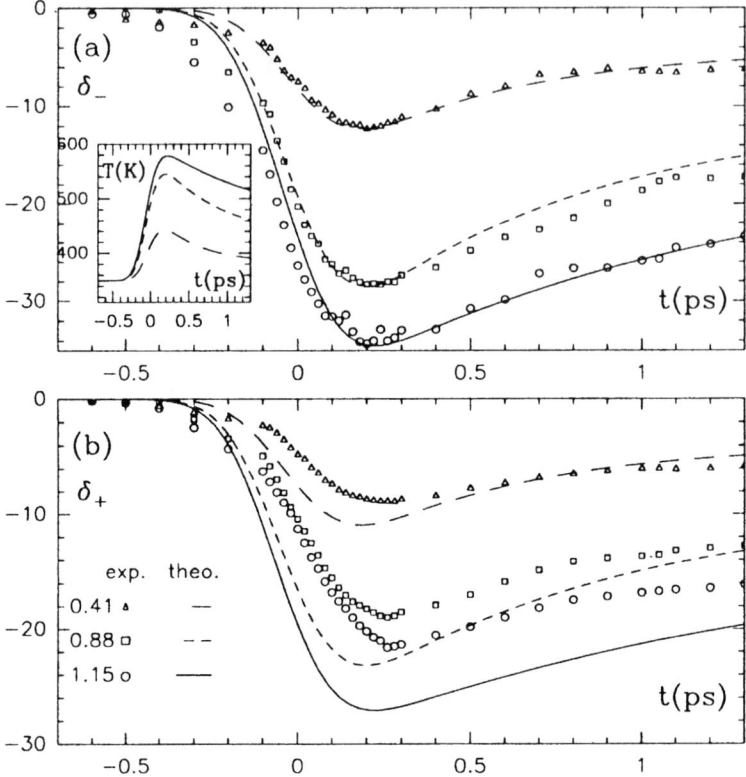

Fig. 6.17. Illustration of the short-time behaviour of the SHG magnetic contrast signal $\delta_-(t) = [\Delta I_-(t) - \Delta I_-(t_0)]/\Delta I_-(t_0)$ and of $\delta_+(t) = [\Delta I_+(t) - \Delta I_+(t_0)]/\Delta I_+(t_0)$. For the electronic temperature $T_{el} = T_{el}(t)$ we use model results shown in the inset. t_0 refers to the time at which the increase of T_{el} begins. The calculated results were obtained using eqns (6.5.6) and (6.5.7) and are compared with experimental ones.[12] δ_- is minimal for $T_{el}(t) \simeq$ max. We use $b \simeq 1.0$. The curves refer to a different fluence as indicated. $M(t)$ is calculated from $M(T_{el})$ and $T_{el} = T_{el}(t)$. The rapid decrease of $\delta_-(t)$ due to the decrease of $M(t)$ caused by the increase of T_{el} due to the hot electrons. After reaching a minimum the quantity $\delta_-(t)$ increases again, since $M(t)$ increases again, see T_{el} and $T_{el} \rightarrow T_{latt}$. Note that the lattice will be somewhat warmer at the time $t \sim$ picoseconds than at the time t_0. Hence $\delta_-(t)$ and $\delta_+(t)$ approach a value which is smaller than the one at t_0. $T_{el}(t)$ refers to Ni with $T_{el,max} \simeq 580\,\text{K}$, $T_{el}(t_0) = T_{latt} \simeq 300\,\text{K}$ and $T_{latt} \simeq T_{el} = 450\,\text{K}$ at $t \gtrsim 2\,\text{ps}$ (see experiments by Hohfeld and Matthias).

While the simple theory using $M = M(T_{el})$ and a temperature-independent ratio

$$b = \frac{|\chi'| M(t_0)}{|\chi_e|} \tag{6.5.8}$$

is in overall fair agreement with experiment, there are interesting revealing discrepancies.[34] In particular, one expects nonequilibrium behaviour for the ultrafast response after laser irradiation has stopped. Then one would have $M(t) \neq M(T_{el}(t))$. Note that for $t < t_{min}$ we find a relatively large discrepancy with experimental results. The somewhat different times at which the minimum is reached for δ_- and δ_+ may signal interesting physics.[30] Also this depends on the width of the laser pulse. Note that for $\alpha \simeq 1$ in our approximate formulae only δ_+ is affected by the temperature dependence of χ' and χ^e. Interestingly, the calculated values for δ_+ at the minimum are too negative, suggesting that $\Delta I_+(t)$ and hence $\chi^e(t)$ increase for increasing temperature and are not constant as assumed. For times $t > t_{min}$, when the electronic temperature approaches the lattice temperature T_{latt} ($T_{el} \to T_{latt}$) for formula $\Delta I_- \propto M(T_{el})$ describes satisfactorily the experiment. Note that δ_- for times $t \sim$ picoseconds is somewhat smaller than $\delta_-(t_0)$ due to the warming up of the lattice by the hot electrons.[35]

Also the simple theory correctly predicts for longer times

$$\frac{\Delta I_-}{\Delta I_+} = \frac{\chi^e \chi'}{|\chi^e|^2 + |\chi'|^2 M^2} M \propto M. \qquad (6.5.9)$$

Note that the different dependences of χ^e and χ' on nonequilibrium and thus on time t and T_{el} and the different decreases of δ_- and δ_+ for $t < t_{min}$ seem very interesting. Further studies, including the determination of the dependence of the dynamics on light frequency $\hbar \omega$, are necessary to understand this. The latter might reveal interesting band-structure effects.[36] Also that thermalization of the electrons occurs over a certain time range might be reflected in $\Delta I_-(T_{el}, M(T_{el})) \propto M(T_{el})$ and ΔI_+.

This demonstrates how the dynamics of nonequilibrium magnetic transition metals can be studied using nonlinear SH and magneto-optics. These experiments support the general theory for the dynamics of magnetism in transition-metals and exhibit magnetic response times ($t \sim J^{-1}$, if $J > V_{so}$) of the order of less than 100 fs or of a few hundred femtoseconds (200–280 fs), long before lattice and electrons have reached equilibrium again. Estimating from ε_F and photon energy $\hbar \omega \simeq 1.6$ eV the relaxation time τ for Ni as of the order of 80 to 300 fs, we predict indeed that $M \to 0$ can occur during a time $t \leq 280$ fs as observed in experiment,[30] and possibly much faster.

In Fig. 6.18 we sketch a further interesting nonequilibrium situation for ferromagnetic transition metals. If T_{el} gets larger than T_c, $T_{el} > T_c$, then the magnetization disappears for a certain time range. Note also that the magnetic response to hot electrons of antiferromagnetic metals like Cr and of ferromagnets, etc. would be interesting.

For rare-earth metals in contrast to transition metals the f electrons constituting the spins may be treated as a separate system coupled to the s, d electrons which get excited. Then, J and spin-orbit interaction V_{so} are larger and one may find for nonequilibrium the situation $T_{el} \neq T_s(t)$, where T_s refers to the temperature of the local spins of the rare-earth atoms, $T_{el} > T_s$, and furthermore that both T_{el} and T_s are different from T_{latt}.[37]

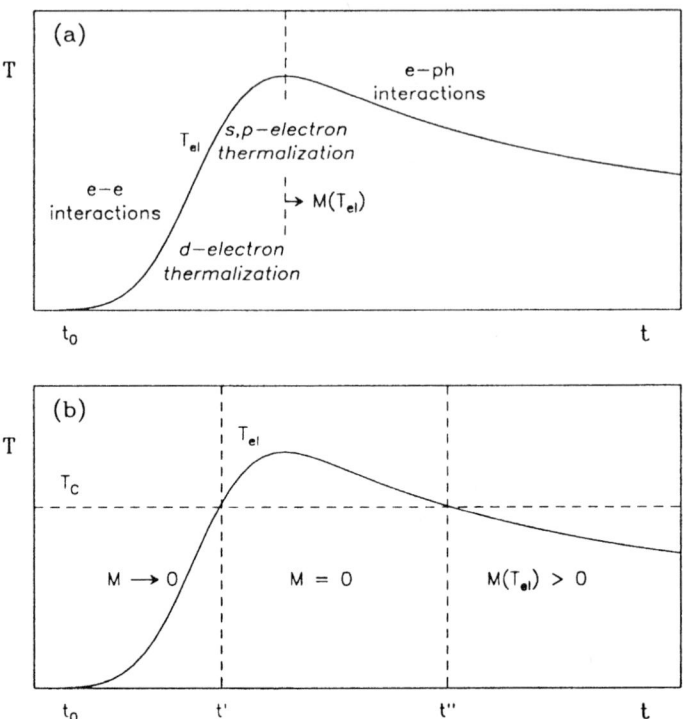

Fig. 6.18. Illustration of dynamics in Ni, for example, due to hot electrons, for a nonequilibrium state. (a) time dependence of the electron temperature T_{el} controlled by electron–electron and electron–lattice interactions. First d and then s electrons thermalize. T_{el}^{max} reflects the interplay of energy distribution over the progressive thermalizing electrons and energy transfer to the lattice $M(t) \to M(T_{el})$ after electron thermalization. (b) If $T_{el} > T_c$, T_c being the Curie temperature, then $M \to 0$ during the time $(t'' - t')$.

Finally, we want to point out an interesting possibility to use nonequilibrium physics to induce the reorientation transition of the magnetization M at the surface of Ni or Fe, for example, due to hot electrons, and which can then be analysed using SHG pump–probe spectroscopy. In Ni, for example, the parallel surface magnetization, $M_{\parallel}$, changes with increasing temperature, due to $M(T)$, to a perpendicular magnetization $M_{\perp}$.[38] Hence, as illustrated in Fig. 6.19, it may be possible to generate optically a *switch*

$$M_{\parallel} \leftrightarrow M_{\perp} \qquad (6.5.10)$$

by a sequence of laser pulses. Each laser pulse will raise T_{el} due to hot electrons and then cause the reorientation transition $M_{\parallel} \to M_{\perp}$, if $T_{el} > T_R$, where T_R is the temperature at which the change $M_{\parallel} \to M_{\perp}$ occurs. Since $T_{el} \to T_{latt}$ (and $T_{latt} < T_R$), the relaxation $M_{\perp} \to M_{\parallel}$ occurs. Then the next light pulse again causes $M_{\parallel} \to M_{\perp}$, and so on. Note that the reorientation involves the magnetic anisotropy coupling (spin–orbit coupling) and thus requires a time $t_{reor} \sim E_{anis}^{-1}$. Here, E_{anis} denotes the magnetic anisotropy energy ($E_{anis} \sim 100\,\mu eV$ for

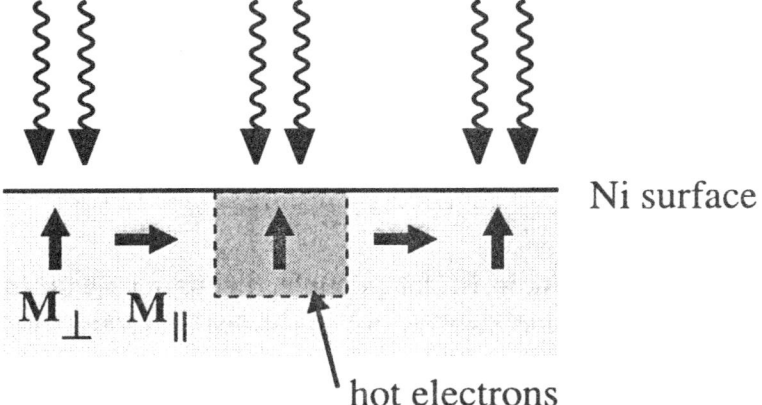

Fig. 6.19. Light-induced magnetic pattern formation due to exciting hot electrons in Ni, for example. If in the irradiated regions the electronic temperature rises such that $T_{el} > T_R$, then the parallel surface magnetization $M_\parallel$ changes at the temperature T_R to the perpendicular magnetization $M_\perp$. The reorientation transition $M_\parallel \to M_\perp$ is achieved by spin–orbit coupling and thus requires the time of the order of $t_{reor} \sim E_{anis}^{-1}$. The cooling from $T_{el} > T_R$ to $T_{el} < T_R$ implies a further time delay.

transition metals). One estimates t_{reor} to be of the order of some hundred picoseconds.

In conclusion, we have identified characteristic times t_{el}, t_μ and t_{reor} for the short-time dynamics of magnetic metals not at equilibrium due to hot electrons. Of course, at times $t \gtrsim$ picoseconds, the spin dynamics may occur for 'cold spins' in an excited warmer lattice. Then the spin–lattice coupling will play an important role for spin relaxation processes occurring on a time scale of some picoseconds approximately $t \sim 100\,\mathrm{ps}$.[39]

The detailed dynamics of the hot electrons is an interesting problem with many aspects. For example, the spin-dependent mean free path of hot electrons needs further analysis. One expects an interesting dependence on the number of excited electrons and on the exchange splitting.[40] SHG including its polarization dependence probing χ^o and χ^e differently is a very useful tool for studying the ultrafast response of electrons and their spins.

It would be interesting to analyse also the dynamics of the many-body Kondo singlet state, of magnetism in heavy-Fermion systems and of superconductivity. The dynamics of the Kondo state is expected to be characterized by the time $t \sim 1/T_K$, where T_K is the Kondo temperature. Hot electrons will destroy the singlet state if $T_{el} > T_K$. Similarly, for heavy Fermions the nonequilibrium hot electrons raising T_{el} will destroy the resonance-like DOS peak at ε_F and thus typical magnetic effects. Again changes occur during a time $t \sim U_{eff}^{-1}$, where U_{eff} is the effective electron correlation responsible for the narrow band of width w and $T_{el} > w$ should destroy the heavy-Fermion characteristics. Finally, it would be interesting to study systems where V_{so} controls angular momentum conservation and thus the magnetic dynamics.

6.6 Outlook

Analysing some problems we have indicated the potential of SHG in metals for determining the electronic structure at interfaces, surfaces or metal–metal interfaces, thin films and multifilm structures. In particular, SHG reveals interface- and film-specific electronic structure and furthermore spin-polarized SHG is capable of determining magnetic anisotropy, atomic-layer-dependent magnetic moments and magnetization. Nonlinear optics and magneto-optics is more than just an alternative to linear optics, but able to analyse uniquely electronic and magnetic structures. Of course, further experimental and theoretical studies are needed to prove that nonlinear optics is a new successful tool offering new possibilities in addition to previous studies. On the experimental side more experiments are needed to demonstrate the particular sensitivity of NOLIMOKE regarding the determination of electronic and atomic structure, for example exhibiting atomic structure changes (fcc → bcc) in thin films and its relationship to corresponding magnetic changes, or revealing reorientation transitions of the magnetization, magnetostriction effects, surface roughness, magnetic structure of multilayer structures and magnetic domain structure. Theoretically the extension of previous SH calculations by determining the dipole matrix elements and the interplay of the electronic energy spectrum and the matrix elements permits to a high degree of accuracy the material-specific analysis of electronic and atomic structure, thus making SHG a real new tool in metal physics.

New studies must show how well SHG is able to analyse also lateral structures. Domain structures, inhomogeneously growing films and clusters deposited on a surface or embedded in a matrix may be examples of such research. Also nonlinear optical studies of magnetism in clusters (the magnetic Mie scattering, etc.) is an interesting problem as well as studies of matter not at equilibrium.

Acknowledgements

We thank R. Knorren, E. Matthias, J. Hohlfeld, W. Hübner, Th. Rasing and J. Kirschner for many helpful discussions.

References

1. W. Hübner and K. H. Bennemann, *Phys. Rev.* B **40**, 5973 (1989).
2. T. A. Luce, W. Hübner, A. Kirilyuk, Th. Rasing, and K. H. Bennemann, *Phys. Rev.* B **57**, 7377 (1998); T. A. Luce, W. Hübner, and K. H. Bennemann, *Phys. Rev. Lett.* **77**, 2810 (1996).
3. W. Hübner and K. H. Bennemann, *Z. Phys.* B **104**, 189 (1997).
4. U. Pustogowa, W. Hübner and K. H. Bennemann, *Phys. Rev.* B **48**, 8607 (1993).
5. A. Kirilyuk, V. Kirilyuk, Th. Rasing, V. Pavlov, and R. V. Pisarev, *J. Magn. Soc. Japan* **20**, 361 (1996).

REFERENCES

6. P. S. Pershan, *Phys. Rev.* **130**, 919 (1963).
7. W. Hübner, K. H. Bennemann, and K. Böhmer, *Phys. Rev.* B **50**, 17597 (1994).
8. P. Guyot-Sionnest, W. Chen, and Y. R. Shen, *Phys. Rev.* B **33**, 1129 (1987); R.-P. Pan, H. D. Wei and Y. R. Shen, *Phys. Rev.* B **39**, 1229 (1989).
9. One may expand the Bloch wavefunctions in terms of Wannier functions centred at atomic sites. Then, performing the summations and integrations in eqn (6.2.8), one will find that approximately only atomic states next to the surface or interface, feeling no inversion symmetry, contribute to SHG.
10. Even the linear Kerr rotation determined by χ_{ij} yields interesting results. Following general ellipsometry arguments, the complex polar Kerr angle is given by

$$\frac{E_y}{E_x} = \frac{\tan\varphi + i\tan\varepsilon}{1 - i\tan\varphi\tan\varepsilon},$$

where φ is the Kerr rotation angle and ε is the ellipticity. E_i are components of the reflected light. Thus,

$$\varphi = \tfrac{1}{2}\arctan\left(\frac{2\,\mathrm{Re}(K)}{1 - |K|^2}\right) + \varphi_0,$$

with $\varphi_0 = 0$ for $|K|^2 \le 1$; $\varphi_0 = 90°$ for $|K|^2 > 1$, $\mathrm{Re}(K) > 0$; and $\varphi_0 = -90°$ for $|K|^2 > 1$, $\mathrm{Re}(K) < 0$. Note that this formula for φ indicates that in principle large Kerr rotations may also result for linear MOKE, as observed for CeSb. For example, $\varphi = 90°$ if $\mathrm{Re}(K) = 0$ and $|\mathrm{Im}(K)| > 1$. Approximately, one has for CeSb that $K = i(\lambda_{so}/\hbar\omega)\{1 + \chi_{xx,i}/L\}\sqrt{(1 + \chi_{xx})}\}^{-1}$. Here, $\chi_{xx,i}$ is the intraband contribution. Thus, larger Kerr rotations may result as was discussed by U. Pustogowa *et al.* Furthermore, $\varphi(M)$ may not always be related to the magnetization M as is usually observed for small Kerr rotations.
11. T. A. Luce, PhD Thesis, Freie Universität Berlin, Arnimallee 14, 14195 Berlin, Germany (1997).
12. C. Carbone, E. Vescovo, O. Rader, W. Gudat, and W. Eberhardt, *Phys. Rev. Lett.* **71**, 2805 (1993); K. Garrison, Y. Chang, and P. D. Johnson, *Phys. Rev. Lett.* **71**, 2801 (1993).
13. M. Straub, R. Vollmer, and J. Kirschner, *Verh. Dtsch. Phys. Ges.* **6**, 1595 (1996), and private communication.
14. R. Mégy, A. Bounouh, Y. Suzuki, P. Beauvillain, P. Bruno, C. Chappart, B. Lecuyer, and P. Veillet, *Phys. Rev.* B **51**, 5586 (1995).
15. Generally, for thick films one expects that the interface contribution to SHG gets less important.
16. M. Straub, R. Vollmer, and J. Kirschner, *Verh. Dtsch. Phys. Ges.* **7**, 818 (1997).
17. P. Bruno, Y. Suzuki, and C. Chappert, *Phys. Rev.* B **53**, 9214 (1996).
18. If both surface and interface contribute nearly equally to SHG, then for example one may write $\langle d|\,p\,|QWS\rangle \approx \langle d|\,p\,|QWS\rangle_s + \langle d|\,p\,|QWS\rangle_i$ where $\langle \cdots \rangle_s$ indicates integration only over the surface regions. Then, for even QWS one has $\langle \cdots \rangle_s = -\langle \cdots \rangle_i$, while for odd QWS both contributions add.
19. Note that the period doubling in the case of contribution to SHG from both surface and interface results from the decomposition of the matrix elements in surface and interface contribution and not from the decomposition of χ into χ^i and χ^s.
20. A. Kirilyuk, Th. Rasing, R. Mégy, and P. Beauvillain, *Phys. Rev. Lett.* **77**, 4608 (1996).

21. Since the regions with large joint density of states are on average at a $k_\perp$ vector of $\frac{7}{10}$th of the Brillouin zone length, this fixes the MOKE period at approximately 7 ML.
22. A lower limit for the MOKE oscillation period is determined by the QWS appearing at E_F. Thus, the oscillation period is rather independent of $\hbar\omega$, although not as strict as in photoemission, since also QWS above E_F can be used as final states.
23. A. Dähn, W. Hübner, and K. H. Bennemann, *Phys. Rev. Lett.* **77**, 3929 (1996).
24. W. Hübner, and K. H. Bennemann, *Phys. Rev.* B **52**, 13411 (1995).
25. Note, that in cases where no SHG signal arises from the interior of the in-plane magnetized domain walls, the SHG yield may result from the domain walls and is then different for Néel and Bloch walls with approximately $M_\perp$. Of course, in a better treatment the continuous change of $M(x, y, z)$ within the domain walls requires additional terms contributing to $P_i = \chi_{ijl}(0)E_j E_l + \chi_{ijlk}E_j E_l M_k + \cdots$, which are of the form $\chi_{ijklm}E_j E_l \nabla_k M_m$, $\chi_{ijklmn}E_j E_l M_k \nabla_m M_n$, etc. see eqn (6.2.3) and P. S. Pershan for a general discussion of the expansion of P_i. Which terms appear in the expansion depends on symmetry.
26. Here, we assume a domain structure with antiferromagnetically ordered nearest-neighbour domains and $M \| x$ within the domain. Note that χ_{yxx} and χ_{yzz} are odd in M. For a detailed analysis see refs 23 and 8. For simplicity we have neglected the global antiferromagnetic-like structure of the domains.
27. Note, that for nearly normal incidence one gets only electric fields $E_x(2\omega)$ and $E_y(2\omega)$ and no nonmagnetic SHG due to inversion symmetry within the (x, y) plane.
28. For example, for $M \perp x$ in the domains one has $I^d_{s \to s} = 0$, since χ_{yyy} is odd in M. However, one has $I^w_{s \to s} \neq 0$.
29. E. Beaurepaire, J. C. Merle, A. Daunois, and J.-Y. Bigot, *Phys. Rev. Lett.* **76**, 4250 (1996).
30. J. Hohlfeld, E. Matthias, R. Knorren, and K. H. Bennemann, *Phys. Rev. Lett.* **78**, 4861 (1997).
31. Note, for simplicity the phase between χ^e and χ^o has been neglected. Also we have assumed that one tensor element dominates in χ_{ijl}. If analysing the polarization dependence of SHG and its time evolution one must of course use $\chi_{ijl}(M)$.
32. Approximately $\Delta I_+ \propto |\chi^e(T)|^2(1 + aM^2)$, with $a(t) = |\chi'|^2/|\chi^e|^2$, $\chi' \equiv \chi_{ijkl}$. Hence, ΔI_+ varies with T_{el}, since the temperature dependence of $\chi^e(T)$ and a is expected to be weaker. This is supported by model calculations: R. Knorren, FU (1997). Note, that the SH intensity I involves Fresnel coefficients which are temperature-dependent: T. A. Luce, W. Hübner, and K. H. Bennemann, *Z. Phys.* B **102**, 223 (1997).
33. W. S. Fann, R. Storz, H. W. K. Tom, and J. Bokor, *Phys. Rev.* B **46**, 13592 (1992).
34. The thermalization of d electrons might be faster than that of s electrons. Therefore, in general for certain situations $M(t)$ might already reach its minimum before $b(t) \sim |\chi'|^2/|\chi^e|^2$ which is responding to the temperature of d and s electrons.
35. One estimates for Ni with $T_{latt} \simeq 300\,\text{K}$ and $T_{el}^{max} \simeq 580\,\text{K}$ that $T_{latt}(t \geq \text{picoseconds}) \simeq 450\,\text{K}$. Then we estimate that $\delta_-(t)$ is lower by 0.05 at times $t \sim$ picoseconds and that $\delta_- \simeq M(t_{min})/M(t_0) - 1 \approx -0.3$ at the minimum. Experimentally one observes corresponding values of 0.02–0.03 at the minimum (see Fig. 6.17).
36. The magnetic response may depend on the excitation energy $\hbar\omega$ and may reveal interesting effects due to the structure in the DOS $N_\sigma(\varepsilon)$. For example, one could excite mainly electrons in the minority band and thus possibly for very short times enhance magnetism until thermalization yields $M(T_{el})$ with $T_{el} > T_{latt}$.

37. The time dependence of $T_{el}(t)$ may be determined approximately also by using the Master equations $c_{el}\dot{T}_{el} = -\alpha_1(T_{el} - T_{latt}) - \alpha_2(T_{el} - T_\mu) + p(t)$ for the electrons, $c_\mu \dot{T}_\mu = -\alpha_2(T_\mu - T_{el}) - \alpha_3(T_\mu - T_{latt})$ for the spins, and $c_{latt}\dot{T} = -\alpha_1(T_{latt} - T_{el}) - \alpha_3(T_{latt} - T_\mu)$. Here, c_{el}, c_μ and c_{latt} denote the specific heat, p is the laser power and α_i are constants. If diffusion is important, then

$$c_{el}\dot{T}_{el} = \frac{\partial}{\partial z}\left(K_{el}\frac{\partial}{\partial z}T_{el}\right) + \cdots,$$

where K_{el} is the thermal diffusion coefficient of the electrons. $T_{el}^{max}(t)$ will be generally determined by the interplay of the electron–electron and the electron–lattice interactions. Note that for nonequilibrium $c(T_{el})$ is generally not given by c_{el} in the ground state.
38. P. Jensen and K. H. Bennemann, *Solid State Commun.* **100**, 585 (1996).
39. W. Hübner and K. H. Benneman, *Phys. Rev.* B **53**, 1 (1996).
40. H. J. Siegmann, *et al.*, private communication; also D. R. Penn, S. P. Apell, and S. M. Girvin, *Phys. Rev.* B **32**, 7753 (1985).

Author Index

Aktsipetrov, O. 43, 134, 201
Allenspach, R. 32
Anisimov, S. I. 222
Argyres, P. N. 9, 10, 54, 372
Aspnes, D. E. 377

Bader, S. 55
Baker, S. K. 368, 370, 372
Bander, M. 26
Barber, B. 293
Beaurepaire, E. 262
Bennemann, K.-H. 42, 133, 148, 151, 169, 170, 177, 264, 267, 429, 437
Bennett, H. S. 10
Bischof, A. 32
Bland, J. A. C. 410
Block, J. H. 46, 50
Bloembergen, N. 133, 137, 268, 322
Bloemen, P. J. H. 165, 178, 264, 301, 310, 326
Bohr, N. 311, 362
Bohren, C. F. 285
Boltzmann, S. 224, 467, 469
Borisov, S. B. 43
Born, M. 283
Brown, F. 44
Bruno, P. 104
Buschow, K. H. I. 55

Callaway, J. 354, 355, 359, 365, 381, 389
Ceperly-Alder, D. M. 400
Christy, R. W. 169, 285, 316
Clemens, W. 165
Cohan, N. V. 348
Cohen, M. C. 340
Conrad, U. 220
Corkum, P. B. 66
Crawford, T.M. 171
Curie, P. 25, 26, 27, 29, 82, 90, 256, 365

Dähn, A. 430
de Jong, W. 214
Debye, P. 358
Detzel, T. 89
Dewitz, J. 430
Dick, B. 136
Driel, van, H. M. 44
Druzinic, R. 430
Dzyaloshinskii, I. E. 336

Eastman, D. 372
Eberhardt, W. 365
Eesley, G. L. 65

Ehrenreich, H. 340, 359, 365
Erickson, R. P. 32
Erskine, J. L. 380, 382
Etteger, van, A. F. 214
Euler, L. 399

Falicov, L. 42
Fann, W. S. 222
Faraday, M. 1, 2, 3, 4, 5, 6, 7, 8, 12, 55, 56, 140, 197, 201, 204, 208, 210
Farle, M. 29,
Feil, H. 325
Ferré, J. 430
Fiebig, M. 192, 211, 213, 336
Franken, P. R. 44
Fresnel, A. 17, 46, 49, 52, 53, 76, 87, 220, 221, 240, 243, 244, 248, 249, 255, 256, 301, 309, 315, 318, 319, 322, 323, 339, 343, 414, 420, 421, 422, 423, 424, 426, 438, 439, 448, 468
Fröhlich, D. 192, 214, 430

Garcia, M. E. 264, 430
Gauss, K. 8
Geerts, W. 410
Geldern, van, P. 110, 214
Gersten, v. J. I. 278
Göppert-Mayer, M. 268
Groot Koerkamp, M. 214
Gros, C. 336
Grosenick, D. 264
Grünberg, P. 103
Güdde, J. 264
Guyot-Sionnest, P. 135, 387
Gyorgy, E. M. 30, 31

Haas, C. 325, 358, 364, 381, 387
Halilov, S.V. 384
Hameka, H. F. 348
Hamilton, J. C. 77
Harrison, W. A. 363, 375
Hathaway, K. B. 389
Hayata, K. 278
Haydock, R. 373
Heinz, T. F. 136
Heisenberg, W. 8, 9, 25, 26, 29
Heskett, D. 50
Hicks, J.M. 52, 66
Hodges, L. 359, 365, 368
Hohenberg, P. 398
Hohlfeld, J. 468, 474
Hua, X. M. 278, 279
Hubbard, J. 311, 331

Hübner, W., 42, 77, 79, 110, 124, 133, 144, 148, 149, 151, 152, 169, 170, 172, 176, 177, 178, 252, 264, 318, 437, 444, 474
Huffmann, D. R. 285
Hulme, H. R. 8, 9

Ishida, H. 46

Jackson, J. D. 294
Jacobsen, J. 48
Jah, S. S. 25
Janner, A. M. 194
Janz, S. 44, 45, 51, 61
Jensen, P. J. 430
Jha, S. S. 44, 268
Jin, Q. 124
Johnson, P. B. 169, 182, 285, 316

Kashuba, A. 32
Kasparian, J. 298
Katayama, T. 106
Kelley, P. L. 268
Kerr, J. 1, 2, 3
Kirilyuk, A. 214
Kirilyuk, V. 214
Kirschner, J. 82, 124, 162, 430, 474
Kittel, C. 9, 104, 144, 350
Kleinmann, L. 304
Knorren, R. 264, 430, 474
Kohlhepp, J. 29
Kohn, W. 398, 399
Koopmans, B. 70, 136, 214, 316, 317, 326
Kondo, J. 473
Koshiba, M. 278
Kosterlitz, J. M. 27
Krakauer, H. 420
Krinchik, G. S. 325, 380, 390

Lessard, A. 430
Li, C. M. 52, 88, 249
Liebsch, A. 45–48, 51–52, 387
Lissberger, P. H. 163
Luce, T. 110, 430, 437
Lybchansky, I. L. 43
Lyubchanskii, I.L. 430

Maiman, J. 42, 44
Manders, F. 214
Mansuripur, M. 56
Matthias, E. 219, 326, 430, 474
Maxwell, J. C. 2, 12, 57, 271, 280, 337
Mermin, M. D. 25, 29
Methfessel, M. 400, 403
Mie, G. A. L. 278, 279, 284, 285, 288, 289, 292, 296, 297, 298, 299, 300, 428, 429, 474
Mills, D. 26, 32

Misemer, D. K. 350, 368, 370
Moog, E. R. 163
Moos, T. H. 365, 377, 430
Moss, D. J. 377
Mössbauer, R. 88
Mott, N. 331
Mueller, F. 359, 371
Müller, J. G. 219
Müller, S. 89
Murphy, R. 45, 378, 379, 387

Nambu, Y. 328, 330
Néel, L. 34, 35, 88, 192, 193, 194, 204, 205, 206, 207, 211, 464, 465, 466
Neumann, J. 337, 338

Onsager, L. 6, 270
Oppeneer, P. M. 325
Östling, D. 278, 279, 297

Pan, R. P. 42, 69, 101, 254, 318,
Papaconstantopoulos, D. A. 310, 366
Pappas, D. P. 30
Pauli, W. 18, 328, 329
Penissard, G. 171
Pershan, P. S. 137, 268, 322, 438
Persson, B. N. J. 345
Pescia, D. 30
Petrocelli, G. 244, 252, 310
Petukhov, A. V. 46, 47, 48, 214
Pin Pan, R.- 133, 144, 147, 151, 152
Pisarev, R. V. 192, 195, 430
Plummer, E. W. 365
Pokrovsky, V. L. 30, 32
Polyakov, A. M. 26
Pustogowa, U. 71, 78, 79, 133, 134, 152, 172, 173, 177, 305, 314, 317, 430

Quail, J. C. 45

Rasing, T. 82, 124, 132, 162, 317, 326, 430, 437, 438, 444, 474
Ready, J. F. 67
Regensburger, H. 124
Reif, S. 43, 50, 62, 75, 78, 96, 133, 134, 152, 157, 160, 388, 430
Rudnick, J. 44
Rustagi, K. C. 225

Schaich, W. E. 46, 51, 52
Schelkunoff, S. A. 293
Schilfgaarde, van, P. 403
Schmalian, J. 430
Schoenes, J. 55
Schoenlein, R. W. 228

AUTHOR INDEX

Sham, L. J. 399
Shen, Y. R. 60, 133, 143, 439, 459
Seitz, F. 377, 386, 387, 443
Sigrist, M. 328
Simon, H. J. 45
Sipe, J. E. 92, 136, 220, 240, 301
Smith, P. V. 368, 370
Sokolowski-Tinten, K. 243
Soleil, J. B. F. 153
Sommerfeld, A. 223
Spierings, G. 43, 133
Stark, J. 365
Stern, E. A. 10, 45, 297
Stolle, R. 214
Straub, M. 124, 161
Suárez, C. 238
Sun, C. K. 222
Suzuki, Y. 104, 166, 410

Thomson, W. 3, 4
Thouless, D. J. 27
Tsui, D. C. 365

Ueda, K. 328
Urbach, L. E. 53
Uspenskii, Y. A. 384

Vaterlaus, A. 427

Veenstra, K.J. 214
Voigt, W. 5, 8, 12, 56, 57, 95
Vollmer, R. 135, 144, 161, 437, 442
Vosko, S. H,. 400

Wagner, H. 25, 29
Wang, C. S. 354, 355, 389
Warke, C. S. 225
Watermann, P. A. 293
Weber, M. G. 45
Weber, W. 165
Weiss, P. 8
Weling, F. 359, 365, 381
Wellershof, S. 219
West, P. 264
Wierenga, H. A. 133, 147, 161, 172, 178, 214, 318
Wigner, E. 377
Wolf, E. 283
Woll, J. 53, 54

Yafet, Y. 30, 31
Yeganeh, M. S. 136
Yeh, C. 56, 293
Ying, Z. 46
Yoshida, S. 104
Yoshino, T. 380

Zak, Y. 12

Subject Index

absorption power 10
  additivity law 18, 25
  adsorbates 47, 81
  Drude behaviour 324, 342, 351, 384, 412, 421
anisotropy 32–8
  bulk 26
  magnetic 26, 27, 28
  SHG 48
  surface 26
  uniaxial 36
anti-ferromagnetism, MSHG 192–4, 336–9
Auger effect 22, 26, 78, 162

bands
  spin split 390
band-structure 108, 109, 311, 360–7, 425, 449, 450, 457
  density-function theory 398–400
  LMTO 400–3
  magneto-optics 403–5
  spin-orbit effects 367, 401–3
  spin-polarized 367–72, 390, 469
Bessel function 280, 294, 363, 400
Bloch wavefunction 10, 204–7, 453, 463–5
  density-function theory 398–400
  LMTO 400–3
  magneto-optics 403–5
  spin-orbit effects 367, 401–3
  spin-polarized 367–72, 390, 469
beam, see light
  direction 353
  incident 15
  reflected 15
birefringence, circular magnetic, see Faraday effect
boundary condition 12, 17, 280
Brillouin zone 111, 357–60, 362, 365–7, 369, 372, 374, 377, 385, 394, 403, 420, 441

conductivity 9, 11, 12
current–current correlation 353
current density operator 10
currents, penetration depth 47

damage thresholds 63
density-function theory, see band-structure
dichroism, magnetic 39
dielectric function 225, 227, 354–5, 418, 421
dielectric theory 3
dipole approximation 275
dipole matrix element 372–7, 456
dipole tensor 6, 7
domains 43, 461–3

antiferromagnetic 211, 212
  Bloch 203, 204
  magnetization 90, 203, 210
  Néel 203, 204
  stripe 30, 32
  structure 209, 211
dynamics 220–2, 419
  electron 253
  hot electrons 220, 419–24
  magnetization 253–61

electric field 9, 58, 71, 342
  circular polarization 6, 321
  nonlinear 71
electromagnetic wave 6, 7
  field 58
electron
  majority 109
  minority 109
ellipticity, light 5, 17, 21, 23–5
exchange interaction 8, 26, 269, 347, 395
  SHG 396

Faraday effect 1, 4, 55, 56
  rotation 5, 6, 7, 8
ferromagnet 89, 105, 377–9
field, electric 6, 7, 15, 17, 57, 59, 321, 322, 323
  harmonic fields 282
  magnetic 5
  Weiss 8
film 103–24
  band-structure 403–7, 411
  garnets 194–200
  growth 82
  Kerr spectra 405–7, 408
  ultrathin 82–91, 438–74
Fresnel coefficient 17, 78, 142, 143, 146, 149, 152, 153, 220, 240, 245, 255, 301, 319–23, 414, 420, 439

generation of light 305; see also SHG
growth, Frank van der Merve 82

Haas-van Alphen effect 358, 364, 381, 387
Hankel function 294, 400
harmonic approximation 6
heating by laser 64, 65, 66, 67
hot electron dynamics, SHG analysis 419–26, 468–73
hybridization, see band-structures
hysteresis 22

SUBJECT INDEX

loop 3, 22, 33–4
MSHG 80–91, 155, 468–71

imaging
  domains 43, 205, 206, 209, 211
  nonlinear magneto-optical 201–12
intensity, SHG 80, 153, 156, 159, 255, 303–5
interaction
  RKKY 104
  dipole 30, 444
  quadrupole 44
interfaces 43, 219, 438–74
Ising model 27, 28, 29

J-dependence
  SHG exchange interaction 396, 397

Kerr angle 17, 21, 24, 57, 60, 81, 134, 150, 321, 323
  ellipticity, *see* ellipticity, light
  frequency dependence 325–6
  linear 54–9
  nonlinear 71–4
Kerr configurations 19, 20, 44, 72, 140
  longitudinal 41, 59, 60, 72, 73–4, 118–19, 142
  polar 44, 72, 75, 141
  transversal 44, 71–2, 142
Kerr effect 2, 17, 56, 355
  linear 12–38, 54–8, 79, 140, 355–6
  macroscopic formalism 12–17
  nonlinear 67–74, 79, 148–51, 355–6
  polarization dependence 86, 87, 142–3, 173, 311–18
Kerr intensity 22
Kerr rotation 4, 17, 21, 73–5, 151
  enhancement 133–4, 319–24
  linear 143, 177
  nonlinear 172–8, 301
Kerr spectra
  frequency dependence 178, 393
  linear 379–80, 382, 393
  nonlinear 377, 384–7, 391–3, 405
  see also spectra
Kerr spectroscopy, *see* Kerr spectra
Kerr susceptibility, *see* SHG susceptibility
Kramers–Kronig relation 9, 11, 12

Lambert–Beer law 163
Langmuir kinetics 49, 91, 164
LEED 23
Legendre polynomial 292
light SH 45, 169
  light incident 58
  polarization 57, 321
  propagation 5, 13, 14, 16
  reflected 58, 301
Liouville equation 225, 341

Lorentz force 6, 7, 55, 59

magnetic anisotropy 32–8
magnetic field 6
magnetic phase transition 25–9
magnetic reorientation 29–32, 37
magnetic resonance 28
magnetism 253, 359, 397
  SHG analysis 253–61, 346–53, 458–67
magnetism, surface 75–82, 160–6, 268, 398, 458–67
  multilayers 167–70
magnetization 26, 27
  2d-magnetization 398
  domains 210
magneto-optical coupling 5
magneto-optical imaging, *see* imaging
magneto-optical Kerr effect 2, 56
  linear 54–66, 140–2
  nonlinear 67–119, 143–211
magneto-optical microscopy 207
magneto-optical rotation, *see* Kerr rotation
magneto optics
  electronic theory 339–46
  linear 9
  nonlinear 132–212, 268, 397
  origin 3–8, 10–19, 20
  see also MSHG
magnetoresistance 103
many-body effects 356–8
Maxwell equations 12, 57, 271–2, 276
medium matrix 16
Mie scattering 291–3
  angular dependence 285–90
Mie theory 278
  angular dependence of intensity 286–91
  linear 278
  magnetic 297–8
  nonlinear 278–96
  scattering profile 279
  THG 297–9
MOKE 42, 54–9, 83, 132, 140
  linear 140–3
  origin 3
MSHG 71–91, 111, 114–15, 118, 134, 143–4, 167–71 256–8, 260
  antiferromagnetic structures 192–4, 458–67
  azimuthal anisotropy 123
  frequency dependence 86
  garnet films 194–201
  multilayers 16, 104, 136, 167–71, 181–91
  polarization 205
  quantum-well states 103–20, 179–91, 445–57
multilayer 16, 132, 136–9

Néel model 35
nonequilibrium physics, SHG 467–73
nonlinear magneto-optics (NOLIMOKE) 133, 143–8, 215, 268–70

SUBJECT INDEX

in antiferromagnets 335–9
classical theory 270–4
electronic theory 268–405
films 160
jellium model 268
multilayers 136–40, 160
oscillator model 274–8
in superconductors 327–36
symmetry analysis 300

Onsager relation 6
optical constants 52
optical transitions 105, 395
  SHG 305, 307, 395
optics 352
oscillations
  MSHG 111, 113
  oscillations of exchange coupling 103
  SHG 111, 114–15, 187, 190, 191, 456–8
oscillator model, see SHG
  classical 276–8

parameter Fletcher–Wohlfarth 359, 362, 371, 391, 415, 420
penetration depth 233–6
permeability
  magnetic 12
perturbation, time dependent 10
  Heitler–Ma theory 348
phase SHG 157–60, 179
phase transition, magnetic 25
  2-dimensional 25, 398
  Kosterlitz–Thouless 27
  XY model 27
plasmon 353
Poisson equation 343
polarization 67–8, 149, 302, 344
  modes 11
  plane 5
  see also SHG polarization dependence 275, 300, 309, 316–17, 458–67
polarized light
  circular 4, 11
  linear 10
  s,p polarized 13–14
pump and probe spectroscopy, SHG 219, 239, 245–9
  MSHG 256–8, 260

quantum size effects 104
  MSHG 103–20, 179–84
  optical transitions 451–52
  SHG oscillations 111, 115, 187
quantum-well states 103–18, 190, 445–54
  magnetic 411–18

recording, magneto-optical 268–9

reflection coefficients 17, 19
reflectivity 52, 228
  frequency dependence 227, 229
  linear 220–38
  metals 219, 228
  pump–probe 232–3
  SHG 239
  temperature dependence 225, 227, 229, 230
  time resolved 219–262
  transient 220, 235–7, 263
reflection 19
refraction
  refractive indices 5, 9, 113, 322
RHEED oscillations 22, 23, 62
remanence, magnetic 28
reorientation, see magnetic reorientation
rotation of light 5, 186

Sandwich systems 106–19
Snell law 58, 139, 143
Schrödinger equation 341, 362, 398, 399, 400
Second harmonic generation (SHG) 106, 107
  spectroscopy 51–3
Second harmonic, SH 49, 50, 426
  azimuthal dependence 93, 94, 95
  bulk 91–4
  multilayers 136–9, 181–91
  p-polarized 48
Stern–Gerlach apparatus 429
Selection rules 353
SHG 61, 107, 113, 132, 163, 166, 186, 191, 239–41
  azimuthal dependence 93
  electronic theory 339–418, 437–59
  frequency dependence 54, 112, 166
  origin 429–7
  oscillator model 274
  polarization dependence 73–5, 120, 173, 251–4, 300–26
  pump-probe spectroscopy 245–9
  reconstruction 89
  surface sensitivity 344
  temperature dependence 243–4, 248–51
  thickness-dependence 163, 171
  time resolved 419–26, 468–73
  wavelength dependence 55, 178
SHG magnetic, see MSHG 42–131, 346–54
SHG metal surfaces 44–53, 135–6, 239–41
  adsorbates 49–51, 95–103
  anisotropy 46–8, 123
  coverage 49, 50, 101, 102, 107, 183, 186
  defects 49, 50
  interface 45, 48, 68
  steps 46
  transition-metal oxides 310–11
SHG of quantum-well states 103–19, 411–17, 445–8, 455
  analysis 68–70, 439–41
  asymmetry 116, 117
  centrosymmetric 47, 219, 440–2

garnet films 195–201
reversal symmetry 6
symmetry, SHG 439–45
SHG oscillator model 274
SMOKE
  surface magneto-optical Kerr effect 3, 20, 21, 25
spectra 377–86, 411
  dependence on magnetization 394–8
spin, electron
  majority 10, 367, 369, 390
  minority 110, 367, 369, 390
spin-orbit interaction 4, 8, 9, 10, 26, 133, 144, 269, 324, 346–51, 398, 401, 402
spin reorientation transition 31–7
superconductivity, *see* nonlinear magneto-optics
superlattice 3, 17, 22, 23
Surface
  anisotropy 46
  magnetism 1–37, 42, 160–7
  steps 32, 35
  terraces 32
Susceptibility 72, 82, 241, 355, 439–42
  asymmetries 71, 72
  electronic theory 339–405
  linear 340–2, 353, 411
  magnetic 29, 144–8, 151–2, 158
  nonlinear 72, 243, 274, 342, 411
  symmetry analysis 68–70, 146, 148
  temperature dependence 242, 244, 249

temperature, two-temperature model 66, 68, 222–3
  electron 224–237
  lattice 224
tensor, magneto-optical 67, 144–51, 157–9
  components 82, 136, 189, 243, 254–5, 309

dielectric 4, 5
even–odd analysis 157–60
phase 155, 158–9, 160
THG 278, 297–9
third-harmonic generation, *see* THG
  angular dependence 299
  time reversal 136
tight-binding theory, *see* SHG 356
transition, magnetic reorientation 29, 30–1, 36
  in-plane 29–30
  inter-band 354
  intra-band 353–4
  perpendicular 29–30
transition matrix elements for SHG 228, 307, 351
  k-dependent 373–7
transmission coefficients 17, 58

Voigt constant 5
Voigt (or Cotton–Mouton) effect 56
Voigt vector 12

wall
  Bloch 205–6
  Néel 205–6
wave
  function 10, 359, 361
  incident 14
  normal modes 5
  propagation 13
  reflected 14
  vector 13, 359, 360

X-rays 9

Zeeman effect 1